THE GALAXY DISK IN COSMOLOGICAL CONTEXT

IAU SYMPOSIUM No. 254

COVER ILLUSTRATION:

The edge-on disk galaxy NGC 4565 (Photo: ESO) on the background of a section of the cosmological "Millennium Simulation" (V. Springel, Max-Planck Institute for Astrophysics, Garching, Germany).

IAU SYMPOSIUM PROCEEDINGS SERIES
2008 EDITORIAL BOARD

Chairman

I.F. CORBETT, IAU Assistant General Secretary
*European Southern Observatory
Karel-Schwarzschild-Strasse 2
D-85748 Garching-bei-München
Germany
icorbett@eso.org*

Advisers

K.A. VAN DER HUCHT, IAU General Secretary,
SRON Netherlands Institute for Space Research, Utrecht, the Netherlands
E.J. DE GEUS, *Dynamic Systems Intelligence B.V., Assen, the Netherlands*
U. GROTHKOPF, *European Southern Observatory, Germany*
M.C. STOREY, *Australia Telescope National Facility, Australia*

Members

IAUS251
SUN KWOK, *Faculty of Science, University of Hong Kong, Hong Kong, China*
IAUS252
LICAI DENG, *National Astronomical Observatories, Chinese Academy of Sciences, Beijing, China*
IAUS253
FREDERIC PONT, *Geneva Observatory, Sauverny, Switzerland*
IAUS254
JOHANNES ANDERSEN, *Niels Bohr Institute, Astronomy Group, Copenhagen University, Denmark*
IAUS255
LESLIE HUNT, *INAF - Istituto di Radioastronomia, Firenze, Italy*
IAUS256
JACOBUS Th. van LOON, *Astrophysics Group, Lennard-Jones Laboratories, Keele University, Staffordshire, UK*
IAUS257
NATCHIMUTHUK GOPALSWAMY, *Solar Systems Exploration Div., NASA Goddard Space Flight Center, MD, USA*
IAUS258
ERIC E. MAMAJEK, *Radio and Geoastronomy Division, Harvard Smithsonian CfA, Cambridge, MA, USA*
IAUS259
KLAUS G. STRASSMEIER, *Astrophysics Institute Potsdam, Potsdam, Germany*

INTERNATIONAL ASTRONOMICAL UNION
UNION ASTRONOMIQUE INTERNATIONALE

THE GALAXY DISK IN COSMOLOGICAL CONTEXT

PROCEEDINGS OF THE 254th SYMPOSIUM OF
THE INTERNATIONAL ASTRONOMICAL UNION
HELD IN COPENHAGEN, DENMARK
JUNE 9–13, 2008

Edited by

JOHANNES ANDERSEN
The Niels Bohr Institute, University of Copenhagen, Denmark

JOSS BLAND-HAWTHORN
School of Physics, University of Sydney, Australia

and

BIRGITTA NORDSTRÖM
The Niels Bohr Institute, University of Copenhagen, Denmark

CAMBRIDGE UNIVERSITY PRESS
The Edinburgh Building, Cambridge CB2 8RU, United Kingdom
32 Avenue of the Americas, New York, NY 10013-2473, USA
477 Williamstown Road, Port Melbourne, VIC 3207, Australia
Ruiz de Alarcón 13, 28014 Madrid, Spain
Dock House, The Waterfront, Cape Town 8001, South Africa

© International Astronomical Union 2009

This book is in copyright. Subject to statutory exception
and to the provisions of relevant collective licensing agreements,
no reproduction of any part may take place without
the written permission of the International Astronomical Union.

First published 2009

Printed in the United Kingdom at the University Press, Cambridge

Typeset in System LaTeX 2_ε

A catalogue record for this book is available from the British Library

Library of Congress Cataloguing in Publication data

ISBN 978 0 521 88985 8 hardback
ISSN 1743–9213

IAU Symposium No. 254 and this volume
are dedicated to
Professor BENGT STRÖMGREN (1908 – 1987)

Photo courtesy of Ole Strömgren

Table of Contents

Conference photo xii

Preface xiii

Organising committees. xv

List of participants. xvii

Introduction

Bengt Strömgren's Approach to the Galaxy 3
    Bengt Gustafsson

Disk Galaxies Throughout Time and Space

Simulations of Disk Galaxy Formation in their Cosmological Context 19
    Simon D. M. White

Disk Galaxies at High Redshift?. 21
    Max Pettini

Spatially resolved dynamics of high-z star forming galaxies. 33
    Reinhard Genzel

Simulating High-Redshift Disk Galaxies: Applications to Long Duration Gamma-Ray Burst Hosts. 35
    Brant E. Robertson

Reconciling the Metallicity Distributions of Gamma-ray Burst, Damped Lyman-α, and Lyman-break Galaxies at $z \approx 3$ 41
    Johan P. U. Fynbo, J. Xavier Prochaska, Jesper Sommer-Larsen, Miroslava Dessauges-Zavadsky, and Palle Møller

Stellar Populations and Dark Matter in the Milky Way Disk and in Local Group Galaxies. 49
    Eva K. Grebel

The mass content of the Sculptor dwarf spheroidal galaxy. 61
    G. Battaglia, A. Helmi, E. Tolstoy, and M. Irwin

Galaxy Interactions, Star Formation History, and Bulgeless Galaxies. 67
    Shardha Jogee

Dark Matter Density in Disk Galaxies. 73
    J. A. Sellwood

Mergers and Disk Survival in ΛCDM. 85
    James S. Bullock, Kyle R. Stewart, and Chris W. Purcell

An 84-μG Magnetic Field in a Galaxy at $Z = 0.692$?. 95
    Arthur M. Wolfe, Regina A. Jorgenson, Timothy Robishaw, Carl Heiles, and Jason X. Prochaska

Abundance Gradient in the Disk of NGC 300 97
  Marija Vlajić, Joss Bland-Hawthorn, and Ken C. Freeman

The Galactic Disk-Halo Transition – Evidence from Stellar Abundances 103
  Poul Erik Nissen and William J. Schuster

Origin, Structure, and Chemical Evolution of Disks

Abundance Gradients and Substructures in Disks 111
  K. C. Freeman

Mapping low-latitude stellar substructure with SEGUE photometry 121
  Jelte T. A. de Jong, Brian Yanny, Hans-Walter Rix, Eric F. Bell, and Andrew E. Dolphin

Cosmic evolution of stellar disk truncations: from $z = 1$ to the Local Universe .. 127
  Ignacio Trujillo, Ruyman Azzollini, Judit Bakos, John Beckman, and Michael Pohlen

Chemically tagging the Galactic disk: abundance patterns of old open clusters.. 133
  G. M. De Silva, K. C. Freeman, and J. Bland-Hawthorn

The Arcturus Moving Group: Its Place in the Galaxy 139
  Mary E. K. Williams, Ken C. Freeman, Amina Helmi, and the RAVE collaboration

The Bulge-disc connection in the Milky Way 145
  James Binney

Stellar abundances tracing the formation of the Galactic Bulge 153
  Beatriz Barbuy, Manuela Zoccali, Sergio Ortolani, Vanessa Hill, Alvio Renzini, Jorge Meléndez, Anita Gómez, Martin Asplund, Dante Minniti, Eduardo Bica, and Alan Alves-Brito

Unveiling the Secrets of the Galactic bulge through stellar abundances in the near-IR: a VLT/Crires project.. 159
  Nils Ryde

Bars in Cuspy Dark Halos... 165
  John Dubinski, Ingo Berentzen, and Isaac Shlosman

Exponential bulges and antitruncated disks in lenticular galaxies............. 173
  Olga K. Sil'chenko

Kinematical & Chemical Characteristics of the Thin and Thick Disks 179
  Rosemary F. G. Wyse

The chemical evolution of the Galactic thick and thin disks 191
  Cristina Chiappini

The chemical fingerprints of the thin and the thick disk 197
  Sofia Feltzing, Sally Oey, and Thomas Bensby

Beryllium and the formation of the Thick Disk and of the Halo 203
  Luca Pasquini, R. Smiljanic, P. Bonifacio, R. Gratton, D. Galli, and S. Randich

The influence of star clusters on galactic disks: new insights in star-formation in
    galaxies 209
    Pavel Kroupa

The initial luminosity and mass functions of Galactic open clusters 221
    Hans Zinnecker, Anatoly E. Piskunov, Nina V. Kharchenko, Siegfried Röser,
    Elena Schilbach, and Ralf-Dieter Scholz

Open Clusters as tracers of the Galactic disk: the Bologna Open Clusters Chemical
    Evolution project 227
    Angela Bragaglia, Eugenio Carretta, Raffaele Gratton, and Monica Tosi

Origin of Star-to-Star Abundance Inhomogeneities in Star Clusters 233
    Jan Palouš, Richard Wünsch, Guillermo Tenorio-Tagle, and Sergyi Silich

Session 3: Accretion and the Interstellar Medium

Warm gas accretion onto the Galaxy 241
    J. Bland-Hawthorn

New evidence for halo gas accretion onto disk galaxies 255
    Filippo Fraternali

Group infall of substructures on to a Milky Way-like dark halo 263
    Yang-Shyang Li and Amina Helmi

Modelling the Disk (three-phase) Interstellar Medium 269
    Gerhard Hensler

Measuring Outer Disk Warps with Optical Spectroscopy 283
    Daniel Christlein and Joss Bland-Hawthorn

Star Formation in Disks: Spiral Arms, Turbulence, and Triggering Mechanisms . 289
    Bruce G. Elmegreen

H I in Galactic Disks 301
    Elias Brinks, Frank Bigiel, Adam Leroy, Fabian Walter, W. J. G. de Blok,
    Ioannis Bagetakos, Antonio Usero, and Robert C. Kennicutt, Jr.

The Molecular Gas Component of Galaxy Disks 307
    Leo Blitz

Disk Stability and Turbulence Generation: Effects of the Stellar Component ... 313
    Woong-Tae Kim

Spiral Arm Tangencies in the Milky Way 319
    Robert A. Benjamin

Session 4: Stars as Drivers and Tracers of Chemical Evolution

Evolution and chemical and dynamical effects of high-mass stars 325
    Georges Meynet, Cristina Chiappini, Cyril Georgy, Marco Pignatari, Raphael
    Hirschi, Sylvia Ekström, and André Maeder

The First Galaxies 337
    Volker Bromm

Chemical enrichment in the early Galaxy 343
    Torgny Karlsson

Halo chemistry and first stars. The chemical composition of the matter in the early Galaxy, from C to Mg 349
    M. Spite, P. Bonifacio, R. Cayrel, F. Spite, P. Francois, H. G. Ludwig, E. Caffau, S. Andrievsky, B. Barbuy, B. Plez, P. Molaro, J. Andersen, T. Beers, E. Depagne, B. Nordström, and F. Primas

Chemical Yields from Supernovae and Hypernovae 355
    Ken'ichi Nomoto, Shinya Wanajo, Yasuomi Kamiya, Nozomu Tominaga, and Hideyuki Umeda

Effects of Supernova Feedback on the Formation of Galaxies 369
    Cecilia Scannapieco, Patricia B. Tissera, Simon D. M. White, and Volker Springel

Chemodynamical simulations of the Milky Way Galaxy..................... 375
    Chiaki Kobayashi

On the chemical evolution of the Milky Way............................... 381
    Nikos Prantzos

Chemical evolution of the Galaxy disk in connection with large-scale winds 393
    Takuji Tsujimoto, Joss Bland-Hawthorn, and Kenneth C. Freeman

Session 5: Disk Galaxy Meets ΛCDM Cosmology

Formation and evolution of disk galaxies.................................. 401
    Joseph Silk

Disk Sizes in a ΛCDM Universe 411
    Qi Guo and Simon White

Cold Dark Matter Substructure and Galactic Disks......................... 417
    Stelios Kazantzidis, Andrew R. Zentner, and James S. Bullock

The Galaxy and its stellar halo – insights from a hybrid cosmological approach 423
    Gabriella De Lucia and Amina Helmi

Numerical simulations of galaxy evolution in cosmological context............ 429
    Marie Martig, Frédéric Bournaud, and Romain Teyssier

Session 6: Surveys, Challenges and Prospects for the Future

The Challenge of Modelling Galactic Disks................................ 437
    Andreas Burkert

Hydrodynamical Adaptive Mesh Refinement Simulations of Disk Galaxies 445
    Brad K. Gibson, Stéphanie Courty, Patricia Sánchez-Blázquez, Romain Teyssier, Elisa L. House, Chris B. Brook, and Daisuke Kawata

Present state and promises to unravel the structure and kinematics of the Milky Way with the RAVE survey 453
M. Steinmetz, A. Siebert, T. Zwitter, and the RAVE collaboration

SEGUE, and the future of large scale surveys of the Galaxy 461
Timothy C. Beers, Young Sun Lee, and Daniela Carollo

Galaxy And Mass Assembly (GAMA) 469
Simon P. Driver and the GAMA team

What will Gaia tell us about the Galactic disk? 475
Coryn A. L. Bailer-Jones

A Roadmap for Delivering the Promise of Gaia 483
T. Prusti, C. Aerts, E. K. Grebel, C. Jordi, S. A. Klioner, L. Lindegren, F. Mignard, S. Randich, and N. A. Walton

The Science of Galaxy Formation 487
Gerard Gilmore

New mechanisms for international coordination of large observing projects 497
Johannes Andersen

Summary, Conclusions and Recommendations 501
Rosemary F. G. Wyse

Poster papers 505

Author index 509

Three other papers presented at the conference but were not submitted for publication in this volume:

Disk chemical evolution with flows
J. Dalcanton

Stellar abundance data for other galaxies
V. Hill

The promise of radial-velocity surveys
C. Rockosi

Preface

Progress in science occurs when theory and observations meet, but do not match. One such battlefield is the disk galaxy: Numerical simulations based on "concordance cosmology" have gone far towards a consistent picture of the formation and evolution of structure in the Universe. But fundamental discrepancies exist between the predictions of such models and the disk galaxies we see in the universe today.

IAU Symposium 254 was held at the Niels Bohr Institute, University of Copenhagen, June 9-13, 2008, to discuss these issues. The time and venue were chosen because 2008 was the 100th anniversary of the birth of Danish astrophysicist Bengt Strömgren, a leading figure in the development of modern astrophysics. His commitment to a physical understanding of our Galaxy - with a focus on its stars and gas and their interplay - laid the foundation for the concept of galactic evolution as we know it today, and we dedicate this IAU Symposium to him. One of our primary goals in organising the meeting was to preserve his comprehensive, yet always focused, approach to the study of Galactic formation and evolution.

It is remarkable the extent to which observations of the near field immediately complement observations of the far field in our pursuit to understand how galaxies evolved from the era of the Hubble Deep Field to the present day. To quote a recent review, "half of all observational cosmology today resides in the near field." This was a theme that ran throughout the meeting.

Bengt Gustafsson started the meeting by recalling Bengt Strömgren's seminal contributions to the field, and Simon White's introductory review crystallised its rationale. The programme then focused on current and future scientific progress in six sessions:

Disk galaxies throughout space and time
Origin, structure, and chemical evolution of disks
Accretion and the interstellar medium
Stars as drivers and tracers of chemical evolution
Disk galaxy meets LambdaCDM cosmology
Surveys, challenges, and prospects for the future

Each session was introduced by 1-2 reviews by prominent experts in each field, followed by shorter invited and contributed papers for a total of 70 oral presentations. Many more oral contributions were proposed than could possibly be accommodated, and a total of 120 posters provided much supplementary material and new results and ideas in most subfields. The poster papers could not be accommodated within the confines of this book, but will be available together with the on-line version of the Proceedings.

Gathering the material for these Proceedings turned out to require patience and persistence. We thank all those authors who returned their manuscripts close to the original deadline and regret that a few did not make the final cut, four months later. For a couple of essential papers, we provide extended abstracts with links to the presentation files. The photos, if not credited otherwise, are by Peter Laursen or other LOC members.

Many of the papers contain beautiful colour illustrations that considerably enhance their information content. Unfortunately, the cost of printing them all would be prohibitive, so we have had to limit the use of colour in this book itself to a few vital cases. All illustrations will, however, appear in full colour in the on-line version of this volume.

By design, the programme emphasised overall understanding rather than individual discoveries, but the great recent strides on both the observational and theoretical fronts were impressive and enlightening. We were gratified to have a large audience (235 participants from 35 countries, 80 of whom were women) of scientists from different observational and theoretical backgrounds, including many PhD students and postdocs. This made for a very varied scientific programme, and the discussions showed how fruitful it is to bring together people from different backgrounds and disciplines. A persistent comment by many participants after the meeting was the new insights obtained by having the topic illuminated from so many different angles, yet with a persistent focus.

The lessons of the Symposium were summarised elegantly by Rosemary Wyse in her concluding review. Rosie also gave us a taste of things to come, not least the armada of new instruments and telescopes that will focus on Galactic evolution.

The social programme had been organised to mark the Strömgren centenary appropriately. Following a welcome reception at the Convocation Hall of the University of Copenhagen on June 8, a reception was offered by the Lord Mayor and City Council of Copenhagen at the City Hall on June 9. In the afternoon of June 11, a reception was held at the former Strömgren residence of honour at the Carlsberg brewery. Guests included the President of the Royal Danish Academy of Sciences (co-host of the Symposium), Prof. Kirsten Hastrup; former IAU President Adriaan Blaauw, a friend of Bengt Strömgren since 1938 and organiser of IAU Symposium No. 1, "Co-ordination of Galactic Research", in 1953(!), and Prof. Helge Kragh, University of Aarhus, who outlined Bengt Strömgren's contributions to Danish science in general. We were delighted that Ole Strömgren, son of Bengt Strömgren, and Hans and Aage Bohr, sons of Niels Bohr (who lived in the mansion 1931-1962) could join us as well.

The conference dinner, with live dance music, was held at the impressive headquarters of the Danish Order of Freemasons, adjacent to the Niels Bohr Institute. Dinner speeches were delivered with gusto by Ole Strömgren and Joss Bland-Hawthorn. Finally, the week was rounded off with a Saturday excursion involving 40 of the participants to the island of Hven, the site of Tycho Brahe's observatories. Public outreach events included public evening talks by meeting participants A.C. Andersen, V. Bromm and J. Silk.

The moral and financial sponsorship of the IAU was of decisive importance for the meeting. Our local sponsors (see next page) contributed substantial additional financial support, which contributed greatly to success of the meeting and enabled us to offer a total of ~60,000 Euro in support for 125 participants, beyond the 14 grants provided from the IAU funds. We thank them all, on behalf of all of us.

Finally, much kind assistance was offered by our LOC colleagues and staff and students at the Niels Bohr Institute (host of the meeting). Lars Lindberg Christensen and Mafalda Martins at the ST-ECF/ESA/ESO kindly provided the design of our web site, poster, and cover picture for this volume. The staff of our conference bureau, Best Destination Partner, were efficient and friendly, and the technical staff of the Panum Institute kept the facilities of the Lundsgaard Auditorium running flawlessly. We thank them all for helping us to create a successful meeting.

November 2008

Johannes Andersen, Joss Bland-Hawthorn and Birgitta Nordström

ORGANISING COMMITTEES

Scientific Organising Committee:

B. Barbuy (Brazil)
J. Bland-Hawthorn (Australia, Co-Chair)
B. Elmegreen (USA)
B. Gustafsson (Sweden)
K. Nomoto (Japan)
D.A. VandenBerg (Canada)
R.F. Wyse (USA)

J. Binney (UK)
V. Bromm (USA)
E. Grebel (Switzerland)
A. Helmi (The Netherlands)
B. Nordström (Denmark, Co-Chair)
S.D. White (Germany)
T. de Zeeuw (The Netherlands)

Local Organising Committee:

J. Andersen (Chair)
J. Knude
K. Pedersen

J.V. Clausen
B. Nordström

Acknowledgements

The coordinating IAU Division for the symposium was Division VII (Galactic System). The proposal was also supported by IAU Divisions IV (Stars) and VI (Interstellar Matter), and by IAU Commissions 25, 28, 29, 30, 33, 45.

On behalf of all participants, the Local Organizing Committee gratefully acknowledges the generous financial and material support received from:

The International Astronomical Union
The Niels Bohr Institute, University of Copenhagen
The Royal Danish Academy of Sciences and Letters
The Danish Natural Science Research Council
The Carlsberg Foundation
The Niels Bohr International Academy
The Foundation of December 29, 1967
The Dark Cosmology Center
The Danish Astrophysics Research School
and
The City of Copenhagen

Remembering Bengt Strömgren in the Pompeii Hall of the former Strömgren residence of honour at Carlsberg. Welcome by the President of the Royal Danish Academy of Sciences and Letters, Prof. Kirsten Hastrup.

Aage Bohr, Ole Strömgren and Adriaan Blaauw at Carlsberg.
Photo: Hans Zinnecker.

List of Participants

Daniel Adén	Lund Observatory	Sweden
Oscar Agertz	University of Zürich	Switzerland
Dinah Allen	University of Hertfordshire	UK
Jin An	Dark Cosmology Centre	Denmark
Anja C. Andersen	Dark Cosmology Centre	Denmark
Jan Marie Andersen	Niels Bohr Institute	Denmark
Johannes Andersen	Niels Bohr Institute	Denmark
Borja Anguiano	Astrophysikalisches Institut Potsdam	Germany
Teresa Antoja	Universitat de Barcelona	Spain
Anna Árnadottir	Lund Observatory	Sweden
Michael Aumer	University of Oxford	UK
Coryn Bailer-Jones	Max Planck Institute for Astronomy, Heidelberg	Germany
Judit Bakos	Instituto de Astrofísica de Canarias	Spain
Dilip G. Banhatti	Madurai-Kamaraj University	India
Beatriz Barbuy	Universidade de São Paulo	Brazil
Marco Barden	University of Innsbruck	Austria
Michael Barker	Institute for Astronomy, Edinburgh	UK
Giuseppina Battaglia	European Southern Observatory	Germany
Timothy C. Beers	Michigan State University	USA
Zahir Bellil	Institut d'Astrophysique, Paris	France
Robert Benjamin	University of Wisconsin-Whitewater	USA
Olivier Bienaymé	Strasbourg Observatory	France
Ilfan Bikmaev	Kazan State University	Russia
James Binney	Oxford University	UK
Joss Bland-Hawthorn	University of Sydney	Australia
Leo Blitz	University of California, Berkeley	USA
Frédéric Bournaud	CEA-Saclay	France
Angela Bragaglia	Osservatorio Astronomico di Bologna	Italy
Elias Brinks	University of Hertfordshire	UK
Karsten Brogaard	European Southern Observatory	Germany
Volker Bromm	University of Texas at Austin	USA
Lars Buchhave	Niels Bohr Institute	Denmark
Sarah Buehler	Royal Observatory, University of Edinburgh	UK
James Bullock	University of California, Irvine	USA
Andreas Burkert	University of München	Germany
Julio A. Carballo-Bello	Instituto de Astrofísica de Canarias	Spain
Luca Casagrande	University of Turku - Tuorla Observatory	Finland
Michael Chabin	Press	USA
Scott Chapman	Institute of Astronomy, Cambridge	UK
Li Chen	Shanghai Astronomical Observatory	China
Cristina Chiappini	Observatoire de Genève	Switzerland
Daniel Christlein	Max Planck Institute for Astrophysics	Germany
Jens Viggo Clausen	Niels Bohr Institute	Denmark
Edoardo Colavitti	Dipartimento di Astronomia di Trieste	Italy
Michelle Collins	Institute of Astronomy, Cambridge	UK
Roberto Costa	IAG/University São Paulo	Brazil
Stephanie Courty	University of Central Lancashire	UK
Katia Cunha	NOAO	USA
Anna Curir	Astronomical Observatory of Turin	Italy
Julianne Dalcanton	University of Washington	USA
Jelte de Jong	Max-Planck-Institut für Astronomie	Germany
Roelof de Jong	Space Telescope Science Institute	USA
Gabriella De Lucia	Max-Planck Institute for Astrophysics	Germany
Serge Demers	Universite de Montreal	Canada
Gayandhi De Silva	European Southern Observatory	Germany
Simon Driver	University of St Andrews	UK
John Dubinski	University of Toronto	Canada
Árdís Elíasdóttir	Dark Cosmology Centre	Denmark
Bruce Elmegreen	IBM, T.J. Watson Laboratory	USA
Claus Fabricius	Institute for Space Studies of Catalonia	Spain

Sofia Feltzing	Lund Observatory	Sweden
David Fernández	Institut de Ciéncies del Cosmos, Barcelona	Spain
Chloé Féron	Dark Cosmology Centre	Denmark
Désirée D.M Ferreira	Dark Cosmology Centre	Denmark
Francesca Figueras	Universitat de Barcelona	Spain
Markus Firnstein	Dr. Remeis-Sternwarte Bamberg	Germany
Chris Flynn	Tuorla Observatory	Finland
Hélène Forest	Collège Ahuntsic, Québec	Canada
Kelly Foyle	Max Planck Institute for Astronomy	Germany
Filippo Fraternali	University of Bologna	Italy
Teddy Frederiksen	Niels Bohr Institute	Denmark
Ken C. Freeman	Mt. Stromlo Observatory, ANU	Australia
Jan Frercks	University of Offenbach	Germany
Eileen Friel	National Science Foundation	USA
Burkhard Fuchs	Astronomisches Rechen-Institut	Germany
Johan Fynbo	Dark Cosmology Centre	Denmark
Christa Gall	Dark Cosmology Centre	Denmark
Esko Gardner	Tuorla Observatory, Turku	Finland
Reinhard Genzel	MPI für Extraterrestrische Physik	Germany
Brad Gibson	University of Central Lancashire	UK
Gerard Gilmore	Cambridge University	UK
Lea Giordano	University of Zürich	Switzerland
Itzhak Goldman	Afeka College	Israel
Eva Grebel	Heidelberg Univ.	Germany
Michel Grenon	Observatiore de Genéve	Switzerland
Preben Grosbøl	European Southern Observatory	Germany
Lisbeth Fogh Grove	Dark Cosmology Centre	Denmark
Frank Grundahl	IFA, University of Aarhus	Denmark
Qi Guo	Max Planck Institute for Astrophysics, Garching	Germany
Bengt Gustafsson	Uppsala University	Sweden
Camilla Juul Hansen	European Southern Observatory	Germany
Steen Hansen	Dark Cosmology Centre	Denmark
Terese Hansen	Niels Bohr Institute	Denmark
Ulrike Heiter	Uppsala University	Sweden
Gerhard Hensler	Instute of Astronomy, University of Vienna	Austria
Stéphane Herbert-Fort	University of Arizona/Steward Observatory	USA
Vanessa Hill	Observatoire de Paris	France
Jens Hjorth	Dark Cosmology Centre, Niels Bohr Institute	Denmark
David Hobbs	Lund Observatory	Sweden
Loren Hoffman	Northwestern University	USA
Berry Holl	Lund Observatory	Sweden
Jinliang Hou	Shanghai Astronomical Observatory	China
Elisa House	Centre for Astrophysics, U. of Central Lancashire	UK
Erik Høg	Niels Bohr Institute	Denmark
Ole Høst	Dark Cosmology Centre	Denmark
Synnøve Irgens-Jensen	Norwegian Research Council	Norway
Shardha Jogee	University of Texas at Austin	USA
Joel Johansson	Royal Institute of Technology	Sweden
Carme Jordi	Dept. of Astronomy, University of Barcelona	Spain
Umesh Joshi	Physical Research Laboratory, Ahmedabad	India
M. Ryan Joung	Princeton University	USA
Andreas Just	Astronomishes Rechen-Institut, Heidelberg	Germany
Xi Kang	Max Planck Institute for Astronomy	Germany
Torgny Karlsson	NORDITA, Stockholm	Sweden
Stelios Kazantzidis	Stanford University	USA
Gyongyi Kerekes	Eötvös University, Dept. of Physics & Astronomy	Hungary
Woong-Tae Kim	Seoul National University	Korea
Rainer Klement	Max-Planck-Institut für Astronomie	Germany
Jens Knude	Niels Bohr Institute	Denmark
Chiaki Kobayashi	Australian National University	Australia

List of Participants

Andreas Korn	IFA, Uppsala University	Sweden
Pavel Kroupa	University of Bonn	Germany
Mark Krumholz	Princeton Univ. / Univ. of California Santa Cruz	USA
Arunas Kucinskas	Institute of Theoretical Physics & Astronomy	Lithuania
David W. Latham	Harvard-Smithsonian Center for Astrophysics	USA
Peter Laursen	Dark Cosmology Centre	Denmark
Giorgos Leloudas	Dark Cosmology Centre	Denmark
Bruno Letarte	California Institute of Technology	USA
Yang-Shyang Li	Kapteyn Astronomical Institute	Netherlands
Lennart Lindegren	Lund Observatory	Sweden
Suzanne Linder	Hamburger Sternwarte	Germany
Thomas Lloyd Evans	University of St Andrews	UK
Dirk Lorenzen	German Public Radio	Germany
Walter Maciel	University of São Paulo	Brazil
Vitalii Makaganiuk	Uppsala Astronomical Observatory	Sweden
Daniele Malesani	Dark Cosmology Centre	Denmark
Luigi Mancini	Department of Physics, University of Salerno	Italy
Michela Mapelli	University of Zürich	Switzerland
Marie Martig	CEA Saclay	France
Cecilia Mateu	Centro de Investigaciones de Astronomía	Venezuela
Lars Mattsson	Uppsala University	Sweden
Søren Meibom	Harvard-Smithsonian Center for Astrophysics	USA
Georges Meynet	Université de Genève	Switzerland
Michal Michalowski	Dark Cosmology Centre	Denmark
Bo Milvang-Jensen	Dark Cosmology Centre	Denmark
Petko Nedialkov	Astronomy Department, Sofia University	Bulgaria
Poul Erik Nissen	IFA, University of Aarhus	Denmark
Ken Nomoto	Univ. Tokyo	Japan
Åke Nordlund	Niels Bohr Institute	Denmark
Birgitta Nordström	Niels Bohr Institute	Denmark
Sally Oey	University of Michigan	USA
Sang Hoon Oh	Seoul National University	Korea
Seungkyung Oh	Seoul National University	Korea
Jan Palouš	Astronomical Institute, Czech Academy of Sciences	Czech Republic
Danuta Paraficz	Dark Cosmology Centre	Denmark
Luca Pasquini	European Southern Observatory	Germany
Kristian Pedersen	Niels Bohr Institute	Denmark
Susana Pedrosa	Institute for Astronomy and Space Physics (IAFE)	Argentina
Jorge Penarrubia	University of Victoria	Canada
Isabel Pérez	Kapteyn Astronomical Institute/ Granada Univ.	Netherlands/Spain
Max Pettini	Cambridge University	UK
Jan Pflamm-Altenburg	University of Bonn	Germany
Franziska Piontek	Astrophysikalisches Institut Potsdam	Germany
Susana Planelles	Universidad de Valencia	Spain
Nedelia Popescu	Astronomical Institute, Romanian Acad. of Science	Romania
Antonio Portas	Centre for Astrophysics Research	UK
Laura Portinari	Tuorla Observatory, University of Turku	Finland
Nikos Prantzos	Institut d'Astrophysique de Paris	France
Timo Prusti	ESA - ESTEC	Netherlands
Ivanio Puerari	INAOE, Puebla	Mexico
Ivan Ramírez	University of Texas at Austin	USA
Jesper Rasmussen	Carnegie Observatories	USA
Abhishek Rawat	IUCAA, Pune	India
Paola Re Fiorentin	University of Ljubljana	Slovenia
Alejandra Recio-Blanco	Observatoire de la Cote d'Azur	France
Bacham E. Reddy	Indian Institute of Astrophysics	India
Yves Revaz	Fed. Technical University, Lausanne	Switzerland
Salvador J. Ribas	Universitat de Barcelona	Spain
Jenny Richardson	Institute for Astronomy, Edinburgh	UK
Linus Riel Petersen	Niels Bohr Institute	Denmark

Signe Riemer-Sørensen	Dark Cosmology Centre	Denmark
Brant Robertson	Kavli Inst. for Cosmological Physics, Univ. of Chicago	USA
Constance Rockosi	Lick Observatory, Univ. of California, Santa Cruz	USA
Joel Roediger	Queen's University, Ontario	Canada
Alessandro Romeo	Onsala Space Observatory	Sweden
Rok Roškar	University of Washington	USA
Christine Ruhland	Max Planck Institute for Astronomy	Germany
Nils Ryde	Lund Observatory	Sweden
Kanak Saha	Indian Institute of Science	India
Stuart Sale	Imperial College London	UK
Laura V. Sales	Kapteyn Astronomical Institute	Netherlands
Cecilia Scannapieco	Max Planck Institute for Astrophysics	Germany
William Schuster	Institute of Astronomy, UNAM	Mexico
Jerry Sellwood	Rutgers University	USA
Zhengyi Shao	Shanghai Astronomical Observatory	China
Olga Silchenko	Sternberg Astronomical Institute of MSU	Russia
Joe Silk	Oxford University	UK
Esteban Silva Villa	Utrecht University	Netherlands
Jennifer Simmerer	Lund Observatory	Sweden
Verne V. Smith	NOAO	USA
Jesper Sollerman	Dark Cosmology Center	Denmark
Jesper Sommer-Larsen	Excellence Cluster Universe/Dark Cosmology Center	Denmark
Linda Sparke	National Science Foundation/Univ. of Wisconsin	USA
Monique Spite	Observatoire de Paris	France
Matthias Steinmetz	Astrophysical Institute Potsdam	Germany
Christiaan Sterken	Free University of Brussels	Belgium
Jesper Storm	Astrophysikalisches Institut Potsdam	Germany
Marianne Takamiya	University of Hawaii, Hilo	USA
Antti Tamm	Tartu Observatory	Estonia
Maria Trinidad Tapia-Peralta	Instituto de Astrofísica de Canarias	Spain
Grazina Tautvaišiene	Institute of Theoretical Physics & Astronomy	Lithuania
Christian Theis	Institute of Astronomy, U. Wien	Austria
Christina Thöne	Dark Cosmology Center	Denmark
Eline Tolstoy	Kapteyn Astronomical Institute, Groningen	Netherlands
Ignacio Trujillo	Instituto de Astrofísica de Canarias	Spain
Takuji Tsujimoto	National Astronomical Observatory of Japan	japan
Eelco van Kampen	University of Innsbruck	Austria
Alvaro Villalobos	Kapteyn Astronomical Institute, Groningen	Netherlands
Marija Vlajic	University of Oxford	UK
Paul Vreeswijk	Dark Cosmology Centre	Denmark
Carl J. Walcher	Institut d'Astrophysique de Paris	France
Darach Watson	Dark Cosmology Centre	Denmark
Simon D. White	Max Planck Institute for Astrophysics, Garching	Germany
Larry Widrow	Queen's University, Kingston	Canada
Mary Williams	Astrophysikalisches Institut Potsdam	Germany
Arthur M. Wolfe	University of California, San Diego	USA
Ted Wyder	California Institute of Technology	USA
Rosemary F. Wyse	Johns Hopkins University	USA
Meng Xiang-Gruess	Institut für Theoretische Physik, Univ. Kiel	Germany
Xiangxiang Xue	Max-Planck-Institute for Astronomy	Germany
Peter Yoachim	University of Texas, Austin	USA
Laimons Zacs	University of Latvia	Latvia
Hans Zinnecker	Astrophysical Institute Potsdam	Germany
Tomaz Zwitter	University of Ljubljana	Slovenia
Anna Önehag	Uppsala University	Sweden

Introduction

Session chair: Birgitta Nordström

Start of the meeting: John Renner Hansen (Director, Niels Bohr Institute), Johannes Andersen, Birgitta Nordström and Bengt Gustafsson.

Bengt Gustafsson delivering his introductory review.

Bengt Strömgren's Approach to the Galaxy

Bengt Gustafsson

[1]Department of Physics & Astronomy, Uppsala University,
Box 515, SE-75120 Uppsala, Sweden
email: bg@astro.uu.se

Abstract. The contributions of Bengt Strömgren to the exploration and understanding of the Galactic Disk are sketched. The question what we can learn from his systematic approach is discussed.

Keywords. Galaxy: disk; stars: fundamental parameters; techniques: photometric; orbituaries, biographies; history and philosophy of astronomy

1. A background

To celebrate the 100th anniversary of Bengt Strömgren's birthday is also to celebrate one century of astrophysics, with remarkable scientific progress. In his lifetime achievement, Bengt Strömgren embodies much of this development. Early well schooled in classical astronomy and heavy numerical calculations, he was exposed to the birth of quantum mechanics and modern atomic physics. This made him contribute fundamentally to stellar and interstellar-matter physics. Later, he added to this new developments in photoelectric stellar photometry, and thus created of these different components a fundamental basis for Galactic exploration. Many contemporary important results in this exploration can be directly traced back to Strömgren's suggestions and inspiration. For this reason, the topic of the present symposium is a natural choice in this year.

2. The early years

Strömgren was born into astronomy in 1908 and achieved, partly through his father who was astronomy professor Elis Strömgren in Copenhagen, a flying start in research with his first published papers (reports on observations of comet Baade) at an age of 14. (For a detailed account of Bengt Strömgren's early career, see Rebsdorf 2005.) He got engaged in an astrometric programme at the transit instrument of the Observatory at Øster Voldgade, resulting in a catalogue of α and δ for 131 northern bright stars. This work was followed by papers on *Die Bestimmung erster parabolischer Bahnen unter Anwendung der Rechenmaschine* from 1925, which was the beginning of his doctoral thesis work. He also published early work on photoelectric registration of stellar meridian transits and on numerical calculations of perturbations of asteroid orbits.

Strömgren got familiar with various aspects of classical astronomy, but also personally familiar with leading astronomers, not the least through his father who had many international contacts. And, then the world was small! Later, in 1932 when Strömgren entered the International Astronomical Union, there were totally about 400 members of the organization. When he left the position as its Secretary General in 1952, this number had increased by about a factor of 1.6, and when he finished as its President in 1973, it had increased by a factor of 5, up to 2000. Today the number is again increased by a factor of 5, close to 10,000. So, the world was different, and it was possible to know most important astronomers. So did Bengt, and this was eased by his social gifts, his profound

Figure 1. Participants at the Copenhagen Conference in 1929 in Auditorium A at the present Niels Bohr Institute. Front row, from left: N. Bohr, R. Kronig, I. Waller, J. Holtsmark, H. Kramers, S. Rosseland, W. Pauli, E. Jordan, P. Ehrenfest and G. Gamow. Second row: L. Rosenfeld (no. 1 from left), O. Klein (2), S. Goudsmit (4), C. Møller (5) and W. Heitler (8). Third row, from left: B. Strömgren (3) and H. Casimir (4). Photo: Niels Bohr Archive.

curiosity concerning what other people were doing, and his extraordinarily good memory, not only for science but also for people.

Strömgren, however, also early got in touch with another, maybe even more important circle: the extensive group of young physicists around Niels Bohr at the newly founded Institute for Theoretical Physics of Copenhagen University, later named after Bohr. There, during the inspiring youth of Quantum Mechanics, Strömgren met a number of leading atomic physicists, among whom, in addition to Niels Bohr, he considered Oskar Klein and the guests Ralph Howard Fowler, Cecilia Payne and Friedrich Hund to be particularly important for him (Strömgren 1983). He got interested in applying the new physics to stellar spectra and stellar modelling in general, and thus turned from a classical astronomer to a modern astrophysicist. He began work on stellar structure, contributed to solving what Arthur Eddington had named "the opacity discrepancy" by computing actual opacity tables, and showed that one could not avoid the conclusion that the Sun and stars contained much more hydrogen than Eddington had assumed earlier (Strömgren 1932). This demonstrates a characteristic feature of Strömgren's theoretical work through the years: he did not only develop the theory but also followed it up with detailed numerical calculations to make the theoretical advances directly applicable.

3. Physics of stars and HII regions

Bengt Strömgren was invited in 1936 by Otto Struve to an assistant professorship at the University of Chicago and Yerkes and McDonald Observatories. Here, he had detailed and continuing discussions with Subrahmanyan Chandrasekhar on problems of stellar structure, and discussed with William W Morgan on spectral classification - Morgan was close to finishing his and Philip Keenan's new two-dimensional classification of stellar

Figure 2. Yerkes Observatory staff in 1936 or 1937. First row, from left: S. Chandrasekhar (no. 2), B. Strömgren (4), G. P. Kuiper (6) and J. Greenstein (8). In second row: G. van Biesbroeck (1) and W.W. Morgan (7). In third row: O. Struve (3). Photo: Yerkes Observatory.

spectra, the MK classification. Strömgren also followed up Otto Struve's discovery of extensive areas in the Milky Way that emitted faintly in the hydrogen Balmer lines, and showed that if a hot early-type star is imbedded in a thin interstellar hydrogen cloud, the transition zone between the ionized gas close to the star and the more distant gas that remains neutral is quite narrow (Strömgren 1939). This research led to one of the three astronomical concepts that carry his name, the Strömgren sphere (the other two being an asteroid and his photometric system). After 18 months in the United States Strömgren returned to Copenhagen to take up a full professorship there, and soon also the directorship of the Copenhagen Observatory after his father.

Strömgren now turned to the solar atmosphere, including the opacity of the negative hydrogen ion, which Rupert Wildt had just found to be significant. Strömgren made a pioneering first model-atmosphere analysis of the solar spectrum and published the result in a Festschrift for his father's 70th birthday (Strömgren 1940). The values of chemical abundances derived agree astonishingly well with present values, within a factor of two for the elements studied. By this work he set a standard for quantitative analysis of stellar spectra, which was inspiringly high. In fact, when he summarised the stellar-atmosphere field at a conference (Strömgren 1955), he had the optimistic vision that an accuracy of one percent should be obtainable in predicted continuous fluxes and line profiles from model atmospheres. Not until recently, half a century later, has that goal been reached, and yet not for all types of stars and all spectral lines.

During the war in isolated Copenhagen, Strömgren took interest in geometrical optics and devised an original Schmidt refractor and calculated optical sine tables, both contributions that he was proud of. After the war he built up an astrophysics school in Copenhagen with young scientists like Jean Claude Pecker, Évry Schatzman and Anne Underhill. He also spent periods in the USA, and finally in 1950 he left Copenhagen to become professor at the University of Chicago and director of Yerkes and McDonald Observatories. It was during these tours to the USA that his work on the evolution of the Galaxy definitely began, with the development of the photometric system, the $uvby\beta$ photometry that is often called Strömgren photometry.

4. The new tools

The dominant kind of detectors used in astronomical photometry during the first half of the century were the photographic plates. Their low efficiency and non-linear response limited the accuracy of the resulting measures. Strömgren saw early new possibilities open up, which would finally drastically change this situation. As he has told in his autobiographical notes (Strömgren 1983), 14 years old he visited Paul Guthnick, at the Berlin-Babelsberg Observatory, who was pioneering photoelectric photometry with a 30 cm refractor equipped with a gas-filled photocell with a potassium cathod. In the early forties Strömgren experimented himself with electrometer vacuum tubes "and was partly successful". He also elaborated various aspects of stellar photometry in two chapters in *Handbuch der Experimentalphysik* (Strömgren 1937) and demonstrated the advantages of filters that were narrower than the traditional broad-band colour filters. In particular, the errors caused by extinction, both interstellar and terrestrial, due to the shift of the effective wavelength of the filters with changing stellar flux distributions, would limit the accuracy at high-precision work.

However, neither did the photographic plates enable such accuracy, nor were such narrow filters accessible. But after the war the situation was totally changed. The photomultiplier tube with its relatively high sensitivity and linear response, though invented already in the thirties, had now been applied to astronomy by Joel Stebbins and Albert Whitford, and their younger collaborators Gerald Kron, Olin Eggen, and Harold Johnson. Strömgren got acquainted with them and proceeded in 1948 at the McDonald Observatory to use the technique to measure the $H\beta$ strength in late-A and early F stars through interference filters, which had also now become available. He used two bands, both about 200 Å wide, one across the hydrogen line, and one beside it. He reported to the IAU General Assembly in 1948: "Very promising results have been obtained by the use of interference filters and photoelectric photometry for spectral classification.... a mean error corresponding to $0^m.006$ (one observation) was determined. It follows that spectral classification... can be effected with an accuracy of, for instance, better than one-tenth of a spectral class in the spectral ranges A5 to F0" (Strömgren 1950). Later, when narrower interference filters were available, he placed one 30 Å band on the line, and two broader comparison bands beside it, in defining what he called the l index (Strömgren 1954). Finally, some years later the Crawford-Strömgren $H\beta$ index was devised with one even narrower and one broad band, both centred on the line.

It is noteworthy, that Strömgren's successful first demonstration of the photoelectric technique for quantitative classification seems to have been a by-product of an attempt to survey interstellar Balmer emission. In general, his different studies in various fields were often linked in a logical and systematic way. However, Strömgren later explicitly ascribed his interest in accurate stellar photometry to his experience, through his model atmosphere work, that hydrogen criteria were indeed the most reliable indicators for temperature and surface gravity for stars hotter than the Sun (Strömgren 1983). Also, the previous photographic stellar spectro-photometry by Daniel Barbier and Daniel Chalonge

(1939), by Bertil Lindblad (1922) and Yngve Öhman (1935), as well as by Robert M. Petrie (Petrie & Maunsell 1949), inspired him to measure the Balmer lines in stellar spectra. These sources of inspiration were obviously significant also for his following attempts to measure the Balmer discontinuity photoelectrically. The first step, which he called the c index, and which was later developed into the c_1 index in the $uvby$ system, was thus established by 1951 (Strömgren 1951a). With his background he found it important to position the ultraviolet u band of the new system with its transmission entirely at shorter wavelengths than the Balmer discontinuity, unlike the location of the U band in the pioneering Six-colour photometry of Stebbins & Whitford (1943) or the corresponding band in the broad-band UBV system (Johnson & Morgan 1951), which was established in parallel with Strömgren's work at Yerkes and McDonald observatories. In addition, already at this stage Strömgren set out to explore the possibilities to establish a two-dimensional classification for the late-type stars, from F8 to K6, by photoelectric photometry together with Kjeld Gyldenkerne (see also Strömgren & Gyldenkerne 1955). Also here, the inspiration came from earlier photographic work in Uppsala and Stockholm by Bertil Lindblad and collaborators (see Lindblad & Stenquist, 1934, and references therein). Strömgren enthusiastically reported probable errors in this spectral classification "of two to four hundreds of a spectral class". Finally, this effort was to lead to the Brorfelde or $gnkmf$ system, and also inspired other narrow-band systems like the DDO system and the Vilnius system.

Strömgren had with his l-c system devised a method for two-dimensional quantitative spectral classification which could provide estimates of stellar temperature and luminosity or surface gravity, a system which would soon be shown to be the most accurate method for stars in the spectral interval B-F. However, in the early fifties a very important discovery was made by Joseph Chamberlain and Lawrence Aller (1951). They analysed spectra the two subdwarfs HD 19445 and HD 140283 and proved them to be significantly metal poor, by a factor of ten or even more as compared with the Sun. This was immediately recognized to be a strong indication that the heavy elements were, indeed, made in stars, and that the gradual build-up of these elements could be traced by measuring the metallicities of stars of various ages and populations. Studying stars of different metallicity should be a way of studying the history of our Galaxy. A photometric system for such purposes obviously also needed a way of estimating metallicity. In order to measure the metallicities quantitatively, Strömgren added an m index to his photometric system. It had three bands with half widths of about 90 Å, one across a region with strong metal lines at 4030 Å, one at 4500 Å and one at 5000 Å, and the index was defined as m=constant - 2.5[log $I(5000)$+log $I(4030)$-2log $I(4500)$]. Three years later, Nancy Roman had demonstrated the existence of a group of F stars with high spatial velocities and very weak metal lines (Roman 1954). Strömgren could show that Nancy Roman's stars stood out from the typical F stars by about 0.1 magnitudes in the m index, more than 10 times the observational error in m, and that the index value correlated with the ultraviolet excess $\delta(U$-$B)$ which was thought to be a metallicity indicator (Strömgren 1958a).

5. The populations

In May 1957, a conference on Stellar Populations was arranged at the Pontifical Academy of Science in the Vatican. This conference turned out to be one of the important ones in the history of 20th century astronomy. The participants were about 20 of the leading astronomers, including Walter Baade, Adrian Blaauw, William Fowler, George Herbig, Fred Hoyle, Georges Lemaître, Bertil Lindblad, W.W. Morgan, Edwin Salpeter,

Figure 3. Participants at the Vatican Conference on Stellar Populations in May 1957. First row, from left: P. Salviucci, B. Lindblad, J.H. Oort, W. Baade, D. O'Connell, G. Armellini, G. Lemaitre, H.A. Brueck. Second row: J.J. Nassau, D. Chalonge, M. Schwarzschild, A. Blaauw, O. Heckmann, F. Hoyle, W.W. Morgan, G.H. Herbig, A.R. Sandage. Third row: E.E. Salpeter, B. Strömgren, L. Spitzer, W.A. Fowler, A.D. Thackeray, V. Preobrajenski, P. Treanor. Photo: Specola Vaticana.

Allan Sandage, Martin Schwarzschild, Lyman Spitzer and Bengt Strömgren. Strömgren presented two papers, one on the composition differences between stellar populations, based on his m-index measurements (Strömgren 1958a), and one on spectrophotometric classification of the population groups, where also the power of the l and c indices was demonstrated (Strömgren 1958b). It is greatly interesting to now, half a century later, read the well minuted discussions at this symposium (O'Connell 1958). Although the concept of stellar populations in the Galaxy had been introduced by Baade already in 1944 in his studies of the stellar content in the Andromeda galaxy and its companion galaxies (Baade 1944), and in fact hinted at by Oort (1926) and Lindblad (1925, 1926, 1927) almost two decades before on kinematical grounds, the full complexity of the concept and its significance for the evolution of the Galaxy were now established at the conference. This was in particular discussed in a presentation at the Symposium by Jan Oort (Oort 1958), who distinguished five different stellar populations: Halo population II, Intermediate population II, Disk population, Intermediate population I and Extreme population I with different characteristics. A great number of questions concerning the origin and interrelations between these various populations were discussed but remained unanswered. A discussion of this conference and its importance for the population concept was presented by Blaauw at the IAU Symp. 164 (Blaauw 1995). The Vatican conference formed a significant basis for Strömgren's continued efforts to study and map the evolution of the Galaxy.

6. The programme

It is clear that Strömgren already the following year had essential parts of his programme ready for Galactic stellar astronomy. In his Halley lecture from 1958 (Strömgren 1958c) he reported about the potential of his lc system: "It would seem that photoelectric narrow-band photometry can lead to information on the initial composition of stars and their ages which may help our efforts to form a picture of the evolution of the Galaxy." He presents the basics of his methods and states, boldly, for B stars: "It is now a simple matter to derive the age of a star from observed values of l and c". For F stars, he claims that ages to a relative accuracy of 10-30% should be possible to obtain: "... such age determinations may prove to be of value. No theoretical mass-age calibration of the $c - l$ diagram for Population I F stars has as yet been carried out, but it is quite feasible to do so". At this stage Strömgren obviously realized that his accurate photometry offered possibilities to measure the location of individual stars within the main-sequence band, which would enable age determinations, and to determine metallicities with relative errors significantly less than a factor of two. This latter accuracy was fully comparable with what one could obtain for (only) bright stars with high-resolution spectrometers and photographic detectors. Not until the development of echelle spectrometers and electronic detector arrays in the seventies could significantly more accurate metallicity determinations be made.

The importance and the possibilities of these achievements within "quantitative spectral classification" were certainly not yet fully understood in the astronomical community. And a number of efforts were needed before the great questions could be attacked. The photometric system had to be optimized for its purpose. Standard stars had to be established, and the system should be tried out and calibrated, and its potential should be explored and verified. Telescope and photometers should be developed for effective surveys. This required hard work and could not be a single man's effort. Collaborators were needed and this required schooling and inspiration, leadership and organization. And, to use a modern word, selling of the ideas, for instance by picking some striking examples demonstrating the power of the approach.

First, the $uvby$ system had to be finally established, with optimum choices of wavelength bands and observations of standards. This was accomplished by 1960 by observation runs at Lick and Palomar observatories, together with Charles Perry and Robert Cameron. The $H\beta$ index was developed at Kitt Peak by David Crawford, and there Strömgren also continued his $uvby$ observations.

Next, the system and its possibilities had to be presented to a broad community. This was done in reviews at symposia, and in prize lectures like the Halley Lecture from 1958 and the George Darwin Lecture (Strömgren 1963a). However, descriptions detailed enough to serve as introductions and even "manuals" for future users of the system, including students, were also needed. Strömgren provided two such descriptions, one in the monograph series *Stars and stellar systems* (Strömgren 1963b) and one particularly important in the then relatively new series of *Annual Reviews of Astronomy & Astrophysics* (Strömgren 1966b).

7. Birth places and worries about metallicity

The key aim of the new system was thus to determine stellar fundamental parameters, effective temperature, surface acceleration of gravity and metallicity, and determine them well enough to be able to deduce stellar ages and masses by comparison to evolutionary tracks. In the mid sixties Strömgren had demonstrated in practice that this was indeed possible, for B, A, and F stars in the main-sequence band. He showed that he was able to obtain an accuracy better than 10% in stellar age, at least on a relative scale (Strömgren 1966a, Kelsall and Strömgren 1965). It is characteristic that this work was based, in addition to Strömgren's photometry, on stellar evolution calculations by Thomas Kelsall in collaboration with himself; he was anxious to have full insight into all critical aspects in his work, and he was certainly competent enough in stellar modelling and its physical and numerical aspects for this to be possible.

For the younger stars he was able to calculate their orbits backwards in time and thus demonstrate that their places of birth were located in spiral patterns according to the C C Lin spiral-arm theory. He reported to the IAU Symposim 31 in 1967: "For the 52 stars of the sample we computed the places of formation, using the Galactic gravitational field adopted by Contopoulos and Strömgren (1965) in their calculation of tables of plane Galactic orbits.... The results... do suggest that the majority of the stars considered were formed in our own spiral arm, and that about 10 to 15% of the stars originated 2 or 3 kpc further away from the Galactic center, i.e. in the Perseus Arm. There is in the sample discussed no evidence for a component contributed by star formation in an inner arm" (Strömgren 1967). Although one could feel that Strömgren sometimes was too optimistic about the accuracy of his results, this was an impressive demonstration of the power of the method. It is also noteworthy that Strömgren had again found a leading specialist, G. Contopoulos in Galactic dynamics, to join in the efforts, but not left the work to calculate orbits to him alone.

For the metallicities based on the m_1 index there was a worry, raised by Bashek (1960), Conti & Deutsch (1966) and others, that the majority of the metal lines affecting the v band were located on the flat part of the curve of growth, and thus sensitive to the microturbulence parameter more than to the metal content of the atmospheres. The issue was then whether the microturbulence parameter varied independently of the fundamental parameters of the stars – effective temperature, surface gravity and metallicity (or, more fundamentally, mass, age, and chemical composition). Strömgren found that the tight correlation between the departure of his observed m_1 from the standard sequence, Δm_1, and spectroscopic [Fe/H] values indicated that the effects of additional microturbulence variations were minor (Strömgren 1966b). A decade later, this view was confirmed by spectrophotometry at high resolution by Poul Erik Nissen and others.

Yet, it must be admitted that the m_1 index was a weak point of the photometric system. Both the Balmer δ line and the blue bands of the Violet CN system are located in the v band, and affect the index for the hotter and cooler stars. More serious is its relatively low sensitivity to metallicity. The alternative to use bands with weaker but more sensitive spectral lines on the linear part of the Curve of Growth does not exist at this spectral resolution for solar-type stars; such methods were, however, later exploited at higher spectral resolution by Nissen, Kjærgaard and others. However, the possibility to go somewhat towards the ultraviolet in the spectrum and then to also include the wings of strong lines like the Ca II H and K lines, as is presently tried with the v band of the SkyMapper project (Murphy et al. 2008), could have been attractive. However, such a bandpass would have mixed the abundances of the α element Ca and Fe. Since these abundances have later been found to vary relative to each other for various stellar populations, that might have led to some confusion when uvby photometry was used for defining samples of stars.

8. The large-scale studies

Strömgren returned to Copenhagen University and Nordita, the Nordic Institute for Theoretical Physics located at the Niels Bohr Institute, in 1967. He systematically pursued his efforts in Galactic exploration by means of his photometric system. New efficient four-channel photometers were built, a Danish 50 cm and a 1.5 m telescope were erected at ESO on La Silla and used a lot for photometry. He engaged a number of young collaborators, like Erik Heyn Olsen, Bent Grønbech and Jens Knude in photometric work, and Poul Erik Nissen in calibrating his photometry by measuring accurate spectroscopic metallicities. He was systematically investigating the potential of his system. When I came to Nordita as a fellow in 1968, Strömgren suggested to me to study the properties of the uvby system for anlysing integrated light from galaxies, with the aim of characterising their stellar population, and in particular to study the possibilities to resolve what was later called the age-metallicity degeneracy in population synthesis. He was also anxious to devise criteria to be able to single out horizontal-branch stars of intermediate population II; he thought of stars like those on the stubby red horizontal branch of the metal-rich globular cluster 47 Tuc, such stars in the Galactic field should be excellent tracers for that population. This required model predictions, and that led me into his old field of stellar model atmospheres.

It should also be mentioned that Strömgren contributed excellent lecture series during these years, both at Nordita/Niels Bohr Institute in its classical Auditorium A, and at the Astronomical Observatory. At the first place he gave series of lectures on stellar evolution and nucleosynthesis, on Galactic dynamics, and also a series on galaxies. These lectures, followed by a wide audience of physicists and astronomers, were dealing with the present research developments and contained excellent summaries and comments on disputed issues. In this art of judging and summarizing recent developments, Strömgren was a true master. These lectures also often reflected Strömgren's own deep research experience. The lecture series given at the Observatory were more operational, dealing with topics like spectrosopy, photometry, and stellar atmospheres. These were very useful as a research education, and in particular valuable for those who were to continue projects along the lines of Strömgren's large programme. It would, in fact, be reasonable to see these lectures as part of the programme itself.

For the F- and early G-type stars, Strömgren was now eager to obtain the distributions of stars on different metallicity within well-defined volumes of space, and to have enough stars to be able to subdivide them into different age bins and space velocities (Strömgren 1969). He accurately mapped the zero-age main sequences of different rich open clusters in order to trace possible differences in helium abundances (Strömgren et al. 1982). And he investigated the distribution of interstellar dust in the solar vicinity by accurately determining the colour excesses of the F stars (Strömgren 1972). These studies had been pioneered by Strömgren earlier but were now, during the seventies and eighties pursued further by his younger collaborators together with him.

It is interesting to see how Strömgren in his late papers, at an age above 75, still worked towards and stressed his great vision to empirically trace the Galactic populations, and their origins and interrelations. He was particularly intrigued by the discovery by Gilmore and Reid (1983) of a "thick disk", with a scale height of more than 1000 pc, as compared with the standard disk with its scale height of about 300 pc, and comprising about 2% of stars in the solar neighbourhood. The physical character of this population, and its relation to "his own" Intermediate Population II, challenged him. In his contribution to IAU Symposium 106, *The Milky Way Galaxy* (Strömgren 1985), he discussed no less than 7 ongoing studies, of which he was involved in most, that were directed towards the clarification of the histories of the Galactic populations. Among these were the photometric surveys of Erik Heyn Olsen, the radial-velocity survey of Johannes Andersen, Michel Mayor and Birgitta Nordstöm, Jens Knude's studies towards the North Galactic pole, and Torben B Andersen's towards the South polar cap. In other late papers he commented on ongoing spectroscopic work by Nissen and collaborators, as well as by Bengt Edvardsson *et al.*, which can also be seen as part of his grand programme.

It was moving to meet him in these years and note his eagerness to learn about the results as soon as they started appearing; a feeling that one is reminded of when reading his late papers reflecting his curiosity, and his understanding that time was now limited. A typical sentence from this time, taken from one of his very last papers (Strömgren 1987), is: "However, as emphasized...a further strengthening of the data base and a revision of the age scale for intermediate population II stars is required before more definitive conclusions can be drawn. We have referred above to the observational and theoretical work now in progress, which within one or two years should lead to the desired clarification." We, who were engaged in this "work now in progress" know that it took longer time, that the picture got even more complex, and that Strömgren, as much as he would have enjoyed seeing the results, would have appreciated the growing complexity.

9. What can we learn?

What can we learn from Strömgren's example, and in particular his approach towards understanding the Galactic disk? In considering this question I first wish to cite Johannes Andersen, who has made the following important remark: "His achievements were so logically designed and systematically pursued that it appeared all natural and self evident to a later generation. How could it be otherwise? Yet, it *was* otherwise until he showed the way."

Andersen stressed Strömgren's logical and systematic approach. Strömgren contemplated what to do, and did it in the right order and right way. In his teaching, he required us to answer a number of questions in the right order: "What do we wish to observe? Why do we observe this? How do we observe, what means do we use? And Where and When do we do this?" Answer the questions is this order, not, as is often done in practice, in the converse order! And even more important, carry out your project according to a systematic plan, which may well take considerable time. His programme was not so rigid that he did not modify it when new discoveries or experiences so required, but he did not redirect the basic aims, nor did he divert from his main track to follow up other sudden ideas, except for some short excursions.

Second, Strömgren through his example and also in his teaching underlined the importance of being methodologically advanced, and to master your tools. This entailed a considerable interest in details. One example, just out of my own recollection; others at this meeting can give many more: When I decided to start calculating blanketed and convective model atmospheres I needed a flexible and rapid method for radiative transfer with scattering. I had happened to come across Feautrier's new method just developed in the *Annales d'astrophysique*, and mentioned that to Strömgren. "Oh yes," he said, "that is a very good one, but beware of the mis-prints in his *Comptes Rendus* publication! They are not so easy to see!" Strömgren kept himself updated, so much that he was used by numerous of us more lazy readers as a living advisory encyclopedia. His encyclopedic knowledge was also evident at seminars and meetings when he would sit in the middle of the front row and always ask a question or make a comment that would enlighten the audience on the real importance of the work being presented. Another characteristic was his willingness to engaging himself in seemingly routine aspects of a project, such as calculating tables, reducing photometric observations or plotting data. He presumably wanted to make sure that everything was correctly done, and digested the information in the data by working on them. I also had the impression that this "handicraft" gave him a certain pleasure in itself, maybe a certain kind of calm that you experience when you do more or less difficult things that you are used to do and master since very long.

In approaching his scientific problems, Strömgren was continuously combining astrophysics and dynamics, developing his already deep insights into most fields of astronomy and exercising his methodological skills. His unusual mastery of all sorts of frontier methodology, from complex numerical calculation to the design of new optical systems, from astrometry to detailed spectroscopy, from quantum-mechanical calculations to reducing observations, from lining up observing programs to making model atmospheres and analysing stellar abundances, is of course not easy to match today. Our mode of operation, in research teams with different specialities represented by different individuals, may be necessary. But Strömgren's mode had its virtues; in particular, it made it possible for him to answer his questions in the right order, starting by defining the problem. He could design his research optimally and follow the problem wherever it brought him methodologically. Maybe we should attempt learning more methods throughout life, and not allow ourselves to become specialists in only one narrow field.

Simultaneously with this keen interest in details, Strömgren was also a grand visionary. Those visions were, however, almost obvious to him. In his autobiographical notes (Strömgren 1983) he states in the end, after having listed a number of planned applications of his new methods: "The connection of the various research goals just mentioned with broader areas of investigation - Galactic spiral structure, star formation, early phases of Galactic evolution - is clear, and I shall not comment on it here." When looking back at his work, however, it is also clear that he succeeded in moving through his interesting landscape of details, with lots of interest and care of these details, without losing track. It is clear from some of his reviews, as well as from his popular articles, that he was what Zygmunt Baumann has called a pilgrim, systematically travelling towards a goal beyond the horizon but on the way cultivating himself, his pupils, and his surroundings in an engaged way (Baumann 1993).

Another aspect gets also very clear when reading his papers. He was very much an empiricist, in spite of his extraordinary theoretical skills. His approach towards Galactic evolution was similar to that of a true historian of nature, or even an archaeologist. In discussing Nobel Prize candidates once, he said: "One should always ask oneself the question whether the scientific world would have been any different if this person had not lived." To him, this pointed more towards people who had done innovative empirical work than those who had come up with fancy speculations first. Incidentally, Strömgren lived up to this criterion; astronomy would have been different and less interesting if he had not lived.

Yet, another characteristic of Strömgren, and his way of working, was his interest in other people, not the least young scientists, and their work. He contributed important inspiration to younger colleagues in Denmark and elsewhere to develop their own specialities and gave them their place in his grand vision, from the astrometrical work of Erik Høg and Lennart Lindegren, leading to the Hipparcos and Gaia satellites to the breakthrough at Nordita and Copenhagen Observatory as regards simulation of convection in stars and the Sun by Åke Nordlund. He once had an appointment with the director at the HC Ørsted Institute, on the other side of the Fælled Park, and I accompanied him there in order to visit the place. He suggested that we should enter a side door of the building, and we came into the workshop, where he immediately engaged in discussion with an engineer. We next proceeded to another floor where he soon was discussing thesis projects with graduate students that he had not met before. During the final, and rather late, approach towards the office of the Director of the Institute, he remarked to me: "It is a good idea to talk to the young fellows first. You tend to learn more from them." In this way, many young scientists have been encouraged by Strömgren's interest and benefited from his advice.

Finally, one could say that Strömgren showed great skill in "being at the right place at the right time", being in Copenhagen during the grand era of Quantum Mechanics, in Chicago and Yerkes Observatory with Otto Struve, Chandrasekhar and W.W. Morgan, being in the Nordic area and in touch with Bertil Lindblad (who served as President of the IAU when Strömgren was Secretary General) who could point out the previous Swedish work on low-resolution spectrophotometric criteria when Strömgren contemplated his photometric system, and being at The Center for Advanced Study at Princeton during its golden age in the fifties and sixties. Maybe one could also say that returning to Europe when ESO developed, and to Denmark and a team of young Danes on long-term contracts, willing to undertake major, systematic studies under his supervision at a time when his programme had reached a stage when such studies were needed, was also a very wise move.

However, this is certainly an argument in retrospect. Maybe, he would under quite different circumstances still have formed an impressive programme which would have been consistent and logical. Also, the very idea that he, in his various environments, was *exposed* to various expertise, scientific and technological developments, is perhaps not the most relevant aspect. He was so scientifically curious and interested that he found people with new good ideas and interacted with them. Then, his extraordinary ability to combine and integrate new patterns of these various ideas, thoughts, methods, facts, into a wholeness made these interactions rewarding for all who met him. We could bring home a message from this, a message which almost seems to be inscribed in the title of his autobiographical notes (Strömgren 1983), *Scientists I have known and some astronomical problems I have met*: If you are open-minded and generous when meeting scientists so you learn to know them and their worlds, you also meet good problems.

10. Final words

Within my limited space and time there has been no possibility to properly review Strömgren's very significant contributions on the organisational level, for Danish astronomy and science in general, as well as for European and international astronomy at large. But in one of his administrative functions, that of Director of Yerkes Observatory in the middle of last century, he gave a speech on the development of astrophysics during the first half of the 20th century, at the 50th anniversary of the Observatory (Strömgren 1951b). He cited Simon Newcomb who, at the dedication of the observatory in 1897, had said: "Slow indeed is progress in the solution of the greatest problems, when measured by what we want to know. Some questions may require centuries, others thousands of years for their answer. And yet never was progress more rapid than during our time. In some directions our astronomers of today are out or sight of those of fifty years ago; we are even gaining heights which, twenty years ago, looked hopeless. Never before had the astronomer so much work, good, hard, yet hopeful work before him as today."

"Another 50 years have passed", said Strömgren, "but Newcomb's comment is as fresh and pertinent today as it was when the Yerkes Observatory first opened its doors." Well, Strömgren and others have, not only literally spoken, opened doors to advanced science to us. We have just to be thankful, enjoy it, and work hard.

References

Baade, W. 1944, The resolution of Messier 32, NGC 205, and the central region of the Andromeda nebula, *ApJ*, 100, 137

Barbier, D. & Chalonge, D. 1939, Remarques préliminaires sur quelques propriétés de la discontinuité de Balmer dans les spectres stellaires, *Ann. Ap.*, 2, 254

Baschek, B. 1960, Abhängigkeit des Strömgrenschen Index m und der Farbindexen U-B, B-V von der Metallhäufigkeit bei sonnenähnlichen Sternen, *Z. Astrophys*, 50, 296

Baumann, Z. 1993, *Postmodern ethics*, Cambridge, MA

Blaauw, A. 1995, Stellar evolution and the population concept after 1950; The Vatican conference, *Stellar populations. Proc IAU Symp 164*, eds. P.C. van der Kruit, G. Gilmore, p. 39

Chamberlain, J. W. & Aller, L. H. 1951, The atmospheres of A-type subdwarfs and 95 Leonis, *ApJ*, 114, 52

Conti, P. S. & Deutsch, A. J. 1966, Color anomalies and metal deficiencies in solar-type disk-population stars, *ApJ*, 145, 742

Contopoulos, G. & Strömgren, B. 1965, Tables of plane Galactic orbits, *Publ. Inst. Space Studies, Goddard Space Flight Center, NASA*

Gilmore, G. & Reid, N. 1983, New light on faint stars. III - Galactic structure towards the South Pole and the Galactic thick disk, *MNRAS*, 202, 1025

Johnson, H. L. & Morgan, W. W. 1951, On the colour-magnitude diagram of the Pleiades, *ApJ*, 114, 522

Kelsall, T. & Strömgren, B. 1965, Calibration of the Hertzsprung-Russell diagram in terms of age and mass for main-sequence B and A stars, *Vistas in Astronomy*, 8, 159

Lindblad, B. 1922, Spectrophotometric methods for determining stellar luminosity, *ApJ*, 55, 85

Lindblad, B. 1925, 1926, *Uppsala Medd. No*, 3, 4, 6, 13

Lindblad, B. 1927, On the state of motion of the Galactic system, *MNRAS*, 87, 553

Lindblad, B. & Stenquist, E. 1934, On the spectrophotometric criteria of stellar luminosity, *Stockholm Obs. Astr. R.*, 11, No. 12

Murphy, S., Keller, S., & Schmidt, B., et al. 2008, SkyMapper and the Southern Sky Survey; a valuable resource for stellar astrophysics, *ASP Conf. Ser.*, XXX, YYY In press, presently http://fr.arxiv.org/abs/0806.1770

O'Connell, D. J. K. 1958, ed., *Stellar Populations. Proceedings of the Conference sponsored by the Pontifical Academy of Science and the Vatican Observatory*, North Holland Publ. Co., Amsterdam, Interscience Publishers, Inc., New York

Öhman, Y. 1935, Some preliminary results from a study of hydrogen absorption for stars in h and χ Persei, *Stockholm Obs. Ann.*, 12, 1

Oort, J. H. 1926, The stars of high velocity, *Groningen Pub.*, 40, 1

Oort, J. H. 1958, Dynamics and evolution of the Galaxy, in so far as relevant to the problem of the populations, in *Stellar Populations. Proceedings of the Conference sponsored by the Pontifical Academy of Science and the Vatican Observatory*, ed. D.J.K. O'Connell, p. 415

Petrie, R. M. & Maunsell, C. D. 1949, The Stark effect in hydrogen as a criterion of luminosity in the early A-type stars, *PASP*, 61, 158

Rebsdorf, S. O. 2005, *The Father, the Son and the Stars. Bengt Strömgren and the History of Twentieth Century Astronomy in Denmark and in the USA*, The Steno Institute, University of Århus

Roman, N. G. 1954, A group of high velocity F-type stars, *AJ*, 59, 507

Stebbins, J. & Whitford, A. E. 1943, Six-color photometry of stars I. The law of space reddening from the colors of O and B stars, *ApJ*, 98, 20

Strömgren, B. 1932, The opacity of stellar matter and the hydrogen content of the stars, *Z.f. Astrophys.*, 7, 222

Strömgren, B. 1937, Aufgaben und Probleme der Astrophotometrie, and Objektive photometrische Methoden, in *Handbuch der Experimentalphysik*,26, p. 321 and 797

Strömgren, B. 1939, The physical state of interstellar hydrogen, *ApJ*, 89, 526

Strömgren, B. 1940, On the chemical composition of the solar atmosphere, *Festschrift für Elis Strömgren. Astronomical papers dedicated to Elis Strömgren*, Einar Munksgaard, Copenhagen, p. 218

Strömgren, B. 1950, Spectrophotometry of stars with interference filters, *IAU Trans*, VII, p. 404

Strömgren, B. 1951a, Spectral classification through photoelectric photometry with interference filters, *AJ*, 56, 142

Strömgren, B. 1951b, On the development of astrophysics during the last half century, in *Proceedings of a topical symposium, commemorating the 50th anniversary of the Yerkes Observatory and half a century of progress in astrophysics*, McGraw-Hill, New York, ed. J.A. Hynek, p. 1

Strömgren, B. 1954, Spectral classification through photoelectric photometry in narrow wavelength regions, *AJ*, 59, 193

Strömgren, B. 1955, Introductory remarks: Stellar atmospheres and line profiles, in *Proc. of the NSF conference on Stellar Atmospheres*, ed. Marshal H. Wrubel, p. 90

Strömgren, B. & Gyldenkerne, K. 1955, Spectral classification of G and K stars through photoelectric photometry with interference filters, *ApJ*, 121, 43

Strömgren, B. 1958a, Composition differences between stellar populations, in *Stellar Populations. Proceedings of the Conference sponsored by the Pontifical Academy of Science and the Vatican Observatory*, ed. D.J.K. O'Connell, p. 245

Strömgren, B. 1958b, Spectrophotometric classification of the population groups, in *Stellar Populations. Proceedings of the Conference sponsored by the Pontifical Academy of Science and the Vatican Observatory*, ed. D.J.K. O'Connell, p. 385

Strömgren, B. 1958c, The composition of stars and their ages, *Observatory*, 78, 137

Strömgren, B. 1963a, Problems of internal constitution and kinematics of main sequence stars, *Quarterly Journal of the RAS*, 4, 8

Strömgren, B. 1963b, Quantitative classification methods, in *Basic Astronomical Data: Stars and stellar systems*, ed. K.A. Strand, University of Chicago Press, Chicago, p. 123

Strömgren, B. 1966a, Age determination for main-sequence B, A, and F stars, in *Stellar evolution. Proceedings of an international conference, Nov. 13-15, 1963, sponsored by the Institute for Space Studies of the Goddard Space Flight Center, NASA*, eds. R.F. Stein, A.G.W. Cameron, Plenum Press, New York, p. 391

Strömgren, B. 1966, Spectral classification through photoelectric narrow-band photometry, *Ann. Rev. A&A*, 4, 433

Strömgren, B. 1967, Places of formation of young and moderately young stars, *IAU Symp.*, 31, 323

Strömgren, B. 1969, Quantitative Spektralklassifikation und ihre Anwendung auf Probleme der Entwicklung der Sterne und der Milchstrasse (Karl-Schwarzschild-Vorlesung), *Mitt. Astron. Ges.*, 27, 15

Strömgren, B. 1972, Interstellar reddening within 200 pc of the Sun, *Quarterly Journal of the RAS*, 13, 153

Strömgren, B., Olsen, E. H., & Gustafsson, B. 1982, Evidence of helium abundance differences between Hyades stars and Coma cluster stars, *PASP*, 94, 5

Strömgren, B. 1983a, Scientists I have known and some astronomical problems I have met, *Ann. Rev. A&A*, 11, p. 1

Strömgren, B. 1985, Star counts, local density and K(z) force, *IAU Symp.*, 106, 153

Strömgren, B. 1987, An investigation of the relations between age, chemical composition and parameters of velocity distribution based on $uvby\beta$ photometry of F stars within 100 pc, *The Galaxy*, eds. G. Gilmore and B. Carswell, D. Reidel Publ. Co., p. 229

Session 1: Disk galaxies throughout time and space

Session chairs: Joe Silk, Árdis Éliasdottir and Steen Hansen

Kirsten Hastrup, President of the Royal Danish Academy of Sciences and Letters, and Birgitta Nordström discussing at Carlsberg. In the background: Hans Bohr (center) and Preben Grosbøl (right)

Simon White, Ken Freeman, Linda Sparke and Gerry Gilmore preparing the meeting at the welcome reception.

Simulations of Disk Galaxy Formation in their Cosmological Context

Simon D.M. White

Max Planck Institute for Astrophysics
Karls-Schwarzschild-Strasse 1
D-85741 Garching, Germany
E-mail: swhite@mpa-Garching.mpg.de

Abstract. Together with the discovery of the accelerated expansion of the present Universe and measurements of large-scale structure at low redshift, observations of the cosmic microwave background have established a standard paradigm in which all cosmic structure grew from small fluctuations generated at very early times in a flat universe which today consists of 72% dark energy, 23.5% dark matter and 4.5% ordinary baryons. The CMB sky provides us with a direct image of this universe when it was 400,000 years old and very nearly uniform. The galaxy formation problem is then to understand how observed galaxies with all their regularity and diversity arose from these very simple initial conditions. Although gravity is the prime driver, many physical processes appear to play an important role in this transformation, and direct numerical simulation has become the principal tool for detailed investigation of the complex and strongly nonlinear interactions between them.

The evolution of structure in the gravitationally dominant Cold Dark Matter distribution can now be simulated in great detail, provided the effects of the baryons are ignored, and there is general consensus for the results on scales relevant to the formation of galaxies like our own. The basic nonlinear units are so-called "dark matter halos", slowly rotating, triaxial, quasi-equilibrium systems with a universal cusped density profile and substantial substructure in the form of a host of much less massive subhalos which are concentrated primarily in their outer regions.

Attempts to include the baryons, and so to model the formation of the visible parts of galaxies, have given much more diverse results. It has been known for 30 years that substantial feedback, presumably from stellar winds and supernovae, is required to prevent overcooling of gas and excessive star formation in the early stages of galaxy assembly. When realistic galaxy formation simulations first became possible in the early 1990's, this problem was immediately confirmed. Without effective feedback, typical halos produced galaxies which were too massive, too concentrated and had too little disk to be consistent with observation.

Simple models for disk formation from the mid 1990's show that the angular momentum predicted for collapsing dark halos is sufficient for them to build a disk population similar to that observed. Direct simulations have repeatedly failed to confirm this picture, however, because nonlinear effects lead to substantial transfer of angular momentum between the various components. In most cases the condensing baryonic material loses angular momentum to the dark matter, and the final galaxy ends up with a disk that is too compact or contains too small a fraction of the stars.

These problems have been reduced as successive generations of simulations have dramatically improved the numerical resolution and have introduced "better" implementations of feedback (i.e. more successful at building disks). Despite this, no high-resolution simulation has so far been able to produce a present-day disk galaxy with a bulge-to-disk mass ratio much less than one in a proper ΛCDM context. Such galaxies are common in the real Universe; our own Milky Way is a good example. The variety of results obtained by different groups show that this issue is very sensitive to *how* star formation and feedback are treated, and all implementations of these processes to date have been much too schematic to be confident of their predictions.

The major outstanding issues I see related to disk galaxies and their formation are the following: Do real disk galaxies have the NFW halos predicted by the ΛCDM cosmology? If not, could the deviations have been produced by the formation of the observed baryonic components, or must the basic structure formation picture be changed? How are Sc and later type galaxies made? Why don't our simulations produce them? What determines which galaxies become barred and which not? Can we demonstrate that secular evolution produces the observed population of (pseudo)bulges from pre-existing disks? How does the observed population of thin disks survive bombardment by substructure and the other transient potential fluctuations expected in ΛCDM halos? Is a better treatment of feedback really the answer? If so, can we demonstrate it using chemical abundances as fossil tracers? And how can we best use observations at high redshift to clarify these formation issues?

Keywords. cosmology: theory, dark matter – galaxies: formation – galaxies: structure – galaxies: evolution – galaxies: halos

---

The presentation made at the meeting is available at:

http://www.mpa-garching.mpg.de/~swhite/IAU254.pdf

Simon White setting the scene for the symposium.

Disk Galaxies at High Redshift?

Max Pettini[1]

[1]Institute of Astronomy, Madingley Road, Cambridge CB3 0HA, UK

Abstract. The successful implementation of integral field near-infrared spectrographs fed by adaptive optics is providing unprecedented views of gas motions within galaxies at redshifts $z = 2-3$, when the universe was forming stars at its peak rate. A complex picture of galaxy kinematics is emerging, with inflows, rotation within sometimes extended and nearly always thick disks, mergers, and galaxy-wide outflows all contributing to the variety of patterns seen. On the computational side, simulations of galaxy formation have reached a level of sophistication which can not only reproduce many of the properties of today's galaxies, but also throws new light on high redshift galaxies which are too faint to be detected directly, such as those giving rise to quasar absorption lines. In this brief review, I summarise recent progress in these areas.

Keywords. galaxies: structure – galaxies: kinematics and dynamics – galaxies: evolution – galaxies: high-redshift – (galaxies:) intergalactic medium

1. Introduction

In the last ten years we have witnessed a remarkable growth in the depth and breadth of galaxy redshift surveys which now stretch out to redshifts beyond $z = 6$, sampling some 90% of the age of the universe. The underlying goal of such endeavours is to follow the process of galaxy formation and subsequent evolution, from the first objects to form at the end of the 'dark ages' all the way to the variety and beauty of the Hubble sequence of present-day galaxies. The kinematics of galaxies at different redshifts are of course a key ingredient of this whole story. In this brief review I shall concentrate on the epochs corresponding to $z = 2-3$, where galaxy kinematics are increasingly drawing the attention of observers and theorists alike, now that a rough picture is taking shape of many of the physical properties of the varied galaxy populations already in place at that time.

The rotation curves of nearby spiral disks which we are all familiar with are assembled mostly from radio observations of atomic and molecular gas extending well beyond the luminous dimensions of galaxies. With the exception of sub-mm detected galaxies and AGN, current observational facilities lack the spatial resolution and sensitivity to conduct analogous measurements at high redshift. Instead, the internal kinematics of galaxies at $z = 2-3$ are being probed via the narrow emission lines, primarily Hα and [O III] $\lambda 5007$, formed in the H II regions of galaxies during a period which saw the peak of the star-formation activity in the cosmic history. At $z > 1$, these nebular lines are redshifted into the near-infrared (near-IR), but are still detectable in the J, H, K atmospheric windows, once the OH emission from the night sky has been carefully subtracted—an example is reproduced in Figure 1.

As we heard at this meeting, Bengt Strömgren was among the first to realise the value of mapping the distribution of ionised gas in the Milky Way via the Balmer lines. Half a century later, the technology available to today's astronomers makes it possible to extend the same quest to the distant universe, probing galaxies only 2–3 Gyr after the Big Bang.

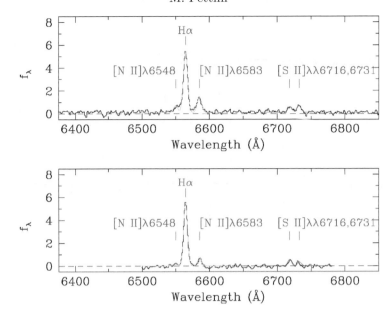

Figure 1. Composite spectra of star-forming galaxies at $z \simeq 2.2$ from the survey by Erb *et al.* (2006). At these redshifts, the emission lines (shown here at their rest-frame wavelengths) are redshifted into the near-IR K band; these spectra were obtained with the NIRSPEC instrument on the Keck II telescope. Pettini & Pagel (2004) proposed that the ratio of the [N II] and Hα emission lines, the $N2$ index, is a convenient approximate measure of the nebular oxygen abundance. Galaxies brighter than $K_s = 20$ (*upper panel*) have a mean ratio [N II]/H$\alpha = 0.25$ which indicates an oxygen abundance $12 + \log{\rm (O/H)} = 8.56$, or $\sim 4/5$ solar, if the local calibration of the $N2$ index with (O/H) applies to these galaxies. Galaxies fainter than $K_s = 20$ (*lower panel*) have [N II]/H$\alpha = 0.13$ and $12 + \log{\rm (O/H)} = 8.39$ or $\sim 1/2$ solar.

2. Near-Infrared Observations of Star-Forming Galaxies at $z = 2 - 3$

2.1. Early Clues

Hints of ordered motions within star-forming galaxies at $z = 2 - 3$ were already present in the first near-IR observations on the ESO VLT with ISAAC, the first single-slit near-IR spectrograph to come on line on an 8–10 m telescope (Moorwood *et al.* 2000; Pettini *et al.* 2001). As the samples increased with the availability of NIRSPEC on the Keck II telescope, cases were commonly encountered where the redshifts of the Hα or [O III] lines varied continuously along the spectrograph slit, indicative of coherent shear which could be interpreted as rotation (Erb *et al.* 2003). However, it soon became apparent (Erb *et al.* 2004) that the velocity spread of the emission depended very sensitively on the seeing conditions under which the galaxies were observed. Furthermore, the lack of prior information on the morphologies of the galaxies targeted made it difficult to interpret such motions as rotation curves (despite some efforts by, for example, Weatherley & Warren 2003), since the slit position angle on the sky was not necessarily aligned with the 'major axes' of the galaxies (see Figure 2).

2.2. Integral Field Observations

This area of study received a tremendous boost with the advent of SINFONI, the integral field near-IR spectrograph on the VLT. This instrument produced the first 'data-cubes' on galaxies at $z = 2$, mapping their kinematics on the plane of the sky with an angular resolution of ~ 0.5 arcsec, sampling spatial scales of a few kpc at these redshifts. An

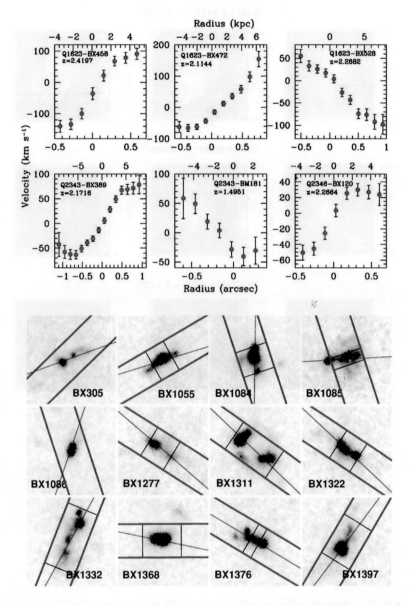

Figure 2. *Upper Panel:* 'Rotation curves' of $z \simeq 2$ galaxies derived from single-slit NIRSPEC observations of the Hα emission line. *Lower Panel:* HST ACS images of selected BX galaxies showing the size and orientation of the NIRSPEC entrance slit (blue lines). The slit is 0.76 arcsec wide, or ~ 6 kpc at the redshifts of the galaxies. (Both figures reproduced from Erb *et al.* 2004).

intensive observational programme described in R. Genzel's contribution to this volume, led to the first reports of galaxy kinematics which convincingly resemble rotation curves. In some of the brighter and more extended galaxies among the BX/BM sample of Steidel *et al.* (2004), the rotation curves could be followed to beyond 10 kpc where they reached circular velocities as high as $v_c \simeq 200\,\mathrm{km\,s^{-1}}$. In these well-developed systems, the implied dynamical masses seem to agree broadly with the stellar masses deduced from model-fitting of the rest-frame UV to near-IR spectral energy distributions, as expected

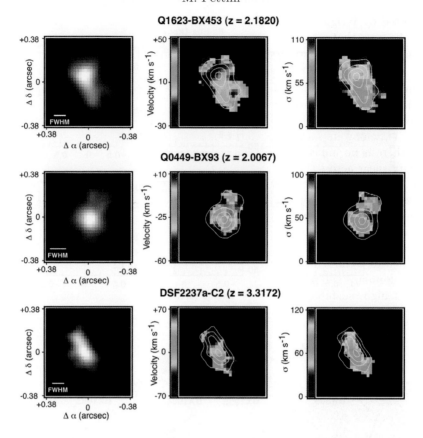

Figure 3. Maps obtained with the OH Suppressing InfraRed Imaging Spectrograph (OSIRIS) on the Keck II telescope (reproduced from Law *et al.* 2007a). For each of the three galaxies illustrated here, the maps show (from left to right): *(a)* the integrated flux in either the Hα or [O III] λ5007 emission line; *(b)* the velocity of the emission line relative to systemic, according to the colour-coding shown on the left; and *(c)* the velocity dispersion. The field of view of 0.75×0.75 arcseconds of each panel corresponds to physical distances of ~ 6 kpc at the galaxies' locations. The resolution of the images is FWHM = $110 - 150$ mas, sampled with 25 mas pixels and shown by the white bar in the lower left-hand corner of the images in the first column.

if the baryons provide most of the gravitational potential in the inner regions of these galaxies.

The successful coupling of laser-guided adaptive optics to the near-IR integral field spectrograph OSIRIS on the Keck II telescope provided the next step forward by affording angular resolutions of 0.1-0.15 arcsec, or ~ 1 kpc at $z = 2-3$. Such fine spatial sampling pushes to the limit the light-gathering power of even 10 m telescopes; consequently, the galaxies initially targeted by Law *et al.* (2007a) were selected from the samples of Steidel *et al.* (2003, 2004) primarily on the basis of their high surface brightness (Law *et al.* 2007b).

Figure 3 shows examples of resolved kinematics in the inner regions (± 3 kpc from the centre) of three of the galaxies studied by Law *et al.* (2007b), illustrating the variety of patterns seen. In some cases the uniform progression from negative to positive velocities along the major axis is highly suggestive of rotation; the $z = 3.3172$ galaxy DSF2237a–C2 shown in the bottom row of Figure 3 provides a good example of what could plausibly be interpreted as a rotating disk with $v_{\rm rot} \simeq 70$ km s^{-1}. An analogous example at $z = 1.5$ has been reported by Wright et al. (2007). If these are indeed disks, they are kinematically 'thick' in the sense that rotational and random velocities are of the same order: $v_{\rm rot}/\sigma \approx 1$, as can be appreciated by comparing the middle and right panels of the bottom row in Figure 3.

A counterexample is provided by the $z = 2.1820$ galaxy Q1623-BX453 (top row of Figure 3). This UV-luminous galaxy is forming stars at a rate of $\sim 80 M_\odot$ yr^{-1} and has already converted (by redshift $z = 2.1820$) more than 50% of its available gas supply into stars, enriching its interstellar medium to near-solar metallicity (Erb et al. 2006). And yet, there is no indication of coherent kinematics in its inner regions, with hardly any velocity shear across the extent of the Hα emission (middle panel of the top row in Figure 3). We are either seeing this galaxy completely face-on, or here we have an example of a galaxy supported not by rotation but by random motions: the width of the resolved Hα emission line indicates velocity dispersions of $\sigma \simeq 60 - 100$ km s^{-1} over most of the galaxy (see right-hand panel of the top row in Figure 3). Q1623-BX453 also provides an illustration of how, with the power of adaptive optics, we are beginning to discern inhomogeneities within the galaxies in the chemical composition of their H II regions (Figure 4). Similar examples of what could be the first indications of radial abundance gradients at high redshifts have been found by Förster Schreiber et al. (2006).

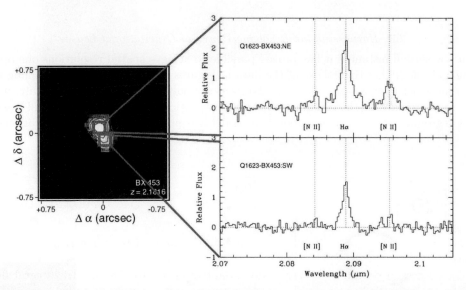

Figure 4. The brighter NE knot of Hα emission in the $z = 2.1820$ galaxy Q1623-BX453 exhibits a high [N II]/Hα ratio indicative of a near-solar oxygen abundance. In the fainter knot, < 2 kpc to the SW, Law et al. 2007a (from which this figure is reproduced) measured a lower value of the $N2$ index, corresponding to O/H $\sim 1/2$ (O/H)$_\odot$.

In some cases, the improved resolution afforded by adaptive optics reveals morphologies and kinematics which are highly suggestive of on-going mergers. Figure 5 is a good example of a galaxy where the velocity shear first detected from single-slit spectra (Pettini et al. 2001) turned out to be due to two distinct 'clumps' (or in this case more likely two galaxies), each with a relatively high internal velocity dispersion ($\sigma \simeq 80$ km s^{-1}) but separated by only a few tens of km s^{-1} in redshift.

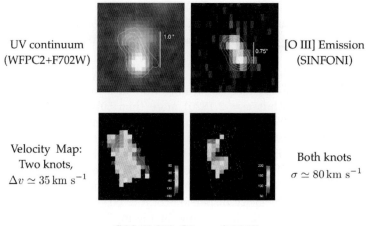

Q0347-383 C5, z = 3.2347

Figure 5. Recent SINFONI observations of the $z = 3.2347$ Lyman break galaxy Q0347−383 C5 by Nesvadba *et al.* 2008 (from which these figures are reproduced) show the source to consist of two merging Lyman break galaxies, separated by only $\sim 5\,\mathrm{kpc}$. Other properties of these two galaxies seem to be typical of the LBG population, with star formation rates $\mathrm{SFR} = 20\text{–}40\,M_\odot\,\mathrm{yr}^{-1}$ and metallicities between solar and $1/5$ solar.

2.3. *Harnessing the Resolving Power of Gravitational Lenses*

In a few special instances it has proved possible to achieve spatial resolutions one order of magnitude higher than those of the data in Figures 3, 4 and 5, thereby sampling the internal structure of galaxies at $z = 2\text{–}3$ on scales of only $\sim 100\,\mathrm{pc}$. Such remarkably sharp views are provided by the combination of adaptive optics and strong gravitational lensing by foreground clusters—with a good lens model it is possible to map the gravitationally

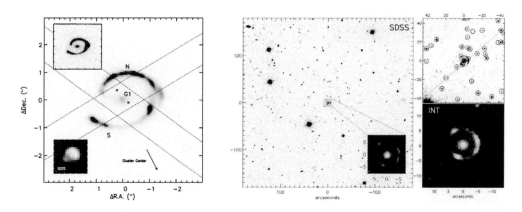

Figure 6. Two examples of recently discovered, strongly lensed, high redshift galaxies. *Left Panel:* the 'Cosmic Eye', a pair of high surface brightness arcs at $z_\mathrm{source} = 3.0735$ bracketing a foreground massive early-type galaxy at $z_\mathrm{lens} = 0.73$, discovered by Smail et al. (2007). *Right Panel:* the 'Cosmic Horseshoe', an almost complete (~ 300 degrees) Einstein ring of diameter 10 arcsec discovered by Belokurov et al. (2007) from a targeted search in Sloan Digital Sky Survey Data Release 5 (SDSS DR5) images. The ring is due to a $z_\mathrm{source} = 2.379$ star-forming galaxy, magnified by a factor of ~ 35 by a $M \simeq 6 \times 10^{12} M_\odot$ foreground galaxy at $z_\mathrm{lens} = 0.444$.

lensed images back onto the source plane at such extraordinary resolution. The number of strongly lensed high-z galaxies has increased significantly in the last couple of years; there are now several examples known which appear as promising as the well studied MS 1512-cB58 (Pettini et al. 2002 and references therein).

Figure 6 shows two such examples where the fortuitous alignments of massive clusters (or a single massive galaxy) with background star-forming galaxies have produced near-complete Einstein rings. Stark et al. (in preparation) have used OSIRIS on the Keck II telescope to map the kinematics of the $z = 3.0735$ galaxy lensed into the two 3 arcsec long arcs which make up the 'Cosmic Eye' (left panel of Figure 6). In the inner ~ 2 kpc, the H II regions of the galaxy follow a rotation pattern which flattens at a projected distance from the centre of $r \sim 1$ kpc to $v_{\rm rot} \sin i = 54$ km s^{-1}. Again, random velocities are comparable to rotation ($v_{\rm rot}/\sigma = 0.9$), so that it is by no means clear which mechanism is the dominant one in providing support for the baryons.

3. Damped Lyman alpha Systems

Long before we had learnt (or had the technological means) to recognise high redshift galaxies directly via their nebular and stellar emission, our knowledge of the 'normal' galaxy population at $z > 1$ relied almost exclusively on the technique of QSO absorption line spectroscopy, illustrated in Figure 7. In particular, Art Wolfe and his collaborators

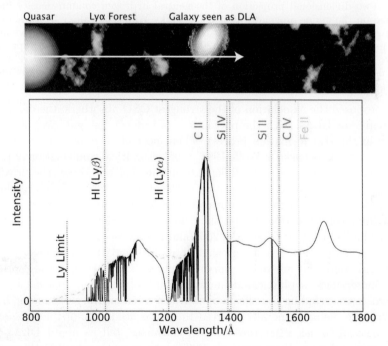

Figure 7. Illustration of the technique of QSO absorption line spectroscopy. The spectra of distant QSOs bear the imprint of intervening diffuse gas, cosmologically distributed along the line of sight, in the form of absorption lines. In the conventional view, low density gas in the intergalactic medium gives rise to the multitude of narrow absorption lines which constitute the Lyman alpha forest. The stronger, and less frequent, damped Lyman alpha lines, indicative of high column densities of neutral gas with $N({\rm H\,I}) \geqslant 2 \times 10^{20}$ atoms cm^{-2}, are thought to be the signposts of intervening galaxies intersected by the line of sight. Such absorption systems, or DLAs, include a variety of metal absorption lines due to the most common astrophysical elements. (Image courtesy of Andrew Pontzen).

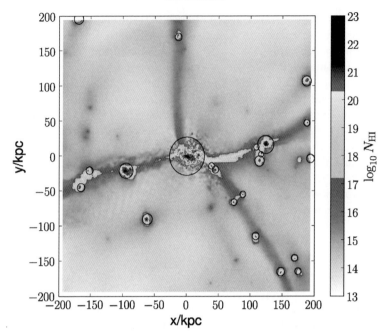

Figure 8. Two dimensional projection of the neutral hydrogen column density in a 400 kpc cube centred on the major progenitor of a $z = 0$ Milky Way-type galaxy in the simulations by Governato et al. (2007). The map is colour coded so that DLAs are dark red, while the lower column density Lyman limit systems (with $20.3 > \log_{10}[N(\text{H\,\textsc{i}})/\text{cm}^{-2}] > 17.2$) appear in green and yellow. The circles indicate the projected positions and virial radii of all dark matter halos with $M > 5 \times 10^8 M_\odot$. All units are physical. (Figure reproduced from Pontzen et al. 2008).

have long proposed the association of the strongest QSO absorbers, the damped Lyman alpha systems—or DLAs, with the progenitors of today's disk galaxies, observed at an early stage in the star formation process, when most of their baryons resided in their interstellar media (Prochaska & Wolfe 1997, 1998). Empirical confirmation of this picture has, however, proved frustratingly difficult to attain. For one thing, the proximity of a bright QSO hinders the direct identification of the absorbers even in high resolution *HST* images and, for another, there is mounting evidence that DLAs sample a wide range of the galaxy luminosity function so that their numbers are dominated by galaxies fainter than L^* (Wolfe, Gawiser, & Prochaska 2005 and references therein).

At this meeting we have heard a great deal about simulations of galaxy formation in a cosmological context (reviewed in Simon White's contribution to these proceedings) which are increasingly sophisticated in resolution, volume, and treatment of relevant physical processes including the effects of star formation on the surrounding interstellar and intergalactic media (generally referred to as the 'feedback' of the star formation). It is only natural to ask what these 'model universes' tell us about DLAs and their association with galaxies. Pontzen et al. (2008) have recently revisited this question by analysing the output of the N-body simulations by the Seattle group (Governato et al. 2007 and references therein) which are among the most successful in reproducing many of the properties of disk galaxies at $z = 0$ (e.g. Brooks et al. 2007; Governato et al. 2008).

Unlike previous simulations specifically aimed at DLAs (e.g. Razoumov et al. 2006; Nagamine et al. 2007 and references therein), there are no free parameters to be adjusted in the analysis by Pontzen et al. (2008)—these authors simply examined the properties of neutral gas in the volumes simulated by Governato et al. (2007) where relevant scaling

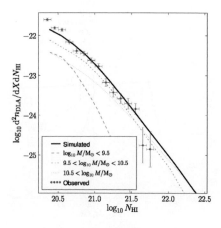 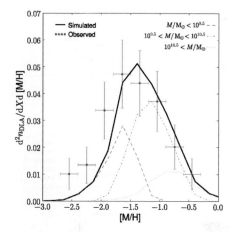

Figure 9. Comparisons between observed and simulated DLAs from the work by Pontzen et al. (2008) from which the figures are reproduced. *Left Panel:* the distribution of DLA column densities reported by Prochaska et al. (2005) from their analysis of SDSS DR5 quasar spectra is well matched by the incidence of DLAs in the volumes generated with the N-body simulations by Governato et al. (2007). *Right Panel:* The observed distribution of DLA metallicities (Prochaska et al. 2007) arises naturally in the Governato et al. (2007) disk galaxy simulations, without recourse to observationally unsubstantiated dust-biasing effects. In both figures the contributions from halos in different mass ranges are indicated separately, as well as their total.

parameters (most importantly the constant of proportionality in the Schmidt law of star formation and the efficiency of supernova energy deposition into the ISM) had already been fixed to the values that best reproduce the characteristics of disk galaxies today. The finding that the model galaxies generated with the code by Governato et al. (2007) do in fact account for many of the observed properties of DLAs without any fine tuning, as Figures 8, 9, and 10 illustrate, then lends support to the view that DLAs are an integral part of the galaxy formation process, and helps us towards the long-sought goal of making a connection between the properties of the absorbers and those of the galaxies producing them (Pettini 2004, 2006).

At $z = 3$, random sightlines through the simulation volumes intersect neutral gas with the same frequency as the observed number of DLAs per unit redshift, implying effective radii of ~ 10 kpc (for column densities of neutral hydrogen $N(\text{H\,\textsc{i}}) \geqslant 2 \times 10^{20}$ cm^{-2}) in galaxies which by $z = 0$ would grow to the mass of the Milky Way (see Figure 8). The most direct observational property of DLAs—their column density distribution, which has changed little since the original census by Wolfe et al. (1986), is reproduced very well (left panel of Figure 9), apart from the highest column density systems ($N(\text{H\,\textsc{i}}) > 5 \times 10^{21}$ cm^{-2}) which are missing (compared to the model predictions) from even the largest quasar surveys, such as the SDSS (Prochaska et al. 2005). Possibly, the gas is mostly molecular and very short-lived (i.e. it turns into stars) when such high surface densities are reached (Schaye 2001).

For the first time, the observed distribution of DLA metallicities is reproduced reasonably well (right panel of Figure 9). Given the substantial uncertainties involved in, for example, the adopted stellar yields of Type II and Type Ia supernovae, the initial stellar mass function, and even the solar abundance scale used for comparison, the agreement may be somewhat fortuitous. However, it is significant that the good match in Figure 9

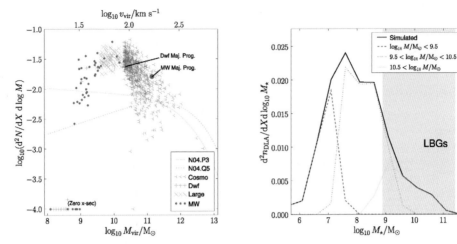

Figure 10. Contributions by different galaxies to the total DLA cross-section on the sky at $z = 3$ in the work by Pontzen et al. (2008) from which the figures are reproduced. *Left Panel:* Relative contributions to the DLA cross-section by halos of different masses (lower x-axis) or virial velocities (top axis). Results from different simulation volumes are colour-coded (see Pontzen et al. 2008 for further details). Halos which do not contain significant column densities of neutral hydrogen, and thus do not give rise to DLAs, are shown artificially at -4 on the y-axis. Also shown for comparison are the relative contributions from galaxies in the earlier simulations by Nagamine et al. (2004). *Right Panel:* Fractional contributions from galaxies of different stellar masses assembled by $z = 3$. Most DLAs arise in galaxies which are less massive and fainter than the Lyman break galaxies surveyed by Steidel and collaborators (Steidel et al. 2003; Shapley et al. 2001).

is achieved without appealing to a 'hidden' population of metal- and dust-rich DLAs which could potentially exclude background QSOs from current, optically-selected, samples. Many previous DLA simulations, which tended to overestimate the frequency of high-metallicity DLAs, have invoked such a scenario to reconcile their predictions with the data, overlooking at times the fact that this supposed source of bias has been shown to be of minor importance by DLA surveys in radio selected QSOs (Ellison et al. 2001, 2005; Jorgenson et al. 2006).

As discussed by Pontzen et al. (2008), the good agreement between theoretical and observed metallicity distributions is achieved naturally in the Governato et al. (2007) simulations because a substantial fraction of the DLA cross-section is produced by halos of relatively low mass, with virial masses $10^{9.5} M_\odot \leqslant M_{\rm vir} \leqslant 10^{10.5} M_\odot$ (see left panel of Figure 10), which have undergone relatively little star formation by $z = 3$ and whose shallow potential wells probably retain only a small fraction of the heavy elements produced (see Julianne Dalcanton's contribution). Such galaxies make up the dominant contribution to the DLA cross-section partly because their interstellar gas is 'puffed up' to larger radii than it would otherwise be the case by the inclusion of feedback in the galaxy formation prescription.

If the simulations by Governato et al. (2007) offer a realistic representation of the connection between DLAs and galaxies, as the analysis by Pontzen et al. (2008) seems to indicate, we can tentatively associate subsets of DLAs chosen according to a particular characteristic with their host halos. For example, Pontzen et al. (2008) find a broad mass-metallicity relation in their simulations, with the least metal-enriched DLAs tracing preferentially low-mass halos (see right panel of Figure 9). Figure 8 gives a visualisation of such halos clustered around the progenitor of a Milky Way-type galaxy. These systems may be some of the best sites where to look for chemical traces of the 'first stars', as well as the unadulterated products of primordial nucleosynthesis (Pettini *et al.* 2008a,b).

4. Conclusions

Continuing advances in observational and computational facilities are giving us clearer views of galaxies at high redshifts than ever before. The recently gained ability to probe the internal kinematics of galaxies at look-back times of more than 75% of the age of the universe is as powerful a demonstration as any that laser-guided adaptive optics has finally come of age and can be successfully used to feed high resolution spectrographs. The picture that is emerging is a complex one. The finding that major mergers were common at $z = 2 - 3$ comes as no surprise. However, there are also many examples of galaxies already exhibiting large-scale rotation, which somewhat remarkably appears not to have been disrupted over a period of several hundred million years, during which the galaxies have accreted a large fraction of their baryonic mass, assembling as much as $M_* \sim 5 \times 10^{10} M_\odot$ in stars (Genzel *et al.* 2006). Clarifying the relative incidence of these different kinematic signatures remains a priority for future observations, given their potential for discriminating between different scenarios of gas accretion and cooling within dark matter halos (e.g. Immeli *et al.* 2004; Dekel & Birnboim 2006).

On the theoretical front, we have yet to develop a unified picture of the different kinematic components of high-z galaxies, which presumably include: *(a)* infalling gas fuelling star formation, *(b)* rotation, *(c)* large velocity dispersions, and *(d)* galaxy-wide outflows of the interstellar medium driven by the energy and momentum released by the star formation activity. In particular, what is the origin of the high turbulence which seems to be present in almost every case, irrespectively of whether a rotation pattern can be discerned or not? Which is the dominant support mechanism in these galaxies? And why is it so difficult to identify the infalling gas, when the observational signatures of the outflows are so obvious? The work by Pontzen *et al.* (2008) strongly suggests that our Milky Way had the characteristics of a damped Lyman alpha system at $z = 3$, but what fraction of the galaxies we see directly at these redshifts evolve to become today's large spirals?

What I took away from this meeting is the high level of interest, both observational and theoretical, in exploring these questions. The goal of bringing together the different pieces of the jig-saw puzzle of galaxy formation is within sight and we can look forward to a time in the not too distant future when the legacy of the work begun by Bengt Strömgren will be fulfilled.

Acknowledgements

It is a pleasure to acknowledge all of my collaborators in the various projects described here; I especially thank David Law and Andrew Pontzen who very generously provided me with much of the material for my presentation. I, like the rest of the participants of this stimulating meeting in the beautiful city of Copenhagen, am indebted to Johannes Andersen and Birgitta Nordström whose unflagging efforts and excellent organisation made the symposium such a success.

References

Belokurov, V., *et al.* 2007, *ApJ*, 671, L9
Brooks, A. M., *et al.* 2007, *ApJ*, 655, L17

Dekel, A. & Birnboim, Y. 2006, *MNRAS*, 368, 2
Ellison, S. L., Yan, L., Hook, I. M., Pettini, M., Wall, J. V., & Shaver, P. 2001, *A&A*, 379, 393
Ellison, S. L., Hall, P. B., & Lira, P. 2005, *AJ*, 130, 1345
Erb, D. K., Shapley, A. E., Steidel, C. C., Pettini, M., Adelberger, K. L., Hunt, M. P., Moorwood, A. F. M., & Cuby, J.-G. 2003, *ApJ*, 591, 101
Erb, D. K., Steidel, C. C., Shapley, A. E., Pettini, M., & Adelberger, K. L. 2004, *ApJ*, 612, 122
Erb, D. K., Steidel, C. C., Shapley, A. E., Pettini, M., Reddy, N. A., & Adelberger, K. L. 2006, *ApJ*, 646, 107
Förster Schreiber, N. M. et al. 2006, *ApJ*, 645, 1062
Genzel, R. et al. 2006, *Nature*, 442, 786
Governato, F., Mayer, L., & Brook, C. 2008, ArXiv e-prints, 801, arXiv:0801.1707
Governato, F., Willman, B., Mayer, L., Brooks, A., Stinson, G., Valenzuela, O., Wadsley, J., & Quinn, T. 2007, *MNRAS*, 374, 1479
Immeli, A., Samland, M., Gerhard, O., & Westera, P. 2004, *A&A*, 413, 547
Jorgenson, R. A., et al. 2006, *ApJ*, 646, 730
Law, D., Steidel, C. C., Erb, D. K, Larkin, J. E., Pettini, M., Shapley, A. E., & Wright, S. A. 2007a, *ApJ*, 669, 929
Law, D. R., Steidel, C. C., Erb, D. K., Pettini, M., Reddy, N. A., Shapley, A. E., Adelberger, K. L., & Simenc, D. J. 2007b, *ApJ*, 656, 1
Moorwood, A. F. M., van der Werf, P. P., Cuby, J. G., & Oliva, E. 2000, *A&A*, 362, 9
Nagamine, K., Springel, V., & Hernquist, L. 2004, *MNRAS*, 348, 421
Nagamine, K., Wolfe, A. M., Hernquist, L., & Springel, V. 2007, *ApJ*, 660, 945
Nesvadba, N. P. H., Lehnert, M., Davies, R., Verma, A., & Eisenhauer, F. 2008, *A&A*, 479, 67
Pettini, M. 2004, in C. Esteban, R. J. García López, A. Herrero & F. Sánchez (eds.), *Cosmochemistry. The melting pot of the elements* (Cambridge: Cambridge University Press), p. 257 (astro-ph/0303272)
Pettini, M. 2006 in V. LeBrun, A. Mazure, S. Arnouts & D. Burgarella (eds.), *The Fabulous Destiny of Galaxies: Bridging Past and Present* (Paris: Frontier Group), p. 319 (astro-ph/0603066)
Pettini, M. & Pagel, B. E. J. 2004, *MNRAS*, 348, L59
Pettini, M., Rix, S. A., Steidel, C. C., Adelberger, K. L., Hunt, M. P., & Shapley, A. E. 2002, *ApJ*, 569, 742
Pettini, M., Shapley, A. E., Steidel, C. C., Cuby, J.-G., Dickinson, M., Moorwood, A. F. M., Adelberger, K. L., & Giavalisco, M. 2001, *ApJ*, 554, 981
Pettini, M., Zych, B. J., Murphy, M. T., Lewis, A., & Steidel, C. C. 2008a, *MNRAS*, in press
Pettini, M., Zych, B. J., Steidel, C. C., & Chaffee, F. H. 2008b, *MNRAS*, 385, 2011
Pontzen, A., Governato, F., Pettini, M., et al. 2008, *MNRAS*, in press
Prochaska, J. X., Herbert-Fort, S., & Wolfe, A. M. 2005, *ApJ*, 635, 123
Prochaska, J. X. & Wolfe, A. M. 1997, *ApJ*, 487, 73
Prochaska, J. X. & Wolfe, A. M. 1998, *ApJ*, 507, 113
Prochaska, J. X., et al. 2007, *ApJ Supp*, 171, 29
Razoumov, A. O., Norman, M. L., Prochaska, J. X., & Wolfe, A. M. 2006, *ApJ*, 645, 55
Schaye, J. 2001, *ApJ*, 562, L95
Shapiro, K. L., et al. 2008, *ApJ*, in press
Shapley, A. E., Steidel, C. C., Adelberger, K. L., Dickinson, M., Giavalisco, M., & Pettini, M. 2001, *ApJ*, 562, 95
Smail, I., et al. 2007, *ApJ*, 654, L33
Steidel, C. C., Adelberger, K. L., Shapley, A. E., Pettini, M., Dickinson, M., & Giavalisco, M. 2003, *ApJ*, 592, 728
Steidel, C. C., Shapley, A. E., Pettini, M., Adelberger, K. L., Erb, D. K., Reddy, N. A. & Hunt, M. P. 2004, *ApJ*, 604, 534
Weatherley, S. J. & Warren, S. J. 2003, *MNRAS*, 345, L29
Wolfe, A. M., Gawiser, E., & Prochaska, J. X. 2005, *ARAA*, 43, 861
Wolfe, A. M., Turnshek, D. A., Smith, H. E., & Cohen, R. D. 1986, *ApJ Supp*, 61, 249
Wright, S. A., et al. 2007, *ApJ*, 658, 78

Spatially resolved dynamics of high-z star forming galaxies

Reinhard Genzel

Max Planck Institute für Extraterrestrische Physik
Giessenbachstrasse, Postfach 1312
D-85748 Garching, Germany
E-mail: genzel@mpe.mpg.de

Abstract. I report on two major programs to study the kinematic properties of galaxies at $z \sim 1.5 - 3$ with spatially resolved spectroscopy for the first time. Using the adaptive optics assisted, integral field spectrometer SINFONI on the ESO VLT, we have observed more than 70 galaxies and find compelling evidence for large, geometrically thick (turbulent), rotating disk galaxies in a majority of the objects that we can spatially resolve. It appears that these star forming disks are driven by continuous, rapid accretion of gas from their dark matter halos, and that their evolution is strongly influenced by internal, secular evolution. In contrast to the 20 submillimeter galaxies that we have investigated with the IRAM Plateau de Bure millimetre interferometer we find strong evidence for compact, major mergers. I discuss the impact of these new observations on our understanding of galaxy evolution in the early Universe.

For the SINS survey we have carried out Hα integral field spectroscopy of well-resolved, UV/optically selected star-forming galaxies at $z \sim 2$ with SINFONI on the ESO VLT. The SINS sample is representative of the majority of massive ($M_* > $ a few $10^{10} M_\odot$) star-forming galaxies at that redshift. Our data obtained with laser guide star assisted adaptive optics in good seeing show the presence of turbulent, rotating star-forming rings/disks in at least a third of the sample, plus central bulge/inner disk components in some of the best cases, whose mass fractions relative to total dynamical mass appears to scale with [NII]/Hα flux ratio and 'star formation' age. Another third of the SINS galaxies show clear signs of kinematic perturbations by a merger, while the last third appear to be compact, 'dispersion' limited systems.

Our interpretation of these data is that the buildup of the central disks and bulges of massive galaxies at $z \sim 2$ can be driven by the early secular evolution of gas-rich 'proto'-disks. High-redshift disks exhibit large random motions. This turbulence may in part be stirred up by the release of gravitational energy in the rapid 'cold' accretion flows along the filaments of the cosmic web. As a result, dynamical friction and viscous processes proceed on a time scale of < 1 Gyr, at least an order of magnitude faster than in disk galaxies at $z \sim 0$. Early secular evolution thus drives gas and stars into the central regions and can build up exponential disks and massive bulges, even without major mergers. Secular evolution along with increased efficiency of star formation at high surface densities may also help to account for the short time scales of the stellar buildup observed in massive galaxies at $z \sim 2$.

Keywords. galaxies: high-redshift – galaxies: formation – galaxies: structure – galaxies: evolution – techniques: high angular resolution

---

The presentation made at the meeting can be downloaded from:
http://www.mpe.mpg.de/ppt/IAU254-RG.ppt

The results discussed here are described in the recent papers by Tacconi et al. (2008), Shapiro et al. (2008), Genzel et al. (2008), and Förster Schreiber et al. (2008).

References

Förster Schreiber, N.M., *et al.* 2008, to be submitted.
Genzel, R., Burkert, A., Bouché, N., *et al.* 2008, *ApJ* 687, 59
Shapiro, K. L., Genzel, R., Förster Schreiber, N. M., *et al.* 2008, *ApJ* 682, 231),
Tacconi, L., Genzel, R., Smail, I., *et al.* 2008, *ApJ* 680, 246

Adriaan Blaauw enjoying his return to the Strömgren residence at Carlsberg.

Simulating High-Redshift Disk Galaxies: Applications to Long Duration Gamma-Ray Burst Hosts

Brant E. Robertson[1,2,3]

[1]Kavli Institute for Cosmological Physics, and
Department of Astronomy and Astrophysics, University of Chicago,
933 East 56th Street, Chicago, IL 60637, USA

[2]Enrico Fermi Institute, 5640 South Ellis Avenue, Chicago, IL 60637, USA

[3]Spitzer Fellow
email: brant@kicp.uchicago.edu

Abstract. The efficiency of star formation governs many observable properties of the cosmological galaxy population, yet many current models of galaxy formation largely ignore the important physics of star formation and the interstellar medium (ISM). Using hydrodynamical simulations of disk galaxies that include a treatment of the molecular ISM and star formation in molecular clouds (Robertson & Kravtsov 2008), we study the influence of star formation efficiency and molecular hydrogen abundance on the properties of high-redshift galaxy populations. In this work, we focus on a model of low-mass, star forming galaxies at $1 \lesssim z \lesssim 2$ that may host long duration gamma-ray bursts (GRBs). Observations of GRB hosts have revealed a population of faint systems with star formation properties that often differ from Lyman-break galaxies (LBGs) and more luminous high-redshift field galaxies. Observed GRB sightlines are deficient in molecular hydrogen, but it is unclear to what degree this deficiency owes to intrinsic properties of the galaxy or the impact the GRB has on its environment. We find that hydrodynamical simulations of low-stellar mass systems at high-redshifts can reproduce the observed star formation rates and efficiencies of GRB host galaxies at redshifts $1 \lesssim z \lesssim 2$. We show that the compact structure of low-mass high-redshift GRB hosts may lead to a molecular ISM fraction of a few tenths, well above that observed in individual GRB sightlines. However, the star formation rates of observed GRB host galaxies imply molecular gas masses of $10^8 - 10^9$ $M_\odot$ similar to those produced in the simulations, and may therefore imply fairly large average H_2 fractions in their ISM.

Keywords. Galaxies:high-redshift, galaxies:ISM, gamma rays: bursts

1. Introduction

To improve the physical description of star formation in hydrodynamical simulations of galaxies, Robertson & Kravtsov (2008) implemented a new model for the ISM that includes low-temperature ($T < 10^4$K) cooling, directly ties the star formation rate to the molecular gas density, and accounts for the destruction of molecular hydrogen by an interstellar radiation field (ISRF) from young stars. They used simulations to study the relation between star formation and the ISM in galaxies and demonstrated that, for the first time, their new model simultaneously reproduces the molecular gas and total gas Kennicutt-Schmidt (KS) relations, the connection between star formation and disk rotation, and the relation between interstellar pressure and the fraction of gas in molecular form (e.g. Wong & Blitz 2002, Blitz & Rosolowsky 2006). The capability of this model to reproduce both the star formation efficiency and molecular abundance of nearby systems makes it useful for simulating low-mass galaxies that have suppressed H_2 abundances (and whose star formation rates would be overestimated in common treatments of star formation based on the KS relation) and high-redshift galaxies whose structural properties may vary substantially from local systems (and may therefore not have the same KS relation normalization). The model should be especially useful for studying low-mass galaxies at high-redshift, such as long duration gamma-ray burst (GRB) host galaxies at $1 \lesssim z \lesssim 2$, which is the focus of this work.

The highly-energetic phenomena known as GRBs were discovered over forty years ago (Klebesadel et al. 1973), but their extragalactic origin was confirmed only in the last decade (e.g., Metzger et al. 1997). Since then, the properties of the cosmological population of galaxies that host GRBs have been increasingly well-studied (e.g., Bloom et al. 2002, Le Floc'h et al. 2006, Prochaska et al. 2006, Berger et al. 2007a,b). Recently, interest in long duration GRB galaxy hosts as possible tracers of the global star formation history of the universe has motivated systematic studies of their star formation efficiencies and stellar masses (Castro Cerón et al. 2008, Savaglio et al. 2008). These studies have found that high-redshift GRB hosts have small stellar masses ($\log M_\star \sim 9.3$) and moderate star formation rates (SFR $\sim 2.5\ M_\odot\ \mathrm{yr}^{-1}$). Compared with other high-redshift galaxy populations, GRB hosts tend to have lower star formation rates at fixed stellar mass compared with Lyman-break galaxies and lower stellar masses at fixed star formation rate compared with field galaxies (for details, see Savaglio et al. 2008).

Spectroscopic studies of GRB sightlines have provided additional information about the post-explosion character of the host galaxy ISM. Tumlinson et al. (2007) failed to detect H_2 in five GRB sightlines and suggested that low metallicity and large far ultraviolet ISRF strengths ($10 - 100\times$ the Milky Way value) were responsible for destroying molecular hydrogen in GRB hosts. They interpreted the lack of vibrationally excited H_2 lines as evidence against the GRB destroying its parent molecular cloud, but noted various caveats to this conclusion such as the parent cloud size or cloud photodissociation before to the GRB. Whalen et al. (2008) used one-dimensional radiative hydrodynamical calculations to show that GRBs can ionize nearby neutral hydrogen, but suggested that an additional ISRF is necessary to remove molecular hydrogen from the nearby ISM. Prochaska et al. (2008) studied NV absorption in GRB sightlines, and argued that if nitrogen ionization by GRB afterglows leads to NV absorption then the observations support a scenario where dense, molecular cloud-like environments serve as the sites of GRBs.

Given the increasingly detailed studies of GRB hosts, their interesting ISM and star formation properties, and their low stellar masses, a theoretical study of GRB host galaxy analogues using hydrodynamical simulations that include a treatment of the molecular ISM is warranted. Below, we present simulations of a model GRB host galaxy that include a prescription for the molecular ISM and star formation in molecular clouds (Robertson & Kravtsov 2008). We use the simulations to examine the star formation efficiency and molecular hydrogen content of galaxies with structural properties similar to those expected for low-mass galaxies at $1 \lesssim z \lesssim 2$. Below, we discuss our methodology and present some initial results.

2. Methodology

To study the properties of long duration GRB host galaxies, we simulate a numerical model of an isolated galaxy using a version of the N-body/Smoothed Particle Hydrodynamics code GADGET (Springel et al. 2001, Springel 2005b) that incorporates a model for the molecular ISM (Robertson & Kravtsov 2008). For details regarding the numerical galaxy models, simulation methodology, and ISM model, we refer the reader to Springel

Figure 1. Simulated long duration Gamma-Ray Burst (GRB) host galaxy analogue at $z \sim 1.5$. Shown is the gas surface density of the GRB host (image intensity), color coded by the median interstellar medium temperature (purple regions have $T < 10^3$ K, while blue regions have $T \gtrsim 10^4$ K). The simulated galaxy has a stellar mass $\log M_\star \approx 9.3$ and a star formation rate SFR $\approx 1.2 M_\odot \text{yr}^{-1}$, similar to high-redshift GRB host galaxies (e.g., Castro Cerón et al. 2008, Savaglio et al. 2008). The simulations include the Robertson & Kravtsov (2008) model of the molecular ISM, enabling a study of the connection between star formation rate, galaxy properties, and H$_2$ abundance in GRB hosts.

et al. (2005a), Robertson et al. (2006a,b), and Robertson & Kravtsov (2008), but a brief summary follows.

The numerical galaxy model is designed to approximate the properties of $1 \lesssim z \lesssim 2$ GRB host galaxies as determined by Savaglio et al. 2008. The stellar disk mass of the system is set to $\log M_\star = 9.3$, with a gas fraction of $f_{\text{gas}} = 0.5$ (appropriate for high-redshift, see Erb et al. 2006), which implies a total virial mass of $\log M_{\text{vir}} = 10.9$ for a typical disk baryon fraction of $f_{\text{b}} = 0.05$. The virial radius is set appropriately for a halo with virial mass M_{vir} at $z \sim 2$. The exponential disk scale length was fixed according to the Mo et al. (1998) formalism, including the adjustment for an effective Navarro et al. (1996) dark matter halo concentration of $c_{\text{NFW}} = 6$ (also appropriate for the chosen virial mass and redshift, see Bullock et al. 2001) and a spin of $\lambda = 0.05$. The density field of the dark matter halo follows the Hernquist (1990) profile, while the velocity fields of the dark matter halo and the exponential stellar disk are set using the Hernquist (1990) distribution function and the epicyclical approximation, respectively. The numerical realizations of the stellar disk, gaseous disk, and dark matter halo are initialized with $N_{\text{disk},\star} = 4 \times 10^5$, $N_{\text{disk,gas}} = 4 \times 10^5$, and $N_{\text{DM}} = 4 \times 10^5$ particles, and are evolved with a gravitational softening of $\epsilon = 70$ pc. The simulation is calculated for a duration of $t \sim 1$ Gyr, or about the time between redshift $z \sim 2$ and $z \sim 1.5$.

The simulation includes a treatment of the physics of the ISM and star formation following the model presented by Robertson & Kravtsov (2008), and interested readers should examine that work for details. The photoionization code CLOUDY (Ferland et al. 1998) is used to tabulate the cooling rate, heating rate, molecular abundance, and related properties of gas as a function of density, temperature, metallicity, and local interstellar radiation field (ISRF) strength. The star formation rate is calculated by converting the

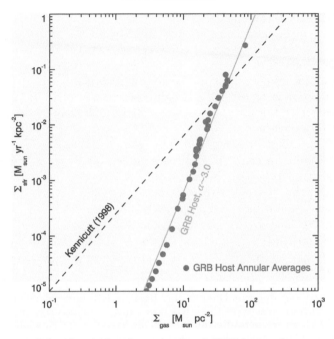

Figure 2. Kennicutt-Schmidt relation for a simulated GRB host galaxy at $z \sim 1.5$. Shown is the star formation rate surface density $\Sigma_{\rm SFR}$ as a function of total gas surface density $\Sigma_{\rm gas}$, measured in annuli (blue dots). The average Kennicutt-Schmidt relation of the GRB host has a steeper power-law index ($\alpha \sim 3.0$, green line) than the disk-averaged relation measured by Kennicutt (1998; $\alpha \sim 1.4$, dashed line), owing to the suppression of H_2 in the galaxy exterior by the interstellar radiation field and the low ISM metallicity (for a detailed discussion, see Robertson & Kravtsov 2008).

molecular gas density to stars on a timescale that scales with the local dynamical time, with an efficiency set to match the star formation efficiency per free fall time in local molecular clouds (e.g., Krumholz & McKee 2005, Krumholz & Tan 2007). The local ISRF spectral shape is fixed to the local Milky Way ISRF inferred by Mathis *et al.* (1983), but the ISRF strength scales with the local star formation rate density (i.e., young, massive stars supply the local ultraviolet radiation field). The abundance of molecular gas tracked using CLOUDY includes the photodissociative and heating effects of this ISRF, and thereby includes a coarse accounting of the regulatory impact of the ISRF on star formation in molecular clouds.

3. Results

Figure 1 shows the gaseous structure of the GRB host galaxy model at $z \sim 1.5$ (after 900 Myr of evolution). The figure shows the gas surface density of the system (image intensity) and the median temperature of the local ISM (purple regions have temperatures $T \lesssim 10^3$K, while blue regions have $T \gtrsim 10^4$K). The system has a rotational velocity of $v_{\rm rot} \approx 100$ km s^{-1} and a disk scale length of $R_{\rm d} \approx 1.5$ kpc. During the simulation the star formation rate of the system varies in the range SFR $\approx 0.5 - 2.5$ $M_\odot$ yr^{-1}, while the specific star formation rate is SFR/$M_\star \approx 0.17 - 1.1$ Gyr^{-1}. These properties are consistent with the properties of high-redshift GRB host galaxies determined by Savaglio *et al.* (2008), who find star formation rates of SFR $\sim 0.1 - 10$ and specific star formation rates of $M_\star$/SFR $\sim 0.1 - 10$ Gyr.

The compactness of the system leads to a dense ISM and a considerable molecular gas fraction. The global, mass-weighted molecular abundance declines from $f_{H2} \sim 0.5$ at $z \sim 2$ to $f_{H2} \sim 0.3$ at $z \sim 1.5$. As a function of radius, the molecular fraction declines from $f_{H2} \sim 1$ near the center of the galaxy to $f_{H2} \sim 0.1 - 0.3$ beyond a disk scale radius. The typical star formation rate-weighted radius of the system is $r_{SFR} \sim 0.8$ kpc, where the ISM molecular fraction is $f_{H2} \sim 0.6$. Hence, if observed high-redshift GRB host galaxies are similar in nature to this simulated system, GRBs will likely occur in molecular-rich regions. While these molecular abundances are consistent with the spectroscopic studies by Prochaska et al. (2008), they are well above the observed H_2 abundance along GRB sightlines (e.g., Tumlinson et al. 2007). In this model, the compact and dense structure of the high-redshift GRB host prevents the diffuse ISRF from suppress the H_2 to levels observed in GRB sightlines. If an ISRF is responsible for suppressing H_2 to observed levels in GRB sightlines, it may be generated by discrete point sources nearby the GRB in a manner not captured by the diffuse ISRF included in these simulations.

We note that in order to supply the star formation efficiency for GRB hosts determined by Savaglio et al. (2008), GRB hosts may need to be fairly molecule rich if their structure is similar to the simulated high-redshift galaxy analogues presented here. Figure 2 shows the total gas Kennicutt-Schmidt (KS) relation for the GRB host galaxy analogue, measured in annuli with a width of $\Delta r = 100$ pc. Plotted is the star formation rate density Σ_{SFR} as a function of the total gas surface density Σ_{gas} (blue points), compared with the mean disk-averaged trend determined by Kennicutt (1998, dashed-line). The central concentration of molecular gas causes the total gas KS relation of the simulated GRB host galaxy analogue to be steeper than the disk-averaged relation. In order to supply the observed star formation rate of SFR $\sim 1-10$ $M_\odot \text{yr}^{-1}$, as this simulated galaxy does, the typical consumption timescales of ~ 100 Myr for molecular gas imply a reservoir of roughly $M_{H2} \sim 0.1 - 1 \times 10^9$ $M_\odot$ (the simulated system has $M_{H2} \sim 2 - 7 \times 10^8$ $M_\odot$ during its evolution). Since observed GRB hosts have stellar masses of only $\log M_\star \sim 9.3$ (Savaglio et al. 2008), the inferred molecular fraction of the ISM should be large even for very gas rich systems.

Overall, we find that under standard assumptions about the mass and redshift scalings of galaxy structure hydrodynamical simulations of disk galaxies with stellar masses of $\log M_\star \sim 9$ that utilize a model for the molecular ISM and star formation in molecular clouds (Robertson & Kravtsov 2008) can reproduce the observed star formation rates and efficiencies of GRB hosts (e.g., Castro Cerón et al. 2008, Savaglio et al. 2008). The star formation in both observed GRB hosts and the simulated GRB host analogue presented here is efficient for their low stellar masses, and to supply the observed range of star formation rates (SFR $\sim 0.1 - 10$ $M_\odot$ yr^{-1}) the molecular gas content such systems may need to be considerable ($f_{H2} \gtrsim 0.1$). While this result is consistent with observations that suggest GRBs occur in dense, potentially molecular-rich regions of the ISM (e.g. Prochaska et al. 2008), more work is needed to reconcile such results with the low molecular abundance observed in GRB sightlines (e.g. Tumlinson et al. 2007) if GRBs cannot efficiently destroy H_2 in the ISM (Whalen et al. 2008).

References

Berger, E., et al. 2007a, *ApJ*, 660, 504
Berger, E., et al. 2007b, *ApJ*, 665, 102
Bloom, J. S., et al. 2002, *AJ*, 123, 1111
Bullock, J. S., et al. 2001, *MNRAS*, 321, 559
Castro Cerón, J. M., et al. 2008, *ApJ*, submitted, arXiv:0803.2235

Erb, D., et al. 2006, *ApJ*, 646,107
Ferland, G., et al. 1998, *PASP*, 110, 761
Hernquist, L. 1990, *ApJ*, 356, 359
Kennicutt, R. C. 1998, *ApJ*, 498, 541
Klebesadel, R. W., et al. 1973, *ApJ* (Letters), 182, L85
Krumholz, M. R. & C. F. McKee. 2005, *ApJ*, 630, 250
Krumholz, M. R. & J. C. Tan. 2007, *ApJ*, 654, 304
Le Floc'h, E., et al. 2006, *ApJ*, 642, 636
Mathis, J. S., et al. 1983, *A&A*, 128, 212
Metzger, M. R., et al. 1997, *Nature*, 387, 878
Mo, H. J., et al. 1998, *ApJ*, 295, 319
Navarro, J., et al. 1996a, *ApJ*, 462, 563
Prochaska, J. X., et al. 2006, *ApJ*, 642, 989
Prochaska, J. X., et al. 2008, *ApJ*, accepted, arXiv:0806.0399
Robertson, B., et al. 2006a, *ApJ*, 641, 21
Robertson, B., et al. 2006b, *ApJ*, 641, 90
Robertson, B. & A. V. Kravtsov . 2008, *ApJ*, 680, 1083
Savaglio, S., et al. 2008, *ApJ*, submitted, arXiv:0803.2718
Springel, V., et al. 2001, *New Astron.*, 6, 79
Springel, V., et al. 2005a, *MNRAS*, 361, 776
Springel, V. 2005b, *MNRAS*, 364, 1105
Tumlinson, J., et al. 2007, *ApJ*, 668, 667
Whalen, D., et al. 2008, *ApJ*, accepted, arXiv:0802.0737

Particpants enjoying the reception at Copenhagen Town Hall.

Reconciling the Metallicity Distributions of Gamma-ray Burst, Damped Lyman-α, and Lyman-break Galaxies at $z \approx 3$

Johan P. U. Fynbo[1], J. Xavier Prochaska[2], Jesper Sommer-Larsen[3], Miroslava Dessauges-Zavadsky[4], and Palle Møller[5]

[1]Dark Cosmology Centre, Niels Bohr Institute, Copenhagen University, Juliane Maries Vej 30, 2100 Copenhagen O, Denmark
email: jfynbo@dark-cosmology.dk

[2]Department of Astronomy and Astrophysics, UCO/Lick Observatory, University of California, 1156 High Street, Santa Cruz, CA 95064, US

[3]Excellence Cluster Universe, Technische Universität München; Boltz-manstr. 2, D-85748 Garching, Germany

[4]Observatoire de Genève, 51 Ch. des Maillettes, 1290 Sauverny, Switzerland

[5]European Southern Observatory, Karl-Scharschild-strasse 2, D-85748 Garching bei München, Germany

Abstract. We test the hypothesis that the host galaxies of long-duration gamma-ray bursts (GRBs) as well as quasar-selected damped Lyman-α (DLA) systems are drawn from the population of UV-selected star-forming, high z galaxies (generally referred to as Lyman-break galaxies). Specifically, we compare the metallicity distributions of the GRB and DLA populations against simple disk models where these galaxies are drawn randomly from the distribution of star-forming galaxies according to their star-formation rate and HI cross-section respectively. We find that it is possible to match both observational distributions assuming very simple and constrained relations between luminosity, metallicity, metallicity gradients and HI sizes. The simple model can be tested by observing the luminosity distribution of GRB host galaxies and by measuring the luminosity and impact parameters of DLA selected galaxies as a function of metallicity. Our results support the expectation that GRB and DLA samples, in contrast with magnitude limited surveys, provide an almost complete census of star-forming galaxies at $z \approx 3$.

Keywords. gamma-rays: bursts – interstellar medium

1. Introduction

The past 10 years has marked the emergence of extensive observational analysis of high redshift ($z > 2$) galaxies. This remarkable and rapid advance was inspired by new technologies in space and ground-base facilities for deep imaging, clever approaches to target selection, and the arrival of 10 m class ground-based telescopes for spectroscopic confirmation. A plethora of classes are now surveyed, each named for the observational technique that selects the galaxies: the Lyα emitters (Hu et al. 1998), the Lyman break galaxies (LBGs, Steidel et al. 2003), sub-mm galaxies (e.g., Chapman et al. 2005), distant red galaxies (van Dokkum et al. 2006), damped Lyα (DLA) systems (Wolfe et al. 2005), extremely red objects (Cimatti et al. 2003), long-duration γ-ray burst (GRB) host galaxies (e.g., Fruchter et al. 2006), MgII absorbers, radio galaxies (Miley & De Breuck 2008), quasar (QSO) host galaxies, etc. Large, dedicated surveys have identified in some cases thousands of these galaxies providing a direct view into the processes of galaxy formation in the young universe.

Because of the significant differences in the sample selection of high z galaxies, there has been a tendency by observers to treat each population separately and/or contrast the populations. However, the various populations will overlap to some extent and it is important to understand how (see also Adelberger et al. 2000; Møller et al. 2002; Fynbo et al. 2003; Reddy et al. 2005).

Of the various galactic populations discovered at $z > 2$ to date, only two offer the opportunity to study the interstellar medium at a precision comparable to the Galaxy and its nearest neighbors: the damped Lyα systems intervening quasar sightlines (QSO-DLA) and the host galaxies of GRBs which exhibit bright afterglows (GRB-DLA). These galaxies are characterized by a bright background source which probes the gas along the sightline to Earth. In the case of QSO-DLA it is the background QSO while for GRB-DLA it is the afterglow of the GRB located within the host galaxy. Hence, the GRB-DLA will not probe the full line-of-sight through the host. The gas in the ISM imprints signatures of the total HI column density, the metal content, the ionization state, the velocity fields, and the molecular fraction along the line-of-sight. Because the galaxies are identified in absorption, there is no formal magnitude limit for the associated stellar populations. In this respect, they may trace a large dynamic range in stellar mass, morphology, star formation rate, etc.

The connection between long-duration GRBs and star-forming galaxies has been empirically established. At large redshift, there is an exclusive coincidence of GRBs with actively star-forming galaxies (e.g., Hogg & Fruchter 1999; Bloom et al. 2002; Fruchter et al. 2006), the majority of which show elevated specific star-formation rates (Christensen et al. 2004). At low z, there is a direct link between GRBs and massive stars via the detection of spatially and temporally coinciding core-collapse supernovae (Hjorth et al. 2003; Stanek et al. 2003, but see also Fynbo et al. 2006). The simplest hypothesis, therefore, is that these galaxies uniformly sample high z galaxies according to star-formation rate, i.e. $f_{GRB} \propto \mathrm{SFR}(L)\phi(L)$, where $\phi(L)$ is the luminosity function.

The link between QSO-DLA and star-forming galaxies is less direct, primarily because the bright background quasar precludes the easy detection of stellar light. Nevertheless, the presence of heavy metals in all QSO-DLAs (and dust in the majority) indicates at least prior star-formation (Prochaska et al. 2003). Furthermore, the observation of CII* absorption suggests heating of the ISM by far-UV photons from ongoing star-formation in at least half of the sample (Wolfe et al. 2003, 2004. Furthermore, a handful of QSO-DLA have been detected in emission and exhibit properties similar to low luminosity LBGs (Møller et al. 2002). In contrast to the GRB-DLA, however, the QSO-DLA are selected according to their covering fraction on the sky, i.e. the probability of detection is the convolution of the HI cross-section with the luminosity function: $f_{DLA} \propto \sigma_{HI}(L)\phi(L)$. While both populations of DLAs may be drawn from the full sample of star-forming galaxies, their distribution functions would only be the same if $\sigma_{HI}(L) \propto SFR(L)$ (see also Chen et al. 2000).

In Fynbo et al. 2008 (F08) we test these ideas by comparing the observed metallicity distributions of QSO-DLA and GRB-DLA with simple predictions based on empirical measurements of star-forming galaxies at $z = 3$. In this paper we give a brief description of the model – for more details we refer to F08. Our study is similar in spirit to the studies of Fynbo et al. (1999) who combined the LBG luminosity function with a Holmberg relation for σ_{HI} to predict the luminosities and impact parameters of QSO-DLA galaxies, Jakobsson et al. (2005) who compared the luminosity distribution function of GRB host galaxies with the LBG luminosity function, and Chen et al. (2005) and Zwaan et al. (2005) who reconciled the properties of local galaxies with the QSO-DLA cross-section and metal abundances.

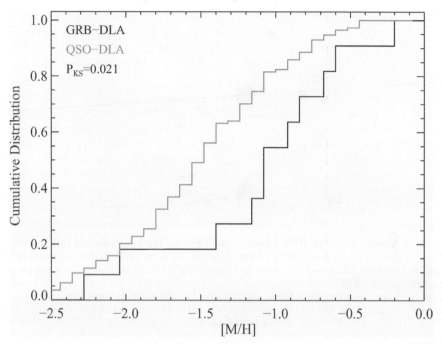

Figure 1. The histograms show the cumulative distribution of QSO-DLA and GRB-DLA metallicities in the statistical samples compiled by Prochaska *et al.* (2003) and Prochaska *et al.* (2007). As seen, the GRB-DLA metallicities are systematically higher than the QSO-DLA metallicities.

2. Analysis and results

Our aim is to model the metallicity distribution function of the GRB-DLA and QSO-DLA shown in Fig. 1. As seen, GRB-DLA systematically have higher metallicities than QSO-DLA (see also Savaglio 2006, Fynbo *et al.* 2006b). Our expectation was that this fact could be due to a combination of two effects: 1) SFR-selection vs. HI cross-section selection causing GRB hosts to on average by brighter and hence more metal rich, and 2) the on average larger impact parameters for QSO-DLA than GRB-DLA, which coupled with metallicity gradients will also shift the GRB-DLA towards lower metallicities. Full details of our model can as mentioned be found in F08. Here we just illustrate the model and repeat the main conclusions. To simulate the distribution of GRB-DLA metallicities we start with the 1700 Åluminosity function for LBGs from Reddy *et al.* 2008. To convert to metallicities we assume a metallicity-luminosity relation with slope 0.2 ($Z \propto (L/L_*)^{0.2}$) as seen locally. We normalise the relation so that it reproduces the metallicities for the LBGs at the bright end of the luminosity function. The simulated distribution based on these few and simple assumptions are in excellent agreement with the observed distribution (see F08). To simulate the distribution of QSO-DLA metallicities we specifically assume that each LBG is embedded in a flat gas disk. Then we assume two further relations: a galaxy luminosity vs. disk size relation and a recipe for assigning metallicity gradients to a galaxy with a given luminosity. Again, the simulated distribution using model parameters that are consistent with observed relations either locally or directly at $z \approx 3$. Fig. 2 and Fig. 3 further illustrate the main elements of our model.

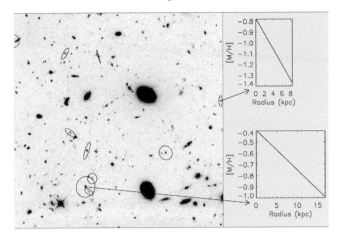

Figure 2. A subfield from the HDF North R-band image. We have selected LBGs with redshift between 2.8 and 3.2 from the catalog of photometric redshifts of Fernández-Soto et al. (1999). Over-plotted on each LBGs is the extent of a randomly inclined HI disk with a radius given by our prescription (See F08) Note that the majority of the total QSO-DLA absorption cross-section is caused by fainter galaxies than those shown here. On the right we show the radial metallicity profile from the centre to the largest radius at which the column density is above the definition of a DLA (N(H) > 2×10^{20} cm^{-2} in our model for two of the galaxies.

3. Shortcomings of the model

Obviously this is a very simple model: $z \approx 3$ galaxies do not all have flat, round gas disks around them, GRBs do not all explode exactly in the centres of their host galaxies, most likely $z \approx 3$ galaxies do not have nice, smooth metallicity gradients, etc. However, these shortcomings are probably of minor importance as long as $z \approx 3$ on average can be described as disks with metallicity gradients and GRBs occur significantly closer to the centres of their hosts than the typical impact parameters for QSO-DLA.

A more serious concern is bias in the observed samples of QSO-DLA and GRB-DLA. It has long been discussed to which extent QSO-DLA samples are biased against dusty (and hence likely metal rich and/or large $\log N(\mathrm{HI})$) systems. Studies of radio selected DLAs (free from dust-bias) have found similar column density and metallicity distributions as for optically selected DLA samples (Ellison et al. 2001, 2004; Ellison, Hall & Lira 2005; Akerman et al. 2005; Jorgenson et al. 2006) showing that any dust bias will be so small that it will not fundamentally change the conclusions about cross-section and metallicity distributions inferred from optically selected surveys.

Concerning GRB-DLA there is no dust bias in the detection of the prompt emission itself as γ-rays are unaffected by dust. However, the requirement of an optical afterglow detection from which the redshift the HI column density and metal columns can be measured does potentially exclude very dusty sightlines. Furthermore, there could be an intrinsic (astrophysical) bias against high metallicity in GRB production. In the collapsar model the limit is estimated to be around 0.3 $Z_\odot$ (Hirschi et al. 2005; Woosley & Heger 2005), but this is very dependent on the as yet poorly understood properties of winds from massive stars (e.g. clumping, Smith 2007). We also note that Wolf & Podsiadlowski (2007) exclude a metallicity cut-off below half the solar value based on statistics of host galaxy luminosities. If true such an intrinsic bias could preferentially exclude massive, dust-obscured starbursts, that typically seem to be enriched above this limit (Swinbank et al. 2004), from the GRB samples. Nevertheless, a few extremely red and luminous GRB hosts have been found (Levan et al. 2006; Berger et al. 2007). So far there are few

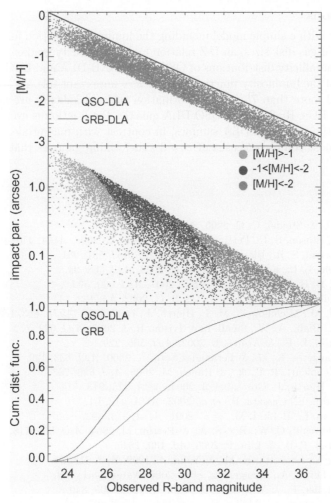

Figure 3. Simulated distributions of luminosity, impact parameter and metallicity (from bottom to top) for QSO-DLA and GRB galaxies at $z = 3$ in our model. In the top panel, QSO-DLA have lower metallicities than GRB hosts at a given R-band magnitude due to the metallicity gradients. In the middle panel impact parameters are lower for fainter QSO-DLA galaxies due to the Holmberg relation and low metallicity QSO-DLA have lower impact parameters due to the luminosity-metallicity relation and the smaller sizes given by the Holmberg relation. In the lower panel QSO-DLA galaxies are fainter than GRB hosts as the selection function for QSO-DLA weights fainter galaxies more than the selection function for GRBs.

examples of GRB sightlines with very large dust columns (for recent examples see Rol et al. 2007; Jaunsen et al. 2008; Tanvir et al. 2008). The near, mid and far-IR properties of GRB host galaxies have been studied by a number of groups (e.g., Chary et al. 2002; Le Floc'h et al. 2003, 2006; Berger et al. 2003; Tanvir et al. 2004; Priddey et al. 2006; Castro Cerón et al. 2007). A few GRB hosts, all at $z < 2$, have tentatively been detected at sub-mm wavelengths, but their inferred UV/optical (bluer, Gorosabel et al. 2003a,b) and dust properties (higher temperatures, Michałowski et al. 2008) are different than those of sub-mm selected galaxies. It is a major goal of ongoing GRB follow-up work to try to build a more complete sample less. For now, the incompleteness of the GRB samples remains a fundamental uncertainty on most conclusions drawn on the issue of GRBs as probes of high-z star-formation.

4. Conclusions

We find that with a simple model including the luminosity function for LBGs, a Holmberg relation for gas disk sizes, an L-Z relation and a metallicity gradient it is possible to reconcile the metallicity distributions of QSO-DLA, GRB-DLA and LBGs. In this model the faint end of the luminosity function plays a very important role. As seen in the lower panel in Fig. 3 more than 75% of star-formation selected galaxies are fainter than the flux limit for LBGs, R=25.5. For QSO-DLA galaxies the fraction is even higher. Hence, in this model the GRB and DLA samples, in contrast with magnitude limited surveys, provide an almost complete census of $z \approx 3$ star-forming galaxies that are not heavily obscured.

References

Adelberger, K. L. & Steidel, C. C. 2000, *ApJ*, 544, 218.
Akerman, C. J., Ellison, S. L., Pettini, M., & Steidel, C. C. 2005, *A&A*, 440, 499.
Berger, E., Kulkarni, S. R., Bloom, J. S., *et al.* 2002, *ApJ*, 581, 981.
Berger, E., Cowie, L. L., Kulkarni, S. R., *et al.* 2003, *ApJ*, 588, 99.
Berger, E., Fox, D. B., Kulkarni, S. R., *et al.* 2007, *ApJ*, 660, 504.
Bloom, J. S., Kulkarni, S. R., & Djorgovski, S. G. 2002, AJ, 123, 1111.
Castro Cerón, J. M., Michałowski, M. J., Hjorth, J., *et al.* 2007, *ApJL*, 653, L85.
Chapman, S. C., Blain, A. W., Smail, I., & Ivison, R. J. 2005, *ApJ*, 622, 772.
Chary, R., Becklin, E. E., & Armus, L. 2002, *ApJ*, 566, 229.
Chen, H.-W., Lanzetta, K. M., & Fernandez-Soto, A. 2000, *ApJ*, 533, 120.
Chen, H.-W., Kennicutt, R. C. Jr., & Rauch, M. 2005, *ApJ*, 620, 703.
Christensen, L., Hjorth, J., Gorosabel, J. 2004, *A&A*, 425, 913.
Cimatti, A., Daddi, E., Cassata, P., *et al.* 2003, *A&A*, 412, L1.
Ellison, S. L., Yan, L., Hook, I. M., *et al.* 2001, *A&A*, 379, 393.
Ellison, S. L. Churchill, C. W., Rix, S. A., & Pettini, M. 2004, *ApJ*, 615, 118.
Ellison, S. L., Hall, P. B., & Lira, P. 2005, *AJ*, 130, 1345.
Fernández-Soto, A., Lanzetta, K. M., & Yahil, A. 1999, *ApJ*, 513, 34.
Fruchter, A. S., Levan, A., Strolger, L., *et al.* 2006, *Nature*, 441, 463.
Fynbo, J. U., Møller, P., & Warren, S. J. 1999, *MNRAS*, 305, 849.
Fynbo, J. P. U., Ledoux, C., Møller, P., *et al.* 2003, *A&A*, 407, 147.
Fynbo, J. P. U., Watson, D., Thöne, C. C., *et al.* 2006a, *Nature*, 444, 1047.
Fynbo, J. P. U., Starling, R. L. C., Ledoux, C., *et al.* 2006b, *A&A*, 451, L47.
Fynbo, J. P. U., Prochaska, J. X., Sommer-Larsen, J., Dessauges-Zavadsky, M., Møller, P. 2008, *ApJ*, in press (F08) (arXiv:0801.3273).
Gorosabel, J., Christensen, L., Hjorth, J. *et al.* 2003a, *A&A*, 400, 127.
Gorosabel, J., Klose, S., Christensen, L., *et al.* 2003b, *A&A*, 409, 123.
Hirschi, R., Meynet, G., & Maeder, A. 2005, *A&A*, 443, 581.
Hjorth, J., Sollerman, J., Møller, P., *et al.* 2003, *Nature*, 423, 847.
Hogg, D. W. & Fruchter, A. S. 1999, *ApJ*, 520, 54.
Hu, E. M., Cowie, L., & McMahon, R. G. 1998, *ApJL*, 502, L99.
Jakobsson, P., Björnsson, G., Fynbo, J. P. U., *et al.* 2005, *MNRAS*, 362, 245.
Jaunsen, A. O., Rol, E., Watson, D. J., *et al.* 2008, *ApJ*, 681, 453
Jorgenson, R. A., Wolfe, A. M., Prochaska, J. X., *et al.* 2006, *ApJ*, 646, 730.
Le Floc'h, E., Duc, P.-A., & Mirabel, I. F. 2003, *A&A*, 400, 499.
Le Floc'h, E., Charmandaris, V., Forrest, W. J., *et al.* 2006, *ApJ*, 642, 636.
Michałowski, M. J, Hjorth, J., Castro Cerón, J. M., Watson, D. 2008, *ApJ*, 672, 817.
Miley, G. & De Breuck, C. 2008, A&ARv, 15, 67.
Møller, P., Warren, S. J., Fall, S. M., Fynbo, J. U. & Jakobsen, P. 2002, *ApJ*, 574, 51.
Priddey, R. S., Tanvir, N. R., Levan, A. J., *et al.* 2006, *MNRAS*, 369, 1189.
Prochaska, J. X., *et al.* 2003, *ApJL*, 595, L9.

Prochaska, J. X., Chen, H.-W., Dessauges-Zavadsky, M., & Bloom, J. S. 2007, *ApJ*, 666, 267.
Reddy, N. A., Erb, D. K., Steidel, C. C., *et al.* 2005, *ApJ*, 633, 748.
Reddy, N. A., Steidel, C. C., Pettini, M., *et al.* 2008, *ApJS*, 175, 48.
Savaglio, S., NJPh, 8, 195.
Stanek, K. Z., Matheson, T., Garnavich, P. M., *et al.* 2003, *ApJL*, 591, L17.
Steidel, C. C., Adelberger, K. L., Shaplet, A. E., *et al.* 2003, *ApJ*, 592, 728.
Tanvir, N. R., Barnard, V. E., Blain, A. W., *et al.* 2004, *MNRAS*, 352, 1073.
Tanvir, N. R., Levan, A. J., Rol, E., *et al.* 2008, *MNRAS*, in press (arXiv:0803.4100).
van Dokkum, P. G., Quadri, R., Marchesini, D., *et al.* 2006, *ApJL*, 638, L59.
Williams, R. E., Blacker, B., Dickinson, M., *et al.* 1996, *AJ*, 112, 1335.
Wolf, C. & Podsiadlowski, P. 2007, *MNRAS*, 375, 1049.
Wolfe, A. M., Gawiser, E., & Prochaska, J. X. 2003, *ApJ*, 593, 235.
Wolfe, A. M., Owk, J. C., Gawiser, E., Prochaska, J. X., & Lopez, S. 2004, *ApJ*, 615, 625.
Wolfe, A. M., Gawiser, E., & Prochaska, J. X. 2005, *ARA&A*, 43, 861.
Woosley, S. E. & Heger, A. 2006, *ApJ*, 637, 914.
Zwaan, M., van der Hulst, J. M., Briggs, F. H., *et al.* 2005, *MNRAS*, 364, 1467.

Johan Fynbo and Lisbeth Fogh Grove relaxing after Johan's talk.

Birgitta Nordström, Jens Viggo Clausen and John Renner Hansen enjoying a light moment during the Town Hall reception. In the background, Torgny Karlsson (left) and Linus Riel Petersen (center).

Anja Andersen, Ole Strömgren and Bengt Gustafsson at the Town Hall.

Stellar Populations and Dark Matter in the Milky Way Disk and in Local Group Galaxies

Eva K. Grebel[1]

[1] Astronomisches Rechen-Institut, Zentrum für Astronomie der Universität Heidelberg, Mönchhofstraße 12–14, D-69120 Heidelberg, Germany
email: grebel@ari.uni-heidelberg.de

Abstract. Our knowledge on the age structure, the chemical evolution, and the kinematics of the Galactic disk has grown substantially during the last years. Recent results on the properties of the stellar populations in the Galactic disk are summarized, and ongoing and future surveys and facilities are discussed. A short overview of recent mass estimates for the Milky Way is presented, and a brief summary of some of the key properties of the Galactic companions is given. The coming decade promises major breakthroughs in understanding our Milky Way, its disk, and the role of its satellites.

Keywords. surveys, Galaxy: abundances, Galaxy: disk, Galaxy: evolution, Galaxy: kinematics and dynamics, Galaxy: structure, galaxies: dwarf, galaxies: evolution, Local Group

1. Stellar Population Surveys of the Galactic Disk

During the last years we have made tremendous progress in learning about the stellar populations contributing to the Galactic disk as well as about the stellar constituents of other galaxies in the Local Group. These advances were made possible by new observational facilities such as space telescopes, large (8 to 10m-class) ground-based telescopes, and surveys with dedicated smaller telescopes. In understanding the Galactic disk, we continue to benefit from the unique heritage of the astrometric Hipparcos mission (Perryman *et al.* 1997) of the European Space Agency (ESA). The next decade looks particularly promising for Galactic astronomy: We will obtain an unprecedentedly deep and detailed picture of the kinematics and evolution of our Galaxy thanks to ESA's Gaia satellite (Perryman *et al.* 2001). This ESA cornerstone mission is scheduled to be launched at the end of 2011. Moreover, studies at many different wavelengths ranging from radio to gamma rays have contributed directly or indirectly to our understanding of the stellar population properties.

1.1. The astrometric Hipparcos mission and the Geneva-Copenhagen Survey

The Hipparcos (High Precision Parallax Collecting Satellite) satellite operated from 1989 to 1993. Hipparcos measured parallaxes for about 120,000 stars with a precision of 1 mas and for about 1 million stars with 30 mas or better. A re-reduction finally yielded the Tycho-2 catalog with 2.5 million stars, encompassing almost all objects down to the 11th magnitude in the V-band (Høg *et al.* 2000). These data include information on positions, motions, luminosities, and colors. The Hipparcos data have provided a basis for more than 1,600 refereed papers to date, revealing the tremendous impact of Hipparcos.

I will highlight here one of the many important studies obtained in part with Hipparcos data: The Geneva-Copenhagen Survey of the solar neighborhood (Nordström *et al.* 2004; Holmberg *et al.* 2007). The Geneva-Copenhagen Survey combines Hipparcos parallaxes,

Tycho-2 proper motions, Strömgren $uvby\beta$ photometry, and radial velocities of more than 14,000 nearby F and G dwarf stars with metallicities and isochrone-based ages.

The Geneva-Copenhagen Survey confirms the G-dwarf problem and shows that radial metallicity gradients exist in the disk. While the mean metallicity of the Galactic disk does not change much with age, a large scatter in metallicity is found at all ages. The thin disk continues to be dynamically heated, and heating from stochastic spiral waves aggravates the kinematic identification of thick disk stars.

1.2. Moving groups and kinematic substructure

Hipparcos as well as other observations have contributed to the detection of kinematic substructure in the Galactic disk. The concept of "moving groups" was first introduced by Eggen, who worked extensively on this topic and identified a number of them (e.g., Eggen 1965). It has since been shown that moving groups can have several possible origins. The most frequent type of moving group consists of the stellar debris of star-forming aggregates (dissolving associations and open clusters). This is particularly common among young moving groups (e.g., López-Santiago et al. 2006), but also old moving groups that appear to be dispersing stellar aggregates have been identified both chemically and kinematically (e.g., HR 1614: De Silva et al. 2007; Feltzing & Holmberg 2000). Here we are witnessing the gradual production of field populations. Other moving groups appear to be due to dynamical resonances in the disk. This seems to be the case for the Hercules stream (Famaey et al. 2007) and also for the Pleiades, Hyades, and Sirius moving groups (Famaey et al. 2008). The third possible origin of moving groups are accretion events: In this case, moving groups are the debris of infalling objects (e.g., Helmi et al. 2006a; Arifyanto & Fuchs 2006). The Arcturus group seems to be one such example (Navarro et al. 2004), and there are also signs of debris streams at larger distances (e.g., Wyse et al. 2006). Ultimately, these detections hint at a possible merger origin of the thick disk.

1.3. Ongoing kinematic surveys: RAVE as an example

The Radial Velocity Experiment (RAVE, 2003–2011) measures the radial velocities and chemical composition of up to 1 million luminous stars in the southern sky (Steinmetz et al. 2006; Zwitter et al. 2008). The data are taken at the 1.2m UK Schmidt Telescope with the 6dF spectrograph of the Anglo-Australian Observatory (Australia). RAVE explores the chemistry and kinematics of our local neighborhood in the Milky Way with spectra in the near-infrared Ca II triplet region (R = 7,500) across a magnitude range of $9 < I < 12$.

Combining RAVE spectra with CORAVEL (Correlation Radial Velocities; used for the Geneva-Copenhagen Survey), Seabroke et al. (2008) searched the local Galactic disk for streams. They were able to exclude nearby crossings of both the Sagittarius stream and the Virgo overdensity. Concentrating on 500 pc around the Sun, Klement et al. (2008) could to retrieve several previously identified moving groups and found one new stream candidate on a radial orbit.

RAVE in combination with ELODIE radial velocities and infrared photometric star counts from 2MASS together with UCAC2 proper motions were used to separate stars of different luminosity classes and measure the scale heights of the thin and the thick disk toward the poles (225 ± 10 pc and 1048 ± 36 pc, respectively). The data indicate that the thick disk cannot have formed through a continuous process from the thin disk (Veltz et al. 2008). With red clump stars towards the South Galactic poles, the inclination of the velocity ellipsoid 1 kpc below the plane was found to be $7.3° \pm 1.8°$ (Siebert et al. 2008). Together with other constraints this argues in favor of a nearly spherical halo and a disk scale length of ~ 2.6 kpc. For further details see the contribution of Steinmetz in these proceedings.

1.4. Recent/ongoing optical and infrared surveys

Vast data sets from ground-based photometric and spectroscopic stellar surveys are either already available or are currently being obtained, and additional survey projects are planned. These data are changing and refining our understanding of the stellar populations in our Galaxy and its neighbors. The past and ongoing surveys are usually carried out with small and medium-sized telescopes, which continue to be immensely useful for survey applications. The observations are often complemented by targeted follow-up studies with large telescopes.

Microlensing surveys such as EROS (Expérience de Recherche d'Objets Sombres, e.g., Derue et al. 2001) MACHO (Massive Compact Halo Objects, e.g., Alcock et al. 1998), and OGLE (Optical Gravitational Lensing Experiment, e.g., Szymanski et al. 1996) had the detection of dark matter candidates as their main goal, but they also yielded valuable data on Galactic structure and on stellar populations in the halo, disk, and bulge, particularly on their variable stars.

The studies of the Quasar Equatorial Survey Team (QUEST) have led to the discovery of detailed substructure in the Galactic halo as traced by RR Lyrae stars (Vivas et al. 2001; Vivas & Zinn 2006; Duffau et al. 2006). The Century Survey Galactic Halo Project relies on blue horizontal branch stars selected from 2MASS and the SDSS (both see below). These data trace the metal-poor thick disk and inner halo, their metallicity distribution functions and their kinematics (Brown et al. 2003, 2005, 2008). Also other surveys including time-domain data are being used to uncover substructure in the Galactic halo, e.g., the Southern Edgeworth-Kuiper Belt Object survey (SEKBO; see Keller et al. 2008 for results).

The Two Micron All Sky Survey (2MASS, Skrutskie et al. 2006) mapped the entire sky in the near-infrared JHK bandpasses using 1.3m telescopes. The 2MASS data base has been of enormous value for numerous stellar and Galactic studies, including, perhaps most spectacularly, the all-sky view of the tidal tails of Sagittarius as traced by M giants (Majewski et al. 2003) and of other overdensities (e.g., Rocha-Pinto et al. 2003, 2004, 2006). Also the IJK Deep Near-Infrared Southern Sky Survey (DENIS, Epchtein et al. 1999), carried out with a 1m telescope at the European Southern Observatory (ESO, La Silla, Chile) has been used for a large number of stellar and Galactic structure studies, for instance, in order to constrain the scale height and the scale length of the Galactic disk and its dust layer (Unavane et al. 1998).

The UKIRT Infrared Deep Sky Survey (UKIDSS; 2005–2012) includes a large-area survey (4000 deg^2) and a Galactic plane survey (GPS, 1800 deg^2; Lucas et al. 2007) in the JHK infrared passbands. The GPS will ultimately contain 2 billion sources and is already proving to be useful in combination with data obtained in other wavelength ranges such as the Isaac Newton Telescope (INT) Photometric Hα Survey of the Northern Galactic Plane (IPHAS), which covers 1800 deg^2 of the plane. The IPHAS point source catalog will contain ~ 80 million sources (Drew et al. 2005) and is a valuable tool for stellar population studies in the disk. The UKIDSS GPS, an excellent resource for Galactic structure studies in the inner Milky Way, was already used recently to refine the position angle and extent of the long Galactic bar (Cabrera-Lavers et al. 2008).

The Spitzer Space Telescope conducted a survey of the Galactic plane in four infrared bands. The data of the Galactic Legacy Infrared Mid-Plane Survey Extraordinaire (GLIMPSE) have provided spectacular results on the spiral structure and bar of our Galaxy as well as detailed information on embedded star-forming regions (see, e.g.,

Benjamin et al. (2003, 2005) and Benjamin, these proceedings). A recent GLIMPSE result is the recognition that our Milky Way appears to have only two major spiral arms as well as several less prominent spiral fragments.

1.5. *The SDSS, SEGUE, and results for the Galactic disk and halo*

Not all of these surveys were initially designed for stellar population studies or for Galactic science. For instance, the Sloan Digital Sky Survey (SDSS) (e.g., York et al. 2000; Grebel 2001; Stoughton et al. 2002) originally had cosmological applications as its main science driver, but it inevitably observed numerous stars of the Galactic thin and thick disk and in the Galactic halo. The SDSS is the largest photometric and spectroscopic sky survey undertaken to date. It uses a dedicated 2.5m telescope at Apache Point Observatory (USA) and is mapping more than one quarter of the sky. The availability of such a large, deep, and homogeneous data set led to, e.g, an improved determination of the vertical disk structure (Chen et al. 2001) and to the discovery of new Galactic substructure (Odenkirchen et al. 2001a, 2003, Rockosi et al. 2002; Newberg et al. 2002, 2003; Yanny et al. 2003, Peñarrubia et al. 2005; Grillmair & Dionatos 2006, Grillmair 2006, Grillmair & Johnson; Belokurov et al. 2006a). The new substructure consisted both of disrupting star clusters as well as new overdensities in the Milky Way such as the Monoceros ring and new tails of the Sagittarius dwarf galaxy.

In recognition of the potential of the SDSS for studies of Galactic structure and evolution, the subsequent continuations of the survey – SDSS-II and SDSS-III – dedicated a substantial amount of survey time to Galactic science projects, in particular, to the Sloan Extension for Galactic Understanding and Exploration (SEGUE, see Newberg et al. 2003 and Rockosi 2005). SEGUE (I+II) will obtain close to 600,000 spectra from $\sim$ 380 to 910 nm ($R = 2000$). APOGEE, the Apache Point Observatory Galactic Evolution Experiment, will take H-band spectra of 100,000 red giants in the bulge and disk with $R \sim 20,000$. SEGUE II and APOGEE are currently the main Galactic SDSS projects. The SDSS is expected to continue until 2014.

The efforts dedicated to Galactic science are leading to substantial progress in our understanding of Galactic structure and chemistry. Jurić et al. (2008) estimated the distances to some 48 million stars in the SDSS and mapped their three-dimensional density distribution from 20 to 100 kpc across 6,500 deg^2. These authors recovered previously known overdensities such as the Monoceros feature of Yanny et al. (2003) and identified two new extended stellar overdensities including the Virgo density enhancement, which extends across 1,000 deg^2. They suggest that the Virgo overdensity may be a dwarf galaxy merging with the disk. All of the detected overdensities are comparatively close to the Galactic disk; the remainder is fairly smooth. Jurić et al. find the Galactic thin and thick disk well-described by exponential disks. The resulting scale lengths and scale heights are 2,600 and 300 pc for the thin disk and 3,600 and 900 pc for the thick disk, respectively. The data support an oblate halo.

Allende Prieto et al. (2006) analyzed spectroscopy and photometry of 22,770 SDSS stars. They show that the metallicity distribution of thick-disk G-dwarf stars at vertical distanced of $1 < |z| < 3$ kpc from the Galactic plane peaks at [Fe/H] ~ -0.7 dex and that it lags behind the local standard of rest (LSR) rotation by 63 km s^{-1}. The authors do not detect a thick disk metallicity gradient across 5 to 14 kpc from the Galactic center, nor do they find a vertical gradient. However, they do find a decline in rotational velocity with increasing distance from the plane. In contrast to the relatively narrow metallicity range of the thick disk, they find a wide range of metallicities in the halo peaking at [Fe/H] ~ -1.4 dex. While most of the star formation in the halo seems to have occurred 11 Gyr or longer ago, some of the thick disk stars appear to be $\geqslant 2$ Gyr younger. Ivezić *et al.* (2008) derive spectroscopic and photometric metallicity estimates for 60,000 F and G main-sequence stars and discuss the metallicity distribution functions. They show that the Monoceros feature is distinct in metallicity and rotates faster than the LSR.

Carollo *et al.* (2007) analyzed SDSS spectra and demonstrate that the Galactic halo consists of two components. The inner halo exhibits an overall prograde rotation with a peak metallicity at [Fe/H] ~ -1.6 dex, while the outer halo shows net retrograde rotation and is more metal-poor (peak metallicity ~ -2.1 dex). This dichotomy and the properties of the outer halo seem to favor an accretion origin. In a structural analysis of ~ 4 million main-sequence turn-off stars, Bell *et al.* (2008) find substantial substructure in the halo, consistent with expectations from a halo built out of the debris of disrupted satellites.

1.6. Other spectroscopic studies of the Galactic disk

The past two decades have seen a large number of high-resolution spectroscopic studies of stars in the Galactic disk and in other Milky Way components. Results of only very few of these studies can be mentioned here. An abundance gradient with Galactocentric radius was found in studies using a variety of tracers including Cepheids, open clusters, and H II regions (Luck *et al.* 2006; Chen *et al.* 2003; Maciel & Costa 2008). These relatively mild gradients tend to become flat at large Galactocentric radii. Individual element abundance ratios indicate that Type II supernovae dominated the enrichment in recent star formation events in the outer disk (Yong *et al.* 2006), in contrast to what is observed in the solar neighborhood. Since older open clusters and field stars reach their lowest metallicities at shorter Galactocentric distances (~ 11 kpc) than the younger Cepheids (~ 14 kpc), the Galactic disk seems to have grown with time (Yong *et al.* 2006). The deviant abundance ratios of a small number of Cepheids – lower [Fe/H] and higher [α/Fe] – may indicate that they formed as a consequence of recent merger events (candidates include the Monoceros feature or Canis Major).

The thick and the thin disk are quite distinct in their abundances (e.g., Bensby *et al.* 2005; Brewer & Carney 2006; Reddy *et al.* 2006). Generally, thick disk stars are enhanced in [α/Fe] as compared to thin disk stars. Only the most metal-poor thin disk stars are α-enhanced; the ratios decline toward solar values for [Fe/H] > -1 dex. In contrast, for the thick disk the decline begins at higher metallicities ([Fe/H] ~ -0.4). Heavy s-process elements are enhanced in thin disk stars relative to thick disk stars, while the thick disk seems to be dominated by Type Ia supernova enrichment. The thick disk does not show variations in its abundance ratios beyond Galactocentric distances of 5 kpc or with vertical distance from the plane (Bensby *et al.* 2005). The differences in nucleosynthetic history support an independent evolutionary history of the thin and the thick disk (Brewer & Carney (2006)).

Meléndez *et al.* (2008) suggest that the local thick disk and the Galactic bulge show the same abundance patterns, indicative of similar chemical enrichment histories, and that they may have formed on similar time scales. Generally, the thick disk is believed to have formed early and rapidly about 12 Gyr ago, and there appears to have been a time delay between the formation of the thick and the thin disk (e.g., Freeman & Bland-Hawthorn 2002). Scenarios for the formation of the thick disk include heating of an early thin disk caused by mergers or interactions, or the formation of the thick disk from the early accretion of satellites. Other scenarios suggest that gas-rich, turbulent disks with giant star-forming clumps (Elmegreen *et al.* 2008) or massive, clustered star formation (Kroupa 2002) lead to disk thickening. There is some transition or overlap in stellar properties between the thick and the thin disk, and Bensby *et al.* (2007) have shown that some stars in the thick disk even reach solar metallicities. They conclude that after pronounced chemical enrichment over a period of ~ 3 Gyr, this process ended some 8–9 Gyr ago. There is also evidence for a gradual transition between the thick disk and the halo beyond distances of 4 kpc above the Galactic plane: Here the mean stellar metallicity is ~ -1.7 dex. The rotational gradient decreases with increasing distance from the plane, and at distances beyond 5 kpc stars show mainly inner-halo properties with a mean [Fe/H] $= -2$ dex (Brown *et al.* 2008).

2. Future Stellar Population Surveys of Galactic Disk

2.1. *Future optical and infrared ground-based imaging surveys*

The VISTA Hemispheric Survey (VHS) is one of the ESO Public Surveys to start 2009. It will contribute to an infrared imaging map of the southern sky (20,000 deg^2) that reaches 4 mag deeper than 2MASS and DENIS. These data will complement optical surveys such as the SSS (see below) and will be particularly useful for structural studies, determinations of the low-mass stellar mass function, etc. The European Galactic Plane Surveys (EGAPS) seek to combine the various ongoing and planned surveys with the Isaac Newton Telescope (INT, Spain) and the VLT Survey Telescope (VST, Chile) to map the complete Galactic plane (10°×360°) down to 21st magnitude in various optical bands. EGAPS comprises individual northern and southern surveys like IPHAS, VPHAS+, and UVEX. The surveys aim mainly at stellar astrophysics.

The Panoramic Survey Telescope & Rapid Response System (Pan-STARRS or PS1; 2008 – 2013; Kaiser 2006) is a northern photometric survey that will reach five times lower luminosities than the SDSS and that will add the time domain. PS1 uses a custom-built 1.8m telescope on Haleakala (USA) with a three-degree imager. PS1 will observe three quarters of the sky. It will be particularly useful for identifying halo substructure, but it will also cover more of the Galactic disk at greater depths than previous surveys. The Southern Sky Survey (SSS, Keller *et al.* 2007) is an automated photometric survey mapping the southern hemisphere with a new 1.3m telescope at Siding Spring Observatory (Australia; to start in 2009). The SSS will provide data for Galactic (sub-)structure, and for variable sources, complementing the northern SDSS and PS1 photometry.

The Large Synoptic Survey Telescope (LSST) is a planned 8.4m telescope with a field of view of 9.6 deg^2. It will map 20,000 deg^2 every three nights using up to six broad-band filters. LSST will reach $r \sim 27.5$. First light is expected in 2014. With its time-lapse imaging, coverage, and depth, LSST will be a fabulous facility for Galactic science.

2.2. *Future kinematic and chemistry surveys*

The Large Sky Area Multi-Fiber Spectroscopic Telescope (LAMOST) is a wide-field spectroscopic survey that will be carried out with a custom-built 4m telescope at Xinglong Station (China). Up to 4,000 stars can be observed simultaneously with a resolution of 1000 – 2000. LAMOST is an ideal facility to map the chemical composition and kinematics of stars in the Milky Way (Zhao *et al.* 2006).

The Wide-Field Multi-Object Spectrograph (WFMOS) is being developed by an international consortium for use with the Gemini and Subaru telescopes. One of the key projects with WFMOS will be Galactic archaeology. The Japanese JASMINE astrometry mission will be complemented by WINERED, a near-infrared high-resolution ($R \sim 20,000$) spectrograph currently under construction. It will provide radial velocities and chemical abundances for Galactic bulge stars.

2.3. *Stellar imaging survey missions from space*

The NASA Explorer mission WISE (Wide-Field Survey Explorer; launch 2010) will map the Milky Way in the infrared during its six-month mission. The data will provide an image of the Milky Ways stellar populations with a much reduced dust extinction, useful for, e.g., Galactic structure studies.

One of the proposed ESA Cosmic Vision missions is EUCLID, which aims at cosmic shear and photo-z measurements (the merger of the DUNE concept of Refregier et al. 2008 and the SPACE concept of Cimatti et al. 2008). EUCLID would deliver diffraction-limited optical and infrared images down to $\sim$ 24th magnitude (wide visual and H-band) of 20,000 deg^2 outside of the Galactic plane. Observations of the plane are under discussion. If approved, this mission (potential launch year: 2018) would deliver an exquisite, deep data set for Galactic structure studies, stellar mass functions, etc.

2.4. Future space astrometry missions

ESA's cornerstone mission Gaia will survey the entire sky down to G $\sim$ 20, covering up to one billion stars. Apart from astrometry accurate to 12 – 25 μas at 15th magnitude and about 100 – 300 μas at G $\sim$ 20, Gaia carries out low-dispersion spectrophotometry with $R = 3$ to 30 nm pixel^{-1} covering a wavelength range from 330 – 1000 nm. From these low-dispersion spectra various photometric passbands can be synthesized. In addition, Gaia obtains spectroscopy with $R = 11,500$ centered on the Ca II triplet. With the wealth of data that Gaia will provide it will truly revolutionize Galactic astronomy, providing six-dimensional phase space information along with physical stellar parameters, metallicities, and to some extent possibly [α/Fe] estimates. What Gaia will be able to accomplish for studies of Galactic structure, of the accretion history of our Milky Way, and other topics is described in Bailer-Jones' contribution in these proceedings and references therein.

The Japanese JASMINE (Japan Astrometry Satellite Mission for INfrared Exploration) satellite is expected to carry out near-infrared astrometry of sources with z-band magnitudes from 6 to 14 with a proper-motion accuracy of $\sim$ 4 μas year^{-1} (Gouda et al. 2006). It may be launched in 2016 or later. It is preceded by a small technical demonstrator mission, Nano-Jasmine (launch planned 2010), which will observe stars brighter than 8th mag. JASMINE will provide an interesting complement to Gaia's optical astrometry particularly in the disk and bulge, as its data will be less affected by dust extinction.

NASA's Space Interferometry Mission (SIM) is a pointed mission (not an all-sky survey) with considerably higher astrometric accuracy than Gaia ($\sim$ 4 μas at V = 20). Due to its measurement technique it will only observe comparatively few selected objects (Unwin et al. 2008). SIM's status is yet to be decided, but if it were to fly it would probably start only in 2016. One of SIM's approved key projects, "Taking Measure of the Milky Way" (PI: Majewski), is designed to provide an accurate map of the mass distribution within our Galaxy.

3. Recent Studies of the Dark Matter Content of the Milky Way

In order to constrain the dark matter content of the Milky Way, one aims at deriving the enclosed mass out to as large Galactocentric distances as possible. Here a few recent results based on stellar populations are presented. Ultimately, these studies seek to determine the circular velocity curve ($V_{circ}(R)$) or the escape velocity curve ($V_{esc}(R)$). The work summarized below does not tell us much about the estimated dark matter content of the Galactic *disk*, but rather of the Milky Way as a function of distance or as a whole.

Smith et al. (2007) measured the local escape speed and used that result to also estimate the total mass of the Milky Way. They use 16 high-velocity stars from the RAVE survey and combined them with 17 high-velocity stars from other surveys, resulting in a sample of 33 stars in total. Smith et al. (2007) find that the local escape velocity lies in the range of (498 < V_{esc} < 608) km s^{-1} with a median V_{esc} = 544 km s^{-1}. Since V_{esc}^2 is much larger than $2V_{circ}^2$ they conclude that there must be a significant amount of mass outside of the solar circle. Adopting different halo models, they derive a virial mass of the Milky Way of $\sim 1.4 \cdot 10^{12}$ $M_\odot$, corresponding to a virial radius of $R_{vir} \sim 305$ kpc and a circular velocity at the virial radius of $V_{circ}(R_{vir}) \sim 142$ km s^{-1}.

Sakamoto et al. (2003) perform a kinematic mass estimate employing (presumably) bound halo objects. They combine radial velocities and, where available, proper motions for 11 Galactic satellite galaxies, 137 globular clusters, and 413 field horizontal branch stars with heliocentric distances of up to 10 kpc in order to constrain the mass of the Milky Way. They find that several high-velocity objects including Leo I, Pal 3, and Draco affect the mass estimate significantly. At the distance of the Large Magellanic Cloud, the resulting Milky Way mass is $\sim 5.5 \cdot 10^{11}$ $M_\odot$. The total Galactic mass is $\sim 2.5 \cdot 10^{12}$ $M_\odot$ when Leo I is included or $\sim 1.8 \cdot 10^{12}$ $M_\odot$ without Leo I.

Battaglia et al. (2005) measure the radial velocity dispersion profile of the halo. They used 240 halo objects with distance and radial velocity measurements, including 24 objects beyond 50 kpc. Their sample includes 44 globular clusters, nine satellites, and a large number of field halo stars (horizontal branch and red giant stars). Battaglia et al. (2005) find that the radial velocity dispersion is almost constant out to 30 kpc with a value of 120 km s^{-1}. Then it decreases smoothly down to 50 km s^{-1} at ~ 120 kpc. The authors conclude that in the case of a constant velocity anisotropy an isothermal profile can be excluded, whereas either a truncated flat halo model ($\sim 1.2 \cdot 10^{12}$ $M_\odot$) or a Navarro, Frenk, & White (NFW) profile with $\sim 8 \cdot 10^{11}$ $M_\odot$ are permitted by the data.

Xue et al. (2008) determine the circular velocity curve of the Milky Way out to 60 kpc using a sample of 2400 blue horizontal branch stars within $|z| \geqslant 4$ kpc and with SDSS photometry and spectroscopy. They derive an enclosed mass at 60 kpc of $\sim 4 \cdot 10^{11}$ $M_\odot$. Assuming an NFW halo profile, they estimate the virial mass of the Milky Way's dark matter halo to be $\sim 10^{12}$ $M_\odot$. This value is lower than previous estimates and questions whether the Galactic satellites are indeed bound. So the question of the total mass and dark matter content of the Milky Way still remain open.

4. Properties of the Satellites of the Milky Way

Our Milky Way is surrounded by a large number of satellites, whose census has been considerably extended in the last few years. While in the previous decades typically four or five new dwarfs were added to the Local Group per decade (van den Bergh 1999), deep imaging surveys recently led to a significant increase of candidate Local Group members. For a more detailed description of the different types of lower-mass dwarf galaxies and their properties see the review by Grebel (2001b). Most of the new galaxies are very faint, gas-deficient dwarf spheroidal (dSph) galaxies that appear to orbit either M31 or the Milky Way (e.g., Zucker et al. 2004, 2006a, 2006b, 2007; Belokurov et al. 2006b, 2007). The Milky Way alone gained at least ten newly recognized companions, and additional ones await confirmation.

The Milky Way has two comparatively massive, gas-rich, nearby neighbors, the Magellanic Clouds. The Clouds are interacting with each other and with the Milky Way, but it is unclear whether they are bound to our Galaxy or on their first passage (Besla et al. 2007). This in turn questions their role in the creation of the warp of the Galactic disk (Weinberg & Blitz 2006). Nonetheless, satellites are of particular interest as interaction partners of the Milky Way and as objects that may ultimately be accreted.

The majority of the Milky Way companions is dominated by old and metal-poor populations. The more luminous dwarf galaxies deviate from this trend however: The Magellanic Clouds have experienced star formation until the present day, the Fornax dSph stopped forming stars only a few hundred million years ago (e.g., Grebel & Stetson 1999), and galaxies like Leo I and Leo II were still active about a billion years ago (e.g., Grebel et al. 2003). While the "classical" dSph galaxies seemed to indicate that the duration of star formation activity and the intermediate-age population fraction correlates with distance from the primary (van den Bergh 1994), no such trend is seen when including the recent, very low-luminosity dwarfs, suggesting that intrinsic properties such as the baryonic mass content may be a more important factor. Nonetheless, spatial variations in star formation are observed even in those dSphs dominated primarily by old populations. While dwarf irregular galaxies show a number of scattered large-scale, long-lived star-forming regions with life times of 10 – 100 Myr (e.g., Grebel & Brandner 1998), in the lower-luminosity dwarfs the somewhat younger and/or somewhat more metal-rich populations are more centrally concentrated (Grebel 1997; Stetson et al. 1998; Hurley-Keller et al. 1999; Harbeck et al. 2001; Koch et al. 2006a), suggesting more extended activity in the central regions where star-forming material could be retained more easily.

In dwarf galaxies with long-lasting star formation histories no simple age-metallicity relation has been found thus far. Leo II, for instance, shows a flat age-metallicity relation over many Gyr (Koch et al. 2007a), which may be indicative of blow-out (see also Lanfranchi & Matteucci 2007) or dilution by gas accretion. The Small Magellanic Cloud shows a complex age-metallicity relation with substantial scatter at any age (Glatt et al. 2008a). Evidence for inhomogeneous enrichment has also been found in other dwarf galaxies (e.g., Kniazev et al. 2005; Koch et al. 2008a, 2008b). Marcolini et al. (2008) present theoretical models describing mechanisms for stochastic enrichment.

All nearby dwarf galaxies studied in detail thus far reveal evidence for old populations (although their fraction may vary), and their main-sequence turn-off ages are even consistent with a common epoch of substantial early Population II star formation (Grebel & Gallagher 2004). (The Small Magellanic Cloud is a possible exception at least as far as its globular clusters are concerned; see, e.g., Glatt et al. 2008b).

There is a growing body of data on detailed element abundance ratios for red giants in nearby dSphs based on high-resolution spectroscopy with 8 to 10m-class telescopes. Generally, dSphs exhibit lower [α/Fe] ratios at a given [Fe/H] than observed in Galactic halo stars of comparable metallicity (e.g., Shetrone et al. 2001; Fulbright 2002; Tolstoy et al. 2003; Sadakane et al. 2004; Geisler et al. 2005; Monaco et al. 2005; Sbordone et al. 2007; Koch et al. 2008). Very few very metal-poor stars with [Fe/H] < -2.5 have been found in dSphs thus far (Helmi et al. 2006b; Koch et al. 2007a), but a recent study succeeded in detecting such objects (Kirby et al. 2008). Both the differences in [α/Fe] ratios and the perceived deficiency in extremely metal-poor stars have been taken as arguments against low-mass dwarf galaxies being *dominant* contributors to the build-up of galaxies like the Milky Way (e.g., Shetrone et al. 2001), although they clearly did contribute (e.g., Carollo et al. 2007). The apparent planar alignment of dwarf satellites (e.g., Lynden-Bell 1982; Koch & Grebel 2006; Pasetto & Chiosi 2007; Metz et al. 2008; and references therein) can be interpreted as an indication of infall and subsequent interactions (e.g., D'Onghia & Lake 2008).

Considering only their present-day numbers, the possible contribution of dSphs to the overall mass of their "parent galaxy" is negligible, since with typical total masses of the order of a few 10^7 M$_\odot$ (e.g., Wilkinson et al. 2004; Koch et al. 2007a, 2007b; Gilmore et al. 2007; Walker et al. 2007; Strigari et al. 2008) they are up to five orders of magnitude less massive than galaxies like the Milky Way. Intriguingly, the above studies indicate that the radial velocity dispersion profiles of dSphs are flat and that these galaxies are likely strongly dark matter dominated, sharing the same mass of a few times 10^7 M$_\odot$ within 600 pc regardless of their baryon content or stellar luminosity.

References

Alcock, C., et al. 1998, ApJ, 492, 190
Allende Prieto, et al. 2006, ApJ, 636, 804
Arifyanto, M. I. & Fuchs, B. 2006, A&A, 449, 533
Battaglia, G., et al. 2005, MNRAS, 364, 433
Belokurov, V., et al. 2006a, ApJ, 642, L137
Belokurov, V., et al. 2006b, ApJ, 647, L111
Belokurov, V., et al. 2007, ApJ, 654, 897
Benjamin, R. A., et al. 2003, PASP, 115, 953
Benjamin, R. A., et al. 2005, ApJ, 630, L149
Bensby, T., Feltzing, S., Lundström, I. & Ilyin, I. 2005, A&A, 433, 185
Bensby, T., Zenn, A. R., Oey, M. S. & Feltzing, S. 2007, ApJ, 663, L13
Besla, G., Kallivayalil, N., Hernquist, L., Robertson, B., Cox, T. J., van der Marel, R. P., & Alcock, C. 2007, ApJ, 668, 949
Brewer, M.-M. & Carney, B. W. 2006, AJ, 131, 431
Brown, W. R., Allende Prieto, C., Beers, T. C., Wilhelm, R., Geller, M. J., Kenyon, S. J., & Kurtz, M. J. 2003, AJ, 126, 1362
Brown, W. R., Geller, M. J., Kenyon, S. J., Kurtz, M. J., Allende Prieto, C., Beers, T. C., & Wilhelm, R. 2005, AJ, 130, 1097
Brown, W. R., Beers, T. C., Wilhelm, R., Allende Prieto, C., Geller, M. J., Kenyon, S. J., & Kurtz, M. J. 2008, AJ, 135, 564
Cabrera-Lavers, A., Gonzalez-Fernandez, C., Garzon, F., Hammersley, P. L., & Lopez-Corredoira, M. 2008, A&A, in press (arXiv:0809.3174)
Carollo, D., et al. 2007, Nature, 450, 1020
Chen, B., et al. 2001, ApJ, 553, 184
Chen, L., Hou, J. L., & Wang, J. J. 2003, AJ, 125, 1397
Cimatti, A., et al. 2008, Experimental Astronomy, 37
Derue, F., et al. 2001, A&A, 373, 126
De Silva, G. M., Freeman, K. C., Bland-Hawthorn, J., Asplund, M., & Bessell, M. S. 2007, AJ, 133, 694
D'Onghia, E. & Lake, G. 2008, ApJ, 686, L61
Drew, J. E., et al. 2005, MNRAS, 362, 753
Duffau, S., Zinn, R., Vivas, A. K., Carraro, G., Méndez, R. A., Winnick, R., & Gallart, C. 2006, ApJ, 636, L97
Eggen, O.J. 1965, in Galactic Structure, eds. A. Blaauw & M. Schmidt (Univ. of Chicago Press), 111
Elmegreen, B. G., Bournaud, F., & Elmegreen, D. M. 2008, ApJ, in press (arXiv:0808.0716)
Epchtein, N., et al. 1999, A&A, 349, 236
Famaey, B., Pont, F., Luri, X., Udry, S., Mayor, M., & Jorissen, A. 2007, A&A, 461, 957
Famaey, B., Siebert, A., & Jorissen, A. 2008, A&A, 483, 453
Feltzing, S. & Holmberg, J. 2000, A&A, 357, 153
Freeman, K. & Bland-Hawthorn, J. 2002, ARA&A, 40, 487
Fulbright, J. P. 2002, AJ, 123, 404
Gilmore, G., et al. 2007, ApJ, 663, 948
Geisler, D., Smith, V. V., Wallerstein, G., Gonzalez, G., & Charbonnel, C. 2005, AJ, 129, 1428
Glatt, K., et al. 2008a, AJ, 136, 1703
Glatt, K., et al. 2008b, AJ, 135, 1106
Gouda, N., et al. 2006, SPIE, 6265, 122
Grebel, E. K. 1997, Reviews in Modern Astronomy, 10, 29
Grebel, E. K. 2001a, Reviews in Modern Astronomy, 14, 223
Grebel, E. K. 2001b, Ap&SSS, 277, 231
Grebel, E. K. & Brandner, W. 1998, The Magellanic Clouds and Other Dwarf Galaxies, eds. T. Richtler & J.M. Braun (Aachen: Shaker Verlag), 151
Grebel, E. K. & Stetson, P.B. 1999, The Stellar Content of the Local Group, IAU Symp. 192, eds. P. Whitelock & R. Cannon (San Francisco: ASP), 17

Grebel, E. K. & Gallagher, J. S., III 2004, *ApJ*, 610, L89
Grebel, E. K., Gallagher, J. S., III, & Harbeck, D. 2003, *AJ*, 125, 1926
Grillmair, C. J. 2006, *ApJ*, 651, L29
Grillmair, C. J. & Dionatos, O. 2006, *ApJ*, 643, L17
Grillmair, C. J. & Johnson, R. 2006, *ApJ*, 639, L17
Harbeck, D., et al. 2001, *AJ*, 122, 3092
Helmi, A., Navarro, J. F., Nordström, B., Holmberg, J., Abadi, M. G., & Steinmetz, M. 2006a, *MNRAS*, 365, 1309
Helmi, A., et al. 2006b, *ApJ*, 651, L121
Høg, E. 2000, *A&A*, 355, L27
Holmberg, J., Nordström, B., & Andersen, J. 2007, *A&A*, 475, 519
Hurley-Keller, D., Mateo, M., & Grebel, E. K. 1999, *ApJ*, 523, L25
Ivezić, Ž., et al. 2008, *ApJ*, 684, 287
Jurić, M., et al. 2008, *ApJ*, 673, 864
Kaiser, N. 2006, in: Large Surveys for Galactic Astronomy, *IAU JD 13*, 7
Keller, S. C., et al. 2007, *PASA*, 24, 1
Keller, S. C., Murphy, S., Prior, S., DaCosta, G., & Schmidt, B. 2008, *ApJ*, 678, 851
Kirby, E. N., Simon, J. D., Geha, M., Guhathakurta, P., & Frebel, A. 2008, *ApJ*, 685, L43
Klement, R., Fuchs, B., & Rix, H.-W. 2008, *ApJ*, 685, 261
Kniazev, A. Y., et al. 2005, *AJ*, 130, 1558
Koch, A. & Grebel, E. K. 2006, *AJ*, 131, 1405
Koch, A., Grebel, E. K., Wyse, R. F. G., Kleyna, J. T., Wilkinson, M. I., Harbeck, D. R., Gilmore, G. F., & Evans, N. W. 2006a, *AJ*, 131, 895
Koch, A., Grebel, E. K., Kleyna, J. T., Wilkinson, M. I., Harbeck, D. R., Gilmore, G. F., Wyse, R. F. G., & Evans, N. W. 2007a, *AJ*, 133, 270
Koch, A., Wilkinson, M. I., Kleyna, J. T., Gilmore, G. F., Grebel, E. K., Mackey, A. D., Evans, N. W., & Wyse, R. F. G. 2007b, *ApJ*, 657, 241
Koch, A., Grebel, E. K., Gilmore, G. F., Wyse, R. F. G., Kleyna, J. T., Harbeck, D. R., Wilkinson, M. I., & Evans, N. W. 2008a, *AJ*, 135, 1580
Koch, A., McWilliam, A., Grebel, E. K., Zucker, D. B., & Belokurov, V. 2008b, *ApJ*, in press (arXiv:0810.0710)
Kroupa, P. 2002, *MNRAS*, 330, 707
Lanfranchi, G. A. & Matteucci, F. 2007, *A&A*, 468, 927
López-Santiago, J., Montes, D., Crespo-Chacón, I., & Fernández-Figueroa, M. J. 2006, *ApJ*, 643, 1160
Lucas, P. W., et al. 2007, *MNRAS*, in press (arXiv:0712.0100)
Luck, R. E., Kovtyukh, V. V., & Andrievsky, S. M. 2006, *AJ*, 132, 902
Lynden-Bell, D. 1982, *Observatory*, 102, 202
Maciel, W. J. & Costa, R. D. D. 2008, arXiv:0806.3443 (these proceedings)
Marcolini, A., D'Ercole, A., Battaglia, G., & Gibson, B. K. 2008, *MNRAS*, 386, 2173
Metz, M., Kroupa, P., & Libeskind, N. I. 2008, *ApJ*, 680, 287
Meléndez, J., et al. 2008, *A&A*, 484, L21
Monaco, L., Bellazzini, M., Bonifacio, P., Ferraro, F. R., Marconi, G., Pancino, E., Sbordone, L., & Zaggia, S. 2005, *A&A*, 441, 141
Navarro, J. F., Helmi, A., & Freeman, K. C. 2004, *ApJ*, 601, L43
Newberg, H. J., et al. 2002, *ApJ*, 569, 245
Newberg, H. J., et al. 2003, *ApJ*, 596, L191
Newberg, H. J., et al. 2003, *BAAS*, 35, 1385
Nordström, B., et al. 2004, *A&A*, 418, 989
Majewski, S. R., Skrutskie, M. F., Weinberg, M. D., & Ostheimer, J. C. 2003, *ApJ*, 599, 1082
Odenkirchen, M., et al. 2001, *ApJ*, 548, L165
Odenkirchen, M., et al. 2003, *AJ*, 126, 2385
Pasetto, S. & Chiosi, C. 2007, *A&A*, 463, 427
Peñarrubia, J., et al. 2005, *ApJ*, 626, 128
Perryman, M. A. C., et al. 1997, *A&A*, 323, L49

Perryman, M. A. C., et al. 2001, A&A, 369, 339
Reddy, B. E., Lambert, D. L., & Allende Prieto, C. 2006, MNRAS, 367, 1329
Refregier, A. et al. 2008, Experimental Astronomy, 26
Rocha-Pinto, H. J., Majewski, S. R., Skrutskie, M. F., & Crane, J. D. 2003, ApJ, 594, L115
Rocha-Pinto, H. J., Majewski, S. R., Skrutskie, M. F., Crane, J. D., & Patterson, R. J. 2004, ApJ, 615, 732
Rocha-Pinto, H. J., Majewski, S. R., Skrutskie, M. F., Patterson, R. J., Nakanishi, H., Muñoz, R. R., & Sofue, Y. 2006, ApJ, 640, L147
Rockosi, C. M., et al. 2002, AJ, 124, 349
Rockosi, C. M. 2005, BAAS, 37, 1404
Sadakane, K., et al. 2004, PASJ, 56, 1041
Sakamoto, T., Chiba, M., & Beers, T. C. 2003, A&A, 397, 899
Sbordone, L., Bonifacio, P., Buonanno, R., Marconi, G., Monaco, L., & Zaggia, S. 2007, A&A, 465, 815
Seabroke, G. M., et al. 2008, MNRAS, 384, 11
Shetrone, M. D., Côté, P., & Sargent, W. L. W. 2001, ApJ, 548, 592
Siebert, A., et al. 2008, MNRAS, 390, 1194
Skrutskie, M. F., et al. 2006, AJ, 131, 1163
Smith, M. C., et al. 2007, MNRAS, 379, 755
Steinmetz, M., et al. 2006, AJ, 132, 1645
Stetson, P. B., Hesser, J. E., & Smecker-Hane, T. A. 1998, PASP, 110, 533
Stoughton, C., et al. 2002, AJ, 123, 485
Strigari, L. E., Bullock, J. S., Kaplinghat, M., Simon, J. D., Geha, M., Willman, B., & Walker, M. G. 2008, Nature, 454, 1096
Szymanski, M., Udalski, A., Kubiak, M., Kaluzny, J., Mateo, M., & Krzeminski, W. 1996, AcA, 46, 1
Tolstoy, E., Venn, K. A., Shetrone, M., Primas, F., Hill, V., Kaufer, A., & Szeifert, T. 2003, AJ, 125, 707
Unavane, M., Gilmore, G., Epchtein, N., Simon, G., Tiphene, D., & de Batz, B. 1998, MNRAS, 295, 119
Unwin, S. C., et al. 2008, PASP, 120, 38
van den Bergh, S. 1994, ApJ, 428, 617
van den Bergh, S. 1999, A&ARv, 9, 273
Veltz, L., et al. 2008, A&A, 480, 753
Vivas, A. K. & Zinn, R. 2006, AJ, 132, 714
Vivas, A. K., et al. 2001, ApJ, 554, L33
Walker, M. G., Mateo, M., Olszewski, E. W., Gnedin, O. Y., Wang, X., Sen, B., & Woodroofe, M. 2007, ApJ, 667, L53
Weinberg, M. D. & Blitz, L. 2006, ApJ, 641, L33
Wilkinson, M. I., Kleyna, J. T., Evans, N. W., Gilmore, G. F., Irwin, M. J., & Grebel, E. K. 2004, ApJ, 611, L21
Wyse, R. F. G., Gilmore, G., Norris, J. E., Wilkinson, M. I., Kleyna, J. T., Koch, A., Evans, N. W., & Grebel, E. K. 2006, ApJ, 639, L13
Xue, X. X., et al. 2008, ApJ, 684, 1143
Yanny, B., et al. 2003, ApJ, 588, 824
Yong, D., Carney, B. W., Teixera de Almeida, M. L., & Pohl, B. L. 2006, AJ, 131, 2256
York, D. G., et al. 2000, AJ, 120, 1579
Zhao, G., Chen, Y.-Q., Shi, J.-R., Liang, Y.-C., Hou, J.-L., Chen, L., Zhang, H.-W., & Li, A.-G. 2006, ChJAA, 6, 265
Zucker, D. B., et al. 2004, ApJ, 612, L121
Zucker, D. B., et al. 2006a, ApJ, 643, L103
Zucker, D. B., et al. 2006b, ApJ, 650, L41
Zucker, D. B., et al. 2007, ApJ, 659, L21
Zwitter, T., et al. 2008, AJ, 136, 421

The mass content of the Sculptor dwarf spheroidal galaxy

G. Battaglia[1,2], A. Helmi[1], E. Tolstoy[1], and M. Irwin[3]

[1] Kapteyn Astronomical Institute, University of Groningen, P.O. Box 800, the Netherlands
[2] European Southern Observatory, K. Schwarzschild-Str. 2, 85748 Garching, Germany
email: gbattagl@eso.org
[3] Institute of Astronomy, Madingley Road, Cambridge CB03 0HA, UK

Abstract. We present a new determination of the mass content of the Sculptor dwarf spheroidal galaxy, based on a novel approach which takes into account the two distinct stellar populations present in this galaxy. This method helps to partially break the well-known mass-anisotropy degeneracy present in the modelling of pressure-supported stellar systems.

Keywords. techniques: spectroscopic, stars: kinematics, galaxies: dwarf, galaxies: individual (Sculptor), dark matter

1. Introduction

The determination of the mass content of galaxies is a fundamental step for our understanding of their formation and evolution. This is particularly true for small systems such as dwarf spheroidal galaxies (dSphs), whose total mass can determine their destiny as being fragile objects, strongly perturbed by supernovae explosions and/or interactions with the environment, or mostly undisturbed objects.

Part of the interest in determining the masses of dSphs has surely been driven by the fact that these galaxies, with mass-to-light (M/L) ratios up to 100s, appear to be the most dark matter (DM) dominated objects known to-date. This makes them potentially good testing grounds for different DM theories of galaxy formation; however at the moment it is still unclear if dSphs inhabit cusped DM profiles, as predicted by cold DM theories of galaxy formation, or cored profiles, typical of warmer kinds of DM. This distinction is often hampered by the degeneracies which are intrinsic to the method generally used to determine the mass of pressure-supported stellar systems, i.e. a Jeans analysis of the line-of-sight (l.o.s.) velocity dispersion obtained considering all stars as a single component embedded in an extended DM halo. This analysis is subject to the well known degeneracy between the mass distribution and the orbital motions of the individual stars in the system (mass-anisotropy degeneracy). Recent observations have shown that some dSphs host multiple stellar populations which have distinct spatial distribution, metallicity and kinematics. The presence of multiple stellar populations requires a modification in the way these systems are dynamically modelled to derive their mass content.

Here we describe the results of our determination of the mass content of the Sculptor (Scl) dSph by adopting a more detailed kinematic modelling than the traditional one, i.e. by considering Scl as two-(stellar) components embedded in an extended DM halo. We first briefly describe and discuss the observational properties of Scl derived from our work (Tolstoy *et al.* 2004; Battaglia 2007; Battaglia *et al.* 2008). We then move to the results of the two-component modelling; we show a clear example of the mass-anisotropy present when modelling Scl as a one-stellar component system, and finally discuss our results. For a more detailed description we refer to Battaglia (2007) and Battaglia et al. (2008).

2. Observed properties of the Sculptor dSph

We acquired extended photometry from the ESO/2.2m WFI and VLT/FLAMES spectra (R $\sim$ 6500) in the Ca II triplet region for hundreds of individual red giant branch (RGB) stars in Scl, out to its nominal tidal radius (Tolstoy et al. 2004: 308 probable members; Battaglia et al. 2008: 470 probable members). We used these data to study the large scale kinematics and metallicity properties of Scl, and the spatial distribution of its stellar populations.

The combined information from the accurate line-of-sight velocities (± 2 km s^{-1}) and metallicities (± 0.15 dex) of RGB stars allowed us to unveil the presence of two distinct stellar populations with different spatial distribution, metallicity and kinematics: the metal rich (MR) stars ([Fe/H]> -1.5) are more centrally concentrated and show colder kinematics than the metal poor (MP) stars ([Fe/H]< -1.7), as can be seen from the line-of-sight (l.o.s.) velocity dispersion profiles in Fig. 1. The MR and MP RGB stars appear to trace, respectively, the red and blue horizontal branch (RHB, BHB) stars. The surface number density profile of RGB stars, representative of the overall stellar population of Scl, is well approximated by a two-component fit, where each component is given by the rescaled best-fitting profile derived from our photometry for the RHB and BHB stars. In contrast, a single profile (be it a Plummer, Sersic or King) does not appear to be a good approximation for the surface number density of RGB stars. Multiple stellar populations are known to exist also in other dSphs, such as Fornax (Battaglia et al. 2006) and Canes Venatici (Ibata et al. 2006), and might be a common feature to this class of objects.

The large spatial coverage and statistics of this spectroscopic data-set has allowed the discovery of another interesting feature, i.e. a velocity gradient of 7.6 km s^{-1} deg^{-1} in the Galactic standard of rest (GSR) † along the projected major axis of Scl. This gradient is likely to due intrinsic rotation, and not to tidal disruption (see Battaglia 2007 and Battaglia et al. 2008 for discussion on this point), and this is the first time that statistically significant rotation is found in a dSph.

Among the other Milky Way (MW) dSphs, a significant velocity gradient is detected in Carina (Muñoz et al. 2006) and a mild one in Leo I (Mateo et al. 2008), but these are interpreted as the result of tidal interaction with the MW. However, velocity gradients are not exclusive of the dSphs satellites of larger galaxies, but are present also in some of the isolated Local Group dSphs, such as Cetus (Lewis et al. 2007) and Tucana (Fraternali et al., in preparation), which are unlikely to have ever interacted with either the MW or M31 and therefore suffered tidal disturbance. This might point to rotation as an intrinsic property of this class of objects and provide insights into the mechanisms that shaped their evolution. This might for instance favour scenarios in which the progenitors of dSphs were rotating systems which were transformed into prevalently pressure supported object by tidal stirring from their host galaxy (e.g., Mayer et al. 2001). It would therefore be interesting to compare the frequency of occurrence of rotation among dSphs and other generally more isolated systems such as dwarf irregular galaxies and transition types. It is possible that more velocity gradients will be discovered in MW dSphs now that larger and more spatially extended velocity surveys are becoming available.

In the analysis presented here we always use rotation-subtracted GSR velocities.

† We use the velocities in the GSR frame to avoid spurious gradients introduced by the component of the Sun and Local Standard of Rest motion's along the l.o.s. to Scl.

3. The Jeans analysis

DSphs generally have low ellipticities and are presumably prevalently pressure supported, and therefore are commonly considered as spherical and with no streaming motions‡. Then, if the system is in dynamical equilibrium, the l.o.s. velocity dispersion predicted by the Jeans equation is (from Binney & Mamon 1982)

$$\sigma_{\mathrm{los}}^2(R) = \frac{2}{\Sigma_*(R)} \int_R^\infty \frac{\rho_*(r)\sigma_{r,*}^2 \, r}{\sqrt{r^2 - R^2}} \left(1 - \beta \frac{R^2}{r^2}\right) dr \qquad (1)$$

where R is the projected radius (on the sky) and r is the 3D radius. The l.o.s. velocity dispersion depends on: the mass surface density $\Sigma_*(R)$ and mass density $\rho_*(r)$ of the tracer, which in our case are the MR and the MP RGB stars; the tracer velocity anisotropy β, defined as $\beta = 1 - \sigma_\theta^2/\sigma_r^2$, and its radial velocity dispersion $\sigma_{r,*}$, which depends on the total mass distribution (for dSphs the contribution due to stellar mass is negligible). A unique solution for the mass profile can be determined from this equation if both the mass distribution of the tracer and $\beta(r)$ are known, although this solution is not guaranteed to produce a phase-space distribution function that is positive everywhere. However, proper motions for individual stars in the system would be necessary to derive $\beta(r)$ and this is not feasible yet. This causes the well-known mass-anisotropy degeneracy, which prevents a distinction amongst different DM profiles. Examples of this are present in the literature: the observed l.o.s. dispersion profiles of dSphs are both compatible with cored DM models, assuming that $\beta = 0$ (e.g., Gilmore et al. 2007), and with cusped profiles, allowing for mildly tangential, constant with radius, β (e.g. Walker et al. 2007). It is therefore advisable to explore several hypotheses for the behaviour of $\beta(r)$ and for the mass distribution when carrying out a Jeans analysis (see below).

In the following we explore the possibility that DM follows a cored (pseudo-isothermal sphere) or a cuspy distribution (NFW profile); we also explore two hypotheses for the behaviour of $\beta(r)$: constant with radius, and following an Osipkov-Merritt (OM) model (isotropic in the central parts and radial in the outskirts). The spatial distributions of the tracers are derived from our photometric data.

Two-components modelling Here we show the results of modelling Scl as a two-(stellar) component system embedded in a DM halo. Eq. 1 is applied separately to the MR and MP stars, assuming that they move in the same DM potential. We explore a range of core radii r_c for the pseudo-isothermal sphere ($r_c = 0.001, 0.05, 0.1, 0.5, 1$ kpc) and of concentrations c for the NFW profile ($c = 20, 25, 30, 35$). By fixing these, each mass model has two free parameters left: the anisotropy and the DM halo mass (enclosed within the last measured point for the isothermal sphere, at 1.3 deg = 1.8 kpc assuming a distance to Scl of 79 kpc, see Mateo 1998; and the virial mass in the case of the NFW model). We compute a χ^2 for the MR and MP components separately (χ_{MR}^2 and χ_{MP}^2, respectively) by comparing the various models to the data. The best-fit is obtained by minimising the sum $\chi^2 = \chi_{\mathrm{MR}}^2 + \chi_{\mathrm{MP}}^2$.

Figure 1a,b shows the models with constant anisotropy, which can be excluded as in this case neither a cored or a cusped model is a good representation of the data ($\chi_{\mathrm{min}}^2 \sim 17$). The situation is different in the hypothesis of an OM velocity anisotropy as shown in Fig. 1c, d). We find that a pseudo-isothermal sphere with a relatively large core ($r_c = 0.5$ kpc, $M(< 1.8\,\mathrm{kpc}) = 3.4 \pm 0.7 \times 10^8$ M$_\odot$) gives an excellent description of the data ($\chi_{\mathrm{min}}^2 = 6.9$). Also an NFW model is statistically consistent with the data

---

‡ We checked that the assumptions of sphericity and absence of streaming motions have a negligible effect on the results.

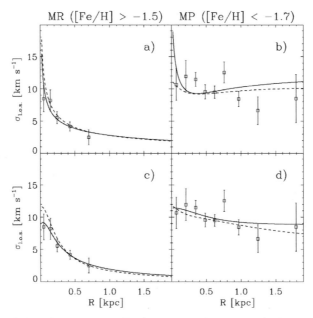

Figure 1. L.o.s. velocity dispersion profile (squares with errorbars), from rotation-subtracted GSR velocities, for the MR (left) and MP (right) RGB stars in Scl. The lines show the best-fitting pseudo-isothermal sphere (solid) and NFW model (dashed) in the hypothesis of $\beta = const$ (a,b) and $\beta = \beta_{\rm OM}$ (c,d). The models have reduced $\chi^2 \sim 1.6$ (a,b), 0.6 (c) and 1 (d).

($c = 20$, virial mass $M_v = 2.2^{+1.0}_{-0.7} \times 10^9$ M$_\odot$, $\chi^2_{\rm min} = 10.8$), but tends to over-predict the central values of the MR velocity dispersion. The mass within the last measured point ($\sim 2.4^{+1.1}_{-0.7} \times 10^8$ M$_\odot$) is consistent with the mass predicted by the best-fitting isothermal sphere model.

One-component modelling Figure 2 shows a clear example of the mass-anisotropy degeneracy when modelling Scl with the traditional one-component analysis. Just by changing the assumption on β (=const. or $= \beta_{\rm OM}$), cored profiles with very different core radii and $M(< 1.8\,{\rm kpc})$ give excellent fit to the data. An excellent fit is also an NFW profile, in the hypothesis of $\beta = const$. These models are practically indistinguishable and all good ($\chi^2 \sim 8$, reduced $\chi^2 \sim 1.1$). No DM model can be favoured nor hypotheses on β.

4. Discussion and conclusions

The combined information of velocity and metallicity from a large sample of FLAMES spectra of individual stars in Scl has allowed us to separate the different kinematics of the two stellar populations in Scl, and to carry out a more detailed kinematic modelling which allows us to partially break the mass-anisotropy degeneracy.

Under the hypotheses explored for the behaviour of the velocity anisotropy $\beta(r)$, the two-component modelling indicates that a cored profile is slightly favoured over a cusped profile, as the latter tends to over-predict the central values (at $R \lesssim 0.1$ kpc) of the MR velocity dispersion. It could be argued that the relatively large central dispersion predicted for the MR stars from the cusped profile could be lowered by allowing the velocity anisotropy of the MR population to be tangential at $R \lesssim 0.1$ kpc. However it should be noted that in order to reproduce the rapid decline of the MR velocity dispersion profile, the MR $\beta(r)$ needs to become very radial already on the scale of 0.3-0.4 kpc, given the observed spatial distribution of MR stars. Observational determinations of β, either

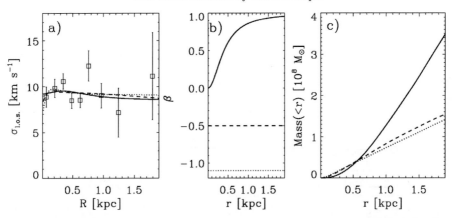

Figure 2. a) Observed l.o.s. velocity dispersion profile, from rotation-subtracted GSR velocities, for all RGB stars in Scl (squares with error-bars), with overlaid the best-fits for a cored DM distribution (solid line: $\beta = \beta_{OM}$ and r_c =0.5 kpc; dotted line: β = const.and r_c =0.05 kpc), and an NFW DM distribution (dashed line: β = const. and c = 35). Panel b) and c) show the velocity anisotropy and mass distribution corresponding to the models in panel a). Note that the model with r_c =0.05 kpc is almost equivalent to a cusp.

from direct measurement or from distribution functions, would be the most desirable (and definitive) solution to this issue.

Strigari et al. (2007) showed that, independently of $\beta(r)$ and the total mass distribution of dSphs, the mass enclosed within 0.6 kpc ($M_{0.6}$) can be derived more accurately than at any other distance from the centre. However, given that the evolution of dSphs is most likely to be governed by their *total* mass and that dynamical analyses provide only the *mass enclosed within the last measured point*, it is important to have reliable determinations of the mass enclosed as farther out as possible. Our analysis shows that, independently of the adopted mass profile, the best fits from our two-component modelling yield consistent values for the mass enclosed within the last measured point, at 1.8kpc.

Finally we would like to comment on the issue that dwarf galaxies might be embedded in haloes of similar mass ($\sim 4 \times 10^7$ $M_\odot$) as has been claimed in the literature. Figure 3(left) shows a revisited version of the mass enclosed within the last measured point versus absolute magnitude plot. The masses plotted are the ones derived in Walker et al. (2007) for a sample of dSphs, because they are derived from the most extended velocity samples and from a homogeneous analysis. For Scl we use our own determination, which is the most accurate so far. The figure shows that the DM mass of Scl is now much larger than the typical DM mass derived so far for the other dSphs. Also Fornax deviates considerably from the relation. In the plot shown here, the masses of Fornax and Sextans were derived from data extending out to $\sim$0.6 and 0.3 tidal radii, respectively. Given that Fornax and Sextans have been recently mapped much farther out (Fornax out to $\sim$ 1 tidal radius: Battaglia et al. 2006; Sextans out to $\sim$ 0.8 tidal radius: Battaglia et al. in preparation), their revised masses are likely to increase quite significantly, and introduce even larger dispersion in this plot. In some cases, $M_{0.6}$ has been used in this kind of analysis, instead than the mass enclosed within the last measured point. However, Fig. 3(right) shows that $M_{0.6}$ is very insensitive to the total mass of the dSph. In fact, NFW haloes of virial masses between 2-40 $\times 10^8$ $M_\odot$(i.e. in the range found by Walker et al. 2007 to fit their samples of dSphs) give remarkably similar values of $M_{0.6}$, probably because at such a small distance there is not much sensitivity to the total mass. Therefore

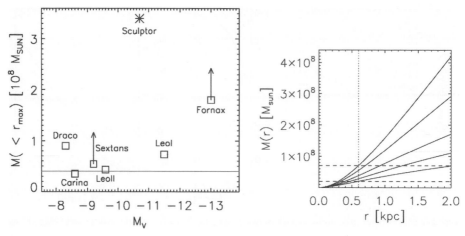

Figure 3. Left: Mass within the last measured point ($r_{\max}$) versus absolute V magnitude, M_V, for MW dSphs. The masses are from Walker et al. 2007 (squares), except for Scl which is from our two-components modelling (asterisk); M_V is from Irwin & Hatzidimitriou (1995). The arrows indicate that the masses measured for Sextans and Fornax are likely to increase given that these objects have been surveyed just to 0.3 and 0.6 tidal radii. The solid line indicates a mass of 4×10^7 $M_\odot$. Right: Mass enclosed within r for NFW halos of virial masses $2,4,8,20,40\times10^8$ $M_\odot$ (solid lines), i.e. the range found by Walker et al. (2007) to fit the observed l.o.s. velocity dispersion profile of their sample of 7 dSphs. Concentrations are derived using the formula in Koch et al. (2007), extrapolated from Jing & Suto (2000). The dashed lines enclose masses between $2\text{-}7\times10^7$ $M_\odot$; the dotted line is placed at a radius of 0.6 kpc.

there is no indication of a common mass scale amongst the dSphs in the Local Group, while perhaps one may draw the conclusion of the existence of a minimum mass for the formation of dSphs.

References

Battaglia, G., et al. 2006, A&A, 459, 423
Battaglia, G. 2007, PhD thesis, Univ. Groningen, The Netherlands, http://irs.ub.rug.nl/ppn/304002712
Battaglia, G., et al. 2008, ApJ (Letters), 681, L13
Binney, J. & Mamon, G. A. 1982, MNRAS, 200, 361
Gilmore, G., et al. 2007, ApJ, 663, 948
Ibata, R., Chapman, S., Irwin, M., Lewis, G., & Martin, N. 2006, MNRAS (Letters), 373, L70
Irwin, M. J. & Hatzidimitriou, D. 1995, MNRAS, 277, 1354
Jing, Y. P. & Suto, Y. 2000, ApJ (Letters), 529, L69
Koch, A. et al. 2007, ApJ, 657, 241
Lewis et al. 2007, MNRAS, 1364, 1370
Mateo, M. L. 1998, ARA&A, 36, 435
Mateo, M., Olszewski, E. W., Walker, M. G. 2008, ApJ, 675, 201
Mayer, L., et al. 2001, ApJ, 559, 754
Muñoz, R. R. et al. 2006, ApJ, 649, 201
Strigari, L. E., et al. 2007, ApJ, 669, 676
Tolstoy, E., et al. 2004, ApJ (Letters), 617, L119
Walker, M. G., et al. 2007, ApJ (Letters), 667, L53

Galaxy Interactions, Star Formation History, and Bulgeless Galaxies

Shardha Jogee[1]

[1]Department of Astronomy, University of Texas at Austin,
1 University Station C1400, Austin, TX 78712-0259
email: sj@astro.as.utexas.edu

Abstract. Hierarchical ΛCDM models provide a successful paradigm for the growth of dark matter on large scales, but they face important challenges in predicting how the baryonic components of galaxies evolve. I present constraints on two aspects of this evolution: (1) The interaction history of galaxies over the last 7 Gyr and the impact of interactions on their star formation properties, based on Jogee et al. (2008a,b); (2) Constraints on the origin of bulges in hierarchical models and the challenge posed in accounting for galaxies with low bulge-to-total ratios, based on Weinzirl, Jogee, Khochar, Burkert, & Kormendy (2008, hereafter WJKBK08).

Keywords. galaxies: evolution, galaxies: interactions, galaxies: structure, galaxies: bulge

1. Galaxy Interactions and their Star Formation over the Last 7 Gyr

The merger history of galaxies impacts the mass assembly (e.g., Dickinson et al. 2003), star formation history, AGN activity (e.g., Springel. et al. 2005b) and structural evolution of galaxies. The merger rate/fraction at $z > 1$ remains highly uncertain, owing to relatively modest volumes and bandpass shifting effects, but with a general trend towards higher merger fractions at higher redshifts Even the merger rate at $z < 1$ has proved hard to robustly measure for a variety of reasons, ranging from small samples in early studies, to different methods on large samples in later studies.

In Jogee et al. (2008a,b), we have performed a complementary and comprehensive observational estimate of the frequency of interacting galaxies over $z \sim 0.24$–0.80 (lookback times of 3–7 Gyr), and the impact of interactions on the star formation (SF) of galaxies over this interval. Our study is based on *HST* ACS, COMBO-17, and Spitzer 24 μm data from the GEMS survey. We use a large sample of ~ 3600 ($M \geqslant 1 \times 10^9\ M_\odot$) galaxies and ~ 790 high mass ($M \geqslant 2.5 \times 10^{10}\ M_\odot$) galaxies for robust number statistics. Two independent methods are used to identify strongly interacting galaxies: a tailored visual classification system complemented with spectrophotometric redshifts and stellar masses, as well as the CAS merger criterion ($A > 0.35$ and $A > S$; Conselice 2003), based on CAS asymmetry A and clumpiness S parameters. This allows one of the most extensive comparisons to date between CAS-based and visual classification results. We set up this visual classification system so as to target interacting systems whose morphology and other properties suggest they are a recent merger of mass ratio $M1/M2 > 1/10$. While many earlier studies focused only on major mergers, we try to constrain the frequency of minor mergers as well, since they dominate the merger rates in ΛCDM models. Some of our results are outlined below.

(1) Among ~ 790 high mass galaxies, *the fraction of visually-classified interacting systems over lookback times of 3–7 Gyr ranges from 9% $\pm$ 5% at $z \sim 0.24$–0.34, to 8% $\pm$ 2% at $z \sim 0.60$–0.80, as averaged over every Gyr bin.(Fig. 1a).* These systems appear to be in merging or post-merger phases, and are candidates for a recent merger of mass

ratio $M1/M2 > 1/10$. Similar results on the interaction fraction are reported by Lotz et al. (2008). The lower limit on the major ($M1/M2 > 1/4$) merger fraction ranges from 1.1% to 3.5% over $z \sim 0.24$–0.80. The corresponding lower limit on the minor ($1/10 \leqslant M1/M2 < 1/4$) merger fraction ranges from 3.6% to 7.5%. This is the first, albeit approximate, empirical estimate of the frequency of minor mergers over the last 7 Gyr.

(2) For an assumed value of ~ 0.5 Gyr for the visibility timescale, it follows that *each massive ($M \geqslant 2.5 \times 10^{10}$ $M_\odot$) galaxy has undergone ~ 0.7 mergers of mass ratio $> 1/10$ over the redshift interval $z \sim 0.24$–0.80.* Of these, we estimate that 1/4 are major mergers, 2/3 are minor mergers, and the rest are ambiguous cases of major or minor mergers. The corresponding merger rate R is a few $\times 10^{-4}$ galaxies Gyr^{-1} Mpc^{-3}. Among ~ 2840 blue cloud galaxies of mass $M \geqslant 1.0 \times 10^9$ $M_\odot$, similar results hold.

(3) We compare our empirical merger rate R for high mass ($M \geqslant 2.5 \times 10^{10}$ $M_\odot$) galaxies to predictions from different ΛCDM-based simulations of galaxy evolution, including the halo occupation distribution (HOD) models of Hopkins et al. (2007); semi-analytic models (SAMs) of Somerville et al. (2008), Bower et al. (2006), and Khochfar & Silk (2006); and smoothed particle hydrodynamics (SPH) cosmological simulations from Maller et al. (2006). To our knowledge, such extensive comparisons have not been attempted to date, and are long overdue. *We find qualitative agreement between the observations and models, with the (major+minor) merger rate from different models bracketing the observed rate, and showing a factor of five dispersion* (Fig. 1b). One can now anticipate that in the near future, improvements in both the observational estimates and model predictions will start to rule out certain merger scenarios and refine our understanding of the merger history of galaxies.

(4) The idea that galaxy interactions generally enhance the star formation rate (SFR) of galaxies is well established from observations (e.g., Joseph & Wright 1985; Kennicutt et al. 1987) and simulations (e.g., Hernquist 1989; Mihos & Hernquist 1994, 1996; Springel, Di Matteo & Hernquist 2005b). However, simulations cannot uniquely predict the factor by which interaction enhance the SF activity of galaxies over the last 7 Gyr, since both the SFR and properties of the remnants in simulations are highly sensitive to the stellar feedback model, the bulge-to-disk (B/D) ratio, the gas mass fractions, and orbital geometry (e.g., Cox et al. 2006; di Matteo et al. 2007). Thus, empirical constraints are needed. Among ~ 3600 intermediate mass ($M \geqslant 1.0 \times 10^9$ $M_\odot$) galaxies, we find that *the average SFR of visibly interacting galaxies is only modestly enhanced compared to non-interacting galaxies over $z \sim 0.24$–0.80* (Fig. 1c). This result is found for SFRs based on UV, UV+IR, and UV+stacked-IR data. This modest enhancement is consistent with the results of di Matteo et al. (2007) based on numerical simulations of several hundred galaxy collisions.

(5) The SF properties of interacting and non-interacting galaxies since $z < 1$ are of great astrophysical interest, given that the cosmic SFR density is claimed to decline by a factor of 4 to 10 since $z \sim 1$ (e.g., Lilly et al. 1996; Ellis et al. 1996; Hopkins 2004; Pérez-González et al. 2005; Le Floc'h et al. 2005). We therefore set quantitative limits on the contribution of obviously interacting systems to the UV-based and UV+IR-based SFR density over $z \sim 0.24$–0.80. Among ~ 3600 intermediate mass ($M \geqslant 1.0 \times 10^9$ $M_\odot$) galaxies, we find that *visibly interacting systems only account for a small fraction (< 30%) of the cosmic SFR density over lookback times of ~ 3–7 Gyr ($z \sim 0.24$–0.80;* Fig. (1d)). Our result is consistent with that of Wolf et al. (2005) over a smaller lookback time interval of ~ 6.2–6.8 Gyr. In effect, our result suggests that *the behavior of the cosmic SFR density over the last 7 Gyr is predominantly shaped by non-interacting galaxies, rather than strongly interacting galaxies.* This suggests that the observed decline in the

cosmic SFR density since $z \sim 0.80$ is largely the result of a shutdown in the SF of non-interacting galaxies.

2. The origin of bulges and the problem of bulgeless galaxies

In ΛCDM models of galaxy evolution, there are in principle three main mechanisms to build bulges of spiral galaxies: major mergers, minor mergers, and secular processes (see WJKBK08 for details). The major merger of two spiral galaxies destroys the disk component and leaves behind a classical bulge, around which a stellar disk forms when hot gas in the halo subsequently cools, settles into a disk, and forms stars. Minor mergers can also grow bulges in several ways. A tidally induced bar and/or direct tidal torques from the companion can drive gas into the inner kpc (e.g., Quinn et al. 1993; Hernquist & Mihos 1995; Jogee 2006 and references therein), where subsequent SF forms a compact high v/σ stellar component, or disky pseudobulge. In addition, the stellar core of the satellite can sink to the central region via dynamical friction. Finally, bulges can also have a secular origin: here, a stellar bar or globally oval structure in a *non-interacting* galaxy drives gas inflow into the inner kpc , where subsequent SF forms a disky pseudobulge (e.g., Kormendy 1993; Jogee 1999; Kormendy & Kennicutt 2004; Jogee, Scoville, & Kenney 2005).

These different mechanisms to form bulges have been postulated for a long time. However, what is still missing is a quantitative assessment of the relative importance of different bulge formation pathways in high and low mass spirals. For instance, although bulges are an integral part of massive present-day spiral galaxies, we still cannot answer the following basic question: do most bulges in massive spirals form via major mergers, minor mergers, or secular processes?

Another thorny issue is the prevalence of bulgeless galaxies. There is rising evidence that bulgeless galaxies are quite common in the local Universe (e.g., Böker et al. 2002; Kautsch et al. 2006; BJM08a; Kormendy & Fisher 2008). Yet, in ΛCDM models of galaxy evolution, most galaxies that had a past major merger at a time when their mass was a fairly large fraction of their present-day mass, are expected to have a significant bulge. So far, no quantitative comparisons have been done between observations and model predictions to assess how serious is the challenge posed by bulgeless galaxies.

In WJKBK08, we attempt one of the first quantitative comparisons of the properties of bulges in a fairly complete sample of high mass ($M_\star \geqslant 1.0 \times 10^{10} M_\odot$) spirals to predictions from ΛCDM-based simulations of galaxy evolution. We derive the bulge-to-total mass ratio (B/T) and bulge Sérsic index n by performing $2D$ bulge-disk-bar decomposition on H-band images of 146 bright, high mass, moderately inclined spirals.

(1) Interestingly, we find that as many as $\sim 56\%$ of high mass spirals have low $n \leqslant 2$ bulges: such bulges exist in barred and unbarred galaxies across all Hubble types (Fig. 2a). Furthermore a striking $\sim 66\%$ of high mass spirals have $B/T \leqslant 0.2$ (Figs. 3a and 3b).

(2) We compare the observed distribution of bulge B/T in high mass spirals to predictions from ΛCDM-based semi-analytical models. In the models, a bulge with $B/T \leqslant 0.2$ can exist in a galaxy with a past major merger, only if the last major merger occurred at $z > 2$ (lookback > 10 Gyr). The predicted fraction of high mass spirals with a past major merger and a bulge with a present-day $B/T \leqslant 0.2$ is *a factor of over fifteen smaller* than the observed fraction ($\sim 66\%$) of high mass spirals with $B/T \leqslant 0.2$ (Fig. 2b). The comparisons *rule out major mergers as the main formation pathway for bulges in high mass spirals*. Contrary to common perception, *bulges built via major mergers seriously fail to account for the bulges present in $\sim 66\%$ of high mass spirals.*

(3) In the models, the majority of low $B/T \leqslant 0.2$ bulges exist in systems that have experienced *only minor mergers, and no major mergers* (Fig. 2b). These bulges can be built via minor mergers and secular processes. So far, we explored one realization of the model focusing on bulges built via satellite stars in minor mergers and find good agreement with the observations. Future models will explore more realistic minor merger scenarios and secular processes.

References

Böker, T. Laine, S., van der Marel, R. P., *et al.* 2002, AJ, 123, 1389
Bower, R. G., Benson, A. J., Malbon, R., *et al.* 2006, MNRAS, 370, 645
Conselice, C. J. 2003, ApJs, 147, 1
Cox, T. J., Jonsson, P., Primack, J. R., & Somerville, R. S. 2006, MNRAS, 373, 1013
Dickinson, M., Papovich, C., Ferguson, H. C., & Budavári, T. 2003, ApJ, 587, 25
Di Matteo, P., Combes, F., Melchior, A.-L., & Semelin, B. 2007, A&A, 468, 6
Hernquist, L. 1989, Nature, 340, 687
Hernquist, L. & Mihos, J. C. 1995, ApJ, 448, 41
Hopkins *et al.* 2007, ApJ, submitted (arXiv:0706.1243)
Jogee, S. 1999, Ph.D. thesis, Yale University
Jogee, S., Shlosman, I., Laine, S., *et al.* 2002, ApJ, 575, 156
Jogee, S., Scoville, N., & Kenney, J. D. P. 2005, ApJ, 630, 837
Jogee, S. 2006, in Physics of Active Galactic Nuclei at all Scales, ed. D. Alloin, R. Johnson, & P. Lira (Berlin: Springer), 143
Jogee, S. *et al.* 2008a, in Formation and Evolution of Galaxy Disks, ed. J. G. Funes, S. J., & E. M. Corsini (San Francisco: ASP), in press (arXiv:0802.3901)
Jogee, S., *et al.* 2008b, ApJ, submitted
Joseph, R. D. & Wright, G. S. 1985, MNRAS, 214, 87
Kautsch, S. J., Grebel, E. K., Barazza, F. D., & Gallagher, J. S., III 2006, A&A, 445, 765
Kennicutt, R. C., Jr., Roettiger, K. A., Keel, W. C., van der Hulst, J. M., & Hummel, E. 1987, AJ, 93, 1011
Khochfar, S. & Burkert, A. 2005, MNRAS, 359, 1379
Khochfar, S., & Silk, J. 2006, MNRAS, 370, 902
Kormendy, J. 1993, in IAU Symposium 153, Galactic Bulges, ed. H. Dejonghe & H. J. Habing (Dordrecht: Kluwer), 209
Kormendy, J. & Kennicutt, R. C. 2004, ARAA, 42, 603
Le Floc'h, E., *et al.* 2005, ApJ, 632, 169
Lilly, S. J., Le Fevre, O., Hammer, F., & Crampton, D. 1996, ApJl, 460, L1
Lotz, J. M., *et al.* 2008, ApJ, 672, 177
Maller, A. H., Katz, N., Kereš, D., Davé, R., & Weinberg, D. H. 2006, ApJ, 647, 763
Mihos, J. C. & Hernquist, L. 1994, ApJ, 437, 611
Mihos, J. C. & Hernquist, L. 1996, ApJ, 464, 641
Navarro, J. F. & Steinmetz, M. 2000, ApJ, 538, 477
Pérez-González, P. G., *et al.* 2005, ApJ, 630, 8
Quinn, P. J., Hernquist, L., & Fullagar, D. P. 1993, ApJ, 403, 74
Rix, H., *et al.* 2004, ApJs, 152, 163
Somerville, R. S., Hopkins, P. F., Cox, T. J., *et al.* 2008, MNRAS, accepted
Somerville, R. S. & Primack, J. R. 1999, MNRAS, 310, 1087
Springel, V. & Hernquist, L. 2005, ApJl, 622, L9
Springel, V., *et al.* 2005a, Nature, 435, 629
Springel, V., Di Matteo, T., & Hernquist, L. 2005b, MNRAS, 361, 776
Steinmetz, M. & Navarro, J. F. 2002, NewA, 7, 155
Weinzirl, T., Jogee, S., Khochfar, S., Burkert, A., & Kormendy, J. 2008, ApJ, submitted (arXiv:0807.0040; WJKBK08)
Wolf, C., *et al.* 2005, ApJ, 630, 771

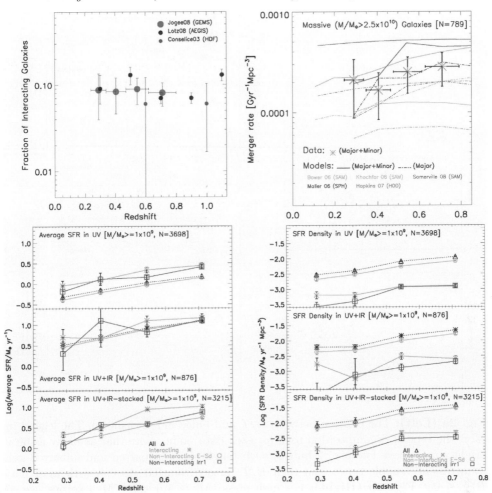

Fig. 1a (Top Left): We show the observed fraction of interacting/merging galaxies from Lotz *et al.* (2008), Jogee *et al.* (2008b), and Conselice (2003). **Fig. 1b (Top Right):** The empirical rate of galaxy mergers with mass ratio $M1/M2 > 1/10$ (orange stars) among high mass galaxies is compared to the rate of (major+minor) mergers (solid lines) predicted by different ΛCDM-based models of galaxy evolution. **Fig. 1c (Lower Left):** The average SFR of interacting and non-interacting galaxies are compared. The average UV-based SFR (top panel; based on 3698 galaxies), average UV+IR-based SFR (middle panel; based on only the 876 galaxies with 24um detections), and average UV+IR-stacked SFR (based on 3215 galaxies with 24um coverage) are shown. In all there cases, the average SFR of interacting galaxies is only modestly enhanced compared to non-interacting E-Sd galaxies over $z \sim 0.24$–0.80 (lookback time ~ 3–7 Gyr). **Fig. 1d (Lower Right):** As in 1c, but now showing the SFR density of galaxies. In all bins, interacting galaxies only contribute a small fraction (typically below 30%) of the total SFR density. [All figures are from Jogee *et al.* (2008b)]

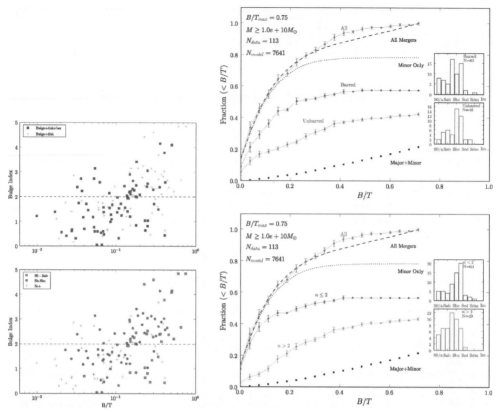

Fig. 2a (Left): The relation between B/T and bulge index is shown. The legend indicates the type of decomposition used for each data point. Note that as many as 60% of bright spirals have low $n \leq 2$ bulges: such bulges exist in barred and unbarred galaxies across all Hubble types, and their B/T ranges from 0.01 to 0.4, with most having $B/T \leq 0.2$. **Fig. 2b (Right):** For high mass ($M_\star \geq 1.0 \times 10^{10} M_\odot$) spirals, we compare the empirical distribution of bulge-to-total mass ratio (B/T) to predictions from ΛCDM-based simulations of galaxy evolution. The y-axis shows the *cumulative* fraction F of galaxies with $B/T \leq$ a given value. The magenta line shows F from the data, while the other two colored lines break this F in terms of bar class (top panel) or bulge n (lower panel). The black dashed line shows F from all model galaxies, while the black dotted line and black dots show the contribution of model galaxies that experienced, respectively, *only past minor mergers* and *both major and minor mergers*. In the models, the fraction ($\sim 3\%$) of high mass spirals, which have undergone a past major merger and host a bulge with $B/T \leq 0.2$ is *a factor of over 15 smaller* than the observed fraction ($\sim 66\%$) of high mass spirals with $B/T \leq 0.2$. Thus, *bulges built via major mergers seriously fail to account for most of the low $B/T \leq 0.2$ bulges present in $\sim 66\%$ high mass spirals.* [All figures are from Weinzirl, Jogee, Khochar, Burkert, & Kormendy (2008)]

Dark Matter Density in Disk Galaxies

J. A. Sellwood

Dept. of Physics & Astronomy, Rutgers University
136 Frelinghuysen Road, Piscataway, NJ 08855
email: sellwood@physics.rutgers.edu

Abstract. I show that the predicted densities of the inner dark matter halos in ΛCDM models of structure formation appear to be higher than estimates from real galaxies and constraints from dynamical friction on bars. This inconsistency would not be a problem for the ΛCDM model if physical processes that are omitted in the collisionless collapse simulations were able to reduce the dark matter density in the inner halos. I review the mechanisms proposed to achieve the needed density reduction.

Keywords. Stellar dynamics, galaxies: halos, dark matter

1. Motivation

I was invited to review secular evolution in disk galaxies. Rather than attempt a very superficial review of this vast topic, I here focus on dynamical friction. Several other possible topics could be included in a review of secular evolution, such as: scattering of disk stars, which I reviewed only recently (Sellwood 2008a); mixing and spreading of disks (e.g. Sellwood & Binney 2002; Roškar *et al.* 2008; Freeman, these proceedings); and the formation of pseudo-bulges (e.g. Kormendy & Kennicutt 2004; Binney, these proceedings).

The current ΛCDM paradigm for galaxy formation (e.g. White, these proceedings) makes specific predictions for the dark matter (DM) densities in halos of galaxies. I first argue that halos of some barred galaxies are inconsistent with this prediction, and then consider whether DM halo densities could be lowered by internal galaxy evolution.

2. Inner Halo Density

Attempts to measure the halo density and its slope in the innermost parts of galaxies are beset by many observational and modeling issues (e.g. Rhee *et al.* 2004; Valenzuela *et al.* 2007), while the predictions from simulations in the same innermost region are still being revised, as shown earlier by White. It therefore makes sense to adopt a more robust measure of central density, such as that proposed by Alam, Bullock & Weinberg (2002). Their parameter, $\Delta_{v/2}$, is a measure of the mean DM density, normalized by the cosmic closure density, interior to the radius at which the circular rotational speed due the DM alone rises to half its maximum value. For those more familiar with halo concentrations, it is useful to note that for the precise NFW (Navarro, Frenk & White 1997) halo form, $\Delta_{v/2} = 672c^3/[\ln(1+c) - c/(1+c)]$, if c is defined where the mean halo density is 200 times the cosmic closure value; thus $\Delta_{v/2} \simeq 10^{5.5}$ for a $c = 9$ halo. However, a further advantage of $\Delta_{v/2}$ is that it is not tied to a specific density profile.

2.1. Prediction

Figure 1, reproduced from Macciò, Dutton & van den Bosch (2008), shows the ΛCDM prediction (shaded) that results when the initial amplitude and spectrum of density

74 J. Sellwood

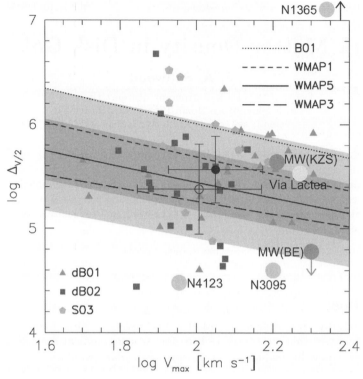

Figure 1. Figure reproduced from Macciò *et al.* (2008, with permission), to which I have added the large labeled points that are described in the text. The shaded regions show the 1− and 2 − σ ranges of the predicted values of $\Delta_{v/2}$, while the lines show the means, as functions of the maximum circular speed from the DM halos. The small colored symbols show various estimates of these parameters for dwarf and LSB galaxies estimated by Macciò *et al.* from data in de Blok, McGaugh & Rubin (2001), de Blok & Bosma (2002), and Swaters *et al.* (2003).

fluctuations match the latest cosmic parameters, as determined by the WMAP team (Komatsu *et al.* 2008). I have added one further predicted point from the *Via Lactea* model (Diemand, Kuhlen & Madau 2007), which is argued to resemble a typical halo that would host a galaxy such as the Milky Way.

2.2. *Data from Galaxies*

Macciò *et al.* plot the small colored triangles, squares and pentagons, which show estimates of $\Delta_{v/2}$ culled from the literature. The data are from dwarf and LSB galaxies, that are believed to be DM dominated and the large black circle indicates their mean in both coordinates, with the error bars indicating the ranges. The open circle with error bars shows a revised mean after subtracting a contribution to the central attraction by the estimated baryonic mass in these galaxies. These authors conclude that these data are consistent with the model predictions.

The large cyan circles are estimates for large galaxies: both NGC 4123 (Weiner, Sellwood & Williams 2001) and NGC 3095 (Weiner 2004) have well-estimated halos that are significantly below the predicted range. The inner density estimated for NGC 1365 (Zanmár Sanchéz *et al.* 2008), on the other hand, is $\Delta_{v/2} \sim 5 \times 10^7$, which is off the top of this plot, though the total halo mass is quite modest; a large uncertainty in the inclination, a possible warp that is very hard to model, together with evidence of a inner disturbance that required us to fit to data only one side of the bar, all conspire to render the inner halo density of this Fornax cluster galaxy quite uncertain.

I also plot two magenta points from different mass models for the Milky Way. The upper point is from Klypin, Zhao & Somerville (2002) while the lower shows the upper bound on the inner halo density estimated by Binney & Evans (2001). Their bound comes from trying to include enough foreground disk stars to match an old estimate (Popowski *et al.* 2001) of the micro-lensing optical depth to the red-clump stars of the Milky Way bulge; current estimates of this optical depth are somewhat lower (Popowski *et al.* 2005), suggesting a reanalysis will allow a higher halo density.

It is important to realize that the creation of a disk through condensation, or inflow, of gas into the centers dark halos must deepen the gravitational potential and cause the halo to contract (e.g. Blumenthal *et al.* 1986; Sellwood & McGaugh 2005). Thus, estimates of the current halo density should be reduced to take account of halo compression. Allowance for compression brings the point for NGC 1365 down by more than one order of magnitude. But making this correction for all the galaxies (which Macció *et al.* did not do for their open circle point) will move all the data points down, including the well-estimated points for NGC 4123 & NGC 3095 that are already uncomfortably low.

The only real difficulties presented by the comparison with the predictions in Fig. 1 arise from two well-determined low points, which could simply turn out to be anomalous. Additional evidence suggesting uncomfortably low DM densities in real galaxies comes from other rotation curve data (e.g. Kassin, de Jong & Weiner 2006) and the difficulty of matching the observed zero point of the Tully-Fisher relation (e.g. Dutton, van den Bosch & Courteau 2008). However, an independent argument, based on the constraints from dynamical friction on bars, also suggests that the DM density in barred galaxies is generally lower than predicted.

2.3. *Bar Slow Down*

Bars in real galaxies are generally believed to be "fast", in that the radius of corotation is generally larger than the semi-major axis of the bar by only a small factor, $\mathcal{R}$. Indications that $1 \lesssim \mathcal{R} \lesssim 1.3$ come from (a) direct measurements in largely gas-free galaxies, summarized by Corsini (2008), (b) models of the gas flow (e.g. Weiner *et al.* 2001; Bissantz, Englmaier & Gerhard 2003), and (c) indirect arguments about the location of dust lanes (e.g. Athanassoula 1992). Rautiainen, Salo & Laurikainen (2008), and others, claim a few counter-examples from indirect evidence, although they concede that they try to match the morphology of the spiral patterns, which may rotate more slowly than the bar.

After some considerable debate, a consensus seems to be emerging that strong bars in galaxies should experience fierce braking unless the halo density is low (Debattista & Sellwood 1998, 2000; O'Neill & Dubinski 2003; Holley-Bockelmann, Weinberg & Katz 2005; Colín, Valenzeula & Klypin 2006). The counter-example claimed by Valenzuela & Klypin (2003) was shown by Sellwood & Debattista (2006) to have resulted from a numerical artifact in their code. The claims of discrepancies by Athanassoula (2003) are merely that weak, or initially slow bars, are less strongly braked, while she also finds that strong, fast bars slow unacceptably in dense halos.

Thus there is little escape from the conclusion by Debattista & Sellwood (2000), that the existence of fast bars in strongly barred galaxies requires a low density of DM in the inner halo. Our original constraint required near maximal disks, although the halo models in that paper were not at all realistic. Figure 2 summarizes the results from a new study of exponential disks embedded in NFW halos, computed using the code described in Sellwood (2003) that has greatly superior dynamic range. When scaled to the Milky

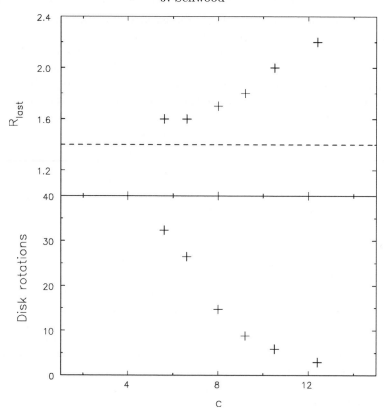

Figure 2. Bars in NFW halos. Above: The value of $\mathcal{R}$ at the time each simulation was stopped. Below: The number of disk rotations before the bar slowed to the point where $\mathcal{R} = 1.4$.

Way, a rotation period at 3 disk scale lengths in these models is 270 Myr. In all cases, the bar becomes slow by the end of the experiment, although the number of disk rotations needed until $\mathcal{R} > 1.4$ increases as the concentration index, c, is reduced. Thus, friction in NFW halos causes bars to become unacceptably slow in a few disk rotations when $c \gtrsim 10$, but on a time scale $\gtrsim 7.5$ Gyr when $c \lesssim 6$ or when the uncompressed $\Delta_{v/2} \lesssim 10^{5.1}$. This conservative bound would exclude well over half the predicted range of halo densities in Fig. 1 for an uncompressed $V_{\max} = 10^{2.25} \simeq 180$ km/s.

In the context of this symposium, it would be nice to test Milky Way models for bar slow-down. The halo and disk in the model tested by Valenzuela & Klypin (2003) were selected from the Klypin *et al.* (2002) models for the Milky Way. The simulations reveal that a very large bar with semi-major axis $\gtrsim 5$ kpc forms quickly, which slows unacceptably within ~ 5 Gyr. Unfortunately, the absence of a realistic bulge in these experiments crucially prevents these results from being regarded as a test of the Klypin *et al.* (2002) MW models, since a bulge should cause a much shorter and faster bar to form.

3. Can the DM Density Be Reduced?

The ΛCDM model would not be challenged if the present-day DM density in galaxies can be reduced by processes that are neglected in cosmic structure formation simulations, which generally follow the dynamics of collisionless collapse only. Four main ideas have been advanced that might achieve the desired density decrease.

3.1. Feedback

This first idea is not a secular effect, and therefore strictly falls outside my assigned topic. However, I discuss it briefly because it should not be omitted from any list of processes that might effect a density reduction.

The basic idea, proposed by Navarro, Eke & Frenk (1997), Binney, Gerhard & Silk (2001), and others is that gas should first collect slowly in a disk at the center of the halo, thereby deepening the gravitational potential well and compressing the halo adiabatically. A burst of star formation would then release so much energy that most of the gas would be blasted back out of the galaxy at very high speed, resulting in a non-adiabatic decompression of the halo, which may possibly result in a net reduction in DM density.

Gnedin & Zhao (2002) present the definitive test of the idea. In their simulations, they slowly grew a disk inside the halo, causing it to compress adiabatically, and then they instantaneously removed the disk. With this artifice, they deliberately set aside all questions of precisely how the star burst could achieve the required outflow, in order simply to test the extreme maximum that any conceivable feedback process could achieve.

They found that the final density of the halo was lower than the initial, confirming that the effect can work, and that mass blasted out from the very center of the potential has greatest effect, presumably because it produces the largest instantaneous change in the gravitational potential. However, density reductions by more than a factor of two required that the disk be unreasonably concentrated, and consequently the baryonic mass has to be blasted out from deep in the potential well.

3.2. Bar-Halo Friction

The same physical process that slows bars, discussed in §2.3, can also reduce the density of the material that takes up the angular momentum, as first reported by Hernquist & Weinberg (1992). This mechanism prompted Weinberg & Katz (2002) to propose the following sequence of events as a means to reconcile ΛCDM halo predictions with bar pattern speed constraints and other data. They argued that a large bar in the gas at an early stage of galaxy formation could reduce the DM density through dynamical friction. The gas bar would then disperse as star formation proceeded, so that were a smaller stellar bar to form later it would not experience much friction. Their idea has been subjected to intensive scrutiny.

Normal Chandrasekhar friction (e.g. Binney & Tremaine 2008, §8.1) is formally invalid in more realistic dynamical systems, such as quasi-spherical halos, because the background particles are bound to the system and will return to interact with the perturber repeatedly. Tremaine & Weinberg (1984) showed that under these circumstances angular momentum exchange occurs at resonances between the motion of the perturber and that of the background particles.

The N-body simulations mentioned in §2.3 generally did not produce a substantial reduction in halo density, despite the presence of strong friction. This could be because the bars were not strong enough, but Weinberg & Katz (2007a,b) argue that the simulations were too crude and that delicate resonances would not be properly mimicked in simulations unless the number of particles exceeds between 10^7 & 10^9, depending on the bar size and strength and the halo mass profile.

Thus two major questions arise: (1) are results from simulations believable? and (2) can realistic bars cause a large density decrease? I addressed both these issues in a recent

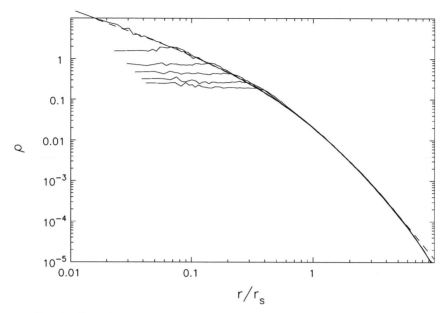

Figure 3. Results from five different experiments with different bar lengths. The dashed line shows the initial profile, while the solid lines show estimates from the particles of the initial (cusped) and final (cored) density profiles from a series of runs with different bar semi-major axes. The final density lines from the lowest to the highest are for bar lengths, $a/r_s = 1$, 0.8, 0.6, 0.4, & 0.2.

paper (Sellwood 2008b). While rigid, ellipsoidal bars are not terribly realistic, I used them deliberately in order to test the analysis and to compare with the simulations presented by Weinberg & Katz. Dubinski (these proceedings) presents results of similar tests using fully self-consistent disks that form bars.

3.2.1. *What N Is Enough?*

Simple convergence tests reveal that experiments with different numbers of particles converge to an invariant time evolution of both the pattern speed and halo density changes at quite modest numbers of halo particles. I report that $N = 10^5$ seemed to be sufficient for a very large bar, while $N \sim 10^6$ was needed for a more realistic bar. I observed no change the results in either case as I increased to $N = 10^8$, or when I employed a spectrum of particle masses in order to concentrate more into the crucial inner halo. I found results for different numbers of particles overlaid each other perfectly, with no evidence for the stochasticity that Weinberg & Katz predicted should result if few particles were in resonance.

I also demonstrated that my simulations did indeed capture resonant responses that converged for the same modest particle numbers. I measured the change in the density of particles $F(L_{\rm res})$, where $L_{\rm res}$ is an angular momentum-like variable that depends on orbit precession frequency. Using this variable, I was able to estimate that some 7% – 20% of halo particles participated in resonant angular momentum exchanges with the bar during a short time interval. This fraction is vastly greater than Weinberg & Katz predicted, because they neglected to take into account the broadening of resonances caused by the evolving bar perturbation, that both grows and slows on an orbital timescale. Athanassoula (2002) and Ceverino & Klypin (2007) also demonstrated the existence of resonances in their simulations.

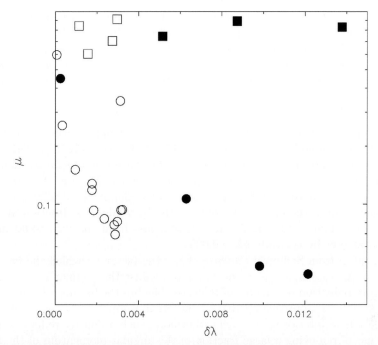

Figure 4. Fractional changes, μ, to $\Delta_{v/2}$ in many experiments. The abscissae show the angular momentum given to the halo, expressed as the usual dimensionless spin parameter. Open circles mark results from experiments in which the density profile of the inner cusp was flattened, while squares indicate experiments where cusp flattening did not occur. Filled symbols show results from experiments in which the moment of inertia of the bar was increased by a factor 5 in all cases except the point at the upper right, where the MoI was increased 10-fold (for more details see Sellwood 2008). The changes to $\Delta_{v/2}$ make no allowance for halo compression.

3.2.2. Density Reduction by Very Strong Bars

Figure 3 shows that a strong bar rotating in a halo within a density cusp ($\rho \propto 1/r$) can flatten the cusp to $\sim 1/3$ bar length. The rigid bar needed to accomplish this must have an axis ratio $a/b \gtrsim 3$, a mass $M_b \gtrsim 30\%$ of the halo mass inside $r = a$.

While cusp flattening is a driven response caused by the slowing bar, it is also a collective effect. I find a much smaller change when I hold fixed the monopole terms of the halo self-gravity. Thus it is dangerous, when studying halo density changes, to include any rigid mass component.

The only simulation I am aware of in which a self-consistent bar flattened the inner density cusp is that reported by Holley-Bockelmann *et al.* (2005). They report a significant density reduction that flattened the cusp to a radius $\sim a/5$. Note that in their model, the initial halo density was not compressed by the inclusion of the disk, since they rederived the halo distribution function that would be in equilibrium in the potential of an uncompressed NFW halo plus the disk.

3.2.3. More Gradual Changes

A number of other simulations in the literature have revealed a modest density reduction caused by angular momentum exchange with a bar in the disk. e.g. Debattista & Sellwood (2000) show a reduction in the halo contribution to the central attraction, and something similar can also be seen in Athanassoula's (2003) simulations. None of these models included very extensive halos.

The inner halo density in fully self-consistent simulations with more extensive and cusped halos can actually rise as the model evolves (Sellwood 2003; Colín et al. 2006). This happens because angular momentum lost by the bar in the disk causes it to contract; the deepening potential of the disk causes further halo compression that overwhelms any density reduction resulting from the angular momentum transferred to the halo.

3.2.4. Angular Momentum Reservoir

A crucial consideration that limits the magnitude of halo density reduction by bar friction is the total angular momentum available in the baryonic disk. Tidal torques in the early universe lead to halos with a log-normal distribution of spin parameters with a mean $\lambda \sim 0.05$, where the dimensionless spin parameter is $\lambda = LE^{1/2}/GM^{5/2}$ as usual. Assuming that the baryons and dark matter are well mixed initially, the fraction of angular momentum in the baryons is equal to the baryonic mass fraction in the galaxy: some 5% – 15%. Thus total angular momentum loss from the disk could increase the halo spin parameter by typically $\delta\lambda \sim 0.005$.

Figure 4, taken from Sellwood (2008b), shows the factor by which the halo density is reduced, $\mu = \Delta_{v/2,\text{fin}}/\Delta_{v/2,\text{init}}$ as the ordinate against the angular momentum gain of halo. A density reduction by a factor of 10 is possible, but the bar must be extreme, having a semi-major axis, a, approaching that of the break radius, r_s, of the NFW profile and a mass $> 30\%$ halo interior to $r = a$. Furthermore, such a density reduction is achieved at the expense of removing a large fraction of the angular momentum of the baryons. It should also be noted that the density changes shown in Fig. 4 also do not take account of any halo compression that might have occurred as the bar and disk formed.

3.3. Baryonic Clumps

El-Zant, Shlosman & Hoffman (2001) proposed that dynamical friction from the halo on moving clumps of dense gas will also transfer energy to the DM and lower its density. They envisaged that baryons would collect into clumps through the Jeans instability as galaxies are assembled and present somewhat simplified calculations of the consequences of energy loss to the halo. The idea was taken up by Mo & Mao (2004), who saw this as a means to erase the cusps in small halos before they merge to make a main galaxy halo, and by Tonini, Lapi & Salucci (2006).

The proposed mechanism has a number of conceptual problems, however. The model assumes that the settling gas clumps maintain their coherence for many dynamical crossing times without colliding with other clumps or being disrupted by star formation, for example. In addition, calculations (e.g. Kaufmann et al. 2006) of the masses of condensing gas clumps suggest they range up to only $\sim 10^6$ $M_\odot$, which is too small to experience strong friction. Larger clumps will probably gather in subhalos, which may get dragged in, but simulations with sub-clumps composed of particles (e.g. Ma & Boylan-Kolchin 2004) indicate that the DM halos of the clumps will be stripped, which simply replaces any DM moved outwards in the halo. Debattista et al. (2008) suggest halo compression is an issue here also, but the essential idea suggested by El-Zant et al. is to displace the DM as the gas settles, which avoids halo compression.

3.3.1. A Direct Test

Setting all these difficulties to one side, Jardel & Sellwood (2008) set out to test the mechanism with N-body simulations. As proposed by El-Zant et al., we divided the entire mass of baryons into N_h equal mass clumps, treated as softened point masses, to which we added isotropic random motion to make their distribution in rough dynamical equilibrium

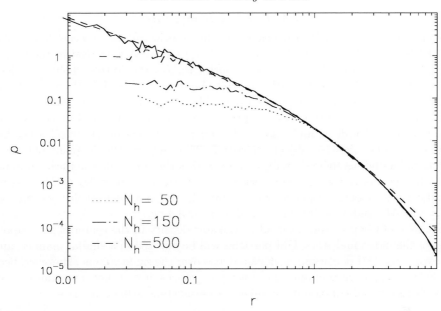

Figure 5. The changes in density caused by the settling of N_h heavy particles with total mass of $0.1 M_{200}$, initially distributed at uniform density within a sphere of radius $4 r_s$ in an NFW halo. The solid line shows the density measured from the particles at the start while the broken lines show the density after 9 Gyr. Another dashed line shows the corresponding theoretical NFW curve.

inside an NFW halo composed of ~ 1 M self-gravitating particles. All particles, both light and heavy, experienced the attraction of all others.

Figure 5 shows results after ~ 9 Gyr, when scaled to a $c = 15$ halo – the timescale would be even longer for less concentrated halos. We find that some density reduction does occur, but the rate at which the density is decreased is considerably slower than El-Zant *et al.* predict. We traced this discrepancy to their use of three times too large a Coulomb logarithm in their calculations.

Fig. 5 also shows that the rate of density reduction rises as the baryon mass is concentrated into fewer, more massive particles. Again our result is consistent with that of Ma & Boylan-Kolchin (2004), who employed a mass spectrum of clumps, and who showed that a much smaller density reduction occurred in a separate simulation that omitted the three heaviest clumps. Thus, if this process is to work on an interesting time-scale, it requires a few gas clumps whose masses exceed 1% of the entire halo.

Mashchenko *et al.* (2006, 2007) argue that the energy input to the halo, mediated by the motion of the mass clumps, can be boosted if the gas is stirred by stellar winds and supernovae – a less extreme form of feedback (*cf.* §3.1). Their simulations of the effect reveal a density reduction in dwarf galaxies. Peirani *et al.* (2008), on the other hand, propose AGN activity to accelerate gas clumps. They present simulations that show the cusp can be flattened to $\sim 0.1 r_s$ with a clump having mass of $\sim 1\%$ of the galaxy mass, being driven outwards from the center to a distance of half the NFW break radius at a speed of 260 km/s. In both these models, it is unclear how the dense material can be accelerated to the required speed (e.g. MacLow & Ferrara 1999).

3.4. *Recoiling/Binary BHs*

In any hierarchical structure-formation model, halos grow through a succession of mergers (e.g. Wechsler *et al.* 2002). If massive black holes (BHs) have formed in the centers of two galaxies that merge, then one expects both BHs to settle to the center of the merged halo and to form a binary pair of BHs in orbit about each other. The physics of the decay of the orbit is interestingly complicated (e.g. Merritt & Milosavljević 2005).

The star density in the centers of elliptical galaxies can be reduced by star scattering as the BH binary hardens, and also by the separate process of BH recoil if the binary encounters another massive object. Merritt & Milosavljević (2005) point out that the star density can be significantly reduced only within the sphere of gravitational influence of the BHs, which extends to $r \sim r_h$, where $r_h = GM_\bullet/\sigma^2$, where $M_\bullet$ is the mass of the BH and σ is the velocity dispersion of the stars. Generous values for disk galaxies might be $M_\bullet = 10^7 M_\odot$ and $\sigma = 100$ km/s, yielding $r_h \simeq 4.4$ pc.

Since none of the processes that affect the star density in the center depend upon the masses of the individual stars, DM particles will be affected in a similar manner, and we must expect the DM density to be depleted also over the same volume. However, because black hole dynamics affects only the inner few parsecs, it can have essentially no effect on bar pattern speed constraints or values of parameters such as $\Delta_{v/2}$.

4. Conclusions

The inner densities of DM halos of galaxies today continue to challenge the ΛCDM model for galaxy formation. Both direct estimates in a few galaxies, and dynamical friction constraints from bar pattern speeds require inner densities lower than predicted in DM-only simulations by at least a factor of a few.

Several processes that are omitted from DM only simulations can reduce the inner halo density. Feedback has a very slight effect unless a large mass of gas is blasted out from the deepest point in the peotential well. Bar friction requires an extreme bar and removes a large fraction of the angular momentum in the baryons. Dynamical friction on massive gas clumps is too slow, unless moving gas clumps exceed $\sim 1\%$ of total baryonic mass. Scattering by merging or recoiling BHs affects only very center. Thus any of the four suggested mechanism needs to be stretched if it is to cause a significant reduction.

Acknowledgements

I thank Victor Debattista for comments on the manuscript and the organizers of the conference for travel support. This work was supported by grants AST-0507323 from the NSF and NNG05GC29G from NASA.

References

Alam, S. M. K., Bullock, J. S., & Weinberg, D. H. 2002, *Ap. J.*, **572**, 34
Athanassoula, E. 1992, *MNRAS*, **259**, 345
Athanassoula, E. 2002, *Ap. J. Lett.*, **569**, L83
Athanassoula, E. 2003, *MNRAS*, **341**, 1179
Binney, J. J. & Evans, N. W. 2001, *MNRAS*, **327**, L27
Binney, J., Gerhard, O., & Silk, J. 2001, *MNRAS*, **321**, 471
Binney, J. & Tremaine, S. 2008, *Galactic Dynamics* 2nd Ed. (Princeton: Princeton University Press)
Bissantz, N., Englmaier, P., & Gerhard, O. 2003, *MNRAS*, **340**, 949
Blumenthal, G. R., Faber, S. M., Flores, R., & Primack, J. R. 1986, *Ap. J.*, **301**, 27
Ceverino, D. & Klypin, A. 2007, *MNRAS*, **379**, 1155

Colín, P., Valenzuela, O., & Klypin, A. 2006, *Ap. J.*, **644**, 687
Corsini, E. M. 2008, in *Formation and Evolution of Galaxy Disks*, eds. J. G. Funes SJ & E. M. Corsini (ASP, to appear)
Debattista, V. P. & Sellwood, J. A. 1998, *Ap. J. Lett.*, **493**, L5
Debattista, V. P. & Sellwood, J. A. 2000, *Ap. J.*, **543**, 704
Debattista, V. P., et al. 2008, *Ap. J.*, **681**, 1076
de Blok, W. J. G., McGaugh, S. S., & Rubin, V. C. 2001, *AJ*, **122**, 2396
de Blok, W. J. G. & Bosma, A. 2002, *A&A*, **385**, 816
Diemand, J., Kuhlen, K., & Madau, P. 2007, *Ap. J.*, **667**, 859
Dutton, A. A., van den Bosch, F. C., & Courteau, S. 2008, in *Formation and Evolution of Galaxy Disks*, eds. J. G. Funes SJ & E. M. Corsini (ASP, to appear) arXiv:0801.1505
El-Zant, A., Shlosman, I., & Hoffman, Y. 2001, *Ap. J.*, **560**, 636
Gnedin, O. Y. & Zhao, H. S., 2002, *MNRAS*, **333**, 299
Hernquist, L. & Weinberg, M. D. 1992, *Ap. J.*, **400**, 80
Holley-Bockelmann, K., Weinberg, M., & Katz, N. 2005, *MNRAS*, **363**, 991
Jardel, J. & Sellwood, J. A. 2008, *Ap. J.*, (submitted)
Kassin, S. A., de Jong, R. S., & Weiner, B. J. 2006, *Ap. J.*, **643**, 804
Kaufmann, T., Mayer, L., Wadsley, J., Stadel, J., & Moore, B. 2006, *MNRAS*, **370**, 1612
Klypin, A., Zhao, H. S., & Somerville, R. S. 2002, *Ap. J.*, **573**, 597
Komatsu, E. et al. 2008, arXiv:0803.0547
Kormendy, J. & Kennicutt, R. C. 2004, *Ann. Rev. Astron. Ap.*, **42**, 603
Ma, C-P. & Boylan-Kolchin, M. 2004, *Phys. Rev. Lett.*, **93**, 21301
Macciò, A. V., Dutton, A. A., & van den Bocsh, F. C. 2008, arXiv:0805.1926
MacLow, M-M. & Ferrara, A. 1999, *Ap. J.*, **513**, 142
Mashchenko, S., Couchman, H. M. P., & Wadsley, J. 2006, *Nature*, **442**, 539
Mashchenko, S., Couchman, H. M. P., & Wadsley, J. 2007, *Science*, **319**, 174
Merritt, D. & Milosavljević, M. 2005, Liv. Rev. Rel., **8**, 8 (astro-ph/0410364)
Mo, J. J. & Mao, S. 2004, *MNRAS*, **353**, 829
Navarro, J. F., Eke, V. R., & Frenk, C. S. 1997, *MNRAS*, **283**, L72
Navarro, J. F., Frenk, C. S., & White, S. D. M. 1997, *Ap. J.*, **490**, 493
O'Neill, J. K. & Dubinski, J. 2003, *MNRAS*, **346**, 251
Peirani, S, Kay, S., & Silk, J. 2008, *A&A*, **479**, 123
Popowski, P. et al. 2001, in *Astrophysical Ages and Times Scales*, eds. T. von Hippel, C. Simpson, & N. Manset, ASP Conference Series **245**, p. 358
Popowski, P. et al. 2005, *Ap. J.*, **631**, 879
Rautiainen, P., Salo, H., & Laurikainen, E. 2008, arXix:0806.0471
Rhee, G., Valenzuela, O., Klypin, A., Holtzman, J., & Moorthy, B. 2004, *Ap. J.*, **617**, 1059
Roškar, R., Debattista, V. P., Stinson, G. S., Quinn, T. R., Kaufmann, T., & Wadsley, J. 2008, *Ap. J. Lett.*, **675**, L65
Sellwood, J. A. 2003, *Ap. J.*, **587**, 638
Sellwood, J. A. 2008a, in *Formation and Evolution of Galaxy Disks*, eds. J. G. Funes S. J. & E. M. Corsini (ASP, to appear) arXiv:0803.1574
Sellwood, J. A. 2008b, *Ap. J.*, **679**, 379
Sellwood, J. A. & Binney, J. J. 2002, *MNRAS*, **336**, 785
Sellwood, J. A. & Debattista, V. P. 2006, *Ap. J.*, **639**, 868
Sellwood, J. A. & McGaugh, S. S. 2005, *Ap. J.*, **634**, 70
Swaters, R. A., Madore, B. F., van den Bosch, F. C., & Balcells, M. 2003, *Ap. J.*, **583**, 732
Tonini, C., Lapi, A., & Salucci, P. 2006, *Ap. J.*, **649**, 591
Tremaine, S. & Weinberg, M. D. 1984, *MNRAS*, **209**, 729
Valenzuela, O. & Klypin, A. 2003, *MNRAS*, **345**, 406
Valenzuela, O., Rhee, G., Klypin, A., Governato, F., Stinson, G., Quinn, T., & Wadsley, J. 2007, *Ap. J.*, **657**, 773
Wechsler, R. H., Bullock, J. S., Primack, J. R., Kravtsov, A. V., & Dekel, A. 2002, *Ap. J.*, **568**, 52
Weinberg, M. D. & Katz, N. 2002, *Ap. J.*, **580**, 627
Weinberg, M. D. & Katz, N. 2007a, *MNRAS*, **375**, 425
Weinberg, M. D. & Katz, N. 2007b, *MNRAS*, **375**, 460
Weiner, B. J., Sellwood, J. A., & Williams, T. B. 2001, *Ap. J.*, **546**, 931

Weiner, B. J. 2004, in IAU Symp. 220, Dark Matter in Galaxies, ed. S. Ryder, D. J. Pisano, M. Walker & K. C. Freeman (Dordrecht: Reidel), p. 35

Zánmar Sánchez, R., Sellwood, J. A., Weiner B. J., & Williams, T. B. 2008, *Ap. J.*, **674**, 797

A packed lecture hall listens to Jerry Sellwood's review.

James Bullock spellbinding his audience.

Mergers and Disk Survival in ΛCDM

James S. Bullock, Kyle R. Stewart, Chris W. Purcell

Center for Cosmology, Department of Physics and Astronomy, The University of California, Irvine, CA 92697 USA

Abstract. Disk galaxies are common in our universe and this is a source of concern for hierarchical formation models like ΛCDM. Here we investigate this issue as motivated by raw merger statistics derived for galaxy-size dark matter halos from ΛCDM simulations. Our analysis shows that a majority ($\sim 70\%$) of galaxy halos with $M_0 = 10^{12} M_\odot$ at $z = 0$ should have accreted at least one object with mass $m > 10^{11} M_\odot \simeq 3 M_{\rm disk}$ over the last 10 Gyr. Mergers involving larger objects $m \gtrsim 3 \times 10^{11} M_\odot$ should have been very rare for Milky-Way size halos today, and this pinpoints $m/M \sim 0.1$ mass-ratio mergers as the most worrying ones for the survival of thin galactic disks. Motivated by these results, we use use high-resolution, dissipationless N-body simulations to study the response of stellar Milky-Way type disks to these common mergers and show that thin disks do not survive the bombardment. The remnant galaxies are roughly three times as thick and twice as kinematically hot as the observed thin disk of the Milky Way. Finally, we evaluate the suggestion that disks may be preserved if the mergers involve gas-rich progenitors. Using empirical measures to assign stellar masses and gas masses to dark matter halos as a function of redshift, we show that the vast majority of large mergers experienced by $10^{12} M_\odot$ halos should be gas-rich ($f_{gas} > 0.5$), suggesting that this is a potentially viable solution to the disk formation conundrum. Moreover, gas-rich mergers should become increasingly rare in more massive halos $> 10^{12.5} M_\odot$, and this suggest that merger gas fractions may play an important role in establishing morphological trends with galaxy luminosity.

Keywords. Cosmology: theory – galaxies: formation – galaxies: evolution

1. Introduction

Roughly 70% of Milky-Way size dark matter halos are believed to host late-type, disk-dominated galaxies (Weinmann *et al.* 2006, van den Bosch *et al.* 2007, Ilbert *et al.* 2006, Choi *et al.* 2007). Conventional wisdom dictates that disk galaxies result from fairly quiescent formation histories, and this has raised concerns about disk formation within hierarchical Cold Dark Matter-based cosmologies (Toth & Ostriker 1992; Wyse 2001; Kormendy *et al.* 2005). Recent evidence for the existence of a sizeable population of cold, rotationally supported disk galaxies at $z \sim 1.6$ ($\sigma/V \sim 0.2$; Wright *et al.* 2008) is particularly striking, given that the fraction of galaxies with recent mergers is expected to be significantly higher at that time (Stewart *et al.* 2008b).

Unfortunately, a real evaluation of the severity of the problem is limited by both theoretical and observational concerns. Theoretically, the process of disk galaxy formation remains very poorly understood in ΛCDM. Though the first-order models envisioned by Mestel (1963), Fall & Efstathiou (1980), Mo, Mao & White (1998) and others provide useful theoretical guides, the formation of disks via a quiescent acquisition of mass is likely not the only channel. Over the last several years, cosmological hydrodynamic simulations have begun to produce galaxies that resemble realistic disk-dominated systems, and in most cases, early mergers have played a role in the disk's formation (Abadi *et al.* 2003; Brook *et al.* 2004; Robertson *et al.* 2005; Governato *et al.* 2007). In particular, the early disks in the simulations of Brook *et al.* (2004) and Robertson *et al.* (2005) originated in gas-rich mergers. Robertson *et al.* (2006) used a suite of focused simulations to show that mergers with gas-fractions larger than $\sim 50\%$ tend to result in disk-dominated remnants and Hopkins et al. (2008) used a larger sample of merger simulations to reach a similar conclusion. Two cautionary notes are in order. First, these results are all subject to the uncertain assumptions associated with modeling 'subgrid' physics in the simulations (ISM pressure, star formation, feedback, etc.). Second, the merger-remnant disks in these simulations tend to be hotter and thicker than the *thin* disk of the Milky Way (Brook et al. 2004). Robertson & Bullock (2008) showed that gas-rich merger remnants are a much closer match to the high-dispersion, rapidly rotating disk galaxies observed by integral field spectroscopy at $z \sim 2$ (Förster Schreiber et al. 2006; Genzel et al. 2006). Gas-rich mergers as an explanation for these high-redshfit disks is further motivated by the expectation that gas fractions should be higher at early times (e.g. Erb et al. 2006).

Observationally, the best quantified thin disk is that of the Milky Way. The thin disk of the Milky Way has a scale height $z_d \simeq 350$ pc (see Jurić et al. 2008 and references therein), a fairly cold stellar velocity dispersion, $\sigma \simeq 40$ km s^{-1}, and contains stars that are as old as 10 Gyr (Nordström et al. 2004). It remains to be determined whether the Milky Way's thin disk is typical of spiral galaxies. This is a vital question. Unfortunately, scale height measurements for a statistical sample of external galaxies remain hindered by the presence of absorbing dust lanes in the disk plane for $\sim L_*$ galaxies (e.g. Yoachim & Dalcanton 2006).

Given the uncertainties associated with the formation of disk galaxies, we might make progress by asking a few focused, conservative questions. First, what is the predicted mass range and frequency of large mergers in galaxy-size halos? Second, can a thin stellar disk survive common mergers, and if not, what does this teach us about disk galaxy formation and/or cosmology? The figures presented below are taken from work described by Stewart et al. (2008a) on halo merger histories, Purcell, Kazantzidis, and Bullock (2008) on stellar disk destruction, and from Stewart et al. (2009, in preparation) on the expected gas fractions of mergers. The simulations in Purcell et al. (2008) were motivated by a program developed in Kazantzidis et al. (2008), which aims to understand the morphological response of disk galaxies to cosmologically-motivated accretion histories.

2. Merger Histories from Cosmological Simulations

As described in Stewart et al. (2008a), our merger trees are derived from an 80 h^{-1} Mpc box ΛCDM simulation. We concentrate specifically on thousands of Milky Way-sized systems, $M_0 \simeq 10^{12} M_\odot$ at $z = 0$. We categorize the accretion of objects as small as $m \simeq 10^{10} h^{-1} M_\odot$ and focus on the infall statistics into main progenitors of $z = 0$ halos as a function of lookback time.

Figure 1 shows a merger tree for a halo of mass $M_0 = 10^{12.5} h^{-1} M_\odot$ at $z = 0$. Time runs from top to bottom and the corresponding redshift for each timestep is shown to the left of each tree. The radii of the circles are proportional to the halo radius $R \sim M^{1/3}$, while the lines show the descendent–progenitor relationship. The color and type of the connecting lines indicate whether the progenitor halo is a field halo (solid black) or a subhalo (dashed red). The most massive progenitor at each timestep — the main progenitor — is plotted in bold down the middle. Once a halo falls within the radius of another halo, it becomes a subhalo and its line-type changes from black solid to red dashed. Figure 1 shows a fairly typical merger history, with a merger of mass $m \simeq 0.1 M_0$ at $z \simeq 0.51$. The merger ratio at the time of the merger was $m/M_z \simeq 0.5$. Note that this large merger does not survive for long as a resolved subhalo — it quickly loses most of its mass via interactions with the center of the halo, which presumably would host a central galaxy.

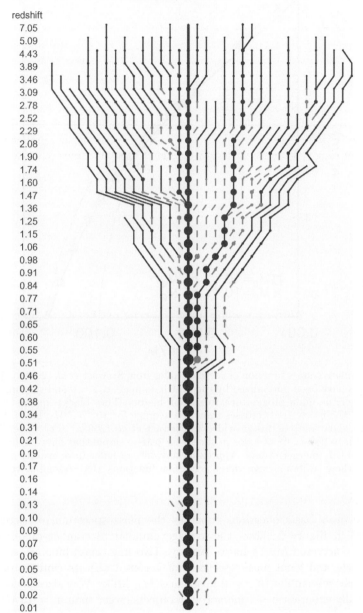

Figure 1. Sample merger tree for a dark matter halo with $z = 0$ mass $M_0 \simeq 3 \times 10^{12} h^{-1} M_\odot$ from Stewart et al. (2008a). Time progresses downward, with the redshift z printed on the left hand side. The bold, vertical line at the center corresponds to the main progenitor, with filled circles proportional to the radius of each halo. The minimum mass halo shown in this diagram has $m = 10^{9.9} h^{-1} M_\odot$. Solid (black) and dashed (red) lines and circles correspond to isolated field halos, or subhalos, respectively. The dashed (red) lines that do not merge with main progenitor represent surviving subhalos at $z = 0$. Note that the halo shown here has a fairly typical merger history, and experiences a merger of mass $m \simeq 0.1 M_0 \simeq 0.5 M_z$ at $z = 0.51$.

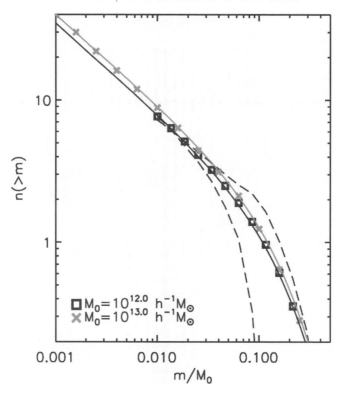

Figure 2. Cumulative mass function of accreted halos from Stewart et al. (2008a). The masses of accreted objects have been normalized by the host halo mass at $z = 0$ and the cumulative count is integrated over the main progenitor's formation history. The (black) squares show the average for $10^{12} h^{-1} M_\odot$ halos; (red) crosses show the average for $10^{13} h^{-1} M_\odot$ halos. Lines through the data points show analytic fits provided in Stewart et al. (2008a). The upper/lower dashed lines indicate the $\sim 25\%/20\%$ of halos in the $10^{12} h^{-1} M_\odot$ sample that have experienced exactly two/zero $m \geqslant 0.1 M_0$ merger events. Approximately 45% of halos have exactly one $m \geqslant 0.1 M_0$ merger event; these systems have mass accretion functions that resemble very closely the average.

Among the most basic questions concerns the mass spectrum of accreted objects. The solid line in Figure 2 shows the average cumulative number of objects of mass greater than m accreted over a halo's history. Two halo mass bins are shown. We see that, on average, the total mass spectrum of accreted objects (integrated over time) is approximately self-similar in $z = 0$ host mass M_0. Milky Way-sized halos with $M_0 \simeq 10^{12} M_\odot$ typically experience ~ 1 merger with objects larger than $m = 0.1 M_0 \simeq 10^{11} M_\odot$, and approximately 7 mergers with objects larger than $m = 0.01 M_0 \simeq 10^{10} M_\odot$ over their histories. Mergers involving objects larger than $m = 0.2 M_0 \simeq 2 \times 10^{11} M_\odot$ should be extremely rare.

Figure 3 shows a particularly important statistical summary for the question of morphological fractions. Specifically we show the fraction of galaxy-sized halos (a bin centered on $M_0 = 10^{12} h^{-1} M_\odot$) that have experienced *at least one* "large" merger within the last t Gyr. The different line types correspond to different absolute mass cuts on the accreted halo, from $m > 0.05 M_0$ to $m > 0.4 M_0$. The lines flatten at high z because the halo main progenitor masses, M_z, become smaller than the mass threshold on m. We find that while fewer than $\sim 10\%$ of Milky Way-sized halos have *ever* experienced a merger with an object large enough to host a sizeable disk galaxy, $(m > 0.4 M_0 \simeq 4 \times 10^{11} M_\odot)$, an

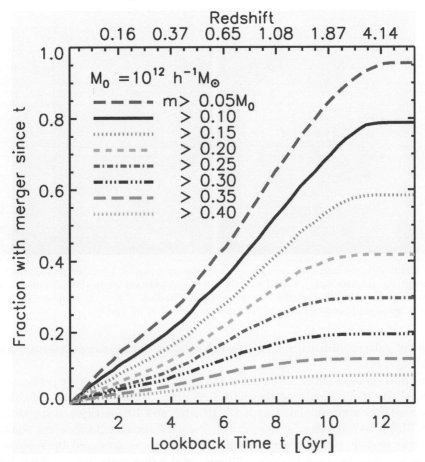

Figure 3. Merger fractions from Stewart *et al.* (2008a). The lines show the fraction of galaxy–sized halos, $M_0 = 10^{12} h^{-1} M_\odot$, that have experienced at least one merger larger than a given mass threshold, m/M_0, since look-back time t.

overwhelming majority ($\sim 95\%$) have accreted an object more massive than the Milky Way's disk ($m > 0.05 M_0 \simeq 5 \times 10^{10} M_\odot$). Approximately 70% of halos have accreted an object larger than $m/M_0 = 0.1$ in the last 10 Gyr. We emphasize that the ratios presented here are relative to the *final* halo mass (m/M_0) not the ratio of the masses just before the merger occurred (m/M_z). As presented, the ratios are quite conservative because halos grow with time $M_z < M_0$ and $m/M_z > m/M_0$. We find that typically, for the mergers we record here, $m/M_z \simeq 2m/M_0$ (Stewart *et al.* 2008a) and that makes the implications for disk survival all the more worrying.

3. Targeted Simulations

Recently, Kazantzidis *et al.* (2008) have investigated the response of galactic disks subject to a ΛCDM-motivated satellite accretion histories and showed that the thin disk component survives, though it is strongly perturbed by the violent gravitational encounters with halo substructure (see also Kazantzidis *et al.*, this proceeding). However, these authors focused on subhalos with masses in the range $0.01 M_0 \lesssim m \lesssim 0.05 M_0$, ignoring the most massive accretion events expected over a galaxy's lifetime. Here, we report on the results of Purcell *et al.* (2008) that expand upon this initiative by investigating the

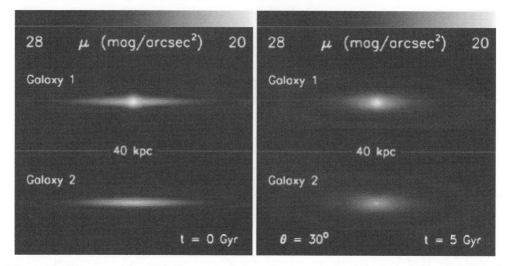

Figure 4. Edge-on surface brightness maps for primary galaxies 1 (upper panels; disk/bulge) and 2 (lower panels; disk only). Initial models ($t = 0$ Gyr) are shown in the left panel, while the final results ($t = 5$ Gyr) for satellite-infall orbital inclination of $\theta = 30°$ appears on the right panel. The associated simulations are discussed in Purcell et al. (2008).

evolution of galactic disk morphology and kinematics during merger events with mass ratio $m/M_0 = 0.1$.

As described in more detail in Purcell et al. (2008), our simulations are performed using the parallel-tree dissipationless code PKDGRAV (Stadel 2001). The host halo, disk, and infalling satellites were simulated with 4×10^6, 10^6 and 10^6 particles, respectively. The primary Milky-Way-analogue system drawn from the set of self-consistent equilibrium models that best fit Galactic observational parameters as produced by Widrow et al. (2008), with a host halo mass of $M_0 = 10^{12} M_\odot$ and a disk mass $M_{\rm disk} = 3.6 \times 10^{10} M_\odot$. We initialize a satellite galaxy with a stellar mass of $2 \times 10^9 M_\odot$ embedded within a dark matter halo of virial mass $m \simeq 0.1 M_{\rm host} = 10^{11} M_\odot$. In the left panel of Figure 4, we show the edge-on surface brightness map for both primary galaxy models, one with a central bulge and one without.

We explore a range of initial orbital parameters assigned to the merging satellite galaxy, motivated by cosmological investigations of substructure mergers (Khochfar & Burkert 2006; Benson 2005). We choose an array of orbital inclination angles ($\theta = 0°, 30°, 60°,$ and $90°$) in order to assess the consequence of this parameter on the evolution and final state of the galactic disk in each case. All simulations are allowed to evolve for a total of 5 Gyr, after which time the subhalo has fully coalesced into the center of the host halo and the stellar disk has relaxed into stability, although there are certainly remnant features in the outer disk and halo that will continue to phase-mix and virialize on a much longer timescale; however, our investigations indicate that the disk-heating process has reached a quasi-steady state by this point in the merger's evolution. The morphological thickening of the initial disk after one typical merger is shown in right panels of Figure 4.

Figure 5 provides a direct comparison of all of our merger remnants to observed properties of the Milky Way. The left panel shows the remnant disk scale heights (derived using two-component fits for a thin and thick disk) alongside the values obtained by Juric et al. (2008). While the thin-disk scale height (z_{thin}) of our initial model agrees well with the Galactic benchmark of $z_{thin} \simeq 0.3$ kpc, the final systems all have thin-disk

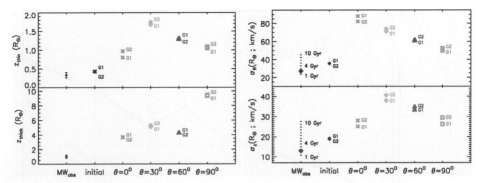

Figure 5. *Left.* The thin- and thick-disk scale heights in the final state for each of our simulated galaxies from Purcell *et al.* (2008), compared to the values derived for the Milky Way. The two panels show the result of a two-component sech^2 fit, with the upper (lower) panel describing the thin (thick) disk's scale height. *Right.* The radial and vertical components of velocity dispersion σ_R and σ_z at the solar neighborhood ($R = 8$ kpc) of our simulated disks, compared to the local values obtained by the Geneva-Copenhagen survey. In each coordinate, the observational spread is marked by a *dotted line* and the dispersion of the sample's median-age stars ($t \sim 2-3$ Gyr) is denoted by a *diamond*.

components with z_{thin} larger by a factor of $\sim 3-5$. The right panel shows remnant disk velocity dispersions (radial and vertical) as measured in disk planes around $R = 8$ kpc compared to the velocity ellipsoid observed in the solar neighborhood (Nordström *et al.* 2004). Following the simulated 1:10 merger, all three components of velocity dispersion are substantially enhanced. None of the remnants are as cold as the Milky Way disk. Note that while the $\theta = 0°$ in-plane accretion produces the least vertical thickening, it produces a huge amount of radial heating, and leaves the remnant disk much hotter than that of the Milky Way. Our conclusion that cosmologically-motivated 1:10 mergers destroy thin stellar disks.

4. Gas-rich Mergers

As discussed in the introduction, the presence of a stabilizing gas component in merger progenitors can potentially alleviate the thin disk disruption we have described. In order to address whether it is plausible that gas-rich mergers occur frequently enough to alleviate the problem, we employ a semi-empirical approach. Specifically, we assign stellar masses and gas masses to halos at each of our merger tree timesteps using empirical relations and then explore the baryonic content of the mergers that occur. For stellar masses, we use the empirical mapping between halo mass and stellar mass advocated by Conryoy & Wechsler (2008). For gas masses we use observational relations between stellar mass and gas mass (Kannappan 2004; McGaugh 2005; Erb *et al.* 2006). The important qualitative trend is that small halos tend to host galaxies with high gas fractions and that gas fractions are inferred to increase in galaxies of a fixed stellar mass at high redshift.

Figure 6 presents an intriguing result from this exploration. The solid black line shows the fraction of halos that have had a merger larger than $m/M_z = 0.3$ since $z = 2$ as a function of the $z = 0$ host mass. (Note that here the merger ratio is the ratio of masses just prior to the merger). We see that a fairly high fraction ($\sim 60\%$) of Milky-Way size halos have experienced a major merger in the last ~ 10 Gyr. Consider, however, the (blue) dotted line, which restricts the merger count to galaxies where *both* of the progenitors are *gas rich* with $f_{\rm gas} = M_{gas}/(M_{gas} + M_*) > 0.5$. We see that the vast majority of the most

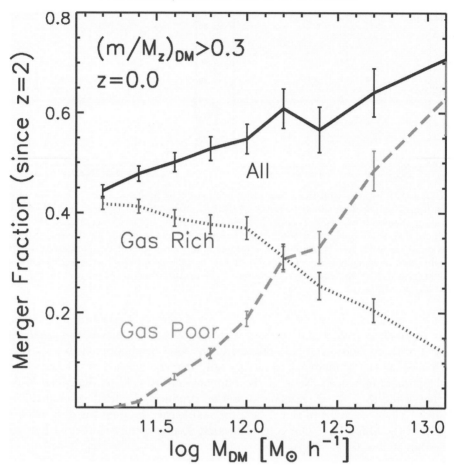

Figure 6. Fraction of dark matter halos of a given mass that have experienced at least one major merger ($m/M_z > 0.3$) since $z = 2$. The solid (black) line is the total DM merger fraction, while the dotted (blue) line *only* includes gas rich major mergers (see text for details), and the dashed (red) line only includes gas poor major mergers. Error bars are Poisson based on number of host halos and the total number of mergers. The figure is modified from Stewart *et al.* (2009, in preparation).

worrying mergers in Milky-Way size halos should have been very gas rich, and that gas-rich major merger become more common in smaller halos. The (red) dashed line shows the fraction mergers that are made up of more gas poor progenitors with $f_{\rm gas} < 0.5$. Not only do these trends provide a possible solution to disk survivability, but they may also provide an interesting clue to the origin of the mass–morphology relation. While dark matter halo merger histories alone show a very weak trend with halo mass, the baryonic content of the mergers should vary significantly with halo mass (and by extension, galaxy luminosity). More massive halos are more likely to experience large, gas-poor mergers, and we expect this to result in a higher fraction of spheroid-dominated galaxies.

Acknowledgements

We thank the conference organizers for an extremely interesting and thought provoking program and we thank our collaborators for allowing us to present our joint work here.

References

Benson, A. J. 2005, MNRAS, 358, 551

Brook, C. B., Kawata, D., Gibson, B. K. & Freeman, K. C. 2004, ApJ, 612, 894

Choi, Y.-Y., Park, C., & Vogeley, M. S. 2007, ApJ, 658, 884

Erb, D. K., Steidel, C. C., Shapley, A. E., Pettini, M., Reddy, N. A., & Adelberger, K. L. 2006, ApJ, 646, 107

Fall, S. M. & Efstathiou, G. 1980, MNRAS, 193, 189

Förster Schreiber, N. M., et al. 2006, ApJ, 645, 1062

Genzel, R., et al. 2006, Nature, 442, 786

Hopkins, P. F., Cox, T. J., Younger, J. D., & Hernquist, L. 2008, arXiv:0806.1739

Ilbert, et al. 2006, A & A, 453, 809

Jurić, M. et al. 2008, ApJ, 673, 864

Kannappan, S. J. 2004, ApJ, 611, L89

Kazantzidis, S., Bullock, J. S., Zentner, A. R., Kravtsov, A. V., & Moustakas, L. A. 2008, ApJ, accepted, arXiv:0708.1949

Khochfar, S. & Burkert, A. 2006, A & A, 445, 403

Kormendy, J. & Fisher, D. B. 2005, in Revista Mexicana de Astronomia y Astrofisica Conference Series, Vol. 23, Revista Mexicana de Astronomia y Astrofisica Conference Series, ed. S. Torres-Peimbert & G. MacAlpine, 101–108

McGaugh, S. S. 2005, ApJ, 632, 859

Mestel, L. 1963, MNRAS, 126, 553

Mo, H. J., Mao, S., & White, S. D. M. 1998, MNRAS, 295, 319

Nordström, B. et al. 2004, A & A, 418, 989

Purcell, C. W., Kazantzidis, S., & Bullock, J. S. 2008, arXiv:0810.2785

Robertson, B. E. & Bullock, J. S. 2008, ApJL, 685, L27

Robertson, B., Bullock, J. S., Cox, T. J., Di Matteo, T., Hernquist, L., Springel, V., & Yoshida, N. 2006, ApJ, 645, 986

Stadel, J. G. 2001, Ph.D. Thesis,

Stewart, K. R., Bullock, J. S., Wechsler, R. H., Maller, A. H., & Zentner, A. R. 2008a, ApJ, 683, 597

Stewart, K. R., Bullock, J. S., Barton, E., & Wechsler, R. H. 2008b, ApJ, submitted

Stewart, K. R., Bullock, J. S. et al. 2009, in preparation

Toth, G. & Ostriker, J. P. 1992, ApJ, 389, 5

van den Bosch, F. C. et al. 2007, MNRAS, 376, 841

Weinmann, S. M., van den Bosch, F. C., Yang, X., & Mo, H. J. 2006, MNRAS, 366, 2

Widrow, L. M., Pym, B., & Dubinski, J. 2008, ApJ, 679, 1239

Wright, S. A., Larkin, J. E., Law, D. R., Steidel, C. C., Shapley, A. E., & Erb, D. K. 2008, arXiv:0810.5599

Wyse, R. F. G. 2001, in Astronomical Society of the Pacific Conference Series, Vol. 230, Galaxy Disks and Disk Galaxies, ed. J. G. Funes, & E. M. Corsini, 71–80

Participants merging and surviving at the Town Hall reception.

55 years of Galaxy symposia: Jan Palouš, Adriaan Blaauw and Bruce Elemgreen at Carlsberg.
Aage Bohr in the background.
Photo: Bruce Elmegreen.

An 84-μG Magnetic Field in a Galaxy at Z=0.692?

Arthur M. Wolfe[1], Regina A. Jorgenson[1], Timothy Robishaw[2], Carl Heiles[2], and Jason X. Prochaska[3]

[1]Dept. of Physics and Center for Astrophysics and Space Sciences
University of California, San Diego
La Jolla, CA 92093-0424, USA
email: awolfe@ucsd.edu, regina@physics.ucsd.edu

[2]Astronomy Department, University of California
Berkeley, CA 94720-3411, USA
email: robishaw@astro.berkeley.edu, heiles@astro.berkeley.edu

[3]UCO-Lick Observatory, University of California, Santa Cruz
Santa Cruz, CA 95464, USA
email: xavier@ucolick.org

1. Abstract

The magnetic field pervading our Galaxy is a crucial constituent of the interstellar medium: it mediates the dynamics of interstellar clouds, the energy density of cosmic rays, and the formation of stars (Beck 2005). The field associated with *ionized* interstellar gas has been determined through observations of pulsars in our Galaxy. Radio-frequency measurements of pulse dispersion and the rotation of the plane of linear polarization, i.e., Faraday rotation, yield an average value $B \approx 3$ μG (Han *et al.* 2006). The possible detection of Faraday rotation of linearly polarized photons emitted by high-redshift quasars (Kronberg *et al.* 2008) suggests similar magnetic fields are present in foreground galaxies with redshifts $z > 1$. As Faraday rotation alone, however, determines neither the magnitude nor the redshift of the magnetic field, the strength of galactic magnetic fields at redshifts $z > 0$ remains uncertain.

Here we report a measurement of a magnetic field of $B \approx 84$ μG in a galaxy at $z = 0.692$, using the same Zeeman-splitting technique that revealed an average value of $B = 6$ μG in the *neutral* interstellar gas of our Galaxy (Heiles *et al.* 2004). This is unexpected, as the leading theory of magnetic field generation, the mean-field dynamo model, predicts large-scale magnetic fields to be weaker in the past, rather than stronger (Parker 1970).

The full text of this paper was published in Nature (Wolfe *et al.* 2008).

References

Beck, R. 2005, *Lect. Notes. Phys.* 664, 41
Han, J. L., Manchester, R. N., Lyne, A. G., Qiao, G. J., & van Straten, W. 2006, *ApJ* 642, 868
Heiles, C. & Troland, T. H. 2004, *ApJ Sup* 151, 271
Kronberg, P. P., Bernet, M. L., Miniati, F., Lilly, S. J., Short, M. B., & Higdon, D. M. 2008, *ApJ* 676, 70
Parker, E. 1970, *ApJ* 160, 383
Wolfe, A. M., Jorgenson, R. A., Robishaw, T., Heiles, C., & Prochaska, J.X. 2008 *Nature* 455, 638

Birgitta Nordström, Jan Palouš and Hans Zinnecker at Carlsberg.
Ole Strömgren and Aage Bohr in the background.

Photo: Bruce Elmegreen.

Marija Vlajic emphasising a point during her lecture.

Abundance Gradient in the Disk of NGC 300

Marija Vlajić[1], Joss Bland-Hawthorn[2] and Ken C. Freeman[3]

[1]Astrophysics, Department of Physics, University of Oxford, Oxford OX1 3RH, UK
email: vlajic@astro.ox.ac.uk

[2]Institute of Astronomy, School of Physics, University of Sydney, Australia
email: jbh@physics.usyd.edu.au

[3]Research School of Astronomy and Astrophysics, Mount Stromlo Observatory,
Australian National University, Australia
email: kcf@mso.anu.edu.au

Abstract. The structure of the outer parts of galactic disks and the nature of their stellar populations are fundamental to our understanding of the formation and evolution of spiral galaxies. Ages and metallicity distributions of stars in the outermost regions of spiral disks provide important clues on how and when the disks are assembled. In our earlier work we trace the extended stellar disk of NGC 300 out to a radius of at least 10 disk scale lengths, with no sign of truncation. We now revisit the outer disk of NGC 300 in order to derive the metallicity distribution of the faint stellar population in its outskirts. We find that predominantly old stellar population in the outer disk exhibits a negative abundance gradient – as predicted by the chemical evolution models – out to about 10 kpc, followed by the metallicity plateau in the outermost disk.

Keywords. galaxies: abundances, galaxies: individual (NGC 300), galaxies: stellar content

1. Introduction

Cosmological simulations of galaxy formation have shown that the faint outer regions of disk galaxies are expected to be replete with signatures of the process of galaxy assembly (e.g. Bullock & Johnston 2005). Due to their long dynamical timescales these regions are able to retain the fossil record from the epoch of galaxy formation in the form of spatial and kinematic distributions, ages and chemical abundances of their stars (Freeman & Bland-Hawthorn 2002). The structure of the outer parts of galactic disks is therefore central to our understanding of the formation and evolution of disk galaxies.

Until recently, it was believed that all galaxy disks undergo truncation near the Holmberg radius (26.5 mag arcsec^{-2}) and that this was telling us something important about the collapsing protocloud of gas that formed the early disk (van der Kruit 1979). This picture was challenged by our earlier finding (Bland-Hawthorn et al. 2005) of a classic exponential disk (Freeman 1970) with no break in NGC 300, out to the distances of 15 kpc, corresponding to about 10 disk scale lengths.

The low surface brightness of galactic outskirts presents a challenge for the studies of the outermost regions of galactic disk. The traditional diffuse light imaging is limited at the levels of ~ 27 mag arcsec^{-2} by various non-stellar sources of radiation. A more powerful approach is to use the star counts to map out the outskirts of spirals. Only by resolving the outer disk stars can we reach the effective surface brightness levels necessary to trace the outer disk structure (Bland-Hawthorn et al. 2005, Ferguson et al. 2007).

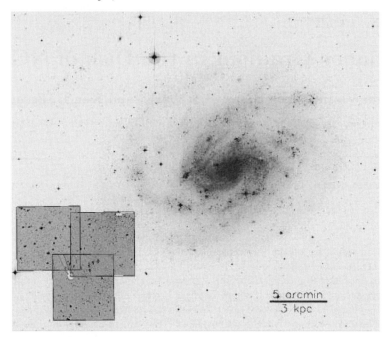

Figure 1. The Digitized Sky Survey (DSS) wide-field image of NGC 300. (North is to the top, east is to the left.) Squares mark the positions of GMOS $5'.5 \times 5'.5$ field of view.

2. Observations and Data Analysis

The deep g' and i' band images of three fields in the outer disk of NGC 300 were obtained with the Gemini Multi Object Spectrograph (GMOS) on Gemini South telescope. The locations of the fields, spanning the galactocentric radii of 7−16 kpc, are shown in Fig. 1. With the total observing time of 34 hours and the average image quality of $0.7''$ and $0.6''$ FWHM in g' and i', respectively, we reach the 50% completeness limit at 26.8 − 27.4 in g' and 26.8 − 27.4 in i' band.

The data were reduced using the standard Gemini/IRAF reduction routines and analyzed with the DAOPHOT/ALLSTAR software suite (Stetson 1987). We performed extensive artificial stars tests to assess the data completeness, as well as derive realistic estimates of the photometric errors and criteria for discarding spurious and non-stellar detections. At these faint depths, careful consideration must be given to the contaminating effect of the background galaxy population. We estimated the contribution of unresolved faint galaxies to our star counts using the observations of the William Herschel Deep Field (WHDF; Metcalfe et al. 2001).

3. Results and Discussion

In Vlajić et al. (2008) we revisit NGC 300 to investigate the faint stellar population in its outskirts. Top panel of Fig. 2 shows the outer disk surface brightness profile of the galaxy derived from the i'-band starcounts of the old red giant branch stars. Surface brightness is calculated in $0.5'$ annuli by summing the flux of all stars in a given annulus and correcting for the radial completeness, number of pixels in the annulus and the pixel scale to obtain surface brightness in the units of mag arcsec^{-2}. We also correct for the light missed due to the fact that we detect only the brightest stars in the outskirts of NGC 300 and the inclination of the galaxy. The surface brightness profile confirms the

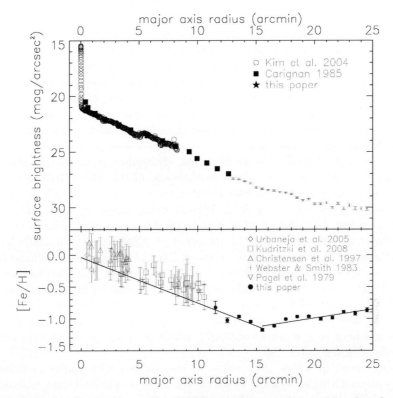

Figure 2. *Top:* Surface brightness profile of NGC 300. Open circles are I-band surface photometry data from Kim *et al.* (2004); the data points have been shifted downward by 1.2 mag. Full squares are B_J surface photometry measurements taken from Carignan (1985) and have been shifted upward by 0.4 mag. The star symbols are our surface brightness measurements derived from star counts. *Bottom:* Metallicity gradient in the disk of NGC 300. Inner disk data points are from the spectroscopic abundance studies listed in the figure. Filled circles are outer disk mean metallicities in 1' bins and solid lines are the linear fits to these values in the regimes $R < 15.5'$ and $R > 15.5'$

earlier finding (Bland-Hawthorn *et al.* 2005) of the extended stellar disk in NGC 300 with the surface brightness declining exponentially out to $\sim 25'$, or about 10 disk scale lengths. We note here that only three other disk have been traced down to the same effective surface brightness levels so far – that of the Galaxy(Ibata *et al.* 2003), M31 (Irwin *et al.* 2005) and M33 (Ferguson *et al.* 2007). Except for the case of a "sub-exponential" disk of M33 the other three disks are either pure exponentials or "super-exponential" and are detected down to the effective surface brightness levels below ~ 30 mag arcsec^{-2}. The existence of stars at these extremely low surface densities (0.01 M$_\odot$ pc^{-2}, Bland-Hawthorn *et al.* 2005) in the outer galactic disk of NGC 300 is puzzling. More insight about the nature of this faint stellar population can be gained by studying the metallicity distribution of the outer disk stars.

We use the position of old red giant branch stars on the color-magnitude diagram, together with the theoretical stellar evolutionary tracks of VandenBerg *et al.* (2006), to derive the metallicity distribution function of the stars in the outer disk of NGC 300. The colors of the stars on the red giant branch are more sensitive to metallicity than to age which, despite the fact that there is an inherent age-metallicity degeneracy and that the procedure assumes coeval star formation, allows for the metallicities to be determined without introducing large uncertainties.

In the bottom panel of Fig. 2 we show the metallicity gradient in the disk of NGC 300. The metallicities of the outermost 14' are calculated as mean metallicities of the stars within 1' wide annuli. We determine the metallicity gradient for stars with $R < 15.5'$ and $R > 15.5'$; these linear fits are shows as solid lines in the figure.

Radial abundance profiles represent one of the most important observational constraints for the models of the evolution of spiral galaxies. Abundance gradients arise naturally in the chemical evolution models as a consequence of the radial dependence of yield, star formation and gas infall; observed abundances exhibit slow negative gradient with large, but uniform, dispersion.

In the outer disk of NGC 300 we find that abundance gradient changes slope at a radius of 15.5' (~ 9.3 kpc). Does this mark the end of the stellar disk and a transition to an old halo? The argument against the disk-halo transition at 15.5' is the relatively high metallicity of the outermost field. Typical stellar halos are more metal-poor than what we find in the outer disk of NGC 300. In the Local Group, the halos of the Galaxy (e.g. Carollo et al. 2007) and M31 (Chapman et al. 2006, Koch et al. 2007) have metallicities of the order of $[Fe/H] = -1.5$, or lower. Even the less massive, almost bulge-less, M33 which is expected to have properties similar to those of NGC 300 has a metal-poor halo with $[Fe/H] = -1.3$ to -1.5, and a more metal-rich disk with the photometric metallicity of $[Fe/H] \approx -0.9$ (McConnachie et al. 2006), similar to the outer disk metallicities we derive. We also note here that stellar halos are usually found to exhibit a negative metallicity gradient (e.g. Koch et al. 2007), which is in disagreement with our positive or flat gradient in the outer disk.

Metallicity trend in which the decline in abundances with radius is followed by a metallicity plateau, similar to what we find in the outer disk of NGC 300, has already been observed in M31 (Worthey et al. 2005) and the Galaxy (Carney et al. 2005, Yong et al. 2005, 2006). The latter authors also find the α-enhanced outer disk with $[\alpha/Fe] \approx 0.2$. We note here that M33 does not exhibit a flattening of its metallicity gradient (Barker et al. 2007). However, this is not surprising given that M33 experiences a break in its light profile at $\sim 35'$ and the surface brightness falls below 31 mag arcsec^{-2} at the radius of 60' (Ferguson et al. 2007), where halo abundances start to dominate.

We present two scenarios to explain the metallicity plateau (or upturn) in the outer disks of spirals.

In the radial mixing scenario, the mechanism described by Sellwood & Binney (2002) is responsible for transporting the stars within the disk while preserving the nearly circular orbits and exponential light profile. Bland-Hawthorn et al. (2005) show that the outer disk of NGC 300 is a high-Q environment, with the value of the Toomre Q parameter of $\sim 5 \pm 2$ and as such is expected to be stable to gravitational perturbations. Scattering of stars by spiral waves (Sellwood & Preto 2002) could then explain the existence of stars in these diffuse regions where no star formation activity is expected. Potentially, stellar migrations could also explain the plateau or upturn we observe in the metallicity gradient of NGC 300. Roškar et al. (2008) reproduce a similar change in trend of the stellar age with radius. However, scattering over such large distances (5 − 8 kpc) would require spiral waves much stronger than those considered by Sellwood & Binney (2002).

The accretion scenario is illustrated in Fig. 3. In this picture, the central spheroid is assembled first in the early Universe ($z > 4$). The first generations of stars that form establish a steep abundance gradient in the spheroid (1), reaching the metallicities of around solar. This is supported by solar metallicities seen in QSOs at high redshifts (Hamann & Ferland 1999, and references therein) and steep abundance gradients observed in the bulge of the Galaxy (e.g. Zoccali et al. 2008). The disk component is established at $z \sim 2$ (Wolfe et al. 2005). The initial disk is metal-poor and with uniform abundance (2). As

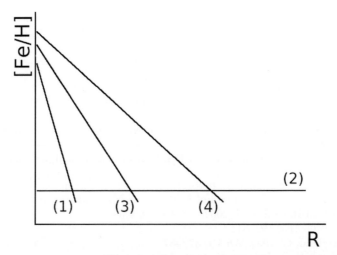

Figure 3. The illustration of the evolution of abundance gradients in disks. (1) Central spheroid is formed at high redshift and intense star formation establishes the steep metallicity gradient. (2) The metal-poor and uniform-metallicity disk is assembled at $z \sim 2$. (3) Star formation sweeps progressively larger radii and enriches the initially low-metallicity regions. (4) As a consequence, the abundance gradient continues to flatten.

the galaxy is built in the inside-out fashion, the extent of star formation migrates towards the outer disk and the parts of the disk that previously had pristine abundances are now forming metals (3). As a consequence, regions of the disk at progressively larger radii become enriched and abundance gradient flattens with time (4). Observations of a variety of chemical gradients in PNe, open clusters, cepheids, OB associations and HII regions (W. Maciel, this volume, and references therein) have shown that older objects exhibit steeper abundance gradients resulting in chemical gradients that become flatter with time. We note however that while some of the chemical evolution models that incorporate inside-out growth of galactic disks were able to reproduce this behavior (e.g. Molla et al. 1997, Hou et al. 2000), others predict steepening of gradient with time (Chiappini et al. 2001). This demonstrates the sensitivity of abundance gradients on adopted form of radial dependence for the star formation and gas infall.

A simple consequence of this picture is that there exist a radius in a galactic disk at which a transition occurs between the star forming disk with a negative abundance gradient and a uniform-abundance initial disk, as observed here in the case of NGC 300, as well as in the disks of the Galaxy (Carney et al. 2005, Yong et al. 2005, 2006) and M31 (Worthey et al. 2005). This is supported by enhanced [α/Fe] ratios and low [Fe/H] both in external gas locally (Collins et al. 2003, 2007) and in the intergalactic medium at high redshift (Prochaska et al. 2003, Dessauges-Zavadsky et al. 2004, 2006).

The accretion scenario suggests that stars today were born roughly in situ, whereas the radial mixing scenario implies that stars largely migrate away from where they were formed. Both of these have consequences for future studies of disk formation.

References

Barker, M. K., Sarajedini, A., Geisler, D., Harding, P., & Schommer, R. 2007, *AJ*, 133, 1125
Bland-Hawthorn, J., Vlajić, M., Freeman, K. C., & Draine, B. T. 2005, *ApJ*, 629, 239
Bullock, J. S. & Johnston, K. V. 2005, *ApJ*, 635, 931
Carignan, C. 1985, *ApJS*, 58, 107
Carney, B. W., Yong, D., Teixera de Almeida, M. L., & Seitzer, P. 2005, *AJ*, 130, 1111

Carollo, D., Beers, T. C., Lee, Y. S., Chiba, M., Norris, J. E., Wilhelm, R., Sivarani, T., Marsteller, B., Munn, J. A., Bailer-Jones, C. A. L., Fiorentin, P. R., & York, D. G. 2007, *Nature*, 450, 1020
Chapman, S. C., Ibata, R., Lewis, G. F., Ferguson, A. M. N., Irwin, M., McConnachie, A., & Tanvir, N. 2006, *ApJ*, 653, 255
Chiappini, C., Matteucci, F., & Romano, D. 2001, *ApJ*, 554, 1044
Collins, J. A., Shull, J. M., & Giroux, M. L. 2003, *ApJ*, 585, 336
—. 2007, *ApJ*, 657, 271
Dessauges-Zavadsky, M., Calura, F., Prochaska, J. X., D'Odorico, S., & Matteucci, F. 2004, *A&A*, 416, 79
Dessauges-Zavadsky, M., Prochaska, J. X., D'Odorico, S., Calura, F., & Matteucci, F. 2006, *A&A*, 445, 93
Ferguson, A., Irwin, M., Chapman, S., Ibata, R., Lewis, G., & Tanvir, N. 2007, in: R. S. de Jong (ed.) *Island Universes, Structure and Evolution of Disk Galaxies*, p. 239
Freeman, K. & Bland-Hawthorn, J. 2002, *ARAA*, 40, 487
Freeman, K. C. 1970, *ApJ*, 160, 811
Hamann, F. & Ferland, G. 1999, *ARAA*, 37, 487
Hou, J. L., Prantzos, N., & Boissier, S. 2000, *A&A*, 362, 921
Ibata, R. A., Irwin, M. J., Lewis, G. F., Ferguson, A. M. N., & Tanvir, N. 2003, *MNRAS* (Letters), 340, 21
Irwin, M. J., Ferguson, A. M. N., Ibata, R. A., Lewis, G. F., & Tanvir, N. R., 2005, *ApJ* (Letters), 628, 105
Kim, S. C., Sung, H., Park, H. S., & Sung, E.-C. 2004, *Chinese Journal of Astronomy and Astrophysics*, 4, 299
Koch, A., Rich, R. M., Reitzel, D. B., Mori, M., Loh, Y.-S., Ibata, R., Martin, N., Chapman, S. C., Ostheimer, J., Majewski, S. R., & Grebel, E. K. 2007, *Astronomische Nachrichten*, 328, 653
McConnachie, A. W., Chapman, S. C., Ibata, R. A., Ferguson, A. M. N., Irwin, M. J., Lewis, G. F., Tanvir, N. R., & Martin, N. 2006, *ApJ* (Letters), 647, 25
Metcalfe, N., Shanks, T., Campos, A., McCracken, H. J., & Fong, R. 2001, *MNRAS*, 323, 795
Molla, M., Ferrini, F., & Diaz, A. I. 1997, *ApJ*, 475, 519
Prochaska, J. X., Gawiser, E., Wolfe, A. M., Castro, S., & Djorgovski, S. G. 2003, *ApJ* (Letters), 595, 9
Roškar, R., Debattista, V. P., Stinson, G. S., Quinn, T. R., Kaufmann, T., & Wadsley, J. 2008, *ApJ* (Letters), 675, 65
Sellwood, J. A. & Binney, J. J. 2002, *MNRAS*, 336, 785
Sellwood, J. A. & Preto, M. 2002, in: E. Athanassoula, A. Bosma, & R. Mujica, (eds.), *Astronomical Society of the Pacific Conference Series, Vol. 275, Disks of Galaxies: Kinematics, Dynamics and Perturbations*, p. 281
Stetson, P. B. 1987, *PASP*, 99, 191
van der Kruit, P. C. 1979, *A&A*, 38, 15
VandenBerg, D. A., Bergbusch, P. A., & Dowler, P. D. 2006, *ApJS*, 162, 375
Vlajić, M., Bland-Hawthorn, J., & Freeman, K. C. 2008, *ApJ submitted*
Wolfe, A. M., Gawiser, E., & Prochaska, J. X. 2005, *ARAA*, 43, 861
Worthey, G., España, A., MacArthur, L. A., & Courteau, S. 2005, *ApJ*, 631, 820
Yong, D., Carney, B. W., & Teixera de Almeida, M. L. 2005, *AJ*, 130, 597
Yong, D., Carney, B. W., Teixera de Almeida, M. L., & Pohl, B. L. 2006, *AJ*, 131, 2256
Zoccali, M., Hill, V., Lecureur, A., Barbuy, B., Renzini, A., Minniti, D., Gomez, A., & Ortolani, S. 2008, *A&A*, 486, 177

The Galactic Disk-Halo Transition – Evidence from Stellar Abundances†

Poul Erik Nissen[1] and William J. Schuster[2]

[1] Department of Physics and Astronomy, University of Aarhus, DK-8000 Aarhus C, Denmark
email: pen@phys.au.dk

[2] Observatorio Astronómico Nacional, Universidad Nacional Autónoma de México, Apartado Postal 877, Ensenada, BC, 22800 México
email: schuster@astrosen.unam.mx

Abstract. New information on the relations between the Galactic disks, the halo, and satellite galaxies is being obtained from elemental abundances of stars having metallicities in the range $-1.5 <$ [Fe/H] < -0.5. The first results for a sample of 26 halo stars and 13 thick-disk stars observed with the ESO VLT/UVES spectrograph are presented. The halo stars fall in two distinct groups: one group (9 stars) has $[\alpha/\text{Fe}] = 0.30 \pm 0.03$ like the thick-disk stars. The other group (17 stars) shows a clearly deviating trend ranging from $[\alpha/\text{Fe}] = 0.20$ at [Fe/H] $= -1.3$ to $[\alpha/\text{Fe}] = 0.08$ at [Fe/H] $= -0.8$. The kinematics of the stars are discussed and the abundance ratios Na/Fe, Ni/Fe, Cu/Fe and Ba/Y are applied to see if the "low-alpha" stars are connected to the thin disk or to Milky Way satellite galaxies. Furthermore, we compare our data with simulations of chemical abundance distributions in hierarchically formed stellar halos in a ΛCDM Universe.

Keywords. stars: abundances, Galaxy: disk, Galaxy: halo, Galaxy: evolution

1. Introduction

As discussed by Venn *et al.* (2004), the chemical signatures of stars in Milky Way dwarf spheroidal (dSph) galaxies are different from the majority of Galactic thin-disk, thick-disk and halo stars. The dSph stars have lower values of α/Fe, where α refers to the abundance of typical alpha-capture elements like Mg, Si, Ca and Ti, than the Galactic stars. Furthermore, the Ba/Y ratio is significantly higher in the dSph stars. Thus, Venn *et al.* conclude that the main components of our Galaxy have not been formed through the merging of dwarf galaxies similar to present-day dSphs. This could be seen as an argument against hierarchical structure formation as predicted in Cold Dark Matter cosmologies. Font *et al.* (2006) have, however, made ΛCDM simulations of the chemical abundance distributions in the the Galactic halo, which show that early accreted dwarf galaxies have chemical properties different from those of present-day satellite galaxies. For the metal-poor part of the Galactic halo, they predict enhanced values of $[\alpha/\text{Fe}]$ corresponding to the ratio calculated for Type II supernovae, but for the metal-rich end of the halo a decline of $[\alpha/\text{Fe}]$ is predicted due to an increasing contribution of iron from Type Ia SNe.

In order to test these recent simulations, we have started a survey of abundance ratios of about 100 halo stars having $-1.5 <$ [Fe/H] < -0.5. This metallicity range also contains thin-disk and thick-disk stars, and hence a comparison of abundances may give new information on the relations between the various Galactic populations.

† Based on observations made with the ESO Telescopes at the Paranal Observatory under programs 65.L-0507, 67.D-0439, 68.D-0094, 68.B-0475, 69.D-0679, 70.D-0474, and 76.B-0133.

The survey is a continuation of a more limited study by Nissen & Schuster (1997), who determined abundances of 13 halo stars and 16 thick disk stars in the metallicity range $-1.3 <$ [Fe/H] < -0.4. Interestingly, 8 of the halo stars turned out to have significantly lower values of [α/Fe]† than the other halo stars. Considering the small number of halo stars studied it is, however, unclear from the work of Nissen & Schuster if the distribution of [α/Fe] in the halo is continuous or bimodal. One of the aims of the present study is to answer this question by observing a larger sample of stars.

2. Selection of stars and observations

The stars have been selected from Schuster et al. (2006), which contains $uvby$-β photometry and complete kinematic data for 1533 high-velocity and metal-poor stars. In order to ensure that a selected star belongs to the halo population, we required that its total space velocity with respect to the Local Standard of Rest (LSR), V_{total}, should be larger than $180 \,\text{km s}^{-1}$. Furthermore, we used the Strömgren indices, $b - y$, m_1 and c_1, to select dwarfs and subgiants with temperatures $5100 < T_{\text{eff}} < 6200 \,\text{K}$ and metallicities $-1.5 <$ [Fe/H] < -0.5. When selecting the stars from the bright end of the Schuster et al. catalogue, these criteria result in a limiting magnitude of $V = 11.1$ for stars that could be reached with the Nordic Optical Telescope (NOT) on La Palma.

The first observations with NOT and its FIbre fed Echelle Spectrograph (FIES) were carried out in May 2008 and resulted in high resolution ($R \sim 45\,000$) spectra with $S/N \sim 150$ for 30 halo stars. The abundance analysis of these data is, however, not yet finished. Instead, we are presenting here results for 26 halo stars that fulfil our selection criteria, and for which spectra obtained with the VLT/UVES spectrograph are available in the ESO science archive. These spectra have $R \sim 60\,000$ and very high signal-to-noise ratios, $S/N > 300$, in the 4800–6500 Å region that we are using. In addition, 13 thick-disk stars with UVES spectra are included. Their atmospheric parameters fall in the same ranges as the halo stars, but their total space velocities relative to the LSR are typical for thick-disk stars ($50 < V_{\text{total}} < 150 \,\text{km s}^{-1}$).

3. Abundance analysis

Abundances of the elements Mg, Si, Na, Ca, Ti, Cr, Ni, Cu, Y and Ba are derived from equivalent widths of weak to medium-strong lines ($5 < EW < 80\,\text{mÅ}$). A grid of 1D MARCS model-atmospheres is used to derive the abundances, and the effect on the electron pressure from variations of [α/Fe] is taken into account. Like Nissen & Schuster (1997), the analysis is performed in a differential way with respect to two bright thick-disk stars, HD 22879 and HD 76932, that are known to have [α/Fe] close to $+0.3$ dex according to several studies (e.g. Reddy et al. 2006). For nearby stars that appear to be unreddened, as judged from the absence of interstellar NaD lines, T_{eff} is derived from the $b - y$ and $V - K$ color indices, and the gravity parameter, $\log g$, is estimated via the Hipparcos parallax. For the more distant and reddened stars, T_{eff} is determined from the excitation balance of Fe I lines, and $\log g$ is derived from the requirement that the difference in Fe abundances derived from Fe II and Fe I lines should be the same as in HD 22879 and HD 76932. The adopted metallicity, [Fe/H], is that derived from Fe II lines.

In order to minimize the dependence of abundance ratios on possible errors in the atmospheric parameters, the abundance ratio of two elements is derived from lines belonging to the same ionization stage, e.g. [Mg/Fe] from Mg I and Fe I lines or [Ba/Y] from

† Defined as [α/Fe] $= \frac{1}{4}$ ([Mg/Fe] + [Si/Fe] + [Ca/Fe] + [Ti/Fe])

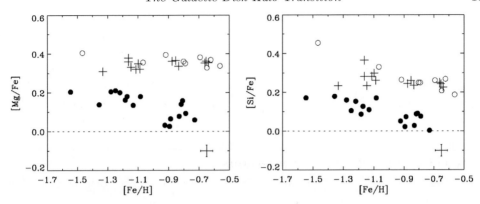

Figure 1. [Mg/Fe] and [Si/Fe] vs. [Fe/H] for the sample of stars with VLT/UVES spectra. Crosses: Thick-disk stars; Open circles: "High-alpha" halo stars; Filled circles: "Low-alpha" halo stars. Typical error bars for the data are shown in the lower right corners of the figures.

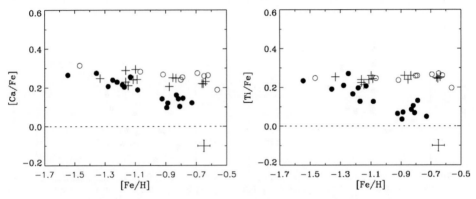

Figure 2. [Ca/Fe] and [Ti/Fe] vs. [Fe/H]. Same symbols as in Fig. 1.

Ba II and Y II lines. Local thermodynamic equilibrium (LTE) is assumed, but due to the limited range of the parameters of our stars, non-LTE effects on the derived *differential* abundances are expected to be small.

4. Results and discussion

Figures 1 and 2 show the derived abundances of the four alpha-capture elements, Mg, Si, Ca and Ti, relative to Fe. As seen, the halo stars fall in two distinct groups: *i*) the "high-alpha" stars that have very near the same [α/Fe] as the thick-disk stars and a remarkably small scatter, ±0.03 dex, around the "plateau" value of [α/Fe] and *ii*) the "low-alpha" stars that show a decreasing trend of [α/Fe] as a function of increasing [Fe/H]. In the cases of Mg and Si the separation of the two groups of halo stars is seen for the whole range $-1.5 <$ [Fe/H] < -0.7 with a maximum separation of ~ 0.25 dex in [Mg/Fe] and ~ 0.20 dex in [Si/Fe]. For Ca and Ti the two groups tend to merge at [Fe/H] $\simeq -1.2$, and the maximum separation is only about 0.12 dex for Ca and about 0.17 dex for Ti.

A possible explanation of these trends is that the "high-alpha" and the thick-disk stars have been formed in regions with a relative high star-formation rate such that only Type II SNe have contributed to the chemical evolution up to [Fe/H] ~ -0.5. The "low-alpha" stars, on the other hand, come from regions with a slower star-formation rate, where

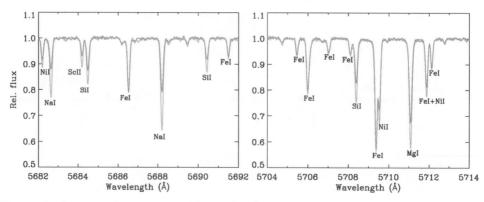

Figure 3. Spectra of two stars with nearly the same atmospheric parameters $T_{\rm eff}$, $\log g$ and [Fe/H]. The spectrum of the "low-alpha" star CD$-45\,3283$ ($T_{\rm eff} = 5603$ K, $\log g = 4.57$, [Fe/H] $= -0.89$, [α/Fe] $= 0.08$) is shown with a thick (red) line, and that of the "high-alpha" star G 159-50 ($T_{\rm eff} = 5648$ K, $\log g = 4.39$, [Fe/H] $= -0.92$, [α/Fe] $= 0.29$) with a lighter (green) line.

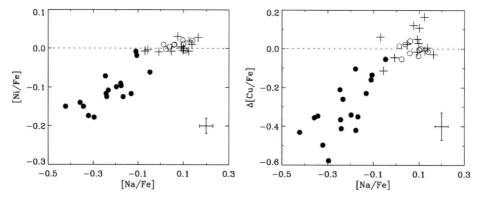

Figure 4. [Ni/Fe] and [Cu/Fe] vs. [Na/Fe] with the same symbols as in Fig. 1. In the case of Cu the deviation of [Cu/Fe] from the relation defined by the thick-disk stars is plotted.

Type Ia SNe have started to contribute with iron at a metallicity [Fe/H] $\simeq -1.5$. The reason for the smaller separation in [Ca/Fe] compared to [Mg/Fe], may be that Mg is almost exclusively produced in Type II SNe, whereas about 25% of Ca originates in Type Ia SNe according to the chemical evolution models of Tsujimoto *et al.* (1995).

The abundance differences between the two halo groups can be seen directly from the observed spectra. Fig. 3 shows two spectral regions for a "high-alpha" and a "low-alpha" star with similar $T_{\rm eff}$, $\log g$ and [Fe/H] values. As seen, the Fe I lines of the two stars have nearly the same strength, whereas the Mg and Si lines are weaker in the "low-alpha" star. The same is the case for Na, Ni and Cu lines and, as shown in Fig. 4, [Ni/Fe] and [Cu/Fe] are well correlated with [Na/Fe]. The reason for these correlations may be that the yields of Na, Ni and Cu depend on the neutron excess in supernovae and that this excess is affected by the α/Fe ratio.

The kinematics of the stars are shown in Fig. 5. As seen from this "Toomre" diagram, both groups of halo stars have an average Galactic rotation velocity close to zero in contrast to the thick-disk stars. Furthermore, the velocity dispersion for the "low-alpha" group is higher than that of the "high-alpha" group. This larger dispersion is mainly caused by larger U velocities, which means that the "low-alpha" stars tend to move on

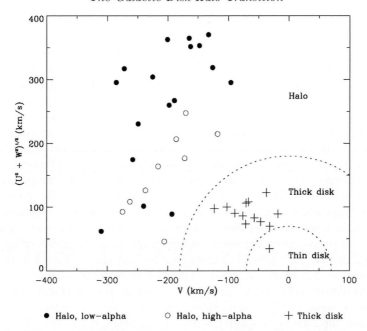

Figure 5. The "Toomre" diagram for the three groups of stars discussed in the present paper. Regions, where stars are most likely to belong to the halo, the thick disk and the thin disk, respectively (see Venn et al. 2004, Fig. 1), are separated by dashed circles corresponding to $V_{\rm total} = 180$ and $70\,{\rm km\,s^{-1}}$.

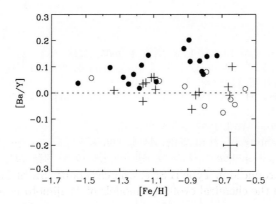

Figure 6. [Ba/Y] vs. [Fe/H] with the same symbols as in Fig. 1

higher energy radial orbits. A larger sample of halo stars is, however, needed before any definitive conclusions about the kinematics of the two groups can be made.

The distribution of the "low-alpha" stars in Figs. 1 and 2 shows a smooth continuation of the α/Fe trend for thin-disk stars, which according to several studies (Reddy et al. 2003, Bensby et al. 2005) stops with $[\alpha/{\rm Fe}] \simeq 0.1$ at $[{\rm Fe/H}] \simeq -0.7$. The thin-disk stars do not share, however, the low [Na/Fe], [Ni/Fe] and [Cu/Fe] values that are being found for the "low-alpha" group. In addition, the "low-alpha" stars have enhanced Ba/Y ratios as shown in Fig. 6. In all these abundance deviations, the "low-alpha" stars resemble stars in dSph galaxies (Venn et al. 2004, Sbordone et al. 2007, Koch et al. 2008) and in the LMC (Pompéia *et al.* 2007), although these satellite galaxies tend to have larger abundance offsets from disk stars than the "low-alpha" stars.

The trend of [Mg/Fe] for the "low-alpha" halo stars in Fig. 1 agrees remarkably well with the trend predicted by Font *et al.* (2006, Fig. 9) from simulations of abundance distributions for a hierarchically formed stellar halo in a ΛCDM Universe. The fact that the "low-alpha" stars do not have quite as low [α/Fe] values as present-day satellite galaxies also agrees with their predictions. On the other hand, the "high-alpha" halo stars are not predicted from the simulations. The existence of this group suggests that the formation of the Galactic halo is more complicated than predicted from the ΛCDM simulations. It remains to be seen if this group of "high-alpha" halo stars can be explained as due to the merger of an exceptionally large satellite or if it is a dissipative component of the Galaxy as suggested by Gratton *et al.* (2003). In the latter case one would expect the group of "high-alpha" stars to have some net rotation. Another problem is that the "low-alpha" halo stars tend to move on high-energy radial orbits that plunge into the Galactic central regions from the outer regions of the halo. As discussed by Gilmore & Wyse (1998), this requires possible parent satellite galaxies to have a very high mean density to provide "low-alpha" stars on orbits with such small perigalactic distances.

In order to obtain more insight into these problems, the elemental abundances of a larger sample of metal-rich halo stars should be studied so that better knowledge of the kinematics and the relative frequency of halo stars belonging to the "high-alpha" and "low-alpha" groups can be obtained. As mentioned earlier, such data is being obtained with the Nordic Optical Telescope.

Acknowledgement. This work has been financially supported from CONACyT project 49434-F.

References

Bensby, T., Feltzing, S., Lundström, I., & Ilyin, I. 2005, *A&A*, 433, 185
Font, A. S., Johnston, K. V., Bullock, J. S., & Robertson, B. E. 2006, *ApJ*, 638, 585
Gilmore, G. & Wyse, R. F. G. 1998, *AJ*, 116, 748
Gratton, R. G., Caretta, E., Desidera, S., *et al.* 2003, *A&A*, 406, 131
Koch, A., Grebel, E. K., Gilmore, G. F. *et al.* 2008, *AJ*, 135, 1580
Nissen, P. E. & Schuster, W. J. 1997, *A&A*, 326, 751
Pompéia, L., Hill, V., Spite, M. *et al.* 2008, *A&A*, 480, 379
Reddy, B. E., Tomkin, J., Lambert, D. L., & Allende Prieto, C. 2003, *MNRAS*, 340, 304
Reddy, B. E., Lambert, D. L., & Allende Prieto, C. 2006, *MNRAS*, 367, 1329
Sbordone, L., Bonifacio, P., Buonanno, R. *et al.* 2007, *A&A*, 465, 815
Schuster, W. J., Moitinho, A., Márquez, A., Parrao, L., & Covarrubias, E. 2006, *A&A*, 445, 939
Tsujimoto, T., Nomoto, K., Yoshii, Y. *et al.* 1995, *MNRAS*, 277, 945
Venn, K. A., Irwin, M., Shetrone, M. D. *et al.* 2004, *AJ*, 128, 1177

Session 2: Origin, Structure, and Chemical Evolution of Disks

Session chairs: Julianne Dalcanton, Burkhard Fuchs, Bacham Reddy and
Eileen Friel

Max Pettini, Ken Freeman, Burkhard Fuchs and Joss Bland-Hawthorn posing outside the auditorium.

Poul Erik Nissen presenting his abundance work.

Abundance Gradients and Substructures in Disks

K. C. Freeman

Research School of Astronomy & Astrophysics
Mount Stromlo Observatory
The Australian National University

Abstract. I will first discuss abundance gradients in the Milky Way and nearby disk galaxies, then the problem of substructures in the Galactic Disk, and finally some new opportunities for investigating substructure in disks.

Keywords. abundances, star clusters, disks, kinematics and dynamics

1. Abundance Gradients

In this section, I will discuss abundance gradients as observed in the Milky Way, M31, and the two similar late-type disk systems M33 and NGC 300. But first some comments on the truncation of disks.

The disk of M33 shows a classical truncation behaviour: out to a radius of about 35 arcmin, the surface brightness profile follows closely the usual exponential decline with radius. Then, at a surface brightness of about 24 I mag arcsec^{-2}, the slope steepens and this steeper exponential continues to the limit of the measurable surface brightness (about 30 mag arcsec^{-2}) (Ferguson *et al.* 2007). This kind of truncation was long believed to be ubiquitous, but it is now known that the surface brightness distribution in the outer regions of disks can take other forms. For example NGC 300, which has a similar appearance and absolute magnitude to M33, continues its exponential decline without any change in slope for about 10 scalelengths to the limit of the surface photometry, again at about 30 mag arcsec^{-2} (Bland-Hawthorn *et al.* 2005). Erwin *et al.* (2005) identified three kinds of outer disk morphology: those which continue as a single exponential (type I, like NGC 300), those which steepen at large radii (type II, the classical truncations, like M33), and those in which the surface brightness profile in the outer disk flattens in slope at large radii (type III).

The classical type II truncations are not understood. Various explanations have been proposed, including angular momentum redistribution by bars and spiral waves, the effect of the star formation threshold at low surface density, the hierarchical accretion process, and bombardment by dark matter subhalos (de Jong *et al.* 2007). Roškar *et al.* (2008) made an SPH simulation of disk formation from cooling gas in an isolated dark halo, including star formation and feedback, which shows the type II break in the surface brightness profile. The break is seeded by a rapid radial decrease in the surface density of cool gas: the break becomes visible within 1 Gyr and gradually moves outwards as the disk grows. The outer (steeper) exponential is fed by secularly redistributed stars from the inner regions, *via* transient disturbances and the Sellwood & Binney (2002) orbit-swapping process, so its stars are relatively old (see Fig. 1).

The Galactic disk shows an abundance gradient, seen very clearly in the abundances of the Galactic cepheids (Luck *et al.* 2006). The mean abundance of these relatively young

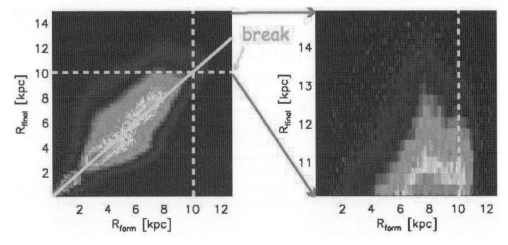

Figure 1. The redistribution of disk stars by the Sellwood & Binney (2002) mechanism: the density distribution of stars in the plane of formation radius R_{form} against final radius R_{final}. The label "break" shows the location of the break in the final surface brightness profile. The right panel shows an expanded view of the region $R_{final} > 10$ kpc. Adapted from Roškar et al. (2008).

stars falls from [Fe/H] = +0.2 at radius R = 5 kpc to about −0.5 at 15 kpc. It is not a simple axisymmetric gradient, however: near the sun, a tongue of relatively metal-rich stars extends out into the $l = 90-180°$ quadrant, with a corresponding tongue of metal-poor stars extending inwards in the third quadrant (see Fig. 2). This asymmetry may be related to noncircular motions induced by the spiral structure.

The Galactic abundance gradient is seen also for older star clusters and red giants (Yong & Carney 2005, Carney & Yong 2005). For the open clusters, with ages between about 1 and 5 Gyr, the abundance gradient appears to bottom out at a radius of about 12 kpc and and abundance of −0.5 (M31 shows a similar outer plateau in its abundance gradient, beginning at a radius of 15 kpc and at the same abundance level: Worthey et al. 2004). These older stars in the outer Galactic disk are α-enhanced, with $[\alpha/Fe] = +0.2$, indicating fairly rapid chemical evolution of the gas in the outer disk from which they formed (unlike the stars of similar age in the solar neighborhood). Comparing the abundance gradients and α-enhancement for these older stars with those for the younger cepheids, it appears that the abundance gradient and the $[\alpha/Fe]$ gradients have flattened with time, tending towards the solar values.

If the outer gas has evolved *in situ*, then the α-enrichment of the 5-Gyr old clusters in the outer galaxy suggests that their chemical evolution was quick: i.e. star formation in the outer disk has been going on for only about 6 Gyr. This is an internal version of down-sizing, depending on surface density rather than total mass. Not only is the star formation timescale longer in the outer Galaxy (Chiappini et al. 2001) but star formation and chemical evolution started later.

A recent study of red giants in the outer regions of NGC 300 (Vlajić et al. 2008) shows an apparent reversal of the abundance gradient in the outer regions. This reversal could be associated with gas accretion. Alternatively, it may result from the orbit-swapping phenomenon mentioned above, with the outermost disk populated by stars scattered out from the more metal-rich inner regions. The similar galaxy M33 shows a clear abundance gradient in its inner exponential component: this gradient appears to flatten in the outer disk where its surface brightness gradient has steepened (Barker et al. 2007). As in the

Milky Way, there is evidence that the abundance gradient in M33 has flattened with time: the gradient appears to be flatter for the HII regions and young stars than for the somewhat older planetary nebulae and red giants (Magrini et al. 2007).

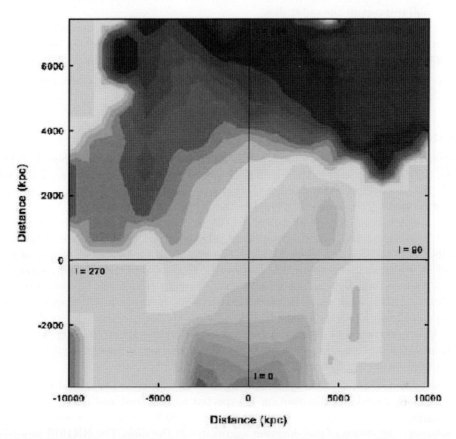

Figure 2. The abundance distribution of cepheids on the galactic plane. The origin is at the sun and the y-axis points towards the anticenter. In the mean, the abundance decreases with radius, but a tongue of metal-richer cepheids extends out into the second quadrant. Adapted from Luck et al. (2006).

In the vertical direction, photometric abundances for a very large sample of Galactic FG stars from the SDSS show that the abundance distribution in the Galactic disk and thick disk is stratified parallel to the Galactic plane for heights $|Z| > 1$ kpc. The Monoceros stream appears as a metal-rich feature extended in Z at a radius of about 15 kpc (Ivezić et al. 2008). The vertical abundance gradient from SDSS agrees well with an independent estimate from thin disk clump giants (Soubiran et al. 2008).

Kinematics for a subsample of the SDSS stars at a mean height $Z \sim 1.1$ kpc show that the azimuthal lag velocity $\langle v_\phi \rangle$ is constant at about 40 km s^{-1} and is uncorrelated with abundance for stars in the abundance range $-0.4 > $ [Fe/H] > -1 (Ivezić et al. 2008). This is unexpected; in the solar neighborhood the correlation of disk kinematics with abundance has been long known: e.g. Strömgren (1986).

To summarize this section on abundance gradients:

• The outer disks of M31, M33, NGC 300 and the Galaxy include a component that is at least several Gyr old

• The abundance gradients in the outer disks of M31 and the Galaxy bottom out at large R, at [Fe/H]= −0.5. The gradient in M33 also flattens at large R and the gradient in NGC 300 reverses in the outer disk

• The abundance gradients in the Galaxy and M33 appear to flatten with time

• The older stars of the outer Galactic disk are α-enhanced, indicating rapid chemical evolution. This α-enhancement is less for the younger stars in the outer disk.

What processes are important for determining the properties of outer disks: gas accretion, tidal effects, orbit swapping ...?

2. Substructures in the Disk

Substructures in the disk are mostly difficult to see in configuration space. Exceptions are the substructures seen in the SDSS data (Jurić et al. 2008, Ivezić et al. 2008) which include the Monoceros feature and two other less promonent disk substructures seen in maps of residuals from a smooth model of the SDSS star counts.

Most disk substructures are identified in (U,V) velocity space as stellar *moving groups*: (U and V are the stellar velocity component relative to the local standard of rest (LSR) towards $l = 0$ and $l = 90°$ respectively). The groups are made up of nearby stars, all around us, which share common U,V motions. The concept of these moving groups goes back to Kapteyn and Eggen. The major nearby moving groups can be seen very clearly in Dehnen's (1999) analysis of Hipparcos motions for nearby stars; it shows the Sirius, Hyades and Pleiades moving groups which dominate the local stellar velocity distribution at low U,V velocities, and also the higher velocity Hercules group (see Fig. 3).

Although the moving groups are not yet well understood, it appears that they are not all produced in the same way:

• some are associated with dynamical resonances (Galactic bar, spiral structure) so their chemical abundance distribution is expected to be broad and fairly typical of the nearby thin and thick disks. The Hercules group is a likely example of a resonance group.

• some are the debris of star-forming aggregates in the disk. The HR1614 group is an example (see below). Such groups may be chemically homogeneous and are potentially very useful for reconstructing the star forming events which built up the Galactic disk.

• Others may be debris of infalling objects, as seen in CDM simulations like those of Abadi et al. (2003). Navarro et al. (2004) proposed that the thick disk Arcturus Group may be an example of this kind of stellar moving group.

The HR1614 group is an example of a group which appears to be the dispersed debris of an old star forming event. Its age is about 2 Gyr and its metallicity [Fe/H] = +0.2. It was studied by Feltzing & Holmberg (2000) who argued for its reality as a relic group. De Silva et al. (2007) measured very precise chemical abundances for many elements in the stars of the HR1614 group, and found a very small spread in the abundances. This is in marked contrast to the stars of the Hercules group studied by Bensby et al. (2007) who found that the Hercules stars appear chemically to be a fairly typical sample of field stars of the thin and thick disk. This supports the view that the Hercules group is dynamically associated with a resonance that affects a wide range of disk stars.

The nature of the Hyades-Sirius complex, which is the dominant feature of the U,V distribution for the nearby stars, remains uncertain. The ages of these stars appears to be less than 1 Gyr: is this substructure the dispersing remnant of a star-forming event,

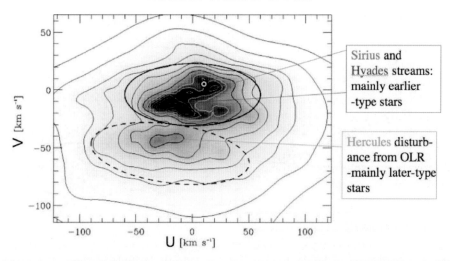

Figure 3. (U,V) distribution of Hipparcos stars, showing the major stellar moving groups which dominate the (U,V) distribution for the nearby stars. Adapted from Dehnen (1999).

or another example of a resonance feature. Quillen & Minchev (2005) showed how Lindblad resonances with spiral density waves can generate velocity space structures that have a morphology at least qualitatively like that of the Hyades-Sirius complex. If the Hyades-Sirius groups are dispersing aggregates passing through the solar neighborhood, then their stars should lie on Lindblad dispersion orbits and therefore on well-defined tracks in the (longitude - radial velocity) plane. Wilson (1988) made a high resolution spectroscopic study of this kind for a large sample of K giants near the galactic plane, and found that the dispersion orbit loci are seen but only in a narrow [Fe/H] range (similar to the abundance of the Hyades cluster), as would be expected for dispersing star forming events (see Fig. 4). On the other hand, Famaey et al. (2008) argue that the stars of the Hyades-Sirius complex include stars with a wide range of ages and are therefore most likely a resonance phenomenon.

To summarize this section on moving groups:

• The HR1614 moving group (age 2 Gyr) appears to be the relic of an old dispersed star-forming event. This is good: we can use such groups to reconstruct the galactic disk. But how has this moving group maintained its dynamical identity for 2 Gyr against the effects of disk heating?

• The Hercules group is chemically inhomogeneous and is probably just a sample of disk stars swept together in velocity space by local dynamical resonances. Mary Williams' new data on the thick disk Arcturus group (Williams, 2008) indicates that it too is probably a resonance phenomenon.

• Many dispersed aggregates will not be recognizable kinematically, because their dynamical identity has been lost. Many kinematical structures are not dispersed aggregates: they are manifestations of resonances with the gravitational fields of bar and spiral wave.

• New resonant substructures are being found in (U,V) space by several authors (eg Fuchs et al., Williams et al.). It seems likely that most of the kinematically defined substructures found in the future will turn out to be associated with resonances. While these are dynamically interesting, they are not much use for probing fossil debris of ancient star forming events. Chemical signatures will probably be more useful for finding fossil debris.

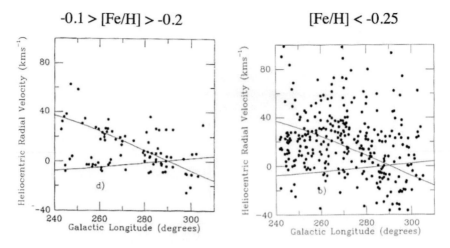

Figure 4. Distribution of bright K giants at low galactic latitude in the radial velocity vs galactic longitude plane. The solid curves show the loci of predicted Lindblad dispersion orbits calculated for the observed (U,V) motions of the Hyades and Sirius moving groups. Stars in the abundance range $-0.1 > [Fe/H] > -0.2$ appear to be clustered around the dispersion orbit loci. Stars with $[Fe/H] < -0.25$ show no such clustering tendency. Adapted from Wilson (1988).

3. New Opportunities for Studying Substructure

- The detailed abundance pattern of the stars in a dispersed star-forming aggregate reflect the chemical evolution of the gas from which the stars formed. These element abundance patterns can be used in chemical tagging, to detect fossil substructure that is no longer dynamically recognizable because of disk heating and orbit-swapping (Freeman & Bland-Hawthorn 2002).
- the hydrostatics of the HI provide a way to detect extended gravitating dark substructures in the disk (Kalberla et al. 2007, O'Brien et al. 2008).

Chemical Tagging: For chemical tagging to work, a few conditions need to be satisfied. Stars must form in large aggregates, and this is believed to be true. The aggregates must be chemically homogeneous, and they must have unique chemical signatures, defined by several elements which do not vary in lockstep from one aggregate to another. Also, a sufficient spread in abundance from aggregate to aggregate is needed, so that their chemical signatures can be distinguished within the accuracy achievable observationally (about 0.05 dex differentially in the element abundances). Testing these chemical conditions was the goal of Gayandhi De Silva's thesis on open cluster and moving group abundances, and they appear to be true (see Fig. 5 and her paper in this volume).

Chemical tagging requires a large high resolution stellar spectroscopic survey of about one million stars (Bland-Hawthorn & Freeman 2004). Such large surveys will be feasible in the relatively near future, with two new instruments that are planned. **WFMOS** is a proposed joint project by Subaru and Gemini. The current concept has 1000 fibers over a 1.5-degree field, feeding spectrometers with a resolution of about 30,000. This instrument would enable a large survey with a signal-to-noise ratio of about 100 per resolution element, down to $V \sim 17$. **HERMES** on the AAT has 400 fibers in a 2-degree field. The resolution is again about 30,000. A large survey is planned down to $V \sim 14$; this magnitude limit matches the fiber density to the typical Galactic stellar density. HERMES has already had its concept design review and is scheduled to begin operation in 2012.

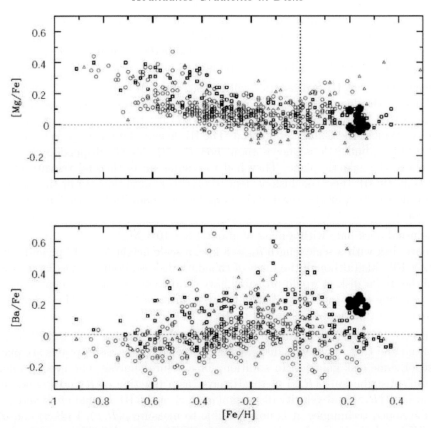

Figure 5. Chemical properties of stars in the HR1614 moving group. The upper panel shows [Mg/Fe] *vs* [Fe/H] abundances for stars of the HR1614 group (large filled circles) and for field stars (small open circles). The lower panel is similar, for [Ba/Fe]. The chemical homogeneity of the HR1614 group stars is evident. Adapted from De Silva *et al.* (2007).

WFMOS and HERMES have an outstanding synergy with GAIA, which will provide precision astrometry for about a billion stars. GAIA is at its best for the relatively bright stars which will be targeted with these two spectrometers. For example, at the V-magnitude limit of 14 planned for the HERMES survey, the expected parallax and proper motion uncertainties are $\sigma_\pi = 10$ μas and $\sigma_\mu = 10$ μas yr^{-1} respectively. These uncertainties correspond to 1% distance errors at a distance of 1 kpc, and transverse velocity errors of 0.7 km s^{-1} at a distance of 15 kpc. The combination of GAIA and HERMES will give accurate distances, element abundances and 3D velocities for most of the survey stars, and isochrone ages for many. It will be possible to construct accurate color-(absolute magnitude) diagrams for the survey stars, providing an independent check that chemically tagged groups of stars do indeed have common ages.

The Hydrostatics of the HI Layer in Disk Galaxies: The hydrostatics of the HI layer in disk galaxies provides a potentially powerful technique to measure the potential gradient $\partial\Phi/\partial z$ to complement the radial potential gradient $\partial\Phi/\partial R$ from rotation data. In combination, the two potential gradients can be used to

- estimate the shape of the dark halo (e.g. Olling & Merrifield 2000)

- settle the maximum disk controversy, because $\partial \Phi / \partial z$ in the inner disk near the galactic plane comes mainly from the stellar disk
- identify concentrations of dark matter in galactic disks, such as might be generated by the accretion of satellite galaxies in circularized orbits

It seems likely that HI hydrostatics will be used more in future, as high spatial resolution HI data with adequate signal-to-noise become available for disk galaxies. To illustrate the power of HI hydrostatics, Kalberla *et al.* (2007) used the HI flaring of the Galactic HI layer to explore the Galactic potential. The HI density data is extends to R = 40 kpc and z = 20 kpc, but little is known about how the HI velocity dispersion changes with height above the galactic plane. They had to assume an isothermal velocity dispersion for the Galactic HI; their conclusions depend on this assumption. To fit the observed HI distribution, they required several dark components, in addition to the known luminous components:

- the usual dark halo with a mass of about 1.8×10^{12} $M_\odot$
- a dark disk with a scale length $h_R = 8$ kpc, a scale height $h_z = 4$ kpc and a mass of about 2×10^{11} $M_\odot$ (although the mass of this dark disk is about 4 times larger than the mass of the stellar disk, it would probably have escaped detection via stellar kinematics).
- a dark ring at radius $R = 13-20$ kpc, with mass of 2.5×10^{10} $M_\odot$, which may be associated with the Monoceros stellar ring.

To use the hydrostatics of the HI layer, one needs to assume that the HI layer is in equilibrium and that its velocity dispersion is isotropic. These assumptions are plausible, at least for some disk galaxies. In addition, for a secure analysis, one needs to know (i) the density distribution $\rho(R, z)$ of the HI layer as a function of R and z, and (ii) how the rotation $V(R, z)$ and velocity dispersion $\sigma(R, z)$ of the HI depend on R and z. Using newly developed techniques, it is now possible to measure $\rho(R, z)$, $V(R, z)$ and $\sigma(R, z)$ for the HI in relatively nearby edge-on galaxies (O'Brien *et al.* 2009).

References

Abadi, M. *et al.* 2003, *ApJ*, 597, 21
Barker, M. *et al.* 2007. *ApJ*, 133, 1138
Bensby,T. *et al.* 2007, *ApJ*, 655, L89
Bland-Hawthorn, J. & Freeman, K. 2004, *PASA*, 21, 110
Bland-Hawthorn, J. *et al.* 2005, *ApJ*, 629, 239
Carney, B. & Yong, D. 2005, *AJ*, 130, 597
Chiappini, C. *et al.* 2001, *ApJ*, 554, 1044
Dehnen, W. 1999, *ApJ*, 524, L35
de Jong, R. *et al.* 2007, , 667, L49
De Silva, G. *et al.* 2007, *AJ*, 133, 694
Erwin, P. *et al.* 2005, *ApJ*, 626, L81
Famaey, B. *et al.* 2008, *A&A*, 483, 453
Feltzing, S. & Holmberg, J. 2000, *A&A*, 357, 153
Ferguson, A. *et al.* 2007, in "Island Universes", *Astrophysics and Space Science Proceedings*, Springer, 239
Freeman, K. & Bland-Hawthorn, J 2002, *ARAA*, 40, 487
Ivezić, Z. *et al.* 2008, astro-ph/0804.3850
Jurić, M. *et al.* 2008, *ApJ*, 673, 864
Kalberla, P. *et al.* 2007, *A&A*, 469, 511
Luck, R.E. *et al.* 2006, *AJ*, 132, 902
Magrini, L. *et al.* 2007, *A&A*, 470, 843
Navarro, K. *et al.* 2004, *ApJ*, 601, L43

O'Brien, J. et al. 2009, in preparation
Olling, J. & Merrifield, M. 2000, *MNRAS*, 311, 361
Quillen, A. & Minchev, I. 2005, *AJ*, 130, 576
Roškar, R. et al. 2008, *ApJ*, 675, L65
Sellwood, J. & Binney, J. 2002, *MNRAS*, 336, 785
Soubiran, C. et al. 2008, *A&A*, 480, 91
Strömgren, B. 1986, in "The Galaxy", NATO ASI Series, (Reidel) Vol 207. Ed
    G. Gilmore & R. Carswell, p 229
Vlajić, M. et al. 2008, in this volume
Williams, M. 2008, in this volume
Wilson, G. 1988, ANU thesis
Yong, D. & Carney, B. 2005, *AJ*

Cecilia Mateu and Jelte de Jong enjoying the welcome reception. In the background, Simon White, Jerry Sellwood and Leo Blitz are deciding the fate of the Galaxy.

Ignacio Trujillo explaining a point during his presentation.

Gayandhi De Silva fielding a question after her talk.

Mapping low-latitude stellar substructure with SEGUE photometry

Jelte T. A. de Jong[1], Brian Yanny[2], Hans-Walter Rix[1], Eric F. Bell[1] and Andrew E. Dolphin[3]

[1] Max-Planck-Institut für Astronomie, Königstuhl 17, 69117 Heidelberg, Germany
email: dejong@mpia.de

[2] Fermi National Accelerator Laboratory, P.O. Box 500, Batavia, IL 60510, United States

[3] Raytheon Corporation, 870 Winter Street, Waltham, MA 02451, United States

Abstract. Encircling the Milky Way at low latitudes, the Low Latitude Stream is a large stellar structure, the origin of which is as yet unknown. As part of the SEGUE survey, several photometric scans have been obtained that cross the Galactic plane, spread over a longitude range of 50° to 203°. These data allow a systematic study of the structure of the Galaxy at low latitudes, where the Low Latitude Stream resides. We apply colour-magnitude diagram fitting techniques to map the stellar (sub)structure in these regions, enabling the detection of overdensities with respect to smooth models. These detections can be used to distinguish between different models of the Low Latitude Stream, and help to shed light on the nature of the system.

Keywords. Galaxy: stellar content — Galaxy: structure

1. Introduction

During the past decade, the structure of the Milky Way (MW) has been mapped in unprecedented detail, owing to deep, wide-field photometric surveys, such as the Sloan Digital Sky Survey (SDSS, York *et al.* 2000) and 2MASS. Apart from an improved understanding of the overall shape and stellar populations of the MW, these data have unveiled a plethora of substructures on all scales. Two examples of large structures are the stellar overdensity towards Canis Major (CMa, Martin *et al.* 2004), and the Low Latitude Stream (LLS), or Monoceros stream, a ring-like structure that seems to encircle the MW at low latitudes (Newberg *et al.* 2002, Ibata *et al.* 2003).

The origin of these structures is as yet unclear, but several hypotheses have been put forward. CMa might be (the remnant of) an accreted dwarf galaxy (e.g. Martin *et al.* 2004, Martínez-Delgado *et al.* 2005, Bellazzini *et al.* 2006, Butler *et al.* 2007, de Jong *et al.* 2007) with the LLS being tidal debris stripped from CMa. Martin *et al.* (2005) and Peñarrubia *et al.* (2005) have shown that dynamical models of such an accretion can indeed reproduce both the CMa overdensity as well as the LLS. On the other hand, as these structures are located at very low Galactic latitudes, it cannot be ruled out that they are intrinsic to the disk itself (e.g. Momany *et al.* 2006). Another explanation for the LLS is that its stars originate in the disk, but have been pulled out of the plane by interactions with satellites (e.g. Kazantzidis *et al.* 2007, Younger *et al.* 2008).

Here we present preliminary results of the application of colour-magnitude diagram (CMD) fitting techniques to photometry at low Galactic latitudes taken as part of the SEGUE survey. These techniques, developed in de Jong *et al.* (2008), allow us to map the 3-D distribution of stars. Fig. 1 demonstrates that the resulting maps of stellar

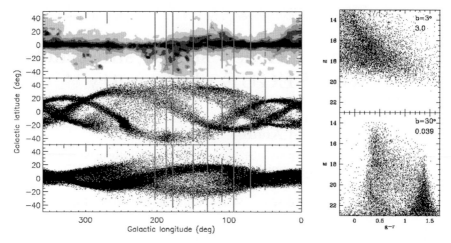

Figure 1. *Left:* Overview of imaging scans used for this analysis, indicated as vertical lines, in Galactic coordinates. The top panel shows reddening according to the dust extinction maps from Schlegel *et al.* (1998), with the gray scale indicating regions with $E(B-V) > 0.1$, 0.25, 0.5 and 1.0 mag. The middle and lower panels show the LLS models of Peñarrubia *et al.* (2005) and Martin *et al.* (2005), respectively. *Right:* CMDs from the scan at $l = 94°$. The top CMD, at $b = 3°$, is heavily extincted (E(B-V) = 3 mag), while the bottom CMD, at $b = 30°$, is typical of the data used for our CMD fitting analysis.

(sub)structure along the SEGUE imaging scans can help to distinguish between different models of the LLS and provide crucial constraints for further modelling endeavours.

2. Data and methods

SEGUE (Sloan Extension for Galactic Understanding and Exploration), an imaging and spectroscopic survey aimed at the study of the MW and its stellar populations, is one of the constituent projects of the extended SDSS survey (SDSS II). The photometric part of the SEGUE survey consists of several 2.5° wide scans going through the Galactic plane (see Fig. 1), allowing a view of the Galaxy at low latitudes. For the CMD-fitting analysis we restrict ourselves to the two most sensitive bands, g and r, which are complete to ~22nd magnitude. Although for the main survey the photometric accuracy is at least 2% down to these limits (Ivezic *et al.* 2004), in crowded regions at low latitudes the accuracy might be worse and the calibration in the data used here is preliminary. In Fig. 1 the coverage of the scans between Galactic latitudes of +50° and −50° is shown. Where possible the SEGUE scans were extended to high latitudes by extracting 2.5° wide strips from SDSS data release 5 (Adelman-McCarthy *et al.* 2007). We de-redden all data using the dust maps from (Schlegel *et al.* 1998), including the correction suggested by Bonifacio *et al.* (2000). Two CMDs from the scan at $l = 94°$ are also shown in Fig. 1. Very close to the plane, the reddening is very high and de-reddening is unable to correct this accurately. As poor de-reddening would limit the accuracy of our results, we avoid regions with reddening higher than E(B-V) = 0.2 mag. The CMD in the bottom right of Fig. 1 is representative of the the data used.

In CMD fitting, observed photometry is compared with models in order to constrain the constituent stellar populations of stellar systems. Traditionally, CMD fitting has been used mostly to determine star formation histories and age-metallicity relations of isolated objects, such as dwarf galaxies and globular clusters. We use the CMD fitting

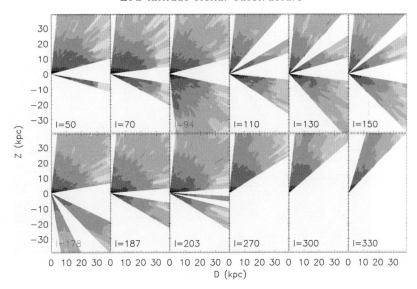

Figure 2. Stellar mass density resulting from the single population fits, as function of distance from the sun along the Galactic plane, and height above or below the plane, in kpc. Each panel shows a different imaging scan, the Galactic longitude of which is listed at the bottom, and darker gray levels correspond to higher densities.

software package MATCH (Dolphin 2001), adapted to solve for distance rather than age-metallicity evolution, to fit the distance distribution of stars along the line-of-sight. A demonstration and a detailed description of this technique and its application to SDSS data can be found in de Jong et al. (2008). Here we shortly discuss the basic approach used in the current analysis.

Observed CMDs are fit with a set of model CMDs, each of which corresponds to a certain age, metallicity and distance. The models are created by populating theoretical isochrones from Girardi et al. (2004) and convolving them with the SDSS photometric errors and completeness (see de Jong et al. 2008). Using maximum-likelihood methods, MATCH determines the best-fitting linear combination of model CMDs, thereby providing the distribution of stars over age-metallicity-distance space. We do this in two ways. First, fitting a narrow colour range of $0.3 < g\text{-}r < 0.8$ and a single combination of age and metallicity ([Fe/H] $= -0.8$, 8 Gyr); this corresponds to measuring the density of main-sequence stars as function of distance modulus. Second, we use a wider colour range ($0.1 < g\text{-}r < 0.8$) that includes turn-off stars, providing information on age and metallicity, and fit for three different stellar populations: a 'thick-disk' population with [Fe/H] $= \sim -0.8$ and $t = 10$ Gyr, a 'halo' population with [Fe/H]~ -1.3 and $t = 13$ Gyr, and a broad 'general' population with [Fe/H] $= \sim -0.8$ and $5 < t < 14$ Gyr.

3. Results

From the fits described above we obtain the stellar mass in bins of constant distance modulus. This can be converted to spatial stellar mass density, giving contour maps such as the ones presented in Fig. 2 for the single population fits. As expected, the density is clearly seen to decrease with increasing distance from the plane and from the Galactic center. Fig. 3 is colour-coded with the type of population, based on the results from the fits with the thick-disk-like (red), halo-like (blue) and very broad (green) populations. It is clear that the thick disk and halo are indeed fit with the appropriate populations,

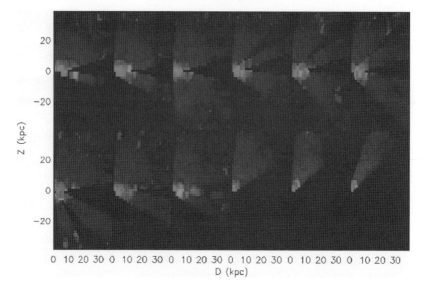

Figure 3. Distribution of stellar populations, following the fit results for three different populations. Each panel is for a different longitude, following the same lay-out as Fig. 2. The thick-disk-like population with [Fe/H] = $\sim$ -0.8 and $t = 10$ Gyr is colour-coded as red, the halo-like population with [Fe/H]$\sim$ -1.3 and $t = 13$ Gyr as blue, and the broad population with [Fe/H] = $\sim$ -0.8 and $5 < t < 14$ Gyr as green.

with the third, broad population mostly visible in the regions where the halo and disk populations have similar densities.

To increase the contrast of any substructures on top of the 'smooth' distribution of stars, a smooth model must be subtracted from the stellar density maps. We assume a model with a double exponential thin and thick disk and an axisymmetric power-law halo. Since the constraints on the disk are limited due to a lack of data at the smallest distances and the masking out of the lowest latitudes, we fix the thin and thick disk scale lengths at 2.6 and 3.6 kpc, respectively, and their scale heights at 0.3 and 1.0 kpc, respectively. Using $R_\odot = 7.6$ kpc (following Vallée (2008)), our best-fit model gives a local thin disk density of 0.08 $M_\odot \mathrm{pc}^{-3}$, local thick disk and halo normalisations of 0.058 and 0.0016, and an almost round ($q = 0.9$) halo with a power-law index of -3. These fit values agree well with previous determinations (see e.g. Siegel *et al.* 2002), although our fit favours a rounder halo than found in previous SDSS studies by Bell *et al.* (2008) and Juric *et al.* (2008). For the fits with three populations the model favours a lower thin disk density and higher normalisations for the thick disk and halo. This is due to the lack of a specific thin disk-like population in the fits. The residuals left after subtracting the best-fitting smooth models from the stellar density maps are shown in Fig. 4. In this figure we have zoomed in to regions close to the plane, where the LLS is expected to be present.

4. Discussion and Conclusions

Colour-magnitude diagram fitting can successfully reproduce the distance distribution of stars. This is corroborated by the fact that the disk and halo model that best fits the density distributions in Fig. 2 gives densities and density ratios between the different model components that are in good agreement with previous determinations from star counts (e.g. Siegel *et al.* 2002 and references therein).

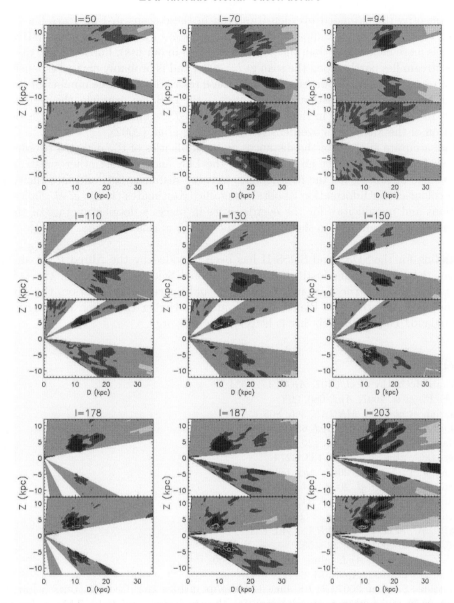

Figure 4. Overdensities at low Galactic latitudes. Residual maps for both single population (upper subpanels) and triple population (lower subpanels) fits for all stripes that cross the Galactic plane. Starting from black, the gray scale levels correspond to areas with residual densities >2, $2>1$, $1>-1$, and <-1 times the model density. White areas contain no information due to an absence of data or too high reddening. Coloured contours show overdensities of >2, $2>1$ times the model density in the individual thick disk-like (red), halo-like (blue) and broad (green) populations.

After subtraction of a smooth Galaxy model, a wealth of substructure becomes visible at low Galactic latitudes (Fig. 4). When searching for pieces of the LLS, possible other overdensities need to be taken into account. The scans at $l=50°$ and $70°$ cross the Hercules-Aquila cloud (Belokurov et al. 2007), a large overdensity extending both above and below the plane to high latitudes at a distances between 10 and 20 kpc.

Consequently, the extended overdensities seen in these scans in Fig. 4 are likely to be related to this structure (it appears slightly more distant in our results because Hercules-Aquila is more metal-poor than the populations used in our fits). Furthermore, the Sagittarius stream lies partly along the scan at $l = 203°$, and is probably partly responsible for the overdensities there. However, even with these restrictions, a large number of overdensities is left, of which at least part corresponds to the LLS. For example, the overdensities at $D = 10$ kpc, $Z = +5$ kpc in the $l = 178°$ and $l = 187°$ scans correspond to the original detection of the Monoceros overdensity by Newberg et al. (2002).

More accurate analysis of the distances and metallicities of the detected overdensities is needed to piece together an overview of the LLS and possible other substructures. One avenue of solving for the degeneracies between distance, age, and metallicity is by using colour-colour information to obtain metallicity estimates (Ivezic et al. 2008). With locations and metallicities for the overdensities in place, a detailed picture of the LLS will emerge that can help to shed light on the nature of this structure.

Funding for the SDSS and SDSS-II has been provided by the Alfred P. Sloan Foundation, the Participating Institutions, the National Science Foundation, the U.S. Department of Energy, the National Aeronautics and Space Administration, the Japanese Monbukagakusho, the Max Planck Society, and the Higher Education Funding Council for England. The SDSS Web Site is http://www.sdss.org/.

References

Adelman-McCarthy et al. 2007, *ApJS*, 172, 634
Bell, E. F., et al. 2007, *ApJ*, 680, 295
Bellazzini, et al. 2006, *MNRAS*, 366, 865
Belokurov, V., et al. 2007, *ApJ*, 657, L89
Bonifacio, P., et al. 2000, *AJ*, 120, 2065
Butler, D. J. et al. 2007, *AJ*, 133, 2274
de Jong, J. T. A., et al. 2007, *ApJ*, 662, 259
de Jong, J. T. A., et al. 2008, *AJ*, 135, 1361
Dolphin, A. E. 2001, *MNRAS*, 332, 91
Girardi, L., et al. 2004, *A&A*, 422, 205
Ibata, R. A., et al. 2003, *MNRAS*, 340, L21
Ivezić, Ž. et al. 2004, *AN*, 325, 583
Ivezić, Ž. et al. 2008, *ApJ*, subm., arXiv:0804.3850
Jurić, M., et al. 2008, *ApJ*, 673, 864
Kazantzidis, S., et al. 2007, *ApJ*, subm. (arXiv:0708.1949)
Martin, N. F., et al. 2004, *MNRAS*, 348, 12
Martin, N. F., et al. 2005, *MNRAS*, 362, 906
Martínez-Delgado, D., et al. 2005, *ApJ*, 633, 205
Momany, Y., et al. 2006, *A&A*, 451, 515
Newberg, H. J., et al. 2002, *ApJ*, 569, 245
Peñarrubia, J., et al. 2005, *ApJ*, 626, 128
Schlegel, D., Finkbeiner, D. & Davis, M. 1998, *ApJ*, 500, 525
Siegel, M. H., et al. 2002, *ApJ*, 578, 151
York, et al. 2000, *AJ*, 120, 1579
Younger, J. D., et al. 2008, *ApJ*, 676, L21

Cosmic evolution of stellar disk truncations: from z = 1 to the Local Universe

Ignacio Trujillo[1], Ruyman Azzollini[1], Judit Bakos[1], John Beckman[1] and Michael Pohlen[2]

[1]Instituto de Astrofísica de Canarias,
C/Vía Láctea s/n, 38205 La Laguna, S/C de Tenerife, Spain
email: trujillo@iac.es,ruyman@iac.es,jbakos@iac.es,jeb@iac.es

[2]Cardiff University, School of Physics & Astronomy,
Cardiff, CF24 3AA, Wales, UK
email: Michael.Pohlen@astro.cf.ac.uk

Abstract. We present our recent results on the cosmic evolution of the outskirts of disk galaxies. In particular we focus on disk–like galaxies with stellar disk truncations. Using UDF, GOODS and SDSS data we show how the position of the break (i.e. a direct estimator of the size of the stellar disk) evolves with time since z∼1. Our findings agree with an evolution on the radial position of the break by a factor of 1.3 ± 0.1 in the last 8 Gyr for galaxies with similar stellar masses. We also present radial color gradients and how they evolve with time. At all redshifts we find a radial inside-out bluing reaching a minimum at the position of the break radius, this minimum is followed by a reddening outwards. Our results constraint several galaxy disk formation models and favour a scenario where stars are formed inside the break radius and are relocated in the outskirts of galaxies through secular processes.

Keywords. galaxies: evolution - galaxies: high-redshift - galaxies: structure - galaxies: formation - galaxies: spiral - galaxies: photometry

1. Introduction

Early studies of the disks of spiral galaxies (Patterson 1940, de Vaucouleurs 1959, Freeman 1970) showed that this component generally follows an exponential radial surface-brightness profile, with a certain scale length, usually taken as the characteristic size of the disk. Freeman (1970) pointed out, though, that not all disks follow this simple exponential law. In fact, a repeatedly reported feature of disks for a representative fraction of the spiral galaxies is that of a truncation of the stellar population at large radii, typically 2-4 exponential scale lengths (see e.g. the review by Pohlen *et al.* 2004).

Several possible break-forming mechanisms have been investigated to explain the truncations. There have been ideas based on maximum angular momentum distribution: van der Kruit (1987) proposed that angular momentum conservation in a collapsing, uniformly rotating cloud naturally gives rise to disk breaks at roughly 4.5 scale radii. van den Bosch (2001) suggested that the breaks are due to angular momentum cut-offs of the cooled gas. On the other hand, breaks have also been attributed to a threshold for star formation (SF), due to changes in the gas density Kennicutt (1989), or to an absence of equilibrium in the cool Interstellar Medium phase (Elmegreen & Parravano 1994, Schaye 2004). More recent models using collisionless N-body simulations, such as that by Debattista *et al.* (2006), demonstrated that the redistribution of angular momentum by spirals during bar formation also produces realistic breaks. In a further elaboration of this idea, Roškar *et al.* (2008) have performed high resolution simulations of the formation of a galaxy embedded in a dark matter halo. In these models, breaks are the result of the interplay between a radial star formation cut-off and redistribution of stellar mass by secular processes. A natural prediction of these models is that the stellar populations present an age minimum in the break position. This prediction could be probed by exploring the color profiles of the galaxies.

Furthermore, addressing the question of how the radial truncation evolves with z is strongly linked to our understanding of how the galactic disks grow and where star formation takes place. Pérez (2004) showed that it is possible to detect stellar truncations even out to z∼1. Using the radial position of the truncation as a direct estimator of the size of the stellar disk, Trujillo & Pohlen (2005) inferred a moderate (∼25%) inside-out growth of disk galaxies since z∼1. An important point, however was missing in the previous analyses: the evolution with redshift of the radial position of the break at a given stellar mass. The stellar mass is a much better parameter to explore the growth of galaxies, since the luminosity evolution of the stellar populations can mimic a size evolution (Trujillo et al. 2004, Trujillo et al. 2006). We present in this contribution a quick summary of our recent findings on the stellar disk truncation origin and its evolution with redshift. The results presented here are based on the following publications: Azzollini et al. (2008a), Azzollini et al. (2008b) and Bakos et al. (2008). Throughout, we assume a flat Λ-dominated cosmology ($\Omega_M = 0.30$, $\Omega_\Lambda = 0.70$, and $H_0 = 70\,\mathrm{km}\,s^{-1}\,Mpc^{-1}$).

2. Color profiles in Local Galaxies

In order to contrain the outer disk formation models, in Bakos et al. (2008), we have explored radial color and stellar surface mass density profiles for a sample of 85 late-type spiral galaxies with available deep (down to ∼27 mag/arcsec^2) SDSS g' and r' band surface brightness profiles (Pohlen & Trujillo 2006). About 90% of the light profiles have been classified as broken exponentials, either exhibiting truncations (Type II galaxies) or antitruncations (Type III galaxies). Their associated color profiles show a significantly different behavior. For the truncated galaxies a radial inside-out bluing reaches a minimum of $(g' - r') = 0.47 \pm 0.02$ mag at the position of the break radius, this minimum is followed by a reddening outwards (see middle row in Fig. 1). The antitruncated galaxies reveal a different behavior. Their break in the light profile resides in a plateau region of the color profile at about $(g' - r') = 0.57 \pm 0.02$.

Using the $(g' - r')$ color (Bell et al. 2003) to calculate the stellar surface mass density profiles reveals a surprising result. The breaks, well established in the light profiles of the Type II galaxies, are almost gone, and the mass profiles resemble now those of the pure exponential Type I galaxies (see bottom row in Fig. 1). This result suggests that the origin of the break in Type II galaxies is more likely due to a radial change in stellar population than being associated to an actual drop in the distribution of mass. The antitruncated galaxies on the other hand preserve to some extent their shape in the stellar mass density profiles.

There are other structural parameters that can be computed to contrain the different formation scenarios. Among these we have estimated the stellar surface mass density at the break for truncated (Type II) galaxies (13.6 ± 1.6 $M_\odot pc^{-2}$) and the same parameter for the antitruncated (Type III) galaxies (9.9 ± 1.3 $M_\odot pc^{-2}$). Finally, we have measured that ∼15% of the total stellar mass in case of truncated galaxies and ∼9% in case of antitruncated galaxies are to be found beyond the measured break radii in the light profiles.

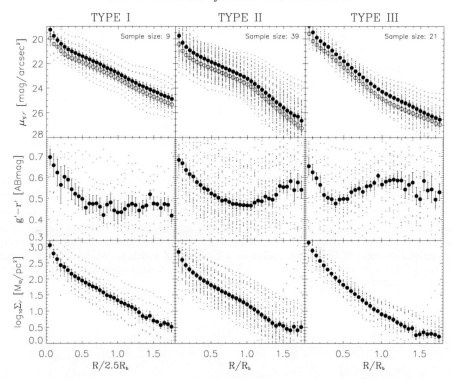

Figure 1. *Upper row*: Averaged, scaled radial surface brightness profiles of 9 Type I (pure exponential profiles), 39 Type II (truncated galaxies) and 21 Type III (antitruncated) galaxies. The filled circles correspond to the r' band mean surface brightness, the open circles to the mean g' band data (Pohlen & Trujillo 2006). The small dots are the individual galaxy profiles in both bands. The surface brightness is corrected for Galactic extinction. — *Middle row*: (g' - r') color gradients. The averaged profile of Type I reaches an asymptotic color value of ~ 0.46 mag being rather constant outwards. Type II profiles have a minimum color of 0.47 ± 0.02 mag at the break position. The mean color profile of Type III has a redder value of about 0.57 ± 0.02 mag at the break. — *Bottom row*: r' band surface mass density profiles obtained using the color to M/L conversion of Bell *et al.* (2003). Note how the significance of the break almost disappears for the Type II (truncated galaxies) case.

3. Stellar disk truncation evolution

In Azzollini *et al.* (2008a), we have conducted the largest systematic search so far for stellar disk truncations in disk-like galaxies at intermediate redshift ($z < 1.1$), using the Great Observatories Origins Deep Survey South (GOODS-S) data from the *Hubble Space Telescope* - ACS. Focusing on Type II galaxies (i.e. downbending profiles) we explore whether the position of the break in the rest-frame B-band radial surface brightness profile (a direct estimator of the extent of the disk where most of the massive star formation is taking place), evolves with time. The number of galaxies under analysis (238 of a total of 505) is an order of magnitude larger than in previous studies. For the first time, we probe the evolution of the break radius for a given stellar mass (a parameter well suited to address evolutionary studies). Our results suggest that, for a given stellar mass, the radial position of the break has increased with cosmic time by a factor 1.3 ± 0.1 between $z \sim 1$ and $z \sim 0$ (see Fig. 2). This is in agreement with a moderate inside-out growth of the disk galaxies in the last ~ 8 Gyr. In the same period of time,

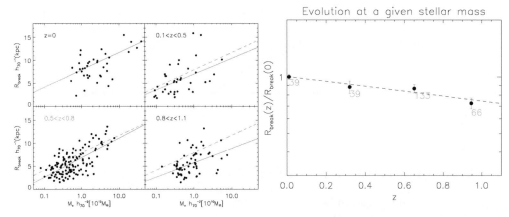

Figure 2. Left: Break Radius of "truncated" galaxies as a function of stellar mass, for 4 ranges of redshift. Local data are from Pohlen & Trujillo (2006) (g'-band results). Right: Size evolution at a given stellar mass of the break radius as a function of redshift. We have found a growth of a factor 1.3 ± 0.1 between $z = 1$ and $z = 0$. The numbers acompanying each point in the right panel give the population of objects which they represent.

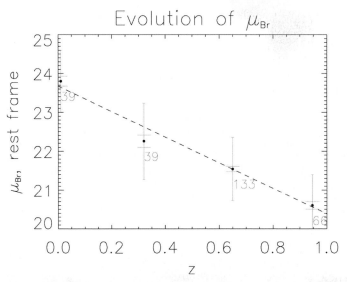

Figure 3. Evolution of the surface brightness at the break for Type II galaxies with redshift. We show the median surface brightness at the break for the distribution of our galaxies. The larger error bars represent the standard deviation of the distributions, while the shorter ones give the error in the median values. The numbers acompanying each point in the right panel give the population of objects which they represent.

the surface brightness level in the rest-frame B-band at which the break takes place has increased by 3.3 ± 0.2 mag/arcsec^2 (a decrease in brightness by a factor of 20.9 ± 4.2).

In Azzollini et al. (2008a) we also find that at a given stellar mass, the scale lengths of the disk in the part inner to the "break" were on average somewhat larger in the past, and have remained more or less constant until recently. This phenomenon could be related to the spatial distribution of star formation, which seems to be rather spread over the disks in the images. So disk galaxies had profiles with a flatter brightness distribution in the inner part of the disk, which has grown in extension, while becoming fainter and "steeper" over time. This is consistent with at least some versions of the inside-out formation scenario for disks.

4. Color profiles in intermediate redshift galaxies

In addition to the evolution on the position of the break in spiral galaxies is important to explore how the color of the surface brightness profiles has evolved with time. This kind of analysis sheds light on when stars formed in different parts of the disk of galaxies, thus giving hints on the stellar mass buildup process.

In Azzollini et al. (2008b) we present deep color profiles for a sample of 415 disk galaxies within the redshift range $0.1 \leqslant z \leqslant 1.1$, and contained in HST ACS imaging of the GOODS-South field. For each galaxy, passband combinations are chosen to obtain, at each redshift, the best possible approximation to the rest-frame $u - g$ color. We find that objects which show a truncation in their stellar disk (type II objects) usually show a

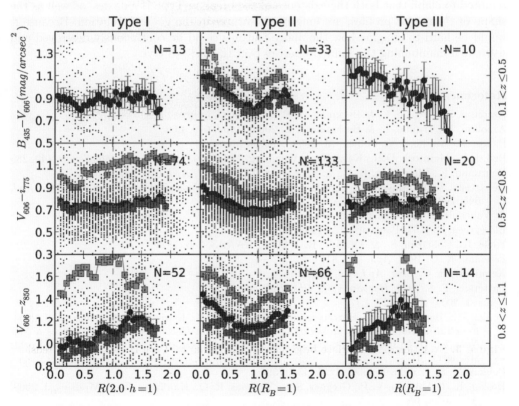

Figure 4. Color profiles of the 415 galaxies under study in Azzollini et al. (2008b). The sample is divided in subsamples according to surface brightness profile type (I or pure exponential profiles, II or truncated galaxies, III or antitruncated, in columns, from left to right) and redshift range (low, mid, or high, in rows, from top to bottom). The colors ($B_{435} - V_{606}, V_{606} - i_{775}, V_{606} - z_{850}$) are chosen as the best proxies to the rest-frame $u - g$ color in each redshift bin. The radii are scaled to the scale radius, R_s, whose definition depends on profile type: $R_s = 2h$ for type I, where h is the scale length of the disk, and it is equal to the break radius, $R_s = R_B$, for types II and III. Small points are individual color profiles. Large black dots are the median color profiles for each subsample, and the error bars give the error in those estimations. The red squares give the median color profile for objects with stellar mass $M_\star > 10^{10} M_\odot$, while the blue squares give the same for objects with $M_\star \leqslant 10^{10} M_\odot$.

minimum in their color profile at the break, or very near to it, with a maximum to minimum amplitude in color of $\leqslant 0.2$ mag/arcsec^2, a feature which is persistent through the explored range of redshifts (i.e., in the last $\sim$8 Gyr and that it is also found in our local sample for comparison (Bakos *et al.* 2008)). This color structure is in qualitative agreement with recent model expectations where the break of the surface brightness profiles is the result of the interplay between a radial star formation cutoff and a redistribution of stellar mass by secular processes (Roškar *et al.* 2008). Our results fit qualitatively their prediction that the youngest stellar population should be found at the break radius, and older (redder) stars must be located beyond that radius. It is not easy to understand how "angular momentum" or "star formation threshold"/"ISM phases" models alone could explain our results. Thus they pose a difficult challenge for these models. However, it will also be necessary to check whether the Roškar *et al.* (2008) models (as well as other available models in the literature like those of Bournaud *et al.* 2007 and Foyle *et al.* 2008) are able to reproduce quantitatively the results shown here.

Combining the results found in Azzollini *et al.* (2008b) and Bakos *et al.* (2008) one is tempted to claim that both the existence of the break in Type II galaxies, as well as the shape of their color profiles, are long lived features in the galaxy evolution. Because it would be hard to imagine how the above features could be continuously destroyed and re-created maintaining the same properties over the last $\sim$8 Gyr.

References

Azzollini, R., Trujillo, I., & Beckman, J. E. 2008a, *ApJ*, in press, arXiv:0805.2259
Azzollini, R., Trujillo, I., & Beckman, J. E. 2008b, *ApJ*, 679, L69
Bakos, J., Trujillo, I., & Pohlen, M. 2008, *ApJ*, in press
Bell, E.F., McIntosh, D.H., Katz, N., & Weinberg, M.D. 2003, *ApJ Supplement Series*, 149, 289
Bournaud, F., Elmegreen, B. G., & Elmegreen, D. M. 2007, *ApJ*, 670, 237
Debattista, V. P., Mayer, L., Carollo, C. M., Moore, B. Wadsley, J., & Quinn, T. 2006, *ApJ*, 645, 209
de Vaucouleurs, G. 1959, *Handb. Phys.*, 53, 311
Elmegreen, B. G. & Parravano, A. 1994, *ApJ*, 435, L121
Foyle, K., Courteau, S., & Thacker, R. J. 2008, *MNRAS*, 386, 1821
Freeman, K. C. 1970, *ApJ*, 160, 811
Kennicutt, R. C. 1989, *ApJ*, 344, 685
Patterson, F. S. 1940, *Harvard Coll. Obs. Bul.*, 914, 9
Pérez, I. 2004, *A&A*, 427, L17
Pohlen, M., Beckman, J. E., Hüttemeister, S., Knapen, J. H., Erwin, P., & Dettmar, R.-J. 2004, *Penetrating Bars through Masks of Cosmic Dust: The Hubble Tuning Fork Strikes a New Note*, ed. D. L. Block, I. Puerari, K. C. Freeman, R. Groess, & E. K. Block (Dordrecht: Springer) 731
Pohlen, M. & Trujillo, I. 2006, *A&A*, 454, 759
Roškar, R., Debattista, V. P., Gregory, S. S., Thomas, R. Q., Kauffman, T., & Wadsley, J. 2008, *ApJ*, 675L, 65
Schaye, J. 2004, *ApJ*, 609, 667
Trujillo, I. *et al.* 2004, *ApJ*, 604, 521
Trujillo, I. & Pohlen, M. 2005, *ApJ*, 630, L17
Trujillo, I. *et al.* 2006, *ApJ*, 650, 18
van den Bosch, F. C. 2001, *MNRAS*, 327, 1334
van der Kruit, P. C. 1987, *A&A*, 173, 59

Chemically tagging the Galactic disk: abundance patterns of old open clusters

G. M. De Silva[1], K. C. Freeman[2] and J. Bland-Hawthorn[3]

[1]European Southern Observatory, Karl-Schwarzschild Str 2. D-85748 Garching, Germany
email: gdesilva@eso.org

[2]Research School of Astronomy and Astrophysics, Mount Stromlo Observatory, Australian National University. ACT 2611. Australia
email: kcf@mso.anu.edu.au

[3]Institute of Astronomy, University of Sydney, NSW 2006, Australia
email: jbh@physics.usyd.edu.au

Abstract. The long term goal of large-scale chemical tagging is to use stellar elemental abundances as a tracer of dispersed substructures of the Galactic disk. The identification of such lost stellar aggregates and exploring their chemical properties will be key in understanding the formation and evolution of the disk. Present day stellar structures such as open clusters and moving groups are the ideal testing grounds for the viability of chemical tagging, as they are believed to be the remnants of the original larger star-forming aggregates. We examine recent high resolution abundance studies of open clusters to explore the various abundance trends and reassess the prospects of large-scale chemical tagging.

Keywords. Galaxy: formation – evolution; open clusters and associations: general

1. Introduction

The aim of chemical tagging (Freeman & Bland-Hawthorn 2002) is to re-construct ancient star-forming aggregates of the Galactic disk, assuming such systems existed from an hierarchical aggregation formation scenario. Observationally the fact that stars are born in rich aggregates numbering hundreds to thousands of stars is supported by many studies from optical, infrared, millimeter and radio surveys (e.g. Carpenter 2000; Meyer *et al.* 2000; Lada & Lada 2003). The existence of stellar aggregates at earlier epochs is also observed in the form of old open clusters, stellar associations and moving groups. Further, theoretical hydrodynamical simulations indicate that star formation occurs in groups, where the original gas cloud undergoes fragmentation preventing contraction onto a single star (e.g. Jappsen *et al.* 2005; Tilley & Pudritz 2004; Larson 1995). Some clusters stay together for billions of years, whereas others become unbound shortly after the initial star-burst, depending on the star formation efficiency. If stars were not born in aggregates, it would be impossible to identify a given star's birth site.

2. Open clusters

Open clusters have historically been used for studying stellar evolution as all stars in a given cluster are coeval. Their key attribute is that they provide a direct time line for investigating change. Young and old open clusters are found in the disk, varying in age from several Myr to over 10 Gyr (see compilation by Dias *et al.* 2002). The older (> 1 Gyr) open clusters are less numerous than their younger counterparts and in general are more massive. These old open clusters are excellent probes of early disk evolution.

They constitute important fossils and are the likely left-overs of the early star-forming aggregates in the disk. Perhaps only the remnant cores are left behind, while the outer stars have dispersed into the disk.

2.1. *Chemical homogeneity*

It is generally assumed that all stars in a cluster, born from the same parent proto-cluster cloud should contain the same abundance patterns. Theoretically, the high levels of supersonic turbulence linked to star formation in giant molecular clouds (McKee & Tan 2002), suggests that the interstellar medium of the gas clouds is well mixed and supports the case for chemically homogenous clusters. Observational evidence show high levels of chemical homogeneity in open clusters. High accuracy differential abundance studies for large samples of Hyades open cluster F-K dwarfs show little or no intrinsic abundance scatter for a range of elements (Paulson *et al.* 2003; De Silva *et al.* 2006). Such internal homogeneity is also observed within old open clusters (e.g. Collinder 261 De Silva *et al.* 2007; Carretta *et al.* 2005, ;). Other old open cluster studies also support the case for internal homogeneity (e.g. Jacobson *et al.* 2007; Bragaglia *et al.* 2008), albeit for only a few stars and larger measurement uncertainties.

The observations of homogeneity in old open clusters show that chemical information is preserved within the stars and effects of any external sources of pollution (e.g. from stellar winds or interactions with ISM) are negligible. Abundance differences may arise within cluster stars at various stellar evolutionary stages, e.g. due to internal mixing of elements during the dredge-up phases in giants. We do not expect main sequence dwarf members of a cluster to show such effects. Further, any internal mixing will only affect the lighter elements synthesized within the stars, while the heavier element abundances should remain at their initial levels. It is, however, interesting to note Pasquini *et al.* (2004)'s study of IC 4651, which shows systematic abundance differences between the main sequence turn-off stars and the giants, although this maybe due to errors in deriving the stellar temperature scales.

2.2. *Cluster sample*

Assuming internal homogeneity holds for most open clusters in the disk, we now compare the different cluster abundance patterns. The mean cluster elemental abundances were taken from several high resolution abundance studies in the literature for clusters with ages greater than the Hyades. The list of clusters and their references are given in Table 1. Figure 1 plots the cluster mean abundance relative to Fe for the various elements studied.

Before the different clusters can be compared we must note that systematic differences are likely to exist due to differences in methodologies and scales. Since the studies are based on different clusters with no overlapping samples, such systematics are difficult to quantify. Where a study included an analysis of a reference star, such as the Sun, we can use the quoted differences as a guide to the expected systematic effects. In other studies, the reference solar abundance levels were simply adopted from past literature sources. This systematic difference arising due to the solar reference values is typically within 0.05 dex.

Systematic uncertainties differ from element to element, depending on how the individual lines were analyzed. We have adopted the published results based on standard LTE analysis to ensure a better comparison, and non-LTE analyses were not included. The employed atomic line data varies between studies and is also gives rise to systematics. In

Table 1. Open cluster sample

Cluster Name	[Fe/H]	Reference
NGC 6253	0.46	Carretta et al. (2007)
NGC 6791	0.47	Carretta et al. (2007)
NGC 7142	0.08	Jacobson et al. (2007)
NGC 6939	0.00	Jacobson et al. (2007)
IC 4756	-0.15	Jacobson et al. (2007)
M 11	0.10	Gonzalez & Wallerstein (2000)
NGC 2324	-0.17	Bragaglia et al. (2008)
NGC 2477	0.07	Bragaglia et al. (2008)
NGC 2660	0.04	Bragaglia et al. (2008)
NGC 3960	0.02	Bragaglia et al. (2008)
Be 32	-0.29	Bragaglia et al. (2008)
NGC 6819	0.09	Bragaglia et al. (2001)
NGC 7789	-0.04	Tautvaišienė et al. (2005)
M 67	-0.03	Tautvaišiene et al. (2000)
NGC 2141	-0.26	Yong et al. (2005)
Be 31	-0.40	Yong et al. (2005)
Be 29	-0.18	Yong et al. (2005)
Be 20	-0.61	Yong et al. (2005)
Tom 2	-0.45	Brown et al. (1996)
Mel 71	-0.30	Brown et al. (1996)
NGC 2243	-0.48	Gratton & Contarini (1994)
Mel 66	-0.38	Gratton & Contarini (1994)
Cr 261	-0.03	De Silva et al. (2007)
Hyades	0.13	De Silva et al. (2006)

differential analyses relative to the Sun or a reference star, the gf values were recalculated for each element and line. Other studies use laboratory measured gf values from various literature sources. Other differences in analyses, such as the use of different model atmospheres and whether the abundance measurements are based on EWs or spectral synthesis also produce systematic variations. Since most of these systematics cannot be accurately quantified, we have not taken them into account and plot the published mean abundance values in Figure 1. The error bars representing the typical measurement errors for the various elements in each study are shown. We refer the reader to the original studies for the individual measurement errors per element per cluster.

Other points to note include the number and type of stars in the different studies. The clusters studied by Bragaglia et al. (2008); Carretta et al. (2007) are based largely on 5-6 red clump stars per cluster, the Yong et al. (2005) and Jacobson et al. (2007) studies on 2-5 red giants per cluster, while other studies are based on main sequence or turn-off dwarfs. As mentioned in section 2.1, we do not expect any effects of stellar evolution for the heavier elements, while the abundances of lighter elements, such as Na, Al, Mg may be affected in the giants by internal mixing. In this case they will no longer represent the cluster's initial abundance levels, which will introduce additional scatter when comparing to dwarf members of other clusters.

2.3. Abundance signatures

Figure 1 shows that different clusters have different elemental abundance patterns. There is also significant scatter for many of the elements. As discussed above, some of this scatter is likely to due systematic uncertainties, although there may be intrinsic variations, especially in elements showing excessive scatter.

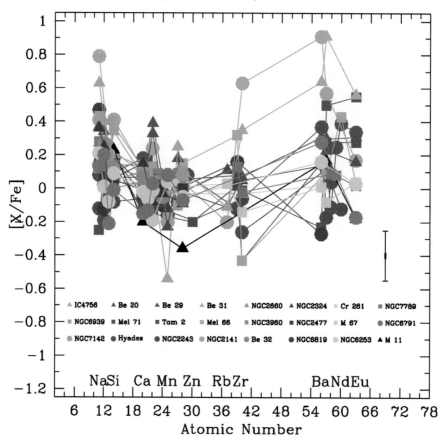

Figure 1. Elemental abundances of old open clusters. Each symbol represents the mean abundance value for individual clusters. The error bars show the typical measurement error. Original references of the cluster data are given in Table 1.

Of the lighter elements, Na has the highest scatter, where most clusters are Na enhanced except for two clusters showing sub-solar levels. The other odd Z and alpha elements show cluster to cluster scatter within 0.15 dex with the average abundance being slightly super-solar. Enhanced alpha elements relative to Fe are indicative of a high rate of star formation where Type II SN dominate over the Type Ia SN. Therefore clusters showing such enhancement is likely to have undergone a phase of rapid star formation in comparison to those which show solar or sub-solar level alpha abundances. Of further interest are the clusters that do not show equal alpha element abundance levels, e.g. with enhanced Si but deficient in Ca. This may represent some form of localized inhomogeneity unique to the time and site of the clusters' formation. Such abundance signatures will play a major role in large scale chemical tagging, when associating field stars to common origins.

The Fe-peak elements, thought to be produced via Type Ia SN, in general show the least scatter with the average abundance close to solar. This is expected given the abundances are plotted relative to Fe and we expect the Fe-peak elements to follow the Fe abundance. Nevertheless it is interesting to note that Ti has a larger scatter. Ti is considered the heaviest of the alpha elements, although by atomic number it falls into the Fe-peak group. It is not considered a pure alpha element either since Type Ia SNe also contribute to its production in addition to the dominant Type II SNe. Among the other Fe-peak elements Mn also shows a higher scatter, however this is dominated by a single cluster which is extremely Mn deficient. Note again that Mn could be synthesized in Type II SN as well as from Type Ia, where the yields are metallicity dependent (McWilliam *et al.* 2003; Shetrone *et al.* 2003). From these examples it is clear that many of the elements behave differently within their groups. They do not necessarily vary in lock-steps and it is likely that various nucleosynthesis processes are at play.

The heavier s- and r- process elements show the largest scatter of all studied elements. Note the number of data points for these neutron capture elements are much less than for the lighter elements, a sign that they are difficult to measure. The mean abundance levels are super-solar for these elements. The exceptions are Zr with clusters showing both super and sub-solar Zr abundance levels, as well as Rb and Ce, which have sub-solar abundances in the clusters although with only few data points. The s-process elements are synthesized in low neutron flux environments, as in AGB stars and mixed into the ISM by stellar winds. The light s-process elements such as Zr seem to show a lower abundance compared to the heavier s-process elements such as Ba. Note however the opposite trend is observed for two clusters. Varying trends between Ba, La and Ce are also seen. The mostly r-process elements such as Nd and Eu, produced in high flux environments during Type II SNe also show various trends among the open cluster abundances. Further the ratio of s- to r- process element abundance varies from cluster to cluster. We can expect this as both the s- and r-processes contribute at different levels to the production of these neutron capture elements. Similar to the alpha and Fe-peak groups, this further demonstrates that the common group elements do not always vary in lock step. Their various abundance levels highlight the different conditions during the formation of the individual clusters.

3. Conclusion

We have used high resolution elemental abundances of old open clusters from the literature to compare the cluster to cluster abundance trends for a large range of elements. We find that different clusters show different abundance levels for a given element, with some elements showing large scatter. Despite systematic uncertainties among the studies which could be the source of much of the abundance scatter, those elements showing a $\sigma > 0.2$ dex is likely to be an indication of real cluster to cluster abundance variations. Further various element to element abundance patterns were seen among the sample, highlighting the decoupled nature of the elements and the existence of chemical signatures unique to the clusters based on their time and site of formation. An *homogenous* high resolution abundance study for a range of elements of the Galactic open cluster population (e.g. the BOCCE project, Bragaglia 2007) will provide much valuable insight to further explore unique chemical signatures. This preliminary look, however, suggests that establishing cluster signatures for large scale chemical tagging of the disk is indeed a viable technique.

References

Bragaglia, A. 2007, ArXiv e-prints, 711, 2171
Bragaglia, A., Sestito, P., Villanova, S., Carretta, E., Randich, S., & Tosi, M. 2008, A & A, 480, 79
Bragaglia, A., Carretta, E., Gratton, R. G., Tosi, M., Bonanno, G., Bruno, P., Calì, A., Claudi, R., Cosentino, R., Desidera, S., Farisato, G., Rebeschini, M., & Scuderi, S. 2001, AJ, 121, 327

Brown, J. A., Wallerstein, G., Geisler, D., & Oke, J. B. 1996, AJ, 112, 1551
Carpenter, J. M. 2000, AJ, 120, 3139
Carretta, E., Bragaglia, A., & Gratton, R. G. 2007, A & A, 473, 129
Carretta, E., Bragaglia, A., Gratton, R. G., & Tosi, M. 2005, A&A, 441, 131
De Silva, G. M., Freeman, K. C., Asplund, M., Bland-Hawthorn, J., Bessell, M. S., & Collet, R. 2007, AJ, 133, 1161
De Silva, G. M., Freeman, K. C., Bland-Hawthorn, J., Asplund, M., & Bessell, M. S. 2007, AJ, 133, 694
De Silva, G. M., Sneden, C., Paulson, D. B., Asplund, M., Bland-Hawthorn, J., Bessell, M. S., & Freeman, K. C. 2006, AJ, 131, 455
Dias, W. S., Alessi, B. S., Moitinho, A., & Lépine, J. R. D. 2002, A& A, 389, 871
Freeman, K. & Bland-Hawthorn, J. 2002, ARAA, 40, 487
Gonzalez, G. & Wallerstein, G. 2000, PASP, 112, 1081
Gratton, R. G. & Contarini, G. 1994, A&A, 283, 911
Jacobson, H. R., Friel, E. D., & Pilachowski, C. A. 2007, AJ, 134, 1216
Jappsen, A.-K., Klessen, R. S., Larson, R. B., Li, Y., & Mac Low, M.-M. 2005, A&A, 435, 611
Lada, C. J. & Lada, E. A., 2003, ARAA, 41, 57
Larson, R. B. 1995, MNRAS, 272, 213
McKee, C. F. & Tan, J. C. 2002, Nature, 416, 59
McWilliam, A., Rich, R. M., & Smecker-Hane, T. A. 2003, ApJl, 592, L21
Meyer, M. R., Adams, F. C., Hillenbrand, L. A., Carpenter, J. M., & Larson, R. B. 2000, Protostars and Planets, 121
Paulson, D. B., Sneden, C., & Cochran, W. D. 2003, AJ, 125, 3185
Pasquini, L., Randich, S., Zoccali, M., Hill, V., Charbonnel, C., & Nordström, B. 2004, A & A, 424, 951
Shetrone, M., Venn, K. A., Tolstoy, E., Primas, F., Hill, V., & Kaufer, A. 2003, AJ, 125, 684
Tautvaišienė, G., Edvardsson, B., Puzeras, E., & Ilyin, I. 2005, A&A, 431, 933
Tautvaišiene, G., Edvardsson, B., Tuominen, I., & Ilyin, I. 2000, A&A, 360, 499
Tilley, D. A. & Pudritz, R. E. 2004, MNRAS, 353, 769
Yong, D., Carney, B. W., & de Almeida, M. L. T. 2005, AJ, 130, 597

The Arcturus Moving Group: Its Place in the Galaxy

Mary E. K. Williams[1,2], Ken C. Freeman[1], Amina Helmi[3] and the RAVE collaboration

[1] Mt Stromlo Observatory, Cotter Road, Weston Creek, ACT 2611, Australia
[2] Astrophysikalisches Institut Potsdam, An der Sternwarte 16, D-14482, Potsdam, Germany
[3] Kapteyn Institute, P.O. Box 800, 9700 AV Groningen, the Netherlands
email: mary@aip.de;
email: kcf@mso.anu.edu.au;
email: ahelmi@astro.rug.nl

Abstract. The Arcturus moving group is a well-populated example of phase space substructure within the disk of our Galaxy. With its large rotational lag ($V = -100$ kms^{-1}), metal poor nature ([Fe/H] ~ -0.6) and significant age (10 Gyr) it belongs to the Galaxy's thick disk. Traditionally regarded as the remains of a dissolved open cluster, it has recently been suggested to be a remnant of a satellite accreted by our Galaxy.

We confirm via further kinematic studies using the Nordstöm et al. (2004), Schuster et al. (2004) and RAdial Velocity Experiment (RAVE) surveys (Steinmetz et al. 2004) the existence of the group, finding it to possibly favour negative U velocities and also possibly a solar-circle phenomenon. We undertook a high-resolution spectroscopic abundance study of Arcturus group members and candidates to investigate the origin of the group. Examining abundance of Fe, Mg, Ca, Ti, Cr, Ni, Zn, Ce, Nd, Sm and Gd for 134 stars we found that the group is chemically similar to disk stars and does not exhibit a clear chemical homogeneity.

The origin of the group still remains unresolved: the chemical results are consistent with a dynamical origin but do not entirely rule out a merger one. Certainly, the Arcturus group provides a challenge to our understanding of the nature and origin of the Galaxy's thick disk.

Keywords. Galaxy: abundances, galaxy: kinematics and dynamics, galaxy: structure

1. Introduction

The Arcturus moving group was discovered by Eggen who over the years gathered a list of stars kinematically associated with Arcturus (Eggen 1971, 1998 and references therein). Eggen based group membership mainly upon the common V space velocity, the component of stellar motion relative to the Local Standard of Rest (LSR) in the direction of rotation. For the Arcturus moving group the stars lag the LSR with $V = \sim -100$ kms^{-1}. While Eggen's analysis was somewhat controversial as he adjusted the stellar parallaxes to return a tight V-velocity relation, *a posteriori* justification was obtained by a tight colour-magnitude relation along an isochrone with age $\tau \gtrsim 10$ Gyr and metallicity [Fe/H] ~ -0.6 (Eggen 1996). The age, metallicity and space velocity identify the group as part of the Galaxy's thick disk.

Eggen introduced the Arcturus group as a dissolved open cluster. In this scenario a single star-forming event creates a cluster of stars which over time dissolves. Eventually, the most distinguishing characteristic of the stars' common origin are their similar space motions. However, recently it was suggested by Navarro (2004) that the Arcturus moving group could be an example of the debris of an accreted satellite in the disk of the Galaxy.

If this indeed is the case, this would be of significance in the debate about the origin of the thick disk; if the Arcturus group is an example of accretion debris this would lend weight to the argument that the thick disk consists mainly of such debris.

The motivation behind this study was therefore to further understand the formation of the Arcturus group in the context of thick disk formation; is the group a remnant of a star-formation event in the disk or accretion debris? Our primary approach was to perform a high resolution spectroscopic abundance program to search for tracers of the origin of the group in the chemistry of the stars. Following this study, we turned to kinematic studies to confirm the group's existence and further define its kinematic properties. In the following we briefly report on the intriguing results of this study, which lead to more questions than answers about the origin of the Arcturus moving group. A more detailed analysis will be presented shortly in Williams et al. (2008).

2. Abundance study

2.1. Candidate Selection

In light of the possible extra-Galactic origin of the Arcturus group there was the possibility that members could have been missed by Eggen as they might not satisfy his strict selection criterion. So in addition to Eggen's list of Arcturus stars we selected group candidates from the RAVE (Steinmetz 2006), Nordström et al. (2004), Beers et al. (2000) and Norris (1986) studies. To develop selection criteria inclusive of both formation scenarios we performed N-body simulations of the formation of the Arcturus group via dissolution of a progenitor in a static Galactic potential with a range of progenitor masses. The simulations follow those presented in Helmi et al. (2006) and Navarro et al. (2004) where a satellite or cluster with an orbit similar to that of the Arcturus group is disrupted in the potential of the Galaxy. The largest of these, the dissolved dwarf spheroidal galaxy, was used to develop the "banana" criterion selecting stars that are near or at the apocentre of their orbits falling within a banana-shaped region in the UV plane around $V = -100$ kms^{-1}. Also, we restricted our study to stars on disk-like orbits with $|W| < 100$ kms^{-1} and with small velocity errors. Some stars from the thick and thin disks were included for comparative purposes. A total of 134 stars from 190 observations were analysed in our study.

2.2. Methodology

Elemental abundances were derived for our candidate stars by performing a Local Thermodynamic Equilibrium (LTE) analysis with the MOOG code (Sneden 1973) on our high resolution, high signal-to-noise UCLES data obtained in three observing periods at the AAT from August 2003 - November 2006. The first of these runs was obtained by Gayandhi de Silva and kindly granted to us for analysis. To measure equivalent widths we used the new DAOSPEC program which automatically fits Gaussian profiles to lines in a spectrum (Pancino and Stetson, in preparation). However, it was necessary to alter the DAOSPEC code with regards to continuum fitting of the spectra; most of our data is in the blue and so very crowded and it was found that DAOSPEC generally set the continuum too low. The DAOSPEC continuum fitting was therefore disabled and hand fits performed. The line list was compiled from the literature utilising laboratory $\log gf$ values where available. Abundances were derived for Fe, Mg, Ca, Ti, Cr, Ni, Zn, Ce, Nd, Sm and Gd for our 134 stars using primarily spectra in the blue. For a subset of these stars we also have red data from which we obtained abundances for Na, Mg, Al and Si. Here we present only those elements or lines for which hyperfine and isotopic splitting need not be considered.

Stellar parameters were calculated using 'physical' and 'spectroscopic' approaches. The former involved calculating T_{eff} and $\log g$ from photometric and astrometric data, while the latter utilises forcing excitation and ionisation balance to derive T_{eff} and $\log g$ respectively. In both cases the microturbulence is found by requiring that there is no dependence of abundance on line strength. In this short report we only include the spectroscopic results as they yielded better agreement in abundance for stars in common with the studies of Reddy et al. (2003, 2006) and Bensby et al. (2003, 2005). The difference in abundance between stars in common with those studies and our own are $<[\text{Fe/H}]_{this\ study}-[\text{Fe/H}]_{Reddy}>=0.09$ with a standard deviation of $\sigma=0.05$, and for $<[\text{Fe/H}]_{this\ study}-[\text{Fe/H}]_{Bensby}>=0.05$ with a standard deviation of $\sigma=0.07$. Other elements have comparable errors. The agreement is good considering the different techniques, line lists and calibrations employed.

2.3. Results

In Section 3 we will see that the Arcturus group is an over-density in a very narrow velocity range around $V = -100$ kms^{-1}, with a standard deviation of only $\sigma_V = 3$ kms^{-1}. We therefore revert here to a selection mimicking Eggen's initial criteria, selecting as Arcturus candidate stars that are ± 10 kms^{-1} from this mean V velocity. Also, to compare the group against the background thick disk stars we include stars with $V < -50$ kms^{-1}. Figure 1 shows selected abundance results for the Arcturus candidates superimposed on the background stars. The results from the Reddy et al. (2003, 2006) and Bensby et al. (2003, 2005) studies are included to further emphasise the general trends of field stars and to increase the number of candidate stars. The largest of the systematic differences between these studies and our results were accounted for by using the zero-point offsets derived from common stars to shift their results to our abundance scale.

We see clearly in these plots that within the limit of our abundance errors there is no distinguishing feature in abundance between the Arcturus group stars and those of the surrounding disk. While Nd suggests a possible clustering of abundance, this is not corroborated by the α- or other elements. Also, the apparently tight CMD relation of Eggen now seems as a sampling from an old population of similar age, i.e., the thick disk, as both the Arcturus candidates and the background stars lie along the reasonably tight isochrone. Employing different selection criteria for the group, such as our banana selection criteria mentioned above or selecting Eggen's original Arcturus group candidates yields the same result: we are unable to find clear distinguishing features in abundance for the Arcturus group when compared to background disk stars.

These results beg the question: *does the Arcturus moving group exist at all as an overdensity in phase space?* So before drawing any further conclusions we turn to kinematic studies to confirm the group's existence and define it kinematically.

3. Kinematic study

While the Geneva-Copenhagen catalog provided a significant number of Arcturus candidates, it is not the ideal source because it has relatively few stars at the high velocity of the group. However, the recent study of solar-neighbourhood metal-poor stars by Schuster et al. (2006) provides a wealth of stars in the thick disk. This study was not available to us at the time of choosing candidates for our abundance study, however we can now use it to investigate the phase-space structure in the vicinity of Arcturus. Note that there are kinematical selection biases in the Schuster data set towards high-proper motion stars which means that stars with a V velocity nearer to the sun ($V \sim 0$) are under-represented. However, the region around the Arcturus group's velocity is not affected by these biases.

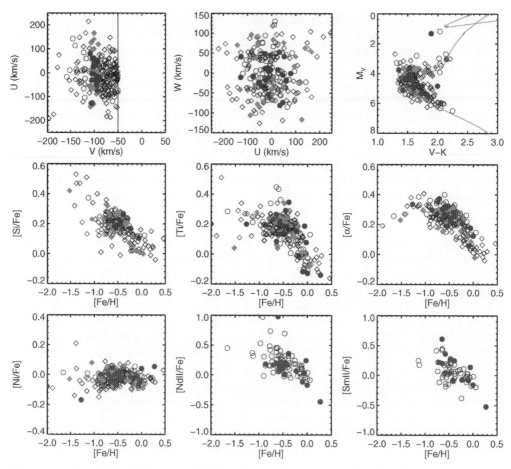

Figure 1. The VU, UW planes, CMD and abundance ratio plots for the Arcturus group candidates (blue circles and red diamonds) selected from stars with $V < -50$ kms^{-1}. Stars from this study are designated by circles, while the amalgamated Reddy and Bensby samples are diamonds. The overlaid Padova isochrone in the CMD has 12.5 Gyr, $Z = 0.006$.

Figure 1 displays plots of kinematic and metallicity plots for the combined Nordström and Schuster data sets where stars in common appear only once. In this diagram we see an over-density at $V \sim -100$ kms^{-1} extending from $0 < $ [Fe/H] < -1. This feature is particularly prominent at lower metallicities. Schuster et al. (2006) interpreted this over-density as the thick disk being split into two components, with the split indicative of the thick disk consisting of merger debris. They link it with the Gilmore, Wyse & Norris (2002) feature but do not associate it to the Arcturus group. We see however that the 'second component' of the thick disk that they identify is clearly at the Arcturus group velocity. Combining their data with the Nordström sample shows that the over-density is perhaps not as localised in [Fe/H] as their results indicated.

We see also that there is a suggestion of a bias towards negative U, which is coincident with the favoured U of the Hercules group. From fits to the generalised histogram in the V-velocity we find that the group is centred around $\mu_V = -102$ kms^{-1} with a very narrow velocity dispersion of $\sigma_V = 2$–3 kms^{-1}.

We have also investigated the Arcturus group in RAVE data, utilising the internal release of October 2007 which contains 191,170 radial velocity measurements as well

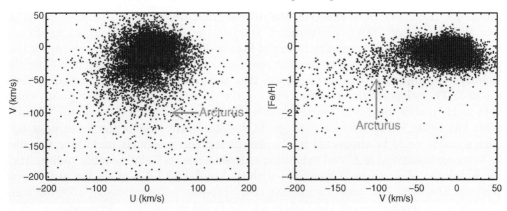

Figure 2. The Arcturus group as seen in the combined Nordström *et al.* (2004) and Schuster *et al.* (2006) data set.

as stellar parameters for a significant number of stars (see Zwitter *et al.* 2008). The stellar parameters enabled us to select those stars associated with the helium-burning red clump. The K-band magnitude of the clump, while being relatively unaffected by extinction, has also been shown observationally to be relatively independent of metallicity and age (Pietrzyński, Gieren & Udalski 2003, Alves 2000). Thus, combining RAVE data with photometric and astrometric data we were able to derive reliable distances and UVW velocites for a sample of 16,000 red clump giants. In this data set we have found that the Arcturus group seems to be more pronounced in the solar circle, i.e., for those with a Galactocentric radius similar to the sun (for full details see Williams *et al.* 2008). If this is indeed the case, such behaviour could be possibly be explained by a dynamical origin of the group.

4. Discussion

Venn *et al.* (2004) demonstrated that current-day satellite dSph galaxies of the Milky Way are α-poor when compared to the Galactic stars of similar [Fe/H], indicating a comparatively slow star formation history (Matteucci 2003). So the similarity between the Arcturus group members and surrounding disk weighs against the accretion debris origin of the group *if we expect the Arcturus progenitor to be similar to current day dSphs*. However, from Tamura *et al.* (2001) we see that a satellite progenitor of Arcturus would have had to have a substantial mass (similar to the LMC) to be enriched to [Fe/H] $= -0.6$. Therefore, it is possible that a such a large dwarf galaxy had a much higher rate of star formation than the remaining satellites of the Milky Way. Also, the simulations of Abadi (2003) of the formation of a Milky Way-like galaxy in the ΛCDM cosmogony gave 60% of the thick disk as tidal debris. So our results showing the Arcturus group as being similar to disk stars does not entirely exclude the possibility that it, and the surrounding thick disk, are tidal debris. However, it would seem remarkable that a disparate group of satellites could conspire to have such similar abundances; a clear picture of the chemical evolution of these satellites to produce the observed abundance patterns has yet to be drawn.

If we instead suppose an *in situ* formation of the thick disk, such as that proposed by Brook *et al.* (2005) from gas-rich mergers, the chemical similarity of the Arcturus group to the surrounding disk favours *in situ* formation of the group as well. However, from our simulations we find that even the largest open clusters (7×10^4 $M_\odot$) could not produce a discernible over-density in the solar neighbourhood after 10 Gyr of evolution (see Williams *et al.* 2008). Furthermore, stars in the Arcturus velocity range do not exhibit the chemical homogeneity expected from a dissolved star forming event (e.g. see de Silva *et al.* 2007). Instead we see that our results are similar to Bensby (2007) for the Hercules moving group where they found that the group is chemically indistinct from the disk. This supports the dynamic origin of the thick/thin disk Hercules group, which is thought to arise due to the Outer Lindblad Resonance with the Galactic bar (Dehnen 2000, Fux 1999). We therefore wonder, could this also be the case for the Arcturus group? This scenario could be supported by the Arcturus group seemingly mimicking Hercules in being asymmetrical in U and exhibiting a dependence on Galactocentric radius. Also, the V velocity of Arcturus is coincident with that of the 6:1 OLR of the bar at the solar position. However, how such a resonance could produce an over-density, affecting stars that spend so much time out the Galaxy's plane is not currently understood.

We are continuing our investigations of the Arcturus group. In our upcoming paper we will present our abundance results in full, as well as including those for extra elements. We will give also present the full kinematic results, comparing them to our simulations of various progenitor scenarios and explore the resonance possibility further. The Arcturus group's origin is an enigma still yet to be solved, but the clues are accumulating.

References

Alves, D. R. 2000, *ApJ*, 539, 732
Beers, T. C., Chiba, M., Yoshii, Y., Platais, I., Hanson, R. B., Fuchs, B., & Rossi, S. 2000, *AJ*, 119, 2866
Bensby, T., Feltzing, S., & Lundström, I. 2003, *A&A*, 410, 527
Bensby, T., Feltzing, S., Lundström, I., & Ilyin, I. 2005, *A&A*, 433, 185
Bensby, T., Oey, M. S., Feltzing, S., & Gustafsson, B. 2007, *ApJ*, 655, L89
Brook, C. B., Gibson, B. K., Martel, H., & Kawata, D. 2005, *ApJ*, 630, 298
Dehnen, W. 2000, *AJ*, 119, 800
Eggen, O. 1971, *PASP*, 83, 271
Eggen, O. 1996, *AJ*, 112, 1595
Eggen, O. 1998, *AJ*, 115, 2397
Fux, R. 2001, *A&AS*, 373, 511
Gilmore, G., Wyse, R.F .G., & Norris, J. E. 2002, *ApJ*, 574, L39
Helmi, A., Navarro, J. F., Nordström, B., Holmberg, J., Abadi, M. G., & Steinmetz, M. 2006, *MNRAS*, 365, 1309
Matteucci, F. 2003, *Ap&SS*, 284, 539
Navarro, J. F., Helmi, A., & Freeman, K. C. 2004, *ApJ*, 601, L43
Nordström, B., Mayor, M., Andersen, J., Holmberg, J., Pont, F., Jørgensen, B. R., Olsen, E. H., Udry, S., & Mowlavi, N. 2004, *A&AS*, 418, 989
Norris, J. 1986, *ApJS*, 61, 667
Pacino & Stetson 2008, in preparation
Pietrzyński, G., Gieren, W., & Udalski, A. 2003, *AJ*, 125, 2494.
Reddy, B. E., Lambert, D. L., & Allende Prieto, C. 2006, *MNRAS*, 367, 1329
Reddy, B. E., Tomkin, J., Lambert, D. L., & Allende Prieto, C. 2003, *MNRAS*, 340, 304
Schuster, W. J., Moitinho, A., Márquez, A., Parrao, L., & Covarrubias, E. 2006, *A&A*, 445, 939
Sneden, C. 1973, PhD thesis, University of Texas, Austin
Steinmetz, M. *et al.* 2006, *AJ*, 132, 1645
Venn, K. A., Irwin, M., Shetrone, M. D., Tout, C. A., Hill, V., & Tolstoy, E. 2004, *AJ*, 128, 1177
Williams, M. E. K *et al.* 2008, in preparation
Zwitter, T. *et al.* 2008, *AJ*, 136, 421

The Bulge-disc connection in the Milky Way

James Binney

Rudolf Peierls Centre for Theoretical Physics,
Keble Road,
Oxford OX1 3NP, UK
email: binney@thphys.ox.ac.uk

Abstract. Bulges come in two flavours – classical and pseudo. The principal characteristics of each flavour are summarised and their impact on discs is considered. Classical bulges probably inhibit the formation of stellar discs. Pseudobulges exchange angular momentum with stars and gas in their companion discs, and also with its embedding dark halo. Since the structure of a pseudobulge depends critically on its angular momentum, these exchanges are expected to modify the bulge. The consequences of this modification are not yet satisfactorily understood. The Galaxy has a pseudobulge. I review the manifestations of its interaction with the disc. More work is needed on the dynamics of gas near the bulge's corotation radius, and on tracing the stellar population in the inner few hundred parsecs of the Galaxy.

Galaxy: kinematics and dynamics; Galaxy: bulge; galaxies: kinematics and dynamics

1. Introduction

In the last decade a consensus has formed that there are two types of bulges: classical bulges, which resemble low-luminosity elliptical galaxies, and "pseudobulges", which do not – for an authoritative case for this distinction see Kormendy & Kennicutt (2004), and for an update on the evidence see Bureau *et al.* (2008). It is believed that pseudobulges form through dynamical instabilities of discs, while classical bulges are products of major mergers. The rapid fluctuations of the gravitational field during a merger randomises the orbits of stars from the progenitor galaxies and violently shocks the progenitors' gas. Stars form rapidly in this gas at a time when there are no circular orbits, so the the stellar system formed is not a centrifugally supported disc but a hot stellar system. Thus classical bulges are highly chaotic stellar systems. A pseudobulge also forms when the gravitational field is unsteady, but the field is not as chaotic as during a major merger, and the pseudobulge inherits a more ordered phase-space structure from its progenitor disc. In particular, it settles to a more rapidly rotating configuration than does a classical bulge, and it is usually triaxial.

In this review I discuss the interactions between bulges of both types and discs, and examine more particularly the case of the Milky Way. The bulge of our Galaxy is still rather mysterious. From our vantage point near the edge of the optical disc, most of it is heavily obscured, and from our edge-on perspective it would be non-trivial to disentangle the bulge's three-dimensional structure even if we had a clear field of view. However, there is every indication that our bulge is a pseudobulge. For this reason alone it would be connected to the disc by history. The connection between bulge and disc can be seen to be ongoing also.

2. Classical bulges

If we define classical bulges to be the analogues of low-luminosity elliptical galaxies, which appear to be axisymmetric (Cappellari *et al.* 2007), then they will be axisymmetric

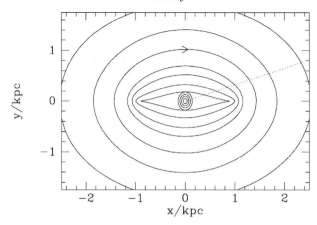

Figure 1. Orbits in a rotating barred potential viewed in the potential's rotating frame. The central orbit family forms the x_2 disc, while the larger, elongated orbits belong to the x_1 family (from Binney & Merrifield 1998).

too. Consequently, we do not expect classical bulges to exchange angular momentum with discs.

Since a thin stellar disc cannot survive a major merger, classical bulges have to be older than any disc that surrounds them. The extraordinary S0 galaxy NGC 4550, which contains two co-spatial discs that rotate in opposite senses (Rubin *et al.* 1992), is a clear indication that galaxies can experience distinct episodes of gas accretion, and that the angular momentum vectors of the gas accreted in the different episodes can be seriously misaligned. Presumably disc galaxies with classical bulges arise when a spheroid that formed in a merger experiences an episode of gas accretion.

Rix & White (1990) showed that many "elliptical" galaxies could host stellar discs. The formation of such faint discs is an ill-understood problem. There are several ways in which the dominant spheroid could profoundly modify the formation of a disc:

• Strong shear associated with the large and fairly constant circular speed generated by the spheroid will inhibit gravitational instability on the gas disc and make any spiral structure tightly wound.

• UV radiation from the spheroid's stars will have a tendency to photo-dissociate molecules, thus increasing the column density of gas required for the transition from atomic to molecular gas that must precede the formation of the cool interstellar cores within which stars form.

• Massive spheroids have a tendency to fill up with gas at the system's virial temperature, $T_{\rm vir}$. This tendency is a consequence of (a) the depth of the gravitational potential well of a massive spheroid, and (b) the large random velocities of the spheroid's stars, which lead to winds from AGB stars colliding at high speed and shocking to $T_{\rm vir}$. The depth of the potential well is significant because it controls whether supernova-heated gas flows out of the galaxy as a wind or accumulates as an X-ray emitting atmosphere. A dense hot atmosphere has the potential to ablate and heat cold gas, thus terminating star formation in that system. I have argued elsewhere (Binney 2004; Nipoti & Binney 2007) that this process is responsible for the cutoff in the galaxy luminosity function at the Schechter luminosity L_* and the transfer of galaxies to the red sequence.

Thus an important and still insufficiently understood area of bulge/disc interaction is the suppression of disc growth by luminous classical bulges.

3. Pseudobulges

Pseudobulges are recognised by their low Sersic indices (their radial surface-bright profiles are closer to exponentials than $r^{1/4}$-laws), fast rotation, and triaxiality (Kormendy & Kennicutt 2004). In earlier-type pseudobulges the mass distribution is sufficiently centrally concentrated to support an inner Lidblad resonance (ILR) (Athanassoula 1992). Inside the ILR the orbits are only mildly non-circular and elongated perpendicular to the bar (Fig. 1). Gas can accumulate on these orbits, and in many systems is dense enough to be support a ring of vigorous star formation. Consequently, pseudobulges, unlike classical bulges, often contain many young stars.

The orbits whose vertical instability drives the formation of a pseudobulge dominate the density at a fair fraction of $R_{\rm CR}$ (Pfenniger & Friedli 1991). Consequently, this is the radial range over which a pseudobulge is vertically thick. Further out, around corotation, in the region of the x_2 disc, the system remains fairly thin vertically (Athanassoula 2005). Thus rather counter-intuitively, important parts of a pseudobulge are disc-like. In particular, the sharp upward rise in the brightness profile of a face-on pseudobulge is more likely to be due to a thin luminous x_2 disc than to a bulge in the classical sense.

Gas that between corotation and the x_2 disc cannot follow nearly circular orbits. At the outer edge of this region gas may flow roughly along x_1 orbit (Fig. 1), but shocks tend to develop in the flow, which deprive the gas of energy, so it drifts inwards at a non-negligible rate, surrendering angular momentum to the stellar bar as it goes. If there is an x_2 disc, the shocks are displaced from the long axis of the bar so as to nearly touch the intersection of the edge of the x_2 disc and the galaxy's minor axis (Athanassoula 1992). In the absence of an x_2 disc, the shocks keep close to the bar's major axis. The shocks are displaced from the major axis in the downstream direction, and the enhancement of the gas density in the post-shock region causes the bar's gravitational field to drain angular momentum from the gas. At optical wavelengths the shocks manifest themselves as dust lanes. The speed with which shocks drain energy and angular momentum from gas in the region $R_{\rm CR} > R > R_{\rm ILR}$ causes the surface density of gas in this region to be small relative to the density outside corotation and in the x_2 disc.

3.1. Impact on spiral structure

Self-sustaining spiral disturbances in a centrifugally supported disc occur either inside their corotation radius (as in the classical WKBJ picture – see Binney & Tremaine 2008, §6.3) or in the immediate vicinity of this radius (Sellwood & Kahn 1991). Since the corotation radius of the bar of a spiral galaxy lies near the ends of the bar, it follows that self-sustaining spiral disturbances rotate less rapidly that the bar; in most of the disc the spiral pattern cannot rotate with the bar (Sellwood & Sparke 1988). However, strong forcing of the disc near the end of the bar can produce a corotating spiral disturbance.

3.2. Exchanges of angular momentum

We have seen that inside $R_{\rm CR}$ gas surrenders angular momentum to the bar. Gas and stars outside $R_{\rm CR}$ tend to take up angular momentum *from* the bar. In particular, gas piles up outside $R_{\rm CR}$ as gas spiralling inward from the outer disc acquires angular momentum from the bar as it approaches $R_{\rm CR}$. Over time stars formed in this region of enhanced gas density form a stellar ring. A further stellar ring often develops further out in the region of the outer Lindblad resonance (OLR), where the orientation of near-circular closed orbits changes from aligned with the bar (inside the OLR) to aligned perpendicular to the bar. These rings have been used to infer the pattern speeds of the bars (Buta & Combes 1996).

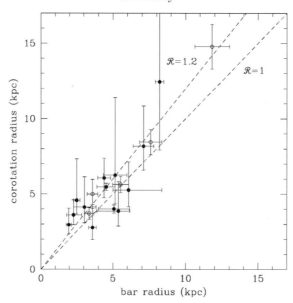

Figure 2. Measurents of corotation radius R_{CR} and bar semi-axis length a from a variety of sources. The straight lines show constant values of $\mathcal{R} = R_{CR}/a$ (From Binney & Tremaine 2008).

3.3. Impact on the bar

A bar consists of orbits that have less angular momentum than circular orbits with the same energy. In this sense, bars are negative-angular momentum perturbations of centrifugally supported discs. The pattern speed of the bar is set by the mean precession speed of the elliptical orbits of its stars. When an orbit of the bar loses angular momentum, it becomes more elongated and precesses more slowly. Consequently, when the bar loses angular momentum, it becomes stronger and slower, and conversely when it gains angular momentum. In particular, gas in the region $R_{OLR} < R < R_{CR}$ weakens and accelerates the bar as it spirals in, while when gas and stars outside R_{CR} take up angular momentum from the bar, they strengthen and slow it. Bars will also lose angular momentum to a slowly on non-rotating dark halo on a timescale that is not long compared to the Hubble time (Weinberg & Tremaine 1984; Debattista & Sellwood 1998).

Numerical simulations of bars forming from unstable discs suggests that all bars are born "fast", that is they have $\mathcal{R} \equiv R_{CR}/a \simeq 1.2$. The empirical evidence that real bars are fast is shown in Fig. 2. This finding suggests that losses of angular momentum by the bar to the dark halo and the disc at $R > R_{CR}$ is balanced by gains of angular momentum from gas that makes it through the barrier at R_{CR}. Certainly there is the possibility of stabilising feedback in that when a bar loses angular momentum, R_{CR} increases, so the non-axisymmetric forces on gas near R_{CR} weaken. Since it is these forces that must be overcome by dissipation if gas is to cross R_{CR}, a decrease in the bar's pattern speed will increase the rate at which gas crosses R_{CR}, and this increase will in turn drive the pattern speed back up.

This proposal, that the pattern speeds of bars are kept up by gas that passes R_{CR}, would appear not to apply to gas-poor S0 galaxies, and in fact most measured pattern speeds are for such systems. Therefore these measurements are currently puzzling.

Bournaud & Combes (2002) propose a more radical explanation. By bringing mass towards the centre, bars endanger their health: once the potential becomes sufficiently centrally concentrated, the bar dissolves (Hasan & Norman 1990; Friedli & Benz 1993).

Perhaps after a bar forms, inspiralling gas builds up outside R_{CR}, while the gas that was initially inside R_{CR} flows to the centre and weakens the bar. Eventually the bar is destroyed and the gas waiting outside R_{CR} flows inwards, and reaches a sufficient surface density to form a new bar. In their numerical simulations Bournaud & Combes (2002) foud that the process of gradual weakening of the bar followed by death and the inward rush of new material can repeat three times in a Hubble time. Each new bar was smaller and faster than the previous one.

4. Case of the Milky Way

There is much evidence that the Milky Way possesses a pseudobulge:
- The COBE/DIRBE near-IR photometry of the bar shows the characteristic "peanut" shape of a pseudobulge.
- Photometry of individual sources, such as clump giants, shows that at longitudes $l \sim 5°$ the bulge is about 0.3 mag closer than at $l \sim -5°$ (Stanek et al. 1997). Thus the bulge is triaxial and and the nearer end is on our left.
- The longitude-velocity plots of both H I and CO show many features indicative of gas flow in a barred potential. The most prominent is the swath of CO emission that runs diagonally across the (l, v) plot for CO, from $(l \simeq 30°, v \simeq 100 \,\mathrm{km\,s^{-1}})$ to the equivalent point on the other side of the origin (Dame et al. 2001). This is the signature of a ring of gas of radius $R_0 \sin(30) \simeq 4\,\mathrm{kpc}$. At $l > 0$ and velocities greater than those of this band there is a marked dearth of gas until l drops to a couple of degrees. This is the signature of the almost empty region between R_{CR} and the edge of the x_2 disc at $\sim R_0 \sin(1.8\,\mathrm{deg}) \simeq 250\,\mathrm{pc}$. At $-1 \lesssim l \lesssim 1.8°$ there is a ridge of emission by CS (which traces very dense gas) at velocities that extend up to $\sim 100\,\mathrm{km\,s^{-1}}$ (Binney et al. 1991). This is the signature of the x_2 disc. Another important feature is the narrow band of strong gas emission that slopes downwards to the right in (l, v) plots, crossing $l = 0$ at $v \simeq -53\,\mathrm{km\,s^{-1}}$ and reaching the curve of tangent velocities at $l \simeq -22°$. This "expanding 3 kpc arm" (van Woerden et al. 1957) is probably associated with the ultraharmonic (4:1) resonance just inside R_{CR}. In this interpretation the arm lies on the near side of the centre. Dame & Thaddeus (2008) have recently identified the counterpart on the far side, and conclude that this feature is satisfyingly symmetrical to the near-side arm. Interestingly they are in the plane $b = 0$ whereas much of the H I on x_1 orbits appears to be in a plane that is significantly inclined to this plane (Liszt & Burton 1980; Ferrière et al. 2007).

The Sun is expected to lie near the OLR of the bar. Kalnajs (1991) pointed out that in this region the local stellar velocity distribution might be expected to be bimodal: at a resonance the orientation of the closed orbits shifts through 90° and near the resonance one would expect to find stars trapped around both kinds of closed orbit, with the result that the velocity distribution would be bimodal. Subsequently Raboud et al. (1998) and Fux (1999) used this effect to explain the distribution of velocities in the Hipparcos catalogue. Dehnen (2000) simulated the effect on the local velocity distribution of the adiabatic growth of the bar potential. He found that the simulated velocity distributions resemble the Hipparcos distribution only if the Sun lies outside the bar's OLR ($R_{OLR} \simeq 0.85 R_0$). The crucial feature is the Hercules stream, which is very prominent in the Hipparcos catalogue and made up of stars that are moving outwards on on orbits that are aligned with the bar. From this analysis Dehnen obtains a value for the pattern speed of the bar $[(1.85 \pm 0.15)\Omega(R_0)]$ that agrees well with independent estimates, for example that obtained by simulating the structure of the (l, v) diagram of molecular gas (Englmaier & Gerhard 1999).

Near-IR star counts from the 2-MASS (Hammersley et al. 2000) and TCS-CAIN catalogues (Cabrera-Lavers et al. 2007) and Spitzer data (Benjamin et al. 2005) show overdensities of stars within 100 pc of the plane at longitudes satisfying $20° < l < 27°$ and distances that imply galactocentric radii 3−4.5 kpc. These features, which include luminous, presumably young stars, seem to lie around or just outside R_{CR} for the bar along a line through the Galactic centre that is inclined to the Sun–centre line by $\sim 43°$. This inclination lies outside the range $30°-15°$ favoured by studies of both the COBE/DIRBE photometry (Binney et al. 1997; Bissantz & Gerhard 2002) and the flow of gas in the bar's potential (Fux 1999; Englmaier & Gerhard 1999). Analyses of the COBE/DIRBE photometry concentrate on the sky at $b \gtrsim 2°$ to minimise the effects of obscuration, so they would not be sensitive to the overdensities in the near-IR star counts.

Hammersley et al. (2000) suggested that these features form "a long thin bar", but it is not clear that the features extend to the centre. It might make more sense dynamically if the features were confined to the corotation region, when they might be created in the disc by the bar's forcing. It would remain puzzling that they lie on the leading edge of the bulge/bar. We still have some way to go before we have a coherent picture of the Galaxy's structure in this region of strongest interaction between the bulge and the disc.

Do we see evidence that the bulge is related to the disc, as the standard theory of pseudobulge formation implies? There does seem to be a strong connection between the bulge and the thick disc. Both systems are old (e.g. McWilliam & Rich 1994; Zoccali et al. 2006) and have similar star-formation histories as diagnosed by the distributions of their stars in the ([O/Fe],[Fe/H]) plane (Meléndez et al. 2008). Haywood (2008) has argued effectively that the chemical evolution of the thin disc follows on naturally from the end point of thick-disc formation. Hence at the moment the data are all consistent with the conjecture that the bulge and thick disc formed from the Galaxy's early disc, and the thin disc has gradually accumulated in the long quiet period that followed. The major caveat that should be made here is that the x_2 disc is clearly the site of vigorous star formation, as attested, for example, by the numerous supernova remnants detected in the central 100 pc at radio continuum wavelengths (LaRosa et al. 2000). In external galaxies similar to the Milky Way central star-forming discs and rings are often prominent and significantly affect our characterisation of the whole bulge. In the case of the Galaxy this region is so highly obscured at optical wavelengths that it has had no impact on our characterisation of the bulge. For example, we say that the bulge is old because the stellar population that is studied a few degrees off the plane is old. When technology allows us to determine the ages and abundances of stars that lie near the plane at $r < 250$ pc, we will probably reassess the bulge. If the x_2 disc has been forming stars throughout the life of the Galaxy, as is perfectly likely, it would be interesting to know how these stars are distributed now.

5. Summary

Bulges fall into two classes: classical and pseudobulges. The impact that a classical bulge has on the associated disc is still speculative, but there is a distinct possibility that a luminous classical bulge will suppress the formation of a co-spatial stellar disc.

Pseudobulges are fashioned out of a disc that has gone bar unstable. Because their patterns rotate, they inevitably exchange angular momentum with any other nearby component.

Bars weaken and speed up when they gain angular momentum and strengthen and slow down when they lose it. It is natural for a bar to lose angular momentum to the disc that lies beyond R_{CR} and to the dark halo. The loss of angular momentum to gas in the surrounding disc leads to a buildup of gas outside $R_{\rm CR}$ and the formation thee of an "inner" stellar ring. "Outer" rings can also form at the outer Lindblad resonance of the bar.

Gas that makes it across the barrier around $R_{\rm CR}$ quickly surrenders most of its angular momentum and settles onto the x_2 disc if one exists. The buildup of gas on the x_2 disc frequently leads to rapid star formation and the formation of a "nuclear" stellar ring.

All the bars that have had their pattern speeds measured to date prove to be "fast" rotators in the sense that their corotation radius lies within ~ 1.2 of their ends. This finding is remarkable because the timescale for bars to gain or lose angular momentum is thought to be significantly shorter than the Hubble time. In spiral galaxies one might conjecture that there is a balance between the angular-momentum gain from gas that slips past $R_{\rm CR}$ and angular-momentum loss to the outer stellar disc and the dark halo. However, this conjecture does not explain why the bars of S0 galaxies, which contain little gas, are also fast. An alternative explanation is that bars destroy themselves by bringing gas in and making the galactic potential more centrally concentrated. After the bar has self-destructed, gas accumulates on circular orbits in the bar's old radial range, and forms a new bar once it has reached a critical mass.

Our Galaxy's bulge displays all the characteristics of a pseudobulge: above the plane we see the typical peanut shape of the thick part of the bar, and in the plane we see a star-forming x_2 disc, a deficiency of gas between this disc and a ring of gas that lies outside $R_{\rm CR}$. We also see features in the local velocity distribution that are attributable to the bar's OLR lying just inside R_0. The first indications that the Galaxy is barred were the presence of gas with large radial velocities along the line of sight to the Galactic centre. The biggest concentration of such gas lies where the "3 kpc" arm crosses the Sun–Centre line. Recently its counterpart on the far side of the Galaxy has been identified. It is remarkably similar to the near-side arm, as we would expect of a feature that is driven by a bar rather than spiral structure. These arms probably lie at the ultraharmonic (4:1) resonance, but this has yet to be securely established.

Although the idea of a Galactic bar is able to account for a large number of observations in a satisfying way, there is still one major puzzle to resolve. This is the status of overdensities in the near-IR starcounts that have been interpreted in terms of a "long-thin bar." These overdensities are securely established at larger longitudes than those at which we expect to see the end of the conventional bar, and photometric distance estimates to these features suggest that they lie along a line through the Galactic centre that is rotated in the direction of Galactic rotation by 15–20 deg with respect to the long axis of the bulge/bar. Since the overdensities are confined to ~ 100 pc of the plane and involve luminous stars, they are presumably associated with gas flow. More work needs to be done on the flow of gas around the Galaxy in the range 5–3 kpc.

The stellar population of the readily observed thick part of the bar is remarkably similar to that of the thick disc. This finding suggests that these two components formed simultaneously from an early, gas-rich and dynamically unstable thin disc. In addition to the old stellar bulge population that has been studied to date, the central bulge must contain a younger population of stars formed in the x_2 disc. Finding these stars and deducing the star-formation history of the x_2 disc is an important task for the future.

References

Athanassoula, E. 1992, MNRAS, 259, 328
Athanassoula, E. 2003, MNRAS, 341, 1179
Athanassoula, E. 2005, MNRAS, 358, 1477

Benjamin, R. A., *et al.* 2005, ApJ, 630, L149
Binney, J., Gerhard, O. E., Stark, A. A, Bally, J., & Uchida, K. I. 1991, MNRAS, 252, 210
Binney, J., Gerhard, O. E., & Spergel, D. 1997, MNRAS, 288, 365
Binney J. & Merrifield, M. 1998, *Galactic Astronomy* (Princeton: Princeton University Press)
Binney, J. 2004, MNRAS, 347, 1093
Binney J. & Tremaine, S. 2008, *Galactic Dynamics* (Princeton: Princeton University Press)
Bissantz, N. & Gerhard, O. E. 2002, MNRAS, 330, 591
Block, D.L., Bournaud, F., Combes, F., Puerari, I. & Buta, R. 2002, A&A, 394, L35
Combes, F. & Sanders, R. H. 1981, A&A, 96, 164
Bournaud, F. & Combes, F. 2002, A&A, 392, 83
Bureau M., Athanassoula, E., & Barbuy, B. 2008, Proc.Intl.Astron.U., 3, IAU Symp. 245
Buta, R. & Combes, F. 1996, FCPh, 17, 95
Cabrera-Lavers, A., Hammersley, P. L., González-Fernández, C., López-Corredoira, M., Garzón, F., & Mahoney, T. J. 2007, A&A, 465, 825
Cappellari, M., & the SAURON collaboration, 2007, MNRAS, 379, 418
Dame, T. M., Hartmann, Dap., & Thaddeus, P. 2001, ApJ, 547, 792
Dame, T. M. & Thaddeus, P. 2008, ApJL, in press (arXiv:0807.1752)
Debattista, V. P. & Sellwood, J. A. 1998, ApJ, 493, L5
Dehnen, W. 2000, AJ, 119, 800
Englmaier, P. & Gerhard, O. 1999, MNRAS, 304, 512
Ferrière, K., Gillard, W. & Jean, P. 2007, A&A, 467, 611
Friedli, D. & Benz, W., A&A, 268, 65
Fux, R. 1999, A&A, 345, 787
Hammersley, P. L., Garzon, F., Mahoney, T. J., Lopez-Corredoira, M., & Torres, M. A. P. 2000, MNRAS, 317, L45
Hasan, H. & Norman, C. A. 1990, ApJ, 361,69
Haywood, M. 2008, arXiv:0805.1822
Kalnajs, A. 1991, in "Dynamics of Disc Galaxies", ed. B. Sundelius (Göteborg: Dept of Astron. Astrophys., Göteborg Univ.), 323
Kormendy, J. & Kennicutt, R. 2004, ARA&A, 42, 603
LaRosa, T. N., Kassim, N. E., Lazio, T. J. W., & Hyman, S. D., 2000, AJ, 119, 207
Liszt, H. S. & Burton, W. B. 1980, ApJ, 236, 779
McWilliam, A. & Rich, R. M. 1994, ApJS, 91, 749
Meléndez, J., *et al.* 2008, A&A, 484, L21
Nipoti, C. & Binney, J. 2007, MNRAS, 382, 1481
Pfenniger D. & Friedli D. 1991, A&A, 252, 75
Raboud, D., Grenon, M., Martinet, L., Fux, R. & Udry, S. 1998, A&A, 335, L61
Raha, N., Sellwood, J. A., James, R. A. & Kahn, F. D., Nature, 352, 411
Rix H.-S. & White S. D. M. 1990, ApJ, 362, 52
Rubin, V. C., Graham, J. A., & Kenney, J. D. P. 1992, ApJ, 394, L9
Sellwood, J. A. & Kahn, F. 1991, MNRAS, 250, 278
Sellwood, J. A. & Sparke, L. S. 1988, MNRAS, 231, P25
Stanek, K. Z., Udalski, A., Szymanski, M., Kaluzny, J., Kubiak, M., Mateo, M. & Krezeminski, W. 1997, ApJ, 477, 163
van Woerden, H., Rougoor, G. W., & Oort, J. H. 1957, Comp. Rend., 244, 1691
Weinberg, M. D. & Tremaine, S. 1984, MNRAS, 209, 729
Zoccali, M., *et al.* 2006, A&A, 457, L1

Stellar abundances tracing the formation of the Galactic Bulge

Beatriz Barbuy[1], Manuela Zoccali[2], Sergio Ortolani[3], Vanessa Hill[4], Alvio Renzini[5], Jorge Meléndez[6], Anita Gómez[4], Martin Asplund[7], Dante Minniti[2], Eduardo Bica[8], and Alan Alves-Brito[1]

[1]Universidade de São Paulo, IAG, Rua do Matão 1226
São Paulo 05508-900, Brazil
email: barbuy@astro.iag.usp.br

[2]Universidad Catolica de Chile, Department of Astronomy & Astrophysics, Casilla 306, Santiago 22, Chile
email: mzoccali@astro.puc.cl, dante@astro.puc.cl

[3]Università di Padova, Dipartimento di Astronomia, Vicolo dell'Osservatorio 2, I-35122 Padova, Italy
email: sergio.ortolani@unipd.it

[4]Observatoire de Paris-Meudon, 92195 Meudon Cedex, France
email: Vanessa.Hill@obspm.fr, anita.gomez@obspm.fr

[5]Osservatorio Astronomico di Padova, Vicolo dell'Osservatorio 5, I-35122 Padova, Italy
email: alvio.renzini@oapd.inaf.it

[6]Centro de Astrofísica da Universidade de Porto, Rua das Estrelas, 4150-762 Porto, Portugal
email: jorge@astro.up.pt

[7]Max Planck Institute for Astrophysics, Postfach 1317, 85741 Garching, Germany
email: asplund@MPA-Garching.MPG.DE

[8]Universidade Federal do Rio Grande do Sul, Departamento de Astronomia, CP 15051, Porto Alegre 91501-970, Brazil e-mail: bica@if.ufrgs.br

Abstract. The metallicity distribution and abundance ratios of the Galactic bulge are reviewed. Issues raised by different groups in recent work, in particular the high metallicity end, a comparison between the oxygen abundances derived from different indicators, the [OI] 630nm and IR OH lines, and the issue of measuring giants vs. dwarfs, are discussed. Finally, abundances in bulge globular clusters are briefly described.

Keywords. Galaxy: abundances, bulge, stellar content, globular clusters

1. Introduction

A debate concerning the nature of the Galactic bulge has been going on over a decade. Kormendy & Kennicutt (2004), Binney (2007; 2008, this conference) consider that classical bulges are the result of mergers and can be found in Sa, Sb spirals, and that these are elliptical galaxies-like that happen to have conspicuous disks, whereas later type spirals have pseudo bulges, formed from disks that develop a bar. Still, there are two different views on this: Sellwood (this conference) for example, forms bulges from bars made of stars, whereas Combes (2007) for example, forms bars from gas. It is then expected that the bulge stellar population should be similar to the inner parts of the disk from which they formed (e.g. Athanassoula 2007).

Observational studies of stellar populations from Colour-Magnitude Diagrams (CMDs) and spectroscopic abundances indicate, so far, that the bulge is old and metal-rich, and α-enhancement indicates early fast formation.

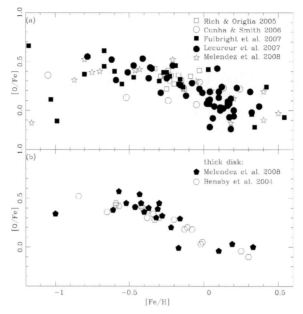

Figure 1. [O/Fe] vs. [Fe/H]: (a) for bulge giants analysed by Rich & Origlia (2005), Cunha & Smith (2006), Zoccali et al. (2006), Fulbright et al. (2007) and Meléndez et al. (2008); (b) for thick disk stars, with dwarfs from Bensby et al. (2004) and giants by Meléndez et al. (2008).

Ortolani et al. (1995), Feltzing & Gilmore (2000), and Zoccali et al. (2003) have shown that the field bulge stars at the Baade's Window are old, since their CMDs are very similar to those of the old metal-rich globular clusters NGC 6528 and NGC 6553, which in turn show similar ages to the old metal-rich thick disk globular cluster 47 Tuc (see Alves-Brito et al. 2005).

Samples of bulge field giants studied by Rich & Origlia (2005), Fulbright et al. (2006, 2007), Cunha & Smith (2006), Zoccali et al. (2006, 2008), Lecureur et al. (2007), Meléndez et al. (2008) include targets from Baade's Window (BW) only, except for Zoccali et al. (2006) and Lecureur et al. (2007) who observed 3 other fields besides BW.

In short, the results so far indicate that the bulge is old, but modellers tend to believe that we are gathering data from parts not much affected by the dynamical evolution of the bar.

2. Abundance ratios

In recent years, abundance ratios for bulge field giants were reported by Fulbright et al. (2007), Cunha & Smith (2006), Zoccali et al. (2006), Lecureur et al. (2007) and Meléndez et al. (2008). In particular the oxygen abundance derived by all these groups indicate an enhancement of oxygen relative to iron as shown in Fig. 1a. The oxygen enhancement is found by these 5 different groups, and these results use both the forbidden [OI] 630nm line and infrared OH lines; therefore the oxygen enhancement appears to be a robust result. Concerning the results by Zoccali et al. (2006) and Lecureur et al. (2007) it is important to recall that O and Mg abundances were derived in four fields, and no difference was found between them.

As for the comparison with the thick disk, Zoccali et al. (2006) compared the bulge data to the thick disk dwarfs by Bensby et al. (2004), whereas Meléndez et al. (2008) compared to thick disk giants analysed in the same way as their sample bulge giants.

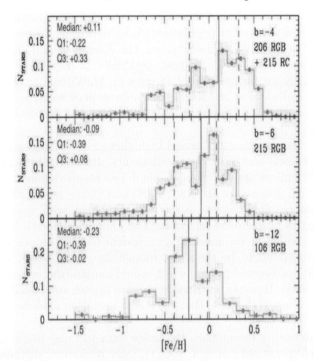

Figure 2. Metallicity distribution in 3 fields along the bulge minor axis (Zoccali et al. 2007; 2008).

Zoccali et al. (2006) found that the bulge is more oxygen-enhanced than the thick disk, whereas Meléndez et al. (2008) found that the oxygen abundances of bulge and thick disk stars overlap each other. The oxygen abundances for thick disk stars are shown in Fig. 1b; it appears that they are not very different, even if one sample concerns dwarfs and the other concerns giants. Therefore, from presently available data, the oxygen abundances of bulge stars are either slightly enhanced, or coincident with the thick disk.

Mg is also enhanced, and shows an overabundance relative to the thick disk (Fulbright et al. 2007; Lecureur et al. 2007). An enhancement is also found for the other α-elements Si, Ca and Ti by Fulbright et al. (2007). However Meléndez et al. (priv. comm.) find in their sample similar abundances of Mg, Si, Ca, Ti for their bulge and thick disk stars.

An issue that will have to be studied in detail as well is that some discrepancies are found between Near-Infrared and optical data, in particular this seems to be clear for Ca abundances.

Given the importance of such studies in terms of their interpretation of bulge formation, a huge effort on this would be necessary to confirm either of the conclusions.

3. Metallicity distribution

Zoccali et al. (2008) present the metallicity distribution for a large sample of about 800 bulge giants, observed in four fields, using FLAMES (Pasquini et al. 2000). The fields are located at: Baade's Window: $1°.14$, $-6°.02$, E(B-V)= 0.46; NGC 6553 field: $5°.25$, $-3°.02$, E(B-V) = 0.70; Field at $-6°$: $0°.21$, $-6°.02$, E(B-V) = 0.48; Blanco field at $-12°$: $0°.00$, $-12°.00$, E(B-V) = 0.20.

In the BW field, two clear peaks are seen around [Fe/H] = +0.3, and -0.25. Progressively along the minor axis, at -6° and -12°, the peak at [Fe/H] = +0.3 tends to disappear as shown in Fig. 2 (Zoccali et al. 2007, 2008). We see therefore a gradient of stellar populations. Kinematical studies of this sample are under way, in order to better understand this most metal-rich stellar population (A. Gómez et al., in preparation).

The high end of the metallicity distribution: The Zoccali et al. (2008) metallicity distributions for the four fields give a highest metallicity at [Fe/H]≈+0.5. The same had been found previously for the samples of 11 stars by McWilliam & Rich (1994), and 27 giants analysed by Fulbright et al. (2006). A recalibration of Sadler et al. (1996)'s results by the latter authors indicated, however that the high end of the metallicity distribution is at [Fe/H]=+1.0.

We believe that the metallicities derived by Sadler et al. (1996), based on indices, are unreliable for the coolest and most metal-rich giants, due to the presence of TiO bands. MgFe, Mg_2 and Fe indices are no longer viable if TiO bands dominate the spectrum.

On the other hand, Johnson et al. (2007, 2008) and Cohen et al. (2008) derived metallicities and abundances for 3 dwarfs in the Galactic bulge, with spectra obtained with high S/N favoured by a microlensing effect. The found [Fe/H]=+0.5 for the 3 dwarfs. They argue that it would be too much of a coincident to have 3 stars at the high metallicity end, so that this might be the mean metallicity of the bulge, and further argue that such stars do not become giants, which would explain the metallicity distribution by Zoccali et al. (2008). However, extreme horizontal branch stars (EHBs) should be seen in the CMDs, which is not the case. Besides, stellar evolution modellers argue that there is no such effect.

Finally, it is interesting to compare the bulge data to bulge-like dwarfs of the solar neighbourhood, which are old (around 10 Gyr), show pericentric distances down to R_{per} ≈ 3.5 kpc, and very low maximum height Z_{max}. We believe that these stars are outer bulge stars accelerated by the bar towards the solar neighbourhood. In Pompeia et al. (2008) it is shown that α-element abundances are intermediate between the bulge and the thick disk.

4. Bulge globular clusters

We have identified a number of Blue Horizontal Branch (BHB) clusters with metallicities around [Fe/H] ≈ −1.0. Within 10°×° and R<4 kpc of the Galactic center, 36 clusters are found, half of them metal-rich ([Fe/H] > − 0.6) and the other half are metal-poor. Around 8 of them show the BHB with intermediate metallicity. From the detailed analyses of 3 of them, namely HP 1 (Barbuy et al. 2006), NGC 6558 (Barbuy et al. 2007), and NGC 6522 (Barbuy et al. 2008, in preparation), the abundance patterns are very similar among them - see Fig. 3, and different from the halo pattern. The unique very metal-poor cluster in the bulge analysed so far, Terzan 4 (Origlia & Rich 2004) shows a similar pattern to the halo.

We suggest that Terzan 4 could be a "normal halo" cluster, whereas the BHB intermediate metallicity ones might be a typical early bulge sample.

5. Planetary Nebulae

A comparison of the latest results on oxygen abundances in bulge giants and bulge Planetary Nebulae is presented in Chiappini et al. (2008). As previously found already, a difference of 0.2 dex with bulge giants showing a higher value, is confirmed. A detailed analysis of the uncertainties involved in the O/H values for stars, and uncertainties in Planetary Nebulae studies, indicates that the results can be considered compatible.

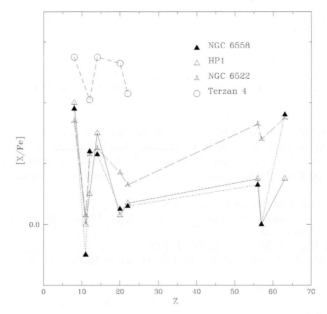

Figure 3. Abundance pattern of metal-poor bulge globular clusters: Terzan 4 with [Fe/H]=-1.6 (Origlia & Rich 2004), and HP 1, NGC 6558, NGC 6522 with [Fe/H] $\approx$ -1.0 (Barbuy et al. 2006, 2007, 2008).

6. Conclusions

Stellar populations studied so far in the bulge are old, as indicated by Colour-Magnitude Diagrams typical of $\sim$0.8M$_\odot$ stars of ages around 12 Gyr. The α-element Oxygen and Magnesium to iron ratios show an enhancement, indicating an early fast chemical enrichment, dominated by SNe type II.

In the very central parts of the Galaxy, within a few pc, a young population is seen, but no sign of such a young component is detected in other fields.

We recognize that more fields should be investigated, in order to have a better coverage in latitude and longitude. On the other hand, there are not that many windows, and reddening is very high towards the more central regions.

Given that we do not see signs of a younger bar population, we suggest that either have we an old bulge, and the peanut-shape bulge/bar, if younger, should be traced through abundance ratios still to be detected, or the bulge and old thick disk should have formed from a bar formed at early times.

Finally, we identify the possibly oldest globular clusters in the Galaxy, showing intermediate metallicity combined with a Blue Horizontal Branch. Should these old globulars have formed early in pre-bulge central parts, or accreted from dwarf galaxies?

References

Alves-Brito, A., Barbuy, B., Ortolani, S., Momany, Y., & Hill, V. et al. 2005, A&A 435, 657
Athanassoula, E. 2007, in *Formation and Evolution of Galaxy Bulges*, IAU Symp. 245, eds. M. Bureau, E. Athanassoula, & B. Barbuy, Cambridge University Press, p. 93
Barbuy, B., Zoccali, M., Ortolani, S., Momany, Y., & Minniti, D. et al. 2006, A&A 449, 349
Barbuy, B., Zoccali, M., Ortolani, S., Hill, V., & Renzini, A., et al. 2007, AJ 134, 1613
Barbuy, B., Zoccali, M., Ortolani, S., Hill, V., & Renzini, A., et al. 2008, A&A in preparation
Bensby, T., Feltzing, S., & Lundström, I. 2004, A&A 415, 155

Binney, J. 2007, in *Formation and Evolution of Galaxy Bulges*, *IAU Symp.* 245 eds. M. Bureau, E. Athanassoula & B. Barbuy, p. 455
Chiappini, C., Górny, S. K., Stasińska, G., & Barbuy, B. 2008, *A&A* in preparation
Cohen, J. G., Huang, W., Udalski, A., Gould, A., & Johnson, J. 2008, *arXiv* 0801.3264
Combes, F. 2007, in *Formation and Evolution of Galaxy Bulges*, *IAU Symp.* 245, eds. M. Bureau, E. Athanassoula, & B. Barbuy, Cambridge University Press, p. 149
Cunha, K. & Smith, V. V. 2006, *ApJ* 651, 491
Feltzing, S. & Gilmore, G. 2000, *A&A*, 355, 949
Fulbright, J. P., McWilliam, A., & Rich, R. M. 2006, *ApJ* 636, 821
Fulbright, J. P., McWilliam, A., & Rich, R. M. 2007, *ApJ* 661, 1152
Johnson, J. A., Scott Gaudi, B., Sumi, T., Bond, I. A., & Gould, A. 2008, *arXiv* 0801.2159
Johnson, J. A., Gal-Yam, A., Douglas, L., Simon, J. D., Udalski, A., & Gould, A. 2007, *ApJ* 655, L33
Kormendy, J. & Kennicutt, R. 2004, *ARA&A* 42, 603
Lecureur, A., Hill, V., Zoccali, M., Barbuy, B., Gómez, A., Minniti, D., Ortolani, S. & Renzini, A. 2007, *A&A* 465, 799
McWilliam, A. & Rich, R. M. 1994, *ApJS* 91, 749
Meléndez, J., Asplund, M., Alves-Brito, A., Cunha, K., & Barbuy, B. *et al.* 2008, A&A, 484, L21
Origlia, L. & Rich, R. M., 2004, *AJ* 127, 3422
Ortolani, S., Renzini, A., Gilmozzi, R., Marconi, G., Barbuy, B., Bica, E., & Rich, R. M., 1995, *Nature* 377, 701
Pasquini, L., Gerardo, A., Allaert, E., Ballester, P., & Biereichel, P. *et al.* 2000, *SPIE* 4008, 129
Pompéia, L., Barbuy, B., Gustafsson, B., & Grenon, M. 2008, *ApJ* in preparation
Rich, R. M. & Origlia, L. 2005, *ApJ* 634, 1293
Rich, R. M., Howard, C., Reitzel, D. B., Zhao, H.-S., & de Propis, R. 2007, in *Formation and Evolution of Galaxy Bulges*, *IAU Symp.* 245, eds. M. Bureau, E. Athanassoula, & B. Barbuy, Cambridge University Press, p. 333
Sadler, E., Rich, R. M., & Terndrup, D. M. 1996, *AJ* 112, 171
Zoccali, M., Renzini, A., Ortolani, S., Greggio, L., Saviane, I., Cassisi, S., Rejkuba, M., Barbuy, B., Rich, R. M., & Bica, E. 2003, *A&A* 399, 931
Zoccali, M., Lecureur, A., Barbuy, B., Hill, V., Renzini, A., Minniti, D., Momany, Y., Gómez, A., & Ortolani, S. 2006, *A&A* 457, L1
Zoccali, M., Lecureur, A., Barbuy, B., Hill, V., & Renzini, A. *et al.* 2007, in *Stellar Populations as Bulding Blocks of Galaxies*, *IAU Symp.* 245, eds. A. Vazdekis & R.F. Peletier, Cambridge University Press, p. 73
Zoccali, M., Hill, V., Lecureur, A., Barbuy, B., & Renzini, A. *et al.* 2008, *A&A* 486, 177

Unveiling the Secrets of the Galactic bulge through stellar abundances in the near-IR: a VLT/Crires project

Nils Ryde[1]

[1]Lund Observatory, Box 43, SE-221 00 Lund, Sweden
email: ryde@astro.lu.se

Abstract. The formation and evolution of the Milky Way bulge can be constrained by studying elemental abundances of bulge stars. Due to the large and variable visual extinction in the line-of-sight towards the bulge, an analysis in the near-IR is preferred. Here, I will present some preliminary results of an on-going project in which elemental abundances, especially those of the C, N, and O elements, of bulge stars are investigated by analysing CRIRES spectra observed with the VLT.

Keywords. stars: abundances, Galaxy: bulge, infrared: stars

1. What Type is the Milky Way bulge?

A bulge is clearly different from its surrounding components: For instance, there are in general, a distinct radial surface brightness peak, specific stellar populations, different kinematics, and less interstellar medium. Although bulges of galaxies form a heterogeneous class of objects, Kormendy & Kennicutt (2004) define two classes of bulges based on the way they are formed: First, the merger-built classical bulges and, second, the secularly evolved ones (pseudo-bulges). These are not clear-cut groups. Galaxies continue to evolve after merger events and can thus exhibit both signatures of a classic bulge and a secularly evolved one. Bulges of the first class are formed through hierarchical clustering and merging events and are similar to elliptical galaxies (bulges of early-type galaxies can therefore be seen as ellipticals in the center of disks). Further, these bulges have had star formation long ago and therefore contain mostly old stars. The second group, the pseudo-bulges, are evolved through slow secular evolution and are important in medium-to-late-type galaxies. Galaxies with pseudo-bulges can not have experienced a merging event for a long time. They are further made slowly out of disk gas and retain a memory of their disk origin, can have ongoing star formation, and exhibit young stars. These two types of bulges give different dynamical and chemical signatures.

The Milky Way is a spiral galaxy, with a pronounced bulge and a small bar, dynamic and transient in nature. The Bulge has a peanut-form profile and its structure and shape have signatures of the secularly evolved pseudo-bulges, but the age of the stellar population(s) and α-element enhancement (O, Mg, Si, Ca, Ti) are signatures of classical bulges. In spite of the fact that the origin, age, and chemical properties of the Bulge remain poorly understood, there is a consensus that there exists a dominant, old, metal-rich population in the Bulge. The bulk of the Bulge's stellar population is thought not to have formed by slow secular evolution. E.g. Zoccali *et al.* (2003) find no trace of a younger population in the Bulge. However, in the Galactic center, within of the order of 100 pc of the geometrical center, there is some evidence of ongoing star formation and a younger population (see e.g. Figer *et al.* (2004) and Barbuy (2002)).

Thus, the formation of the Milky Way bulge is clearly not well understood and its classification is inconclusive. However, the different formation scenarios can be constrained by abundance surveys. From stellar populations and abundance analyses it can be investigated which process dominated the star formation: whether stars formed rapidly a long time ago during hierarchical clustering of galaxies (as a classical bulge), or through secular evolution (such as defines a pseudo-bulge). The α-element composition (e.g. O, Mg, Ca, etc.) relative to iron as a function of the metallicity, [Fe/H], can infer star-formation rates (SFR) and initial-mass functions (IMF) from stellar compositions. A shallower IMF will increase the number α-element producing stars thus leading to higher [α/Fe] values. A faster enrichment due to a high star-formation rate will keep the over-abundance of the α elements at a high value also at higher metallicity. Different populations may show different behaviours.

2. Advantages of a Spectral Analysis in the Near-IR

The Bulge has remained fairly unexplored, mainly because of the large optical obscuration due to dust in the line-of-sight toward the Galactic center. A spectroscopic investigation of elemental abundances based on near-IR spectra alleviates this problem. The Bulge is more accessible in the IR than in the optical for multiple reasons (Ryde *et al.* 2005). The most important is of course the smaller interstellar extinction in the IR ($A_K \sim 0.1 \times A_V$; Cardelli *et al.* 1989). Furthermore, red stars stand out the most in the near-IR not only because they are brightest there. Admittedly, the dust extinction decreases monotonically with wavelength, which favors longer wavelengths, but interstellar dust radiates strongly in the mid- and far-IR, thus favoring the near-IR. Furthermore, in the thermal infrared (i.e. beyond approximately 2.3 μm) the telluric sky starts to shine due to its intrinsic temperature, making observations increasingly more difficult. Thus, this leaves us with the J, H, and K bands for an optimal spectroscopic study. The near-IR is also preferred for analysis of abundances, due to the fact that the absorption spectra are less crowded with lines, that fewer lines are blended, and that it is easier to find portions of the spectrum which can be used to define a continuum compared to wavelength regions in the optical spectral window. Since the transitions occur within the electronic ground-state, the assumption of *Local Thermodynamic Equilibrium (LTE)* in the analysis of the molecules is probably valid (Hinkle & Lambert 1975), which simplifies the analysis dramatically. Moreover, in the Rayleigh-Jeans regime, the intensity is less sensitive to temperature variations. This means that the effects of, for example, effective-temperature uncertainties or surface inhomogeneities on line strengths should be smaller in the IR. Clearly, the near-IR is the optimal spectroscopic region to work in, in order to get a handle on the Milky Way bulge through an abundance analysis.

Recently, a few studies of elemental abundances of bulge stars using near-IR spectra at high resolution have been done, see for instance Meléndez *et al.* (2008), Cunha *et al.* (2007), Cunha & Smith (2006), and Meléndez *et al.* (2003).

3. Drawbacks of a Spectral Analysis in the Near-IR

A general drawback of a spectral analysis in the near-IR is that there are much fewer atomic and ionic lines. The ones that exist often originate from highly excited levels in metals, which also complicates an interpretation. Furthermore, many lines are not properly identified and/or lack known atomic line-strengths, which are needed in an abundance analysis. There are, however, many molecular lines in this region which can be used. Note, though, the lack of signatures from several molecules such as TiO and ZrO in the near-IR. Furthermore, even though large advances have been made when it comes to the technology for recording near-infrared light, existing spectrometers are still much less effective than optical ones, one of the main reasons being the lack of cross-dispersion. Finally, determining the stellar parameters based only on near-IR spectra is difficult.

4. Unveiling the secrets of the Galactic bulge: an infrared spectroscopic study of Bulge giants

We† are engaged in an on-going VLT project (080.D-0675), in which we are analysing near-IR spectra recorded at a spectral resolution of $R = 60,000$ with the CRIRES spectrometer (Moorwood 2005; Käufl et al. 2006) on the *Very Large Telescope, VLT*, see also Ryde et al. (2007). The aim of the project is to draw firm conclusions on the formation history of the Galactic bulge by obtaining precise abundances of C, N, and O in addition to α elements (Mg, Si, S, Ca, and Ti), and a few other elements for a well-chosen sample of stars, sampling different stages of the chemical enrichment history and different parts of the Bulge. Here, we present the first three stars, chosen from Arp (1965), namely Arp 4329, Arp 4203, and Arp 1322 (see Table 1) in Baade's Window. The H-magnitudes of the stars range from $H = 9.2 - 11.1$ and the exposure times from $100 - 300$ s. The spectra will be shown and presented in more detail in Ryde et al. (2008, in prep.).

The CRIRES spectrometer records approximately a three times broader wavelength range compared to, for instance, the Phoenix spectrometer (Hinkle et al. 1998, 2003), which is an important advantage. Note, however, that the signal-to-noise over the CRIRES wavelength range may vary by a factor of two. In the range we have observed, we have identified many CO, CN, and OH lines, 10 Si, 5 Ti, 4 S, 5 Ni, one Cr, and many Fe lines. The modelling of the atmospheres, the determination of the line strengths, a discussion on the sensitivity of the molecular lines caused by *(i)* changes in the C, N, O abundances by 0.1 dex, and *(ii)* uncertainties in the stellar parameters, will be presented in Ryde et al. (2008, in prep.). The abundances of Mg, Si, Al. and Na are important for the calculation of the model atmosphere since these elements are important electron donors and thereby affect the electron pressure in the atmosphere. Also, the continuous opacity which directly affects line strengths, is due to H^- free-free, which depends on the electron pressure. In Figure 1, the relative importance of these different elements as electron donors are shown versus the Rosseland optical depth in the model atmosphere of the bulge star Arp 1322. Therefore, it is important to know these abundances, also when investigating other elements. For a further discussion, see Ryde et al. (2008 in prep.).

The resulting abundances are presented Table 2. A good agreement is found between our near-IR results and the abundances from Fulbright et al (2007), an analysis based on optical spectra. This is reassuring for future analysis of stars for which only near-IR spectra will exist. It should be noted that a general problem with determining elemental abundances for stars with only near-IR spectra is, however, the temperature sensitivity of the molecular lines and the difficulties in the determination of the stellar parameters.

5. Conclusions

Stellar surface abundances in Bulge stars, especially those of the C, N, and O elements, can be extensively studied in the near-IR, due to lower extinction. Here, we have shown

† B. Edvardsson, B. Gustafsson (PI), A. Alves Brito, M. Asplund, B. Barbuy, K. Eriksson, V. Hill, K.. Hinkle, S. Johansson, H.-U. Käufl, D. L. Lambert, A. Lecureur, J. Melendez, D. Minniti, S. Ortolani, F. Primas, A. Renzini, N. Ryde, V. V. Smith, and M. Zoccali

Table 1. Stellar parameters for the three stars presented here. The parameters are taken from Fulbright *et al.* (2007).

Star	$T_{\rm eff}$ [K]	$\log g$ cgs	[Fe/H]	$\xi_{\rm micro}$ [km s^{-1}]
Arp1322	4106	0.89	−0.23	1.6
Arp4203	3902	0.51	−1.25	1.9
Arp4329	4197	1.29	−0.90	1.5

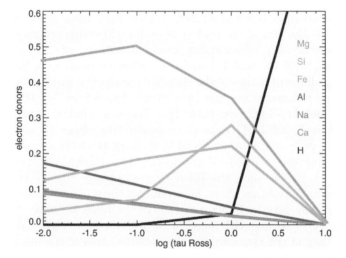

Figure 1. Elements contributing to the electron pressure in the model atmosphere of the Bulge star Arp 1322. The electron pressure affects the continuous opacitiy which directly affects the equivalent widths of spectral lines. At 1.5 μm the continuum is formed at $\log \tau_{\rm Ross} \sim 0.5$, and the lines of the molecules are formed further out, namely at $\log \tau_{\rm Ross}^{\rm CO} \sim 0.2$, $\log \tau_{\rm Ross}^{\rm CN} \sim 0.1$, and $\log \tau_{\rm Ross}^{\rm OH} \sim -2.4$.

a good agreement between near-IR and optically determined abundances in stars in Baade's window, stars which can be observed in both wavelength ranges. It will be very important to extend the analysis to other regions of the Galactic bulge, such as in the galactic plane, in order to get a proper handle on its formation and evolution. In these regions the optical extinction is large which only permits observations in the near-IR. Near-IR, high-spectral-resolution spectroscopy offers a promising methodology to study the whole Bulge to give clues to its formation and evolution.

References

Arp, H., 1965, ApJ 141, 43
Barbuy, B., 2002, in J. J. Claria, D. Garcia Lambas, & H. Levato (eds.), Revista Mexicana de Astronomia y Astrofisica Conference Series, Vol. 14 of Revista Mexicana de Astronomia y Astrofisica, vol. 27, p. 29
Cardelli, J. A., Clayton, G. C., & Mathis, J. S., 1989, ApJ 345, 245
Cunha, K., Sellgren, K., Smith, V. V., et al., 2007, ApJ 669, 1011
Cunha, K. & Smith, V. V., 2006, ApJ 651, 491
Figer, D. F., Rich, R. M., Kim, S. S., Morris, M., & Serabyn, E., 2004, ApJ 601, 319
Fulbright, J. P., McWilliam, A., & Rich, R. M., 2006, ApJ 636, 821
Fulbright, J. P., McWilliam, A., & Rich, R. M., 2007, ApJ 661, 1152

Table 2. Abundances of C, N, O, Ti, Si, S, Cr, Ni, and Fe for Arcturus and three bulge giants as determined from our near-IR spectra observed with VLT/CRIRES. As a comparison the abundances for the same stars determined by Fulbright et al. (2006, 2007) are also provided.

Star	Ref.	$\log \varepsilon(\text{C})^{a}$ [dex]	$\log \varepsilon(\text{N})$ [dex]	$\log \varepsilon(\text{O})$ [dex]	$\log \varepsilon(\text{Ti})$ [dex]	$\log \varepsilon(\text{Si})$ [dex]	$\log \varepsilon(\text{S})$ [dex]	$\log \varepsilon(\text{Cr})$ [dex]	$\log \varepsilon(\text{Ni})$ [dex]	$\log \varepsilon(\text{Fe})$ [dex]
Arcturus	this work	8.06	7.67	8.76	4.68	7.35	6.94	5.17	5.78	7.00
	Fulbright et al.	–	–	8.67	4.68	7.39	–	–	–	6.95
	difference	–	–	0.09	0.00	−0.04	–	–	–	0.05
Arp 4203	this work	6.62	7.70	7.71	3.98	6.75	6.26	4.28	5.12	6.25
	Fulbright et al.	–	–	7.55	4.03	6.82	–	–	–	6.20
	difference	–	–	0.16	−0.05	−0.07	–	–	–	0.05
Arp 4329	this work	7.42	7.23	8.25	4.33	7.15	6.81	4.77	5.41	6.60
	Fulbright et al.	–	–	8.16	4.30	7.14	–	–	–	6.55
	difference	–	–	0.09	0.02	0.01	–	–	–	0.05
Arp 1322	this work	7.93	8.10	8.61	4.87	7.52	7.35	5.62	6.02	7.33
	Fulbright et al.	–	–	8.80	4.84	7.43	–	–	–	7.22
	difference	–	–	−0.19	0.03	0.09	–	–	–	0.11

Notes:
[a] $\log \varepsilon(\text{X}) = \log n_\text{X}/n_\text{H} + 12$, where $\log n_\text{X}$ is the number density of element X.

Hinkle, K. H., Blum, R. D., Joyce, R. R., et al., 2003, in P. Guhathakurta (ed.), Discoveries and Research Prospects from 6- to 10-Meter-Class Telescopes II. Edited by Guhathakurta, Puragra, Vol. 4834 of Proc. SPIE, p. 353

Hinkle, K. H., Cuberly, R. W., Gaughan, N. A., et al., 1998, SPIE 3354, 810

Hinkle, K. H. & Lambert, D. L., 1975, MNRAS 170, 447

Käufl, H. U., Amico, P., Ballester, P., et al., 2006, The Messenger 126, 32

Kormendy, J. & Kennicutt, Jr., R. C., 2004, ARA&A 42, 603

Meléndez, J., Asplund, M., Alves-Brito, A., Cunha, K., Barbuy, B., Bessell, M. S., Chiappini, C., Freeman, K. C., Ramírez, I., Smith, V. V., & Yong, D., 2008, A&A 484, L21

Meléndez, J., Barbuy, B., Bica, E., et al., 2003, A&A 411, 417

Moorwood, A., 2005, in H. U. Käufl, R. Siebenmorgen, & A. F. M. Moorwood (eds.), High Resolution Infrared Spectroscopy in Astronomy, p. 15

Ryde, N., Edvardsson, B., Gustafsson, B., & Käufl, H.-U., 2007, in A. Vazdekis & R. F. Peletier (eds.), IAU Symposium, Vol. 241 of IAU Symposium, p. 260

Ryde, N., Gustafsson, B., Eriksson, K., & Wahlin, R., 2005, in H. U. Käufl, R. Siebenmorgen, & A. F. M. Moorwood (eds.), High Resolution Infrared Spectroscopy in Astronomy, p. 365

Zoccali, M., Renzini, A., Ortolani, S., Greggio, L., Saviane, I., Cassisi, S., Rejkuba, M., Barbuy, B., Rich, R. M., & Bica, E., 2003, A&A 399, 931

James Binney clarifies the final key point in his review.

Nils Ryde presenting his paper.

Bars in Cuspy Dark Halos

John Dubinski[1], Ingo Berentzen[2] and Isaac Shlosman[3,4]

[1] Department of Astronomy and Astrophysics, University of Toronto,
50 St. George Street, Toronto, ON M5S 3H4, Canada
email: dubinski@astro.utoronto.ca

[2] Astronomisches Rechen-Institut,
Mönchhofstr. 12-14 69120, Heidelberg, Germany
email: iberent@ari.uni-heidelberg.de

[3] JILA, University of Colorado, Boulder, CO 80309-0440, USA; and

[4] Department of Physics and Astronomy, University of Kentucky,
Lexington, KY 40506-0055, USA
email: shlosman@pa.uky.edu

Abstract. We examine the bar instability in models with an exponential disk and a cuspy NFW-like dark matter (DM) halo inspired by cosmological simulations. Bar evolution is studied as a function of numerical resolution in a sequence of models spanning $10^4 - 10^8$ DM particles - including a multi-mass model with an effective resolution of 10^{10}. The goal is to find convergence in dynamical behaviour. We characterize the bar growth, the buckling instability, pattern speed decay through resonant transfer of angular momentum, and possible destruction of the DM halo cusp. Overall, most characteristics converge in behaviour for halos containing more than 10^7 particles in detail. Notably, the formation of the bar does not destroy the density cusp in this case. These higher resolution simulations clearly illustrate the importance of discrete resonances in transporting angular momentum from the bar to the halo.

Keywords. galaxies: spiral, structure, evolution; dark matter, methods: n-body simulations, stellar dynamics

1. Introduction

The bar instability in a cold gravitating disk plays a major role in a spiral galaxy's dynamical evolution. At least 2/3 of spiral galaxies host bars (Knappen *et al.* 2000) and the fraction has not evolved significantly since $z \sim 1$ (Jogee *et al.* 2004; Sheth *et al.* 2008). As models of galaxy formation become more sophisticated and reveal complex dynamical behaviour, it is important to understand the details of different physical processes that shape their morphology as well as to verify that numerical resolution is in fact adequate to follow their evolution. The bar-halo interaction is the driving mechanism in disk galaxy evolution. As a bar churns through the DM halo with a pattern speed Ω_b resonant interactions with halo orbits – a form of dynamical friction – transfer angular momentum from the bar to the halo and cause it to spin down (Tremaine & Weinberg 1984). This process was first pointed out by Lynden-Bell & Kalnajs (1972) and has been studied in models with idealized rigid bars (Weinberg 1985; Hernquist & Weinberg 1992; Weinberg & Katz 2002; Weinberg & Katz 2007) as well as in models in which a stellar bar forms self-consistently in an unstable disk (e.g., Sellwood 1980; Debattista & Sellwood 1998; O'Neill & Dubinski 2003; Holley-Bockelmann *et al.* 2005; Martinez-Valpuesta *et al.* 2006). There has been some concern that the process is too efficient, leading to bars that are much smaller than their corotation radii and so discrepant with observed bar galaxies (Debattista & Sellwood 2000). More recent studies with greater resolution suggest that the bars tend to lengthen moving out to their co-rotation radii as they slow down and so perhaps they are not inconsistent with reality (O'Neill & Dubinski 2003; Martinez-Valpuesta et al. 2006). Weinberg & Katz (2002) hold a cautious view that lower resolution simulations can lead to spurious results because of the diffusive nature of noise that may move orbits into and out of resonances artificially while insufficient particle numbers may also underpopulate the resonant regions of phase space. While most current work to date has used $\sim 10^6$ particles, they claim that as many as 10^8 DM particles (or more) may be needed to simulate the resonant process with N-body methods.

In this study, we attempt to clear up the inconsistencies of current work and address the problem of numerical resolution hoping to converge to the correct physical behaviour. We present a series of bar-unstable disk+halo N-body models with increasing resolution spanning a range $N_h = 10^4 - 10^8$ DM particles with $N_d = 1.8 \times 10^{3-7}$ disk particles. One further simulation uses a multi-mass method that increases the halo particle number density by 200× in the halo centre so giving an effective $N_h \sim 10^{10}$. The mass model is constructed from a 3-integral distribution function (Widrow & Dubinski 2005) describing an exponential disk embedded within a tidally truncated NFW halo and is similar to the model studied by Martinez-Valpuesta et al. (2006). Natural units for the simulations are $D = 10$ kpc, $M = 10^{11}$ M$_\odot$, $V = 208$ km/s, and $T = 47.2$ Myr. We discuss results both in simulation and physical units throughout and clarify when necessary. (For further details on the models and simulations see Dubinski, Berentzen & Shlosman 2008). Animations of the simulations are available for viewing at the URL: www.cita.utoronto.ca/~dubinski/IAU254 along with higher-resolution figures.

Animation 1 shows the evolution of the disks in six models with increasing resolution in face-on and edge-on views. The lowest resolution simulations with $N_h \leqslant 10^5$ are clearly deficient and either lose the bar or suffer from heating affects. At higher resolution, the behaviour is similar exhibiting the buckling instability and relaxation to a bar in quasi-equilibrium that gradually lengthens and slows down. Animations 2 & 3 show the multi-mass model with $N_h = 10^8$ in an inertial and corotating frame and illustrate how the bar grows from noise from the inside out saturating as a thin bar on reaching the corotation radius and then evolving into a fatter bar with a peanut-shaped bulge after the buckling instability.

2. Bar Growth and Pattern Speed Evolution

We study bar growth using the normalized Fourier amplitude $|A_2|$ of the $m = 2$ disturbance in the plane of the disk within a fixed radius $R < 0.5$ (5 kpc). Figure 1 shown the exponential growth of $|A_2|$ before saturation. Higher resolution simulations reach saturation at later times. This time delay is the result of the lower amplitude of the Poisson fluctuations that seed the bar. Since the instability grows from these fluctuations if takes longer for them to saturate if the initial amplitude is lower. We estimate the time delays from the peak in $|A_2|$ and synchronize the simulations for comparison of various evolutionary characteristics.

We can also use A_2 to measure the phase angle of the $m = 2$ mode and so estimate the pattern speed by taking the difference between subsequent snapshots. Figure 2 shows the pattern speed evolution as a function of numerical resolution. There is some scatter in behaviour that can be accounted for from the expected variance introduced by different initial conditions but overall the agreement is good. The highest resolution simulations

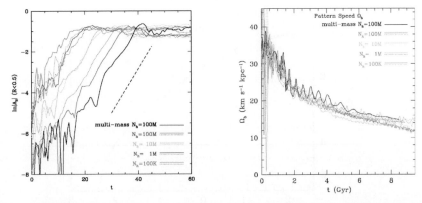

Figure 1. Initial growth of the $m = 2$ Fourier component $|A_2|$ for stars with $R < 0.5$ for two model sequences using $N_d = 18K, 180K, 1.8M, 18M$ with $N_h = 100K, 1M, 10M, 100M$ respectively. The $\ln|A_2|$ grows approximately linearly with time independent of the choice of N_d and N_h showing the exponential growth of the bar mode. The dashed line shows an exponential timescale that is approximately $\tau = 8$ (370 Myr). Since the bar grows from the Poisson noise within the disk then we expect the noise amplitude to be proportional to $N^{1/2}$ so larger simulations will saturate at later times.

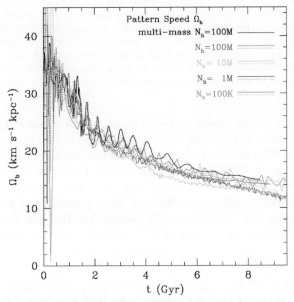

Figure 2. Evolution of the pattern speed Ω_b for two model sequences at different resolution. The curves have been shifted in time so that the bar growth evolution is coincident with the 1M particle model.

show a modulation of the pattern speed at a frequency close to Ω_b itself. This probably indicates an interaction between the bar and the gravitational wake in the halo that only shows up with sufficient numerical resolution.

3. Halo Density Profile

We also measured the evolution of the halo density profile over the course of the simulation. Previous work has shown both preservation and destruction of the density

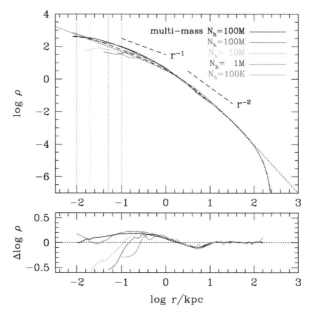

Figure 3. A comparison of density profiles at $t = 7.08$ Gyr for different halo particle numbers N_h. We also show the initial density profile (dashed line) and the best fit NFW model curve (dotted line) to the initial profile over the range $0 < r < 100$ kpc. The NFW parameters for the fit are $r_s = 4.3$ kpc, $v_{max} = 160$ km s^{-1}, where v_{max} is the maximum circular velocity at $r = 2.16 r_s = 9.3$ kpc. Note that this halo is more concentrated than the typical galactic dark matter halos in cosmological simulations to model the contraction expected during dissipative galaxy formation. The dotted vertical lines show the softening length ϵ used at different resolutions.

cusp so we focus on the processes in the central regions. Only studies using the self-consistent field N-body method (SCF) lead to a core (Holley-Bockelmann *et al.* 2005; Weinberg & Katz 2007b) and there has been some concern about numerical instabilities that arise in those methods (Selwood 2003). Figure 3 exhibits the final density profile along with the difference from the initial profile as a function of halo particle number. The profiles agree well to within the limit of their gravitational softening radius and show convergent behaviour. The cusp is not destroyed in this case but rather the central density increases modestly, by a factor of 2, as a result of the bar evolution, following its buckling, which leads to an increase in the central stellar density (Sellwood 2003).

The bar that forms in this simulation has a similar mass ratio $M_b/M_{halo} \approx 0.6$ but is much thicker than the fiducial rigid bar simulation in Weinberg *et al.* (2007b) that destroys the cusp. Their study also included thick bars which did not destroy the cusps and the bar that forms here overlaps with those in their study. We conclude that there is no direct contradiction with the most recent results of Weinberg *et al.* (2007b) but would argue that the thicker bar models are probably more relevant to real galaxies. Thin bars do not persist for long before responding to the buckling instability and so the rigid bar approximation is not applicable over a Hubble time and probably not relevant to most galaxies (Martinez-Valpuesta & Shlosman 2004).

4. Orbital Resonances

The bar slowdown is the result of dynamical friction that leads to angular momentum transport to the DM halo. The process is due to resonant interactions between the rotating bar's pattern speed and the halo particles' azimuthal and radial orbital frequencies. When $l_1\Omega_r + l_2\Omega_\phi = m\Omega_b$ then orbital resonances occur and halo particles torque or are torqued by the bar leading to a change in angular momentum. We can estimate the orbital frequencies at a specific time by freezing the potential and integrating the orbits of particles in this potential rotating with the pattern speed at that time (e.g., Athanassoula 2002; Martinez-Valpuesta et al. 2006). Spectral analysis can then be applied to the orbital time series to determine the fundamental orbital frequencies Ω_r and Ω_ϕ (Binney & Spergel 1982). (Note we label these frequencies with the usual epicyclic variables $\kappa \equiv \Omega_r$ and $\Omega \equiv \Omega_\phi$ in our figures below.) The dimensionless frequency $\eta = (\Omega - \Omega_b)/\kappa$ is a useful way to characterize resonances since the values $\eta = 1/2, 0, -1/2$ correspond to the inner Lindblad, corotation, and outer Lindblad resonances for $m = 2$. Further negative half-integer values correspond to higher order resonances that can also absorb the angular momentum.

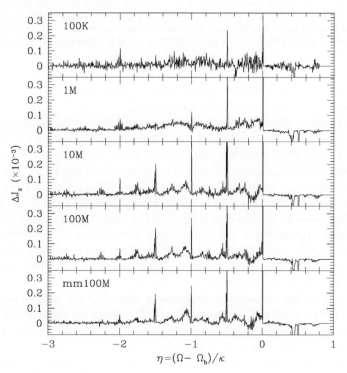

Figure 4. The distribution in the change in the z angular momentum ΔJ_z between $t = 100$ (4.7 Gyr) and $t = 150$ (7 Gyr) plotted as a function of the dimensionless frequency $\eta = (\Omega - \Omega_b)/\kappa$ measured at $t = 100$. The spikes at $\eta = 0.5, 0.0, -0.5$ correspond to the ILR, COR, and OLR respectively while other spikes refer to higher order resonances. This plot shows how halo particles in resonant orbits are the main sink of angular momentum. The detailed distributions are converging for $N_h > 10^7$ particles while lower resolution simulation miss some of the higher order resonances.

Figure 4 shows the change in the z component of the angular momentum J_z for particles binned as a function η between $t = 100$ and $t = 150$ at different numerical resolution. The spikes at the half-integer values reveal the resonances. Most angular momentum is transferred through the corotation resonance though it is also appears to be transferred at other frequencies, but more randomly – this difference is probably the result of particles in resonance before or after $t = 100$. The detailed behaviour of the distributions seem to

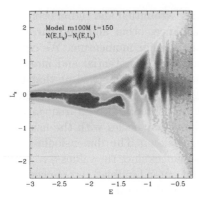 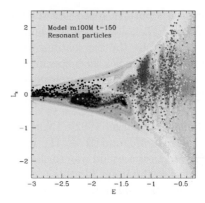

Figure 5. Change in particle number density in (E, L_z) space between $t = 0$ and $t = 150$ (7.0 Gyr) for the $N_h = 10^8$ single mass model. The resonant regions show up clearly as peaks (red regions) in phase space in the left panel. In the right panel, we overplot the (E, L_z) coordinates of a random subset of particles located at discrete resonances at $t = 150$ within $\delta\eta = \pm 0.05$ (black-ILR-$\eta = 0.5$, red-COR-$\eta = 0.0$, green-OLR-$\eta = -0.5$, blue-$\eta = -1.0$, magenta-$\eta = -1.5$, and cyan-$\eta = -2.0$.

converge for $N_h \geqslant 10^7$. At lower resolutions, the resonant spikes have a smaller amplitude and higher order resonant interactions are missing.

Finally, we examine the change in halo phase space density by computing the particle number density in (E, L_z) space and computing the difference between $t = 0$ and $t = 150$ in a similar way to Holley-Bockelmann *et al.* (2005). In this way, we clearly see the resonant regions visible as discrete islands of particle overdensity in (E, L_z) space (Fig. 5). We can also overplot the values of (E, L_z) for the particles found in the resonant spikes in the analysis to see where they lie in phase space. The right panel of Figure 5 clearly shows that these islands are directly related to the discrete resonances extracted in our spectral analysis. Animation 4 describes the time evolution of the differential number density in phase space and reveals how the resonances move through a large fraction of the halo mass. By counting particles in resonant peaks at different times we estimate that roughly 30% of the halo particles are affected.

We conclude that the resonances are broader than thought and so simulations with more than 1M halo particles do a reasonable job of tracking bar evolution. However, a look at the distribution of orbital frequencies reveals that higher order resonances are missed at lower resolution with less than 10M particles. This effect could account for the different rate of angular momentum loss at higher resolution. Future studies should examine the bar instability self-consistently using the same initial conditions with different N-body methods to resolve current inconsistent results on the cusp/core evolution of DM halos.

References

Athanassoula, E. 2002, *ApJ (Letters)*, 569, L83
Binney, J. & Spergel, D. 1982, *ApJ*, 252, 308
Debattista, V. P. & Sellwood, J. A. 1998, *ApJ (Letters)*, 493, L5+
—. 2000, *ApJ*, 543, 704
Dubinski, J. 1996, New Astronomy, 1, 133
Dubinski, J., Berentzen, I., & Shlosman, I. 2008, in prep.
Hernquist, L. & Weinberg, M. D. 1992, *ApJ*, 400, 80
Holley-Bockelmann, K., Weinberg, M., & Katz, N. 2005, *MNRAS*, 363, 991
Jogee, S., *et al.* 2004, *ApJ (Letters)*, 615, L105

Knapen, J. H., Shlosman, I., & Peletier, R. F. 2000, *ApJ*, 529, 93
Lynden-Bell, D. & Kalnajs, A. J. 1972, *MNRAS*, 157, 1
Martinez-Valpuesta, I. & Shlosman, I. 2004, *ApJ (Letters)*, 613, L29
Martinez-Valpuesta, I., Shlosman, I., & Heller, C. 2006, *ApJ*, 637, 214
O'Neill, J. K. & Dubinski, J. 2003, *MNRAS*, 346, 251
Sellwood, J. A. 1980, *A&A*, 89, 296
—. 2003, *ApJ*, 587, 638
Sheth, K. *et al.* 2008, *ApJ*, 675, 1141
Tremaine, S. & Weinberg, M. D. 1984, *MNRAS*, 209, 729
Weinberg, M. D. 1985, *MNRAS*, 213, 451
Weinberg, M. D. & Katz, N. 2002, *ApJ*, 580, 627
—. 2007a, *MNRAS*, 375, 425
—. 2007b, *MNRAS*, 375, 460
Widrow, L. M. & Dubinski, J. 2005, *ApJ*, 631, 838

John Dubinski explaining a fine point in his simulations.

Beatriz Barbuy illustrating the view of the Bulge through Baade's window.

Olga Sil'chenko fielding a question after her talk, with chairman Burkhard Fuchs.

Exponential bulges and antitruncated disks in lenticular galaxies

Olga K. Sil'chenko[1]

[1]Sternberg Astronomical Institute, Moscow 119991, Russia
email: olga@sai.msu.su

Abstract. The presence of exponential bulges and anti-truncated disks has been noticed in many lenticular galaxies. In fact, it could be expected because the very formation of S0 galaxies includes various processes of secular evolution. We discuss how to distinguish between a pseudobulge and an anti-truncated disk, and also what particular mechanisms may be responsible for the formation of anti-truncated disks. Some clear examples of lenticular galaxies with a multi-tier exponential stellar structure are presented, – among them two central group giant S0s seen face-on and perfectly axisymmetric.

Keywords. Galaxies: elliptical and lenticular, cD; galaxies: bulges; galaxies: disks; galaxies: evolution

1. Introduction

Bulges have been traditionally thought to have de Vaucouleurs' brightness profiles just as elliptical galaxies (e.g. Freeman 1970). However John Kormendy (Kormendy 1982a, Kormendy 1982b, Kormendy 1993) has proved that there exists a type of bulges named 'pseudobulges' which resemble disks from the dynamical point of view: they are rather cold and demonstrate fast rotation. Despite their dynamical properties, they are bulges from the geometrical point of view: they are 'fat' and have rather large scaleheights. Pseudobulges are thought to be products of secular evolution; but if it is so, then dynamical simulations (e.g. Pfenniger & Friedli 1991) predict that they must have exponential brightness profiles. And indeed, small bulges of late-type spirals which could be easily made by secular evolution, practically all have exponential brightness profiles (e.g. Andredakis *et al.* 1995, Graham 2001).

Not only small bulges in late-type spirals have exponential brightness profiles. Lenticular galaxies whose bulges usually are not small at all typically have mean Sersic coefficients lower than Sa galaxies, often with $n < 2$: the Sersic parameter n peaks at the Sa–Sb morphological type and falls further toward S0s (Graham 2001, Laurikainen *et al.* 2005, Laurikainen *et al.* 2007). In fact, it is quite natural because the event of a galaxy transformation to S0 from a spiral must include various processes of secular evolution, resulting in radial matter re-distribution and reshaping of the bulge.

But galactic disks outside the bulges can also consist of several exponential segments. Now it becomes clear that so called anti-truncated disks, consisting of two exponential segments with the outer scalelength larger than the inner one, may be the dominant type of galactic disks among some types of galaxies (Erwin *et al.* 2008a). We reported finding a few large nearby spiral galaxies with such two-tiers disks during the last 10 years. A significant population of such disks has been found where the brightness profiles reach 27th or 28th magnitudes per square arcsecond (Pohlen & Trujillo 2006). Does it mean that the outer segments of anti-truncated disks represent always low surface-brightness (LSB) disks? The statistics by Erwin et al. (Erwin et al. 2008b) for barred galaxies says so. Among unbarred spiral galaxies, we find only one galaxy, NGC 5533, that has a LSB outer disk (Sil'chenko et al. 1998). Other galaxies (NGC 7217, Sil'chenko & Afanasiev 2000; NGC 615, Sil'chenko et al. 2001; NGC 4138, Afanasiev & Sil'chenko 2002; NGC 7742, Sil'chenko & Moiseev 2006) have on the contrary quite normal outer disks as compared with the reference value of the central B surface brightness of 21.7 (Freeman 1970); and simultaneously they have compact bright inner disks.

Can we distinguish the compact inner disk from a pseudobulge? Yes, by using a multi-variant approach. For example, in NGC 7742 all spirals and current star formation are confined to the inner exponential disk, and the outer one is smooth and featureless (Sil'chenko & Moiseev 2006), so we can conclude that the inner exponential stellar component is dynamically cold and cannot be a bulge. But the key property is a visible geometry of the stellar component with the exponential profile. To be a disk, it must be thin. For disks inclined to the line of sight, a good check is isophotal analysis. The outer disk is always assumed to be thin: then its isophote axis ratio characterizes the cosine of the inclination. If the inner component has the same visible axis ratio as the outer one and if two disks are coplanar, the inner structure cannot be a spheroid, it must be a disk. For face-on galaxies, this approach does not work: their isophotes are always round independent of the scaleheights. But for face-on galaxies we can use kinematical data and estimate their thickness by measuring vertical stellar velocity dispersions.

We have now started a program of studying systematically multi-tier (anti-truncated) exponential structures in early-type, presumably lenticular galaxies. The study will include photometric as well as spectral observations. Some first results are presented in this talk.

2. Observations

The photometric observations which we discuss here have been made with the focal reducer SCORPIO of the Russian 6m telescope (Afanasiev & Moiseev 2005) in the direct-image mode. An EEV 42-40 CCD detector with 2048 × 2048 pixels was used, binning 2 × 2 pixels. The field of view was about 6 arcminutes, the scale 0.35 arcsec per binned pixel. The photometric observations were undertaken on August 21, 2007, with a seeing of about 2″. We have exposed 5 fields in the NGC 80 group and 6 fields in the NGC 524 group, in the B and V filters. The exposure times were selected in accordance with the surface brightness of the targets observed; for example, we exposed NGC 524 itself during 60 sec in the B-filter and during 30 sec in the V-filter. As a flat field, we used the exposures of the twilight sky. The calibration onto the standard Johnson BV system has been made by using multi-aperture photoelectric data collected by HYPERLEDA for NGC 80 and NGC 524.

3. Photometric structure of the central group S0 galaxies

The two central group galaxies under consideration are typical giant lenticular galaxies, with blue absolute magnitudes of ~ -21.6 (HYPERLEDA). Both are very red, $(B-V)_e = 1.07$, and are seen face-on, $b/a > 0.9$.

We have calculated azimuthally averaged surface brightness profiles for NGC 524 in two filters, B and V. The data are rather precise, and we trace the profiles up to $R = 80''$, or about 10 kpc from the center, with an accuracy better than 0.01 mag. At larger radii

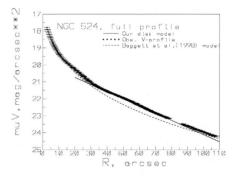

Figure 1. The V-band azimuthally-averaged brightness profile of NGC 524 obtained by us with the reducer SCORPIO at the 6m telescope; the model decomposition proposed by Baggett et al. (1998) is overlaid. One can see that the high-accuracy data do not agree with a single de Vaucouleurs' bulge model; instead at least two exponential components are needed.

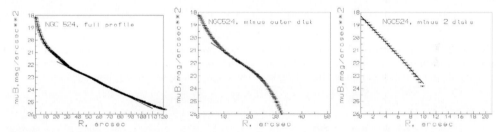

Figure 2. The B-band azimuthally-averaged brightness profile of NGC 524 obtained by us with the reducer SCORPIO at the 6m telescope can be decomposed step-by-step into three exponential components.

the accuracy is worse due to bright stars projected onto the galaxy. Comparing our V-profile with the model decomposition proposed by Baggett et al. (Baggett et al. 1998) (Fig. 1), i.e. with a single de Vaucouleurs' bulge approximation, shows that their model is inappropriate. Instead we see at least two exponential components: the 'outer' one in the radius range $R = 35'' - 90''$ and the 'inner' one in the radius range $R = 10'' - 25''$.

We fit the outer part of the surface brightness profile by an exponential law, construct a 2D model image of this disk and subtract it from the observed full image. For the residual image, we calculate again the azimuthally averaged surface brightness profile, fit its outer part by an exponential law, construct the 2D image of the inner disk and subtract it from the first-step residual image. Interestingly, the brightness profile of the second-step residual image which is safely traced up to $10''$, is also exponential (Fig. 2)! Figure 2 presents all steps of our decomposition procedure, and in Table 1 we give the parameters of the exponential stellar substructures obtained by this procedure.

Table 1. Parameters of the photometric components of NGC 524 approximated by an exponential law

Component	Radius range of the fit	μ_0, mag/$\square''$	h''	h, kpc
		B-filter		
Outer disk	$34'' - 90''$	20.906 ± 0.007	26.07 ± 0.06	3.02
Inner disk	$10'' - 23''$	19.67 ± 0.01	7.67 ± 0.05	0.89
Bulge	$1'' - 8''$	18.25 ± 0.01	2.04 ± 0.01	0.24

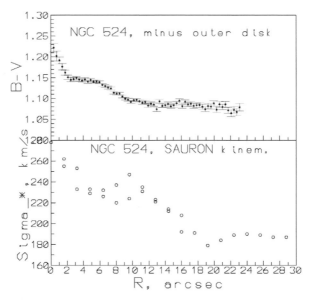

Figure 3. Comparison of the azimuthally averaged profiles in the $B-V$ colour, calculated from the residual image of NGC 524 after subtraction of the outer disk (upper), and the stellar velocity dispersion (lower), showing a break at the edge of a dynamically hot bulge.

Figure 3 presents the $B-V$ colour profile of the first-step residual image (the inner disk). It reveals a net colour difference between the inner disk, $R > 10''$, and the 'bulge', $R < 10''$. This colour difference, $\Delta(B-V) = 0.07$, may result from a metallicity difference of 0.2 dex at an old stellar age ($T = 12$ Gyr), or alternatively from an age difference of 8 Gyr at a metallicity of [Z/H]= +0.1 (Worthey 1994, and also his WEBsite, Dial-a-Galaxy option), the inner disk being younger or/and more metal-poor.

We have compared the colour profile of the residual image with the stellar velocity dispersion profile similarly averaged in the rings over the map obtained from SAURON data (Emsellem et al. 2004, see also the WEBsite of the SAURON project) (Fig. 3). The profiles of the colour and of the stellar velocity dispersion are qualitatively similar! Certainly, we see a transition from the (exponential) bulge to the inner disk at $R \approx 10''$. Preliminary estimates of the scaleheight of the inner disk in NGC 524, by treating the measured line-of-sight stellar velocity dispersion as a vertical one, give a value of about 1.5 kpc; a typical value for a thick stellar disk as expected in a lenticular galaxy.

A very similar surface brightness profile, consisting of two exponential disks and a very compact bulge, has been observed by us earlier in the giant lenticular galaxy NGC 80, which is also located at the center of a rich X-ray group (Sil'chenko et al. 2003). Our new observations certainly confirm this multi-tier structure. In both galaxies the outer stellar disks are quite normal as regards their scalelengths or their central surface brightnesses, and the inner disks are compact and bright. Both galaxies are seen face-on and are certainly unbarred; the low ellipticity of their isophotes over the full radial extension proves that the galaxies are strictly axisymmetric. If the exponential inner stellar structures have been formed by secular evolution, where are the signatures of the main 'driver' of secular evolution, of a bar?

4. What can be the mechanisms for forming anti-truncated disks in lenticular galaxies?

It seems clear that anti-truncated disks are the result of radial matter re-distribution in a disk galaxy, and the very event of re-distribution must be rather fast and discrete. Several candidate mechanisms can be proposed. Younger et al. (Younger et al. 2007) simulate a minor merger and obtain an anti-truncated stellar disk in the merger remnant, mainly due to stellar diffusion from the inner part of the initial spiral galaxy into an outer region. In such models the outer part of the multi-tier stellar disk in the merger remnant seems to be a low surface brightness disk.

Some years ago I proposed another mechanism where a transient interaction, due to, say, a passage of a rather massive galaxy, provokes intense gas inflow and results in the gas concentrating in the very inner part of a galaxy with a subsequent star formation burst. In this model, the inner disk must be brighter and more compact than the usual large-scale stellar disks of spiral galaxies. We may expect that a statistical study of the parameters of inner and outer exponential disks in the anti-truncated galaxies would help to select a model.

In the group NGC 80, besides the central galaxy, we have analysed brightness profiles in more 10 lenticular galaxies, with absolute blue magnitudes from −17 to −20. Among those, 7 S0s appear to possess two-tier exponential disks. Of these, three have compact inner disks and normal outer disks, while four have normal inner disks and LSB outer disks. Together with NGC 80 itself and with NGC 524, with their compact bright inner disks and extended normal ones, we obtain half-to-half preliminary statistics, which implies that the origin of the multi-tier exponential disks may be different in different galaxies.

5. Conclusions

If we assume that multi-tier exponential profiles are formed by secular evolution of galactic disks, the best place to search for them would be lenticular galaxies. Lenticular galaxies have had to reform their stellar disks during their secular transformation from S to S0; hence S0s must be the hosts of both multi-tier disks and pseudobulges. To select a particular mechanism for forming anti-truncated disks, we must have a better knowledge of the statistics of their properties, which still does not cover representative samples. Among the possible alternatives to the origin of multi-tier exponential profiles are minor mergers or tides (but the same alternative exists for the origin of S0s outside clusters!). To choose, we need to know the ratio between compact+normal and normal+LSB stellar disk combinations.

Secular redistribution of stars and other matter in disks is thought to be caused by bars: bars are usually generated by any interaction and even without interactions – by intrinsic instabilities. But almost all our galaxies with multi-tier disks are unbarred; the face-on S0s NGC 524 and NGC 80 are perfectly axisymmetric. And the samples of Erwin et al. (Erwin et al. 2008a, Erwin et al. 2008b) imply the same conclusion: among barred galaxies, the anti-truncated disks contribute one third of all, among unbarred – more than 50%. This inconsistency is a complete puzzle yet.

Acknowledgements

The 6m telescope is operated under the financial support of Science and Education Ministry of Russia (registration number 01-43). During our data analysis we used the Lyon-Meudon Extragalactic Database (HYPERLEDA) supplied by the LEDA team at the CRAL-Observatoire de Lyon (France), and the NASA/IPAC Extragalactic Database

(NED) operated by the Jet Propulsion Laboratory, California Institute of Technology under contract with the National Aeronautics and Space Administration. The study of multi-tier galactic disks is supported by the grant of the Russian Foundation for Basic Research (RFBR), no. 07-02-00229a. My attendance at IAU Symposium no. 254 was made possible by an IAU grant.

References

Afanasiev, V. L. & Sil'chenko, O. K. 2002, *AJ* 124, 706
Afanasiev, V. L. & Moiseev, A. V. 2005, *Astronomy Letters* 31, 194
Andredakis, Y. C., Peletier, R. F., & Balcells, M. 1995, *MNRAS* 275, 874
Baggett, W. E., Baggett, S. M., & Anderson, K. S. J. 1998, *AJ* 116, 1626
Emsellem, E., Cappellari, M., Peletier, R. F., *et al.* 2004, *MNRAS* 352, 721
Erwin, P., Pohlen, M., Beckman, J. E., Gutiérrez, L., & Aladro, R. 2008, in: J. H. Knapen, T.J. Mahoney, & A. Vazdekis (eds.), *Pathways through an Eclectic Universe. ASP Conf. Ser. v. 390* (San Francisco), p. 251
Erwin, P., Pohlen, M., & Beckman, J. E. 2008, *AJ* 135, 20
Freeman, K. C. 1970, *ApJ* 160, 811
Graham, A. W. 2001, *MNRAS* 326, 543
Kormendy, J. 1982, in: L. Martinet & M. Mayor (eds.), *Morphology and Dynamics of Galaxies, Twelfth Advanced Course of the Swiss Society of Astronomy and Astrophysics* (Sauverny: Geneva Obs.), p. 113
Kormendy, J. 1982, *ApJ* 257, 75
Kormendy, J. 1993, in: H. Habing & H. Dejonghe (eds.), *Galactic Bulges. IAU Symp. 153* (Dordrecht: Kluwer), p. 209
Laurikainen, E., Salo, H., & Buta, R. 2005, *MNRAS* 362, 1319
Laurikainen, E., Salo, H., Buta, R., & Knapen, J. H. 2007, *MNRAS* 381, 401
Pfenniger, D. & Friedli, D. 1991, *A&A* 252, 75
Pohlen, M. & Trujillo, I. 2006, *A&A* 454, 759
Sil'chenko, O. K., Burenkov, A. N., & Vlasyuk, V. V. 1998, *New Astronomy* 3, 15
Sil'chenko, O. K. & Afanasiev, V. L. 2000, *A&A* 364, 479
Sil'chenko, O. K., Vlasyuk, V. V., & Alvarado, F. 2001, *AJ* 121, 2499
Sil'chenko, O. K., Koposov, S. E., Vlasyuk, V. V., & Spiridonova, O. I. 2003, *Astronomy Reports* 47, 88
Sil'chenko, O. K. & Moiseev, A. V. 2006, *AJ* 131, 1336
Worthey, G. 1994, *ApJS* 95, 107
Younger, J. D., Cox, T. J., Seth, A. C., & Hernquist, L. 2007, *ApJ* 670, 269

Kinematical & Chemical Characteristics of the Thin and Thick Disks

Rosemary F. G. Wyse

Department of Physics & Astronomy, Johns Hopkins University,
Baltimore, MD 21218, USA
email: wyse@pha.jhu.edu

Abstract. I discuss how the chemical abundance distributions, kinematics and age distributions of stars in the thin and thick disks of the Galaxy can be used to decipher the merger history of the Milky Way, a typical large galaxy. The observational evidence points to a rather quiescent past merging history, unusual in the context of the 'consensus' cold-dark-matter cosmology favoured from observations of structure on scales larger than individual galaxies.

Keywords. Galaxy: disk, Galaxy: evolution, Galaxy: formation, Galaxy: stellar content, Galaxy: structure; cosmology: dark matter

1. Introduction: Formation of Thin and Thick Disks in ΛCDM

Dissipational collapse of a gas-rich system is an important ingredient in establishing the thin disks so prevalent today. In the context of hierarchical clustering scenarios, such as ΛCDM, gaseous mergers are required to produce disks (Zurek, Quinn & Salmon 1988; Robertson et al. 2006). Centrifugally supported, extended disks also are only produced if angular momentum is largely conserved during collapse within the dark halo (Fall & Efstathiou 1980; Mo, Mao & White 1998). However, angular-momentum transport and evolution, for example due to gravitational torques and tidal effects, are natural during the mergers inherent in ΛCDM, leading to low angular momentum, very concentrated disks (Navarro, Frenk & White 1995). The typical merger history (of dark haloes) is fixed by the dark matter power spectrum, so with ΛCDM additional baryonic processes are introduced to implement 'feedback', to both suppress early star formation and prevent dissipation, delaying disk formation until after the epoch of most active merging (cf. Simon White's talk). Later mergers into the disk will heat the thin disk, with minor mergers producing a thick stellar disk (some thin stellar disk component persists, e.g. Kazantzidis et al. 2007) and also driving gas into the central regions to build up a bulge (Mihos & Hernquist 1996).

During a merger, orbital energy is absorbed into the internal degrees of freedom of the merging systems, and orbital angular momentum is redistributed, some absorbed by the larger system, and some being lost to the system. The evolution of a satellite, and of its orbit, as it merges with, and is assimilated by, a larger system depends on the relative masses (the dynamical friction timescale on which the satellite sinks to the center scales like the mass ratio, being shorter for more massive satellites), on the relative density profiles of the satellite and host (denser satellites can survive tidal effects, leading to mass loss and disruption, closer to the center of the larger system), and on the initial orbital parameters (e.g. sense of rotation, inclination angle to the plane of the host, peri-Galacticon and orbital eccentricity). The effect of the satellite(s) on a pre-existing stellar disk is also a sensitive function of the satellite's properties and orbit. The mechanism by which a thin disk is heated by the merging process is a combination of local deposition of orbital energy from the satellite(s), local scattering (in which azimuthal streaming motions are transformed into random motions), and resonant excitation of modes in the disk, providing heating on a more global scale (e.g. Quinn & Goodman 1986; Tóth & Ostriker 1992; Sellwood, Nelson & Tremaine 1998).

Early simulations focussed on the impact of one minor merger (e.g. Quinn, Hernquist & Fullagar 1993; Velazquez & White 1999). These produced a plausible thick disk, similar in structure to that of the Milky Way (Gilmore & Reid 1982; see also Gilmore & Wyse 1985 for kinematics and an order-of-magnitude estimation of the mass of satellite needed), from a merger of a robust (dense) satellite with a mass ratio to the stellar *disk* (not to the total mass) of 10–20%, for a range of initial orbital parameters. Simulations of the merging of cosmologically motivated ensembles of satellites are more relevant, and also find significant heating of a pre-existing stellar disk, over an extended period of time (e.g. Hayashi & Chiba 2006; Kazantzidis *et al.* 2007, also this volume). The satellites in these later simulations have a orbital distribution similar to that of subhaloes identified in dissipationless ΛCDM cosmological simulations, with a mean ratio of initial apocenter to pericenter distance of around 6:1 (e.g. Ghigna *et al.* 2000; Diemand, Kühlen & Madau 2007). The distribution of pericenter distances is also important, since satellites with pericenters that are significantly beyond the disk (larger than say 10 disk scale lengths, see Fig. 2 of Hayashi & Chiba 2006) do not couple effectively to the disk. In addition to realistic orbits, the internal density profiles of the satellites are critical, since fluffy satellites are disrupted early and provide little heating (Huang & Carlberg 1997). High-resolution, N-body simulations of the formation of the dark halo of a 'Milky Way galaxy' with the CDM power spectrum have demonstrated that a significant population of substructure is indeed dense enough to persist and survive many pericenter passages. The shapes of the mass- and velocity-functions of subsystems are reasonably independent of redshift and at $z = 0$ are well-established as power laws (see convergence tests in Reed *et al.* 2005; S. White's talk at this conference). The amplitudes depend on resolution, with the number of satellite dark haloes still increasing with increased resolution (S. White, these proceedings); 'overmerging', particularly in the central regions of the larger host galaxy halo, can artificially reduce substructure. Indeed, the population of subhaloes within the analogue of the solar neighborhood is not yet well-established, even in pure dark-matter simulations (the addition of baryons will increase central densities). However, the present generation of (baryon-free) simulations imply that robust satellites penetrate the region of the disk (e.g. Diemand, Kühlen & Madau 2007). Simulations which model gas physics and star formation also find that thick (and thin) disks are produced. For example, the bulk of the younger stars (ages less than 8 Gyr) in the thick disk in the simulation of Abadi *et al.* (2003) are produced by heating of the pre-existing stellar disk (we return to the older stars in section 2 below).

The most massive satellite provides the greatest heating, with the scaling such that the increase in scaleheight (or, equivalently, in the square of the vertical velocity dispersion) is proportional to the square of the mass of the satellite (Hayashi & Chiba 2006). For an ensemble of satellites with the differential mass function seen in CDM simulations, namely a power-law with slope close to -2 (e.g. Diemand, Kühlen & Madau 2007), the cumulative heating is also dominated by the most massive systems (see also White 2000). The substructure distribution at earlier times contains more satellites of higher mass, since these are more affected by dynamical friction, which brings them into the central regions where they are more effectively stripped of mass from their outer parts (e.g. Zentner *et al.* 2005). These massive satellites, after this shriking of their orbits, can be more effective at heating the thin disk, prior to their demise. Thus at early times the satellite distribution is more concentrated than is the host dark matter halo, while at the present day it has evolved to be less concentrated. One must allow for this evolution of the mass function and orbital parameters, rather than simply adopting a surviving satellite retinue from the end-point of a simulation, which minimizes the predicted overall heating (as found by e.g. Font et al. 2001). Indeed following the full merging history is preferable. Such analyses imply that late (after redshifts of unity) heating of thin stellar disks seems to be inevitable in ΛCDM (e.g. Abadi et al. 2003; Stewart et al. 2008; Kazantzidis, this volume). The current models (Stewart et al. 2008) show that over the last 10Gyr, fully 95% of galaxy haloes of present total mass 10^{12} M$_\odot$ have accreted a system of mass equal to that of the present-disk (5×10^{10}M$_\odot$ – their simulation resolution limit is 10^{10} M$_\odot$); the mass ratio of the substructure to that of the stellar disk at the epoch of accretion is the more important ratio for disk heating and such a large mass is highly likely to cause severe heating. More numerous, lower-mass mergers are also expected, continuing to late epochs.

Gas physics can also play a role in the formation of the thick disk, in terms of slow settling to the disk plane (Burkert, Truran & Hensler 1992) or a starburst in a rapidly changing potential such as a gas-rich merger (Robertson et al. 2006; Brook et al. 2007). The latter mechanism has some observational support at high redshift (see Elmegreen's and Genzler's contributions to this volume). One must of course also take account of adiabatic compression and heating of an existing thick disk by subsequent slow accretion of gas to buildup the thin disk (cf. Tóth & Ostriker 1992; Elmegreen & Elmegreen 2006). Here I will focus on the – apparently inevitable – late heating of thin stellar disks, as a probe of ΛCDM. The important issues are the predicted chemical abundance and age distributions of stars in the thin and thick disks, given a typical merger history, and how they compare with the observations.

2. Evidence for Minor Mergers in the Past

Satellites that are accreted are in general only partially assimilated, with 'shredded satellite' debris maintaining some coordinate-space coherence for a few orbits, kinematic coherence for longer, and with persistent stellar population signatures, in terms of their chemical abundances and stellar age distributions, allowing identification over a Hubble time. In general one can expect satellite debris to be deposited along the (evolving) orbit of its center-of-mass, leading to a prediction that former-satellite member stars will populate the thick disk – halo interface at a range of Galactocentric radii (see e.g. Fig. 19 of Huang & Carlberg 1997, Fig. 9 of Abadi et al. 2003, Fig. 3 of Meza et al. 2005), again with details depending on the satellite mass, internal density distribution and initial orbit. Indeed, it has been predicted that a large fraction of the old stars in the thick and thin disks consists not of stars formed *in situ* but rather stars accreted from satellites many Gyr after they were formed (Abadi et al. 2003), with debris from each satellite populating a different radial range, and the parent satellite having been brought to a circular orbit prior to mass loss. This late (redshifts $z < 1$) accretion of old stars, directly into the disks, on high angular momentum orbits, would allow reconciliation of the delayed formation of disks, required in ΛCDM as discussed above, with the presence of old 'disk' stars.

Only high mass, dense (robust) satellites can experience efficient circularization of their initial orbits through dynamical friction. Accretion of such objects into the plane of the disk should also cause heating of the thin disk that has formed by the epoch of their accretion, leading to a thick disk component with a stellar age distribution that reflects the thin disk star formation history up to that epoch. The derived star-formation history of the (local) thin disk is fairly smooth and continuous from the earliest times, corresponding to the lookback time of $\sim 10 - 12$ Gyr that equals the ages of the oldest stars in the thin disk (Binney, Dehnen & Bertelli 2000). Taking Abadi *et al.* as an example, with a significant accretion event at $z = 0.73$, i.e. a look-back time of ~ 7 Gyr, one expects stars in the thick disk as young as 7 Gyr, rather than a uniformly old population. This is indeed what they find in that simulation (their Fig. 8).

Note that the Monoceros Stream (Newberg *et al.* 2002) and Canis Major overdensity have been interpreted in terms of the in-plane accretion of a dwarf galaxy (Peñarrubia *et al.* 2005). The null detection of an associated over-density of RR Lyrae stars by Mateu *et al.* (poster this conference) would be unexpected in this scenario, but is consistent with dynamical instabilities – warp, flare, spiral arms – in the outer disk. †

2.1. *The Age Distribution of Stars in the Thick Disk*

All available observations are consistent with a dominant old age for the stars of the thick disk of the Milky Way, where 'old' means as old as the globular clusters of the same metallicity (e.g. 47 Tuc), or at least 10 Gyr, and probably 12 Gyr. The most reliable evidence comes from looking at the turnoff for *in situ* thick disk stars, several thin disk scale heights above the plane (to minimize contamination by outlier thin disk stars), as a function of metallicity. Samples analysed in this way show very few stars bluer than the 10-12 Gyr turn-off colour at a given metallicity, for both thick disk and stellar halo (Gilmore, Wyse & Jones 1995 and references therein).

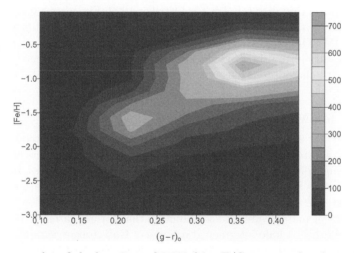

Figure 1. Contour plot of the locations of 8,600 faint F/G stars in the plane of de-reddened colour and metallicity. The rather abrupt turn-offs of each of the thick disk and halo are apparent, with few younger stars, which, if present, would occupy the upper left portion of the plane.

The data for 8,600 faint F/G dwarfs with metallicities from intermediate-resolution spectroscopy (obtained with the AAT/AAΩ multi-object spectrograph) are shown in Fig. 1 (Wyse, Gilmore & Norris, in preparation), where the dominant turn-off is seen. The turn-off colour of the thick disk (and the bluer turn-off colour of the more metal-poor stellar halo) from the full Sloan Digital Sky survey imaging data, covering a significant fraction of the sky, also implies this old age for both thick disk and stellar halo, more globally across the Galaxy (e.g. Ivezic *et al.* 2008). Local samples – making use of Strömgren photometry – are more susceptible to contamination by thin disk stars with extreme kinematics, such as due to three-body interactions, and require careful analysis to isolate a

† 'The Monoceros Stream' was first identified by Corlin (1920; his Table 2) as a local moving group. While clearly a different feature, this first Monoceros Stream may hold lessons for the interpretation of the current Stream.

clean thick disk sub-sample. With this caveat, such samples are in general consistent with this dominant old age (e.g. Strömgren 1987; Nordström et al. 2004; Reddy et al. 2006; Schuster et al. 2006).

Such a high value for the age of the dominant stellar population of the thick disk has major implications for the minor merging history of the Milky Way. As noted above, stars of all ages are found in the local thin disk, with a derived star formation history that extends back to earliest times. Thus, if the thick disk originated through merger-induced heating of a pre-existing thin stellar disk, the last significant (defined as having a mass ratio to the disk of $\sim 20\%$, and surviving to interact with the disk) dissipationless merger can be dated by the age distribution of stars in thick disk: if the typical thick disk star is old, then the last such merger was long ago, with an age greater than 10 Gyr setting a limit of no significant merger activity and heating since a redshift of $\gtrsim 2$. This scenario also requires there to have been an extended thin stellar disk in place at $z \sim 2$ (even allowing for some radial mixing subsequently).

Mergers do not only heat thin disks, but can also drive gas and stars into the central regions to build-up the bulge. It may then be no coincidence that the old age of the dominant stellar population in the bulge of the Milky Way, again 10–12 Gyr (e.g. Zoccali et al. 2003; Feltzing & Gilmore 2001), provides a consistent limit on merger activity. As noted previously (Wyse 2001), it could be that the merger that created the thick disk induced gas inflow and an associated starburst to form the (bar/)bulge *in situ*.

2.2. *What fractions of the disks can be direct stellar accretion from satellites?*

2.2.1. *Kinematic Constraints*

Interloper stars, accreted during the merger with a satellite galaxy, should be identifiable, probably with different kinematics, chemical abundances, age and spatial distributions from stars formed *in situ*. It might be remembered that the Sagittarius dwarf spheroidal was discovered serendipitously during a spectroscopic survey of the Milky Way bulge (Ibata, Gilmore & Irwin 1994) by the distinct kinematics and colour distributions of its member stars. In the scenario whereby the thick disk results from heating of a pre-existing thin disk by minor mergers, the 'shredded-satellite' stars should be distinguishable from 'heated thin-disk stars'.

Such satellite debris was identified with 'thick disk' stars, observed several kiloparsecs above the thin disk plane, in two lines-of-sight at longitude $\sim 270°$, on orbits of significantly lower angular momentum than the standard thick disk star – with a lag in mean azimuthal streaming of ~ 100 km/s behind the Sun, compared to the standard thick disk lag of ~ 40 km/s (Gilmore, Wyse & Norris 2002). A similar population was found among the Galactic field stars along lines of sight to dwarf spheroidal galaxies, at widely separated lines-of-sight (Wyse et al. 2006). The radial velocity histogram from the subset of our AAT/AAΩ data at $\ell \sim 270°$, where the line-of-sight velocity has a significant contribution from the azimuthal streaming, is shown in Fig. 2 (Wyse, Gilmore & Norris, in preparation). There is clearly again a significant population with a rotational lag of ~ 100 km/s. These stars have a broad range of metallicity (derived from the Ca II K line), down to -3 dex. A full analysis is underway.

While we interpreted our results in terms of discontinuous kinematics distinguishing 'shredded-satellite stars' from 'heated thin-disk stars' (true thick disk in this picture), others with similar quality spectroscopic data have modelled their velocities by smooth gradients as a function of height from the disk plane (e.g. Chiba & Beers (2000) with a similar sample size of around one thousand stars). The Sloan Digital Sky Survey photometric data for $\sim 60,000$ F/G stars at the NGP ($b > 80°$) have been analysed together with proper motions by Ivezic et al. (2008), using photometric metallicity determinations

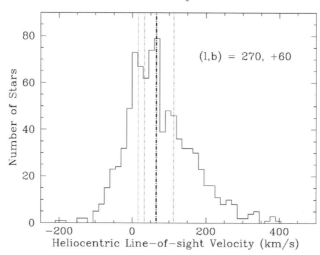

Figure 2. Line-of-sight velocity histogram for the ~ 900 faint F/G stars at $\ell = 270°$ from our wide-area spectroscopic survey with AAΩ (some 12,000 stars in total). The predicted mean velocities for the standard (old) thin disk, thick disk and stellar halo are shown as fainter vertical dot-dashed lines (left to right), while the heavier vertical dot-dash line is the predicted mean for a component lagging behind the Sun by 100 km/s. This clearly matches the peak velocity.

to derive distances and hence tangential velocities (decomposable into velocity towards the Galactic center and azimuthal streaming velocity). They deduce a steep gradient in rotational lag with Z-height for the thick disk, similar in amplitude to that of Chiba & Beers (2000), namely ~ 30 km s^{-1}/kpc. It will be interesting to try alternative models to describe the data, while noting that the heated thin-disk stars in the minor-merger simulations of Hayashi & Chiba (2006) shows a vertical gradient in rotational velocity of comparable amplitude.

2.2.2. Age Constraints

The surviving satellites in the Local Group all contain old stars, consistent with star formation in all small galaxies being initiated around the epoch of reionization (e.g. Hernandez, Gilmore & Valls-Gabaud 2000; Dolphin 2002). Most of the luminous satellites had an extended and fairly continuous star-formation history, albeit non-monotonic, and contain a dominant intermediate-age population, contrasting with the dominant old ages seen in the bulge, thick disk and stellar halo of the Milky Way. While satellites accreted early will therefore contain stars with the same age distribution of the non-thin-disk components of the Milky Way, satellites accreted later may be expected to contain significantly younger stars. Accretion of typical luminous satellites to form more than a few percent of the stellar halo is then limited to epochs prior to a look-back time of ~ 10 Gyr, or again a redshift of ~ 2 (Unavane, Wyse & Gilmore 1996). The similar old age of the thick disk stars produces similar constraints. Systems that contain uniformly old stellar populations, such as the Ursa Minor dSph, could of course be assimilated into the Galaxy at any epoch and would not be distinguishable by the stellar age distribution (or their stellar mass function; Wyse et al. 2002), but only a small fraction of the stars in dwarf galaxies now are as old as the stars in the Ursa Minor dSph, and the stellar mass of the Ursa Minor dSph, $\sim 10^6$ M$_\odot$, is a tiny fraction of even the stellar halo.

2.2.3. Overall Metallicity Constraints

The metallicity distribution of the local thick disk is distinct from any of the other stellar components, but of course the tails overlap (see e.g. Wyse & Gilmore 1995). The mean metallicity of the local thick disk, expressed as an iron abundance, is around one-quarter of the Solar value. The luminosity-metallicity relation for galaxies in the Local Group (e.g. Mateo 1998) implies that only the most luminous satellites can self-enrich to this value, suggesting that the thick-disk stars formed in a system of relatively deep potential well. The Large Magellanic Cloud has enriched to this level, and the metallicity distribution of the inner disk of the LMC (Cole, Smecker-Hane & Gallagher 2000) is similar to that of the local thick disk. Further, the total stellar masses are comparable. However, the LMC had a much slower enrichment history than did the (progenitor of) the thick disk, and reached [Fe/H] ~ -0.6 dex only ~ 5 Gyr ago (Hill et al. 2000). The rapid enrichment of the thick disk points to a high early star formation rate, and chemical evolution in a system significantly more massive than the LMC, suggestive that the overall Milky Way potential was already established at $z \sim 2$, and the bulk of the star formation was *in situ*.

2.2.4. Elemental Abundance Constraints

Stars of different masses synthesize and eject different elements, on different timescales, so that elemental abundances contain much more information than does overall metallicity. The latter is an integral over past star formation and chemical enrichment, while the former reflects the ratio of recent star formation rate (through enrichment by Type II supernovae, which evolve on timescales of $\sim 10^7$ yr after birth of the progenitor star, faster than the typical duration of star formation) to past star formation (e.g. through Type Ia supernovae, which evolve on timescales of several times 10^8 years, up to a Hubble time, after the birth of the progenitor stars). Massive stars, ending their lives as core-collapse (Type II) supernovae, create and eject intermediate-mass elements, in particular those synthesized by the addition of successive helium nuclei, and known collectively as the 'alpha-elements'. The $r-$process elements are also created in the high neutron-flux environments of Type II supernovae. Stars that are formed early in a star-forming event, prior to significant Type Ia supernovae activity, (not necessarily early in absolute terms) will be enriched by only Type II supernovae. Provided there is good mixing of ejecta, and a massive enough star-formation event for the massive-star Initial Mass Function (IMF) to be fully sampled, the interstellar medium from which these early stars form will be enriched with a characteristic ratio of α-elements to iron. This characteristic ratio is set by the massive-star IMF, since the mass of iron that is produced is essentially independent of progenitor mass, while the mass of α-elements produced is a steeply increasing function of progenitor mass, independent of progenitor metallicity (see e.g. Fig. 1 in Gibson 1998; Kobayashi et al. 2006). Thus if the massive-star IMF were biased towards more massive stars (remembering the relevant range is ~ 8 M$_\odot$ to ~ 100 M$_\odot$), stars enriched by the resulting Type II supernovae only would show a higher value of [α/Fe]. As discussed in Wyse & Gilmore (1992), IMF slopes that have been proposed (for various reasons) predict values of this 'Type II plateau' that can differ by greater than 0.3 dex, certainly an observable effect. Such differences have not been observed, providing strong evidence against a variable IMF.

Type Ia supernovae are produced by accretion onto a massive white dwarf in a binary system and each produce a relatively large mass of iron, and a small mass in α-elements. The signature of the incorporation of the ejecta from Type Ia in the element ratios of long-lived low mass stars is a lower value of [α/Fe] than the Type II plateau. Irrespective of the details of the model, the minimum delay time after star formation, for a Type Ia

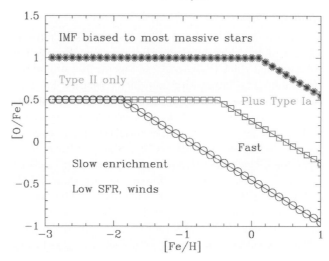

Figure 3. Schematic elemental ratio pattern for self-enriching star-forming regions. Stars formed at early times in the star-forming event are pre-enriched by only Type II supernovae and with good mixing of ejecta into the interstellar medium, this pre-enrichment has a fixed value of [O/Fe]. The asterisks represent an IMF biased towards the most massive stars, while the open squares and circles represent a normal IMF. The values of the 'Type II plateau' reflect the differences in massive-star IMF.

supernova, is given by the time taken for an ~ 8 M$_\odot$ star (the most massive progenitor in this case) to become a white dwarf, accrete sufficient material to exceed the Chandrasekhar mass, and then explode. This is the origin of the several times 10^8 yr delay noted above; the shortest timescale for incorporation of significant iron into the ISM and the next generation of stars is usually estimated as $\sim 10^9$ yr. Lower-mass progenitors take longer to end up as white dwarfs, and different binary systems can range tremendously in their evolution and accretion times, leading to a long tail in delay times (see e.g. Matteucci & Greggio 1986; Smecker & Wyse 1991). The enrichment by Type Ia supernovae is set by a delay *time,* and the iron abundance corresponding to that time depends on the efficiency of chemical enrichment. The rate of chemical enrichment depends on the star formation rate, gas flows, and on the ability of the system to retain metals. In the absence of a mechanism to remove individual elements preferentially in a wind, none of these will modify the value of the Type II plateau, but will modify the iron abundance at which the downturn from this plateau appears.

The situation is illustrated schematically in Fig. 3 (modified from Wyse & Gilmore 1993). With a fixed IMF, the value of [α/Fe] is fixed, for stars that form early, and for a normal IMF that value is $\sim +0.3$. Thus one expects the metal-poor stars in any self-enriching system to show such values. Of course, if there are subsequent bursts of star formation, so that Type II supernovae dominate again, newly forming stars in that burst will also show these enhanced values of [α/Fe] (e.g. Gilmore & Wyse 1991 for models; Koch *et al.* 2008 for application to the Carina dSph).

It is clear that elemental abundance patterns reflect the IMF and star formation histories of the star-forming systems. In particular, systems like the dwarf spheroidal galaxies, with inefficient enrichment over extended periods, should show low values of [α/Fe] at low values of iron (Unavane, Wyse & Gilmore 1996), as observed (Venn *et al.* 2004). Each (surviving) satellite galaxy in the Local Group has its own star-formation and enrichment history, leading to the expectations of a unique pattern in elemental abundances for each system. The realisation of this expectation is demonstrated by Geisler *et al.* (2007; their

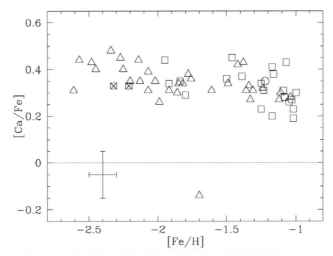

Figure 4. Elemental abundance ratios for the α-element Calcium, in metal-poor stars ([Fe/H] < -1 dex) selected from the RAVE catalogue on the basis of disk-like kinematics. The stars are assigned to populations using several criteria, based on the refined stellar parameters from high-resolution spectra. Circles are thin-disk stars, squares are thick-disk stars and triangles are halo stars. The crossed- square symbols at [Fe/H] < -2 dex represent two stars with uncertain thick disk–halo designation at present. The thick and thin disks clearly extend to low iron abundances, and those stars have enhanced [α/Fe], unlike the bulk of stars in satellite galaxies, which have values of [α/Fe] < 0.2 at these metallicities. We expect to double this sample this observing season.

Fig. 12), where the different loci in the [α/Fe]–[Fe/H] plane of Galactic stellar populations and of each dwarf satellite is rather dramatic. The vast majority of stars in dwarf galaxies now all lie below the Galactic populations in this plane, with $0.2 > [\alpha/Fe] > -0.3$. The distinct patterns of stars in different satellites means that one should be able to identify candidate 'interloper' stars from a given system, from their joint kinematics and elemental abundance distributions, provided their parent system formed stars for longer than $\sim 10^9$ yr, and the stars were accreted subsequently (see Nissen & Schuster, this volume, for an interesting discussion of 'low-alpha' stars with halo kinematics).

Thus late accretion of satellite galaxies into the plane of the disk may be expected to produce disk stars with low iron abundance and low [α/Fe]. The RAVE spectroscopic survey of apparently bright stars (Steinmetz *et al.* 2006, also this volume) provides an ideal sample in which to look for candidates: stars with low metallicity but disk-like kinematics. We (Ruchti *et al.*, in preparation) have initiated a programme to obtain follow-up high resolution spectroscopy of such candidates, using the AAT, Magellan and the ESO 2.2m telescope. These spectra provide improved stellar parameters and elemental abundances. Reflecting the magnitude-limited nature of the RAVE selection function, most of the candidates are giants, with distances in the range of 500 pc to a few kpc. Several criteria, with different dependences on the distance estimates and kinematics (radial velocities and proper motions), are used to assign each star to a given population. This assignment is probabilistic in nature and so cannot be definitive for any one star; large samples are needed and we have been awarded further observing time this semester to obtain a statistically significant sample. Elemental abundances are derived using the methodology of Fulbright (2000) and our results thus far for metal-poor stars are illustrated in Fig. 4. We find that the thick disk extends to at least [Fe/H] $= -2$ dex, and the thin disk below -1 dex. These metal-poor disk stars do not have the low ratios of [α/Fe] of the bulk of the population in dwarf galaxies. This limits the accretion of stars from satellite galaxies, into the disks, to have occurred only very early, and is not obviously consistent with late accretion, at epochs $z < 1$, as has been proposed (Abadi et al. 2003).

These data also serve to illustrate another unexpected and important point: the small scatter about the 'Type II plateau', even at the lowest abundances where only a handful of supernovae suffice to provide the enrichment (see also Cohen et al. 2004; Cayrel et al. 2004; Spite, this volume). The different star-forming regions that were the basis for the major stellar components of the Milky Way were apparently enriched by stars with a fixed (massive-star) IMF, and were well enough mixed, and massive enough, to average out the different yields from supernovae of different progenitor masses. The separation between different components, and lack of scatter, within any one component (e.g. Reddy, Lambert & Allende-Prieto 2006; Bensby et al. 2007a), around the downturn signalling the contributions of Type Ia supernovae, at least in samples probing a few kiloparsec of the solar circle, is hard to understand if many distinct small subsystems contribute.

2.3. *Substructure and Mergers into the Thin Disk*

As noted, it has been suggested that direct accretion of satellites on high angular momentum orbits could provide old stars in the thin disk, and perhaps explain the Monoceros Stream. However, the thin disk is clearly far from a smooth, equilibrium system which could provide a well-understood background population against which to define substructure. Similarly to the situation in external galaxies, spiral arms in the thin disk can cause significant disturbances in stellar kinematics and positions that persist long after the perturbation has passed (e.g. de Simone et al. 2004), and resonant scattering can lead to significant radial mixing (Sellwood 2008 and references therein; Roškar et al. 2008). Resonances with the bar in the Milky Way can also induce coherent motions (Dehnen & Binney 1998). These effects can all give rise to 'moving groups' within which there is a large range of stellar age and metallicity, consisting of stars that were not born together, but have been perturbed to move together. Observational evidence has been provided by detailed analysis of space motions (e.g. Famaey et al. 2005) and elemental abundances (Bensby et al. 2007b; Williams et al. this volume). Indeed, it appears that all disk substructure, including the Monoceros Stream (Momany et al. 2006), can be produced by dynamical effects in the disk.

3. Cosmological Context

The available stellar population evidence implies that the Milky Way galaxy had a quiet history, with no significant mergers with dark matter haloes since a redshift of ~ 2, where 'significant' means $\sim 20\%$ mass ratio to the disk, robust satellites, on non-circular orbits that take them into the region of the disk. This lack of merging is apparently unusual in ΛCDM, where subsystems are typically more concentrated than their hosts, and typically indeed on radially biased orbits. External disk galaxies also often have a thick disk component, and this component is again old (Mould 2005; Dalcanton, Seth & Yoachim 2007).

Gentle accretion must have dominated the mass build-up of the Milky Way, with fluffy, gas-rich systems being the primary means of matter infall. Late mergers are clearly contributing to the outer stellar halo, e.g. the Sagittarius dwarf spheroidal. Late accretion is perhaps building up the gaseous disk, in the form of high velocity clouds. With accretion dominated by gas-rich systems, most star formation will have occurred *in situ*, consistent with the high metallicity of the bulge, and of the thick and thin disks, the components that dominate the stellar mass of the Milky Way. The first detections of CO emission lines in massive disk galaxies at redshifts $z \sim 1.5$ also imply *in situ* star formation (Daddi et al. 2008), as inferred for the Milky Way (and thus satisfying the Copernican Principle).

However, we do still lack important knowledge of the large-scale stellar populations of the Galaxy, including the thick and thin disks far from the Sun. Our knowledge of how the different stellar components are connected is also far from satisfactory. There is a continuing need for large-scale spectroscopic studies, at both medium resolution, for broad kinematics and metallicities across the Galaxy, and high-resolution, for detailed elemental abundances and tracing streams. There is also a need for comprehensive surveys of M31 and M33, to place the results for the Milky Way in a better context. The proposed Gemini/Subaru MOS instrument WFMOS will play an important part in this endeavour.

Bengt Strömgren emphasised the role played by the stellar populations of the Milky Way Galaxy in guiding our models of galaxy formation. That this remains true is testament to his legacy.

Acknowledgements

I thank the organisers for financial support, the Aspen Center for Physics for a stimulating environment, Greg Ruchti for help with Fig. 1 and him and other collaborators for allowing me to show results in preparation.

References

Abadi, M., Navarro, J., Steinmetz, M., & Eke, V. 2003, *ApJ*, 597, 21
Bensby, T., Zenn, A., Oey, S., & Feltzing, S. 2007a, *ApJ*, 663, L13
Bensby, T., Oey, S., Feltzing, S., & Gustaffson, B. 2007b, *ApJ*, 655, L89
Binney, J., Dehnen, W., & Bertelli, G. 2000, *MNRAS*, 318, 658
Brook, C. et al. 2007, *ApJ*, 658, 60
Burkert, A., Truran, J., & Hensler, G. 1992, *ApJ*, 391, 651
Cayrel, R. et al. 2004, *A&A*, 416, 1117
Chiba, M. & Beers, T.C. 2000, *AJ*, 119, 2843
Cohen, J. G. et al. 2004, *ApJ*, 612, 1107
Cole, A., Smecker-Hane, T., & Gallagher, J. 2000, *AJ*, 120, 1808
Corlin, A. 1920, *AJ*, 33, 113
Daddi, E. et al. 2008, *ApJ*, 673, 21
Dalcanton, J., Seth, A. C., & Yoachim, P. 2007, in: R. de Jong (ed.) *Island Universes*, (Dordrecht: Springer), p. 29
Diemand, J. et al. 2008, *Nature*, in press (arXiv:0805.1244)
Diemand, J., Kühlen, M. & Madau, P. 2007, *ApJ*, 667, 859 (erratum: 2008, *ApJ*, 680, 25)
Dehnen, W. & Binney, J. 1998, *MNRAS*, 298, 387
Dolphin, A. E. 2002, *MNRAS*, 332, 91
Elmegreen, B. & Elmegreen, D.M. 2006, *ApJ*, 650, 644
Fall, S. M. & Efstathiou, G.P. 1980, *MNRAS*, 193, 189
Famaey, B. et al. 2005, *A&A*, 430, 165
Feltzing, S. & Gilmore, G. 2000, *A&A*, 355, 949
Font, A., Navarro, J., Stadel, J., & Quinn, T. 2001, *ApJ*, 563, L1
Fulbright, J. 2000, *AJ*, 120, 1841
Geisler, D., Wallerstein, G., Smith, V., & Casetti-Dinescu, D. 2007, *PASP*, 119, 939
Gibson, B. K. 1998, *ApJ*, 501, 675
Gilmore, G. & Reid, I. N. 1983, *MNRAS*, 202, 1025
Gilmore, G. & Wyse, R. F. G. 1985, *AJ*, 90, 2015
Gilmore, G. & Wyse, R. F. G. 1991, *ApJ*, 367, L55
Gilmore, G., Wyse, R. F. G., & Jones, J. B. 1995, *AJ*, 109, 1095
Gilmore, G., Wyse, R. F. G., & Norris, J. E. 2002, *ApJ*, 574, L39

Hayashi, H. & Chiba, M. 2006, *PASJ*, 58, 835
Hernandez, X., Gilmore, G., & Valls-Gabaud, D 2000, *MNRAS*, 317, 831
Hill, V. *et al.* 2000, *A&A*, 364, L19
Huang, S. & Carlberg, R. 1997, *ApJ*, 480, 503
Ibata, R., Gilmore, G., & Irwin, M. 1994, *Nature*, 370, 194
Ivezic, Z. *et al.* 2008, *ApJ*, in press (arXiv:0804.3850)
Kazantzidis, S. *et al.* 2007, *ApJ*, in press (arXiv:0708.1949)
Kobayashi, C. *et al.* 2006, *ApJ*, 653, 1145
Koch, A. *et al.* 2008, *AJ*, 135, 1580
Madau, P., Diemand, J. & Kühlen, M. 2008, *ApJ*, 679, 1260
Mateo, M. 1998, *ARAA*, 36, 435
Matteucci, F & Greggio, L. 1986, *A&A*, 154, 279
Meza, A., Navarro, J. F., Abadi, M. G., & Steinmetz, M. 2005, *MNRAS*, 359, 93
Mihos, J. C. & Hernquist, L. 1996, *ApJ*, 464, 641
Mo, H., Mao, S. & White, S. D. M. 1998, *MNRAS*, 295, 319
Momany, Y. *et al.* 2006, *A&A*, 451, 515
Mould, J. 2005, *AJ*, 129, 698
Navarro, J., Frenk, C. S. & White, S. D. M. 1995, *MNRAS*, 275, 56
Newberg, H. *et al.* 2002, *ApJ*, 569, 245
Nordström, B. *et al.* 2004, *A&A*, 418, 989
Peñarrubia, J. *et al.* 2005, *ApJ*, 626, 128
Quinn, P. & Goodman, J. 1986, *ApJ*, 309, 472
Quinn, P., Hernquist, L. & Fullagar, D. 1993, *ApJ*, 403, 74
Reddy, B., Lambert, D. & Allende Prieto, C. 2006, *MNRAS*, 367, 1329
Reed, D., *et al.* 2005, *MNRAS*, 359, 1537
Robertson, B. *et al.* 2006, *ApJ*, 645, 986
Roškar, R. *et al.* 2008, *ApJL*, accepted (arXiv:0808.0206)
Schuster, W., *et al.* 2006, *A&A*, 445, 939
Sellwood, J. A. 2008, in: E. M. Corsini & J. G. Funes (eds.), *Formation and Evolution of Galaxy Disks*, (San Francisco: ASP) (arXiv:0803.1574)
Sellwood, J. A., Nelson, R. W. & Tremaine, S., 1998, *ApJ*, 506, 590
de Simone, R., Wu, X. & Tremaine, S. 2004, *MNRAS*, 350, 627
Smecker, T. & Wyse, R. F. G. 1991, *ApJ*, 372, 448
Steinmetz, M., *et al.* (the RAVE collaboration) 2006, *AJ*, 132, 1645
Stewart, K., *et al.* 2008, *ApJ*, accepted (arXiv0711.5027)
Strömgren, B. 1987, in: G. Gilmore & B. Carswell (eds.), *The Galaxy*, (Reidel: Dordrecht), p. 229
Tóth, G. & Ostriker, J. P. 1992, *ApJ*, 389, 5
Unavane, M., Wyse, R. F. G. & Gilmore, G. 1996, *MNRAS*, 278, 727
Velazquez, H. & White, S. D. M. 1999, *MNRAS*, 304, 254
Venn, K. *et al.* 2004, *AJ*, 128, 1177
White, S. D. M. 2000, presentation at ITP conference *Galaxy Formation and Evolution*, http://online.itp.ucsb.edu/online/galaxy_c00/white
Wyse, R. F. G. & Gilmore, G. 1992, *AJ*, 104, 144
Wyse, R. F. G. & Gilmore, G. 1993, in: G. H. Smith & J. P. Brodie (eds.), ASP Conf. Ser. 48, *The Globular Cluster – Galaxy Connection*, (San Francisco: ASP), p. 727
Wyse, R. F. G. & Gilmore, G. 1995, *AJ*, 110, 2771
Wyse, R. F. G. *et al.* 2002, *New Astr*, 7, 395
Wyse, R. F. G. *et al.* 2006, *ApJ*, 639, L13
Wyse, R. F. G. 2001, in: J. G. Funes & E. M. Corsini (eds.), ASP Conf. Ser. 230, *Galaxy Disks and Disk Galaxies*, (San Francisco: ASP), p. 71
Zentner, A. *et al.* 2005, *ApJ*, 624, 505
Zoccali, M., *et al.* 2003, *A&A*, 399, 931
Zurek, W. H., Quinn, P. J., & Salmon, J. K. 1988, *ApJ*, 330, 519

The chemical evolution of the Galactic thick and thin disks

Cristina Chiappini[1,2]

[1] Geneva Observatory, Geneva University, 51 Chemin des Mailletes, Sauverny, CH1290, Switzerland, email: `Cristina.Chiappini@obs.unige.ch`

[2] Osservatorio Astronomico di Trieste - INAF, Via G. B. Tiepolo 11, Trieste, Italy

Abstract. Recent data have revealed a clear distinction between the abundance patterns of the Milky Way (MW) thick and thin disks, suggesting a different origin for each of these components. In this work we first review the main ideas on the formation of the thin disk. From chemical evolution arguments we show that the thin disk should have formed on a long timescale. We also show clear signs that the local stellar samples are contaminated by stars coming from inner radii. We then check what would have to be changed in such a model in order to explain the observables in the thick disk. We find that a model in which the thick disk forms on a much shorter timescale than thin disk and with a star formation efficiency of around a factor of 10 larger than that in the thin disk can account for the observed abundance ratio shifts of several elements between thick and thin disk stars. Moreover, the lack of scatter in the abundance ratio patterns of both the thick and thin disks suggest both components to have been formed in situ by gas accretion and not by mergers of smaller stellar systems. Especially for the thick disk, this last constraint becomes a strong one if its metallicity distribution extends to, at least, solar. Finally, we briefly discuss the interplay between present deuterium abundance and present infall rates in connection with the thin disk evolution.

Keywords. Galaxy: abundances, Galaxy: disk, Galaxy: evolution, stars: abundances

1. Introduction

Since Gilmore & Reid (1983) it is known that our Galaxy has not only a thin disk, but also a thick disk component. Moreover, it has been found that thick disks are common in many other spiral galaxies (e.g. Yoachim & Dalcanton 2008). For our Galaxy, there are striking differences between these two components, namely: a) thick disk stars are old (10-12 Gyrs) whereas thin disk stars are younger than $\sim$10 Gyrs; b) although they overlap in metallicity, their abundance patterns (i.e. [X/Fe] vs. [Fe/H]) show important differences for X = α-elements, Mn, among other elements (see Feltzing, this volume), while being similar for most of the iron-peak elements; c) the thin disk is still forming stars whereas in the thick disk there is no more star formation or gas.

The question of how the thick and thin disk formed is intimately connected to the more general question of how galaxies form. As showed in this meeting one way to approach this problem is via galaxy formation simulations, starting with a cosmological model and computing the formation of galaxies in large scale. As beautifully summarized by S. White (this conference) this approach has led to an enormous progress, but still faces several problems such as uncertainties related to *feedback* and resolution. A stringent constraint upon models of galaxy formation is that of their chemical composition, which is directly linked to *feedback*. Here we use an alternative and complementary approach which has often been named the *Archaeological Approach*: we start from the present properties of the thick and thin disks in our Galaxy, for which detailed abundances are measured, and infer their past history.

2. The thin disk

A robust result from chemical evolution models is that a closed-box model or a model in which metal-poor gas is accreted into the thin disk on very short timescales are not only incompatible with the G-dwarf metallicity distribution, but also with other observables such as the present star formation rate and the present deuterium abundance.

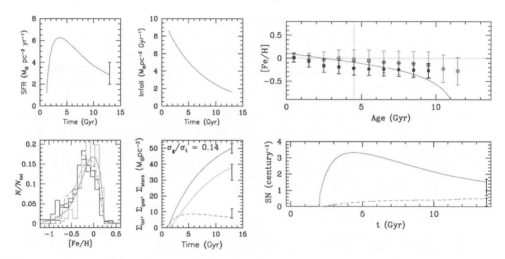

Figure 1. Best model predictions for the thin disk at the solar vicinity (see Chiappini et al. 1997). The vertical bars in some plots mark the observed value of the different quantities at present time at the solar vicinity. The data shown as histograms are from Rocha-Pinto & Maciel (1996), Kotoneva et al. (2002) and Jørgensen (2000). For further details, see text.

Fig. 1 shows our best model predictions for the thin disk at the solar vicinity. The main input parameters are the infall rate $f(t)$ and the law for star formation, $\psi(t)$. We assume $f(t) = Ae^{(-t/\tau)}$ (where $\tau = 7$Gyrs) and $\psi(t) = B(R)\Sigma_g^{1.5}$, where B(R) is related to the star formation efficiency and the total mass density at each radii and Σ_g is the gas surface mass density (see details in Chiappini et al. 1997). Note that the 7 Gyr timescale is compatible with recent cosmologically motivated infall (e.g. Naab & Ostriker 2006, Colavitti et al. 2008). These assumptions ensure a good agreement with several observables.

Fig. 1 shows the most important ones, namely (from upper left to bottom right): a) the star formation history resulting from the adoption of the infall rate (b), c) the age metallicity relation (AMR) compared with recent data by Soubiran et al. (2008 - squares) and da Silva et al. (2006 - dots); d) the G-dwarf metallicity distribution as measured by different groups (histograms) compared with the prediction of the second infall of Chiappini et al. (1997 - dashed line) and with the similar model present here, only for thin disk (solid line); e) the predicted variation of the total (solid), stellar (dotted) and gas (dashed) surface mass densities along ∼11 Gyrs of evolution and f) the type II (solid) and Ia (dashed) supernovae rates. In addition, the model shown in Fig. 1 reproduces the solar and present-time abundances of the 28 elements included in our code.

Although we find good agreement for most of the observables, the model predicts a flatter AMR than observed, especially at larger ages. This is expected to happen for the following two reasons: a) the AMR for objects older than 10 Gyrs most probably includes thick-disk objects. In the case of Soubiran et al. thick-disk candidates were removed from their sample, which explains the better agreement of our model with this data-set; b) contamination of local samples with old metal-rich stars born in the inner regions of the disk (see Roskar, this conference). The latter effect seems to also happen. Fig. 2 (left panel) shows that although our model can well explain the [Fe/O] trend with metallicity (which means that our enrichment timescales are correct), it stops at [O/H]∼0.1 dex. This is because, as said before, this model reproduces both the solar and the ISM oxygen abundances which are well constrained by observations.

Recent results have shown oxygen in young stars of the solar vicinity to be solar (see Chiappini *et al.* in prep and references therein). The similarity in the abundances of the Sun (4.5 Gyrs ago) and the ISM (now) can be understood in the framework of our model as due to the quite inefficient star formation rate in the last 4.5 Gyrs, leading to only a mild enrichment of the ISM since the formation of the Sun. However, the thin disk data shown in Fig. 2 extends up to at least [O/H]∼+0.3 dex. We interpret this discrepancy as a clear indication that the local samples adopted to constrain chemical evolution models are contaminated by stars born at inner radii (see Grenon 1999).

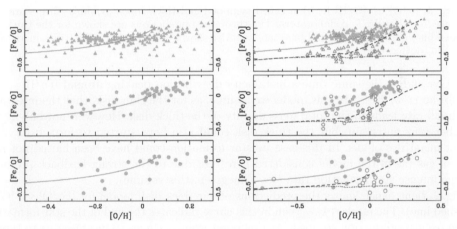

Figure 2. [Fe/O] vs. [O/H]. Left panel: Filled symbols show thin disk stars observed by different groups using different oxygen lines and thus suffering from different uncertainties (Ramirez *et al.* (2007) – upper panel; Bensby *et al.* (2004) – middle panel and Mélendez *et al.* (2008) – lower panel). The solid line shows our best thin-disk model predictions (in both panels). Right panel: shows both thin disk (filled symbols) and thick disk stars (open symbols) measured by the same author to avoid systematic shifts. Our thick disk model with (dashed) and without (dotted) the contribution of SNIa is also shown.

3. The thick disk

For the thick disk, fewer constraints are available. Thick disk stars are older than ∼10 Gyrs, and the metallicity distribution of this galactic component has a peak around [Fe/H]= −0.5, extending from ∼ −1.5 to solar or above (the exact form of the metallicity distribution is still very uncertain, with large differences among different authors). The thick disk stars comprise 4 to 15% of the mass of the thin disk (Jurić *et al.* 2008).

The above properties support the idea that the star formation history of the thick disk was radically different from the one of the thin disk, being peaked towards earlier ages, hence implying a shorter timescale for the formation of this component in contrast to the extended formation of the thin disk. Here we present a thick-disk model similar to the one of the thin-disk but with two main differences: a much shorter infall timescale $\tau = 0.4$ Gyr and a star formation efficiency larger than the one in the thin disk by a factor of 10. In this way we manage to get almost all stars in this component to have ages above

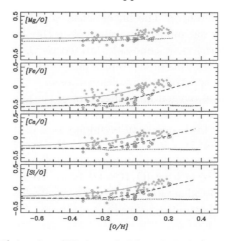

Figure 3. [α/O] vs. [O/H] for thin (filled symbols) and thick (open symbols) disk stars. The data are from Bensby & collaborators. The theoretical predictions are shown for the thick disk (dashed line) and thin disk (solid line) models. Also shown is the case of a thick disk model with the SNIa (dotted line).

∼10 Gyr and at the same time a metallicity distribution peaking around [Fe/H]= −0.5 (Soubiran *et al.* 2008). In this model we assume, as for the thin disk, that the integral of the infall rate leads to the total mass density of the thick disk. However, not only is this number very uncertain, but the thick disk could also have started to form from much larger quantities of gas. In this case its star formation could have been halted not by a lack of gas (as in the model considered here) but due to the strong feedback produced by the intense star formation rate (see Elmegreen, this volume).

Fig. 2, right panel, shows the predictions of our thick-disk model for [Fe/O] vs. [O/H] (dashed line). The marked *knee* seen in this curve indicates the end of the star formation, when oxygen production stopped. At this point iron continues to increase due to type Ia supernovae (whose progenitors are low and intermediate-mass stars born before the halt in star formation). Also shown is a model without the SNIa contribution (dotted line) which obviously produces no up-turn. Notice that the reason for the up-turn seen in the thin disk curve is completely different. In this case, it is the larger contribution of SNIa with respect to SNII that increases the [Fe/O] ratio with time (*Time-delay model*, see Matteucci 2001). Good agreement is also obtained for a large number of other abundance ratios such as Mg/O, Ca/O, Si/O (see Fig. 3), Mn/O, among others (see Chiappini *et al.* in prep.). As clearly seen in Fig. 3, there is not only a simple shift between thick and thin-disk stars, but the differences among the two samples is a function of metallicity and it is well reproduced by our model. It can also be seen that the scatter in the abundance ratios is small.

A crucial point is that the abundance ratios of thick disk stars shown here exhibit trends with metallicity and are not just constant plateaus. However, it should be said that these trends become more evident if the thick disk really extends above solar metallicities. If this is the case, these trends can be interpreted as an indication of the coherent formation of the thick disk, challenging the idea that this component was formed from accreted stellar building blocks and favouring scenarios where the thick disk had an *in situ* formation, via gas accretion (similar to the one envisaged by Bournaud *et al.* 2007). Unfortunately, it is currently still debated if the stars with [O/H] > 0 are really thick disk stars or are again stars coming from other galactic components such as thin-disk (other radii) or even bulge stars (Bensby *et al.* 2007, Ramirez *et al.* 2007).

Finally, we notice a certain coincidence between the thick disk properties we infer here from our *Archaeological Approach* and the properties of the recently discovered disks at redshift ~2 (Pettini, this conference). These objects also seem to show high star formation efficiencies and to be actively forming their stars at times compatible with the star formation history we infer here for the thick disk.

4. Infall and deuterium in the thin disk

Up to now we have discussed models for the solar vicinity where most of the observational constraints are found. To satisfy other constraints such as the radial profiles of gas and star formation rate along the thin disk as well as the abundance gradients, we assume that the thin disk formed *inside out* (understood as shorter infall timescales in the inner regions relative to the outer ones - Chiappini et al. 2001). By integrating over several rings, we obtain the total present star formation and infall rates in the thin disk. Here we show the results for the model of Chiappini et al. (2001) (see Fig. 4). We see that this model predicts an infall rate of around 1 $M_\odot$/yr at ~5 Gyrs ago and a present rate of ~0.4 $M_\odot$/yr†. Current estimates of the rate of infalling metal poor gas into the disk are ~0.2 $M_\odot$/yr (Peek et al. 2007), but some extra contribution could still come from the Magellanic Clouds or other hidden sources (Fraternali, this conference).

For the integrated star formation rate in the disk we predict that it was around a factor of 5 larger ~8 Gyrs ago, than at present time. Interestingly, Bell (2008) found a similar result by studying the star formation rates obtained with Spitzer for spiral galaxies at redshifts <1 (lookback times corresponding to ~8 Gyrs ago).

As mentioned before, infall is a solution not only to the G-dwarf problem, but also helps to explain the present quantities of deuterium in the ISM. Deuterium is a very sensitive chemical marker of the gas consumption in a given locale. All of the deuterium in the Universe was created in the Big Bang. Stars destroy deuterium, so mass-loss from stars reduces the D abundance in the ISM, while accretion of pristine material from intergalactic space increases it. Thanks to infall, chemical evolution models for the solar vicinity that reproduce the major observational constraints predict only moderate D depletion, by at most a factor of 1.5 (e.g. Tosi et al. 1998, Chiappini et al. 2002). This was considered to be in good agreement with the observed values in the ISM.

With the new FUSE results, showing a large scatter in [D/H] in the ISM, the interpretation of the data becomes complex (see Steigman et al. 2007 for a critical discussion). Linsky et al. (2006) prefer the largest values of D (the low ones attributed to depletion into dust grains) and conclude that $[D/H] = 2.31 \pm 0.24 \times 10^{-5}$, which is smaller than the primordial value by only a factor ~1.2, in conflict with the predictions of chemical evolution models. According to Steigman et al. (2007) the real value can be slightly larger and still compatible with current models (Romano et al. 2006).

A possible way to alleviate the problem is to imagine larger quantities of infalling gas in the thin disk at present time. Right now this seems not to be possible because, as discussed above, the predicted infall rates are already on the limit of the observed values. On the other hand, Fraternali (this conference) presented some examples where the real infall rates are probably much larger than what is estimated from observed infalling gas clouds. If this was the case also for the MW, we would immediately alleviate the deuterium problem. It remains to be seen if larger quantities of infalling gas would still

† In that model the age of the thin disk was ~12 Gyrs. If we assume the disk is only 10 Gyrs old, then the present infall rate will be slightly larger, although the exact value depends on the adopted disk scale-length and total central mass density.

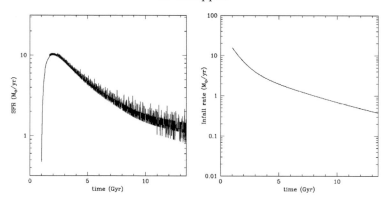

Figure 4. Predicted star formation and infall rates in the whole thin disk. This figure was obtained by integrating the several 2kpc rings of the model of Chiappini et al. (2001).

lead to models in good agreement with constraints such as the present gas surface density and present star formation rate.

References

Bell, E. 2008, *Formation and Evolution of Galaxy Disks*, eds. J. G. Funes & E. M. Corsini
Bensby, T., Zenn, A. R., Oey, M. S., & Feltzing, S. 2007, *ApJ*, 663, 13
Bensby, T., Feltzing, S., & Lundström, I. 2004, *A&A*, 415, 155
Bournaud, F., Elmegreen, B. G., & Elmegreen, D. M. 2007, *ApJ*, 670, 237
Chiappini, C., Renda, A., & Matteucci, F. 2002, *A&A*, 395, 789
Chiappini, C., Matteucci, F., & Romano, D. 2001, *ApJ*, 554, 1044
Chiappini, C., Matteucci, F., & Gratton, R. 1997, *ApJ*, 477, 765
Colavitti, E., Matteucci, F., & Murante, G. 2008, *A&A*, 483, 401
da Silva, L., Girardi, L., Pasquini, L. et al. 2006, *A&A*, 458, 609
Gilmore, G. & Reid, I. N. 1983, *MNRAS*, 202, 1025
Grenon, M. 1999, *Astrophys. Space Science*, 265, 331
Haywood, M. 2006, *MNRAS*, 371, 1760
Jørgensen, B. R. 2000, *A&A*, 363, 947
Jurić, M., Ivezic, Z., Brooks, A. et al. 2008 *astro-ph/0510520*
Kotoneva, E., Flynn, C., Chiappini, C., & Matteucci, F. 2002, *MNRAS*, 336, 879
Linsky, J. L., Draine, B. T., Moos, H. W. et al. 2006, *ApJ*, 647, 347
Matteucci, F., 2001 *The chemical evolution of the Galaxy*, Kluwer
Meléndez, J., Asplund, M., Alves-Brito, A. et al. 2008, *A&A*, 484, L21
Naab, T. & Ostriker, J. P. 2006, *MNRAS*, 366, 899
Peek, J. E. G., Putman, M. E., & Sommer-Larsen, J. 2008, *ApJ*, 674, 227
Ramirez, I., Allende Prieto, C., & Lambert, D. L. 2007, *A&A*, 465, 271
Romano, D., Tosi, M., Chiappini, C., & Matteucci, M. 2006, *MNRAS*, 369, 295
Rocha-Pinto, H. J. & Maciel, W. J. 1996, *MNRAS*, 279, 447
Soubiran, C., Bienaymé, O., Mishenina, T. V., & Kovtyukh, V. V. 2008, *A&A*, 480, 91
Steigman, G., Romano, D., & Tosi, M. 2007, *MNRAS*, 378, 576
Tosi, M., Steigman, G., Matteucci, F., & Chiappini, C. 1998, *ApJ*, 498, 226
Yoachim, P. & Dalcanton, J. J. 2008, *ApJ*, 682, 1004

The chemical fingerprints of the thin and the thick disk

Sofia Feltzing[1], Sally Oey[2] and Thomas Bensby[3]

[1]Lund Observatory,
Box 43, SE-221 00 Lund, Sweden
email: sofia@astro.lu.se

[2]University of Michigan
830 Dennison Building, Ann Arbor, MI, USA 48109-1042, USA

[3]European Southern Observatory
Alonso de Cordova 3107, Vitacura, Casilla 19001, Santiago 19, Chile
email: tbensby@eso.org

Abstract. The past history and origin of the different Galactic stellar populations are manifested in their different chemical abundance patterns. We obtained new elemental abundances for 553 F and G dwarf stars, to more accurately quantify these patterns for the thin and thick disks. However, the exact definition of disk membership is not straightforward. Stars that have a high likelihood of belonging to the thin disk show different abundance patterns from those for the thick disk. In contrast, we show that stars for the Hercules Stream do *not* show unique abundance patterns, but rather follow those of the thin and thick disks. This strongly suggests that the Hercules Stream is a feature induced by internal dynamics within the Galaxy rather than the remnant of an accreted satellite.

Keywords. stars: abundances, stars: kinematics, Galaxy: abundances, Galaxy: disk, Galaxy: formation

1. Chemical fingerprints of the Milky Way stellar disks

That the Milky Way has two stellar disks became clear when the existence of the thick disk was confirmed through star counts (Gilmore & Reid 1983). The existence of the two stellar populations in the Milky Way disks with partly overlapping properties has since been confirmed in a number of studies, notably those on elemental abundances and ages (e.g., Gratton *et al.* 2003, Soubiran & Girard 2005, Reddy *et al.* 2003, Reddy *et al.* 2006). In general, the thick disk is observed to be older and less metal-rich than the thin disk.

Of particular concern to us in the context of galactic chemical evolution is the fact that these two disk components appear to have distinct abundance patterns. Here we will, as an example, discuss two studies of elemental abundances in F and G dwarf stars in the solar neighbourhood: Fuhrmann (2008) and Bensby *et al.* (2004). These two studies nicely illustrate some of the best established properties of the thin and the thick disks.

Fuhrmann (2008) studied the trends for magnesium relative to iron and found that his volume-limited sample ($d < 25$ pc) of late F and early G dwarf stars with respect to the elemental abundances split into two major groups. One of these shows an overabundance of Mg relative to Fe of about +0.4 dex, presenting a plateau that extends from an Fe abundance of -1 to -0.3 dex. Fuhrmann associates these stars with the Milky Way thick disk. The other major group consists of stars that are less enhanced in Mg relative to Fe, and these stars also show a shallow decline in [Mg/Fe] as [Fe/H] increases. Fuhrmann also found that the stars with less enhancement of Mg were younger than the stars with

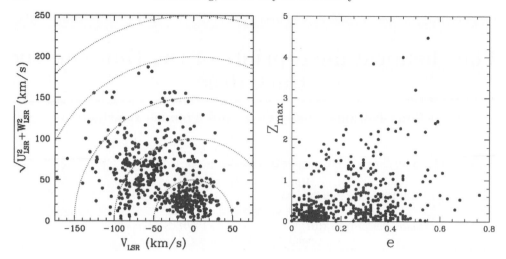

Figure 1. Left hand panel: Toomre diagram showing the kinematic properties for the sub--sample of five hundred stars. **Right hand panel:** Two orbital parameters, maximum height above the Galactic plane in kpc ($Z_{\max}$) and ellipticity of the orbit (e) for the same stars. The orbital parameters are taken from the calculations in Nordström et al. (2004).

higher Mg enhancement. He identifies these stars with the thin disk. Stars that do not fall onto either of the two trends were dubbed "transition" stars by Fuhrmann.

Bensby et al. (2004), on the other hand, employed kinematic criteria to define two samples of F and G dwarf stars: one in which the stars have a much higher probability to be thin than thick disk stars, and one in which they have a much higher probability to be thick than thin disk stars. For a full description of how the stars were selected see Bensby et al. (2003). For these stars, we obtained oxygen abundances from the forbidden line at 630.0 nm. Also for these samples, a clear distinction was found between the abundance trends traced by the thin and thick disk stars, respectively, such that the thick disk stars are more enhanced in O relative to Fe than the thin disk is.

Hence, we may conclude that at least one of the chemical fingerprints for the thin and the thick disks is this difference in enhancement in α-elements relative to iron. Furthermore, this trend appears to be quite insensitive to the actual definition of thin vs thick disk stars: both a volume-complete, as well as kinematically selected samples, show the same overall picture, even though the details might differ.

However, even though it appears to be reasonably easy to kinematically distinguish the two major components of the stellar disk, the thin and the thick disk, it is clear that the stars in the solar neighbourhood are not smoothly distributed in velocity space, (Nordström et al. 2004). Indeed, several streams and so-called moving groups have been identified. The origin of these kinematic groupings is, however, not altogether clear; are they dissolved stellar clusters, or accreted satellites, or do they result from various dynamical process such as resonances with the Galactic bar (e.g. Dehnen 2000). Studies of the chemical fingerprints of such kinematic structures will help to settle these questions, as we show below (see also Freeman & Bland-Hawthorn 2002).

2. A new sample of stars

We have now extended our studies to include about 900 F and G dwarf stars. The full sample consists of several sub-samples that were each defined to study a particular question, e.g., the question of how metal-rich the thick disk can be (Bensby et al. 2007b).

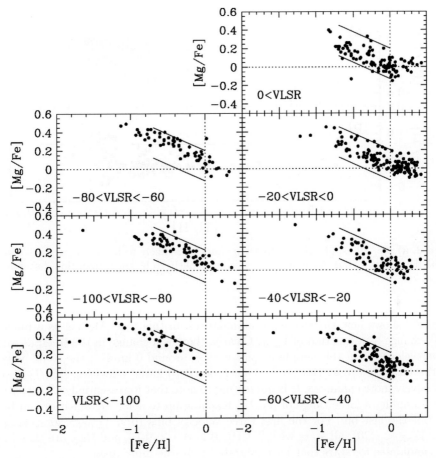

Figure 2. [Mg/Fe] vs [Fe/H] for samples with different $V_{\rm LSR}$ velocities, as indicated. The solid lines are to guide the eye and have been fitted to trace the upper and lower boundaries of the full sample.

For these stars we have obtained high-resolution, high S/N spectra using several spectrographs (Bensby *et al.* in prep.). Normally, S/N is above 250 and the resolution, for a significant fraction of the stars 60,000 or higher. In this contribution we present the first results for a subset of ∼500 of these stars. The elemental abundances and stellar parameters are derived using a methodology based on that used in Bensby *et al.* (2003). We base our $\log g$ and age determinations on the *Hipparcos* parallaxes, and the effective temperatures on excitation balance (Bensby *et al.*, in prep.). However, we now include enhancement of α-elements in the model atmospheres and have developed a more automated approach to the determination of stellar parameters based on a large grid of model atmospheres. The kinematic properties and some orbital parameters of this sub-sample are shown in Fig. 1. The orbital parameters are taken from Nordström *et al.* (2004).

2.1. *Elemental abundances and kinematics are intimately connected*

Here, we investigate how the elemental abundance trends change as a function of $V_{\rm LSR}$, where $V_{\rm LSR}$ is the velocity in the direction of rotation around the centre of the Galaxy, $U_{\rm LSR}$ is the radial velocity, and $W_{\rm LSR}$ the vertical velocity. Fig. 1 shows the distribution of velocities for our sample.

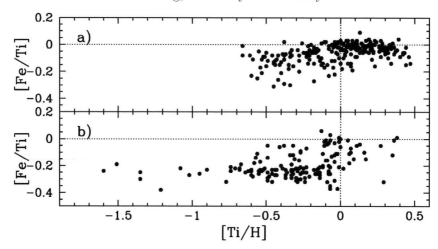

Figure 3. a) [Fe/Ti] vs. [Ti/H] for stars ten times more likely to be thin than thick disk stars. b) [Fe/Ti] vs. [Ti/H] for stars ten times more likely to be thick than thin disk stars.

Figure 2 shows how the elemental abundances, in this case [Mg/Fe] as a function of [Fe/H], change as a function of $V_{\rm LSR}$. Clearly, there is a change in the trend as we move to lower velocities. In this simplistic plot we see, around 0 km s^{-1}, the trend associated with the local thin disk. However, as we proceed to lower velocities, [Mg/Fe] becomes progressively more enhanced. It is interesting to note that for essentially all velocity bins, we are sampling a large range of [Fe/H], reaching up to solar metallicities and beyond, in all but the last bin. For the first three velocity bins ($0 < V_{\rm LSR}$, $-20 < V_{\rm LSR} < 0$, $-40 < V_{\rm LSR} < -20$), all stars with [Fe/H]>0 have $e < 0.15$, and they are therefore not likely candidates for stars that have migrated from the inner disk.

It is interesting to note that in, e.g., the bin with $-80 < V_{\rm LSR} < -60$, the metal-rich stars are more elevated in their [Mg/Fe]. Here, most stars with [Fe/H]>0 have orbits with higher ellipticity, i.e., they are probing the disk interior to the solar circle. This could thus potentially be a signature of a more rapid star formation process, leading to higher levels of [Mg/Fe], in the inner disk. However, this interpretation is tentative and requires more investigation.

2.2. *Abundance trends in kinematically defined samples of the thin and the thick disk*

We now proceed to look at the thin and thick disk defined using the kinematic criteria developed by Bensby *et al.* (2003). For the thick disk, we chose stars that are ten times more likely to belong to the thick than thin disk, and for the thin disk stars, ones that are ten times more likely to belong to the thin than thick disk. Figure 3 shows the resulting abundance trends for [Fe/Ti] vs. [Ti/H]. Ti is here preferred as the reference element instead of the commonly used Fe, as Ti is mainly produced in SN II, while Fe is produced in both SN II and SN Ia.

The two stellar samples show different abundance trends. The thin disk sample has a sharp, upper envelope that defines a gentle curve. The thick disk sample has a clear "bottom", a floor that defines the highest level of α-enhancement in the stars. This plateau ranges from [Ti/H]=−1.5 to solar Ti values. At the solar value [Fe/Ti] increases rapidly.

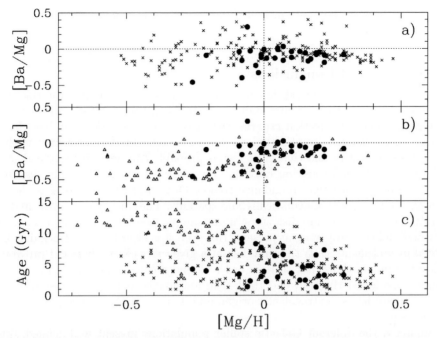

Figure 4. Elemental abundances and ages for the Hercules stream. We only show data for stars that have internal uncertainties in the age determinations of less than 3 Gyr. **a)** [Ba/Mg] vs [Mg/H] for thin disk (×) and Hercules stars (•). **b)** [Ba/Mg] vs [Mg/H] for thick disk (△) and Hercules stars (•). **c)** Stellar ages as a function of [Mg/H]. Symbols as in a) and b).

3. The Hercules stream

The existence of unique abundance patterns for the thin and thick disks allows us to test whether local kinematic features are comprised of stars that are members of the thin and thick disks, or whether they might be remnants of accreted satellite galaxies. Famaey et al. (2005) found that the Hercules stream of stars makes up about 6% of the stars in the solar neighbourhood, and that they have a net drift of $\sim$40 km s^{-1} directed radially away from the Galactic centre. The stream is also lagging behind the Local Standard of Rest by about the same amount as the thick disk. We used this kinematic information to isolate 60 F and G dwarf stars as likely members of the Hercules stream. For these stars, we obtained high-resolution ($R \sim 65,000$), high S/N ($\geqslant$250) spectra with the MIKE spectrograph on Magellan. The selection of stars, data reduction and subsequent abundance analysis are detailed in Bensby et al. (2007a) and Feltzing et al. (in prep.).

We find that the Hercules stream does not consist of stars with a unique metallicity, nor with any other unique feature. Instead, the members of the Hercules stream fall on the trends of both the thin and thick disks, indicating that they are only members of the underlying general disk populations. Dehnen (2000) indeed showed that the Hercules stream might be the result of the dynamical interaction between the local disk and the bar. Our data seem to support this model.

In fact, it appears more and more likely that the majority of the moving groups and stellar streams do not originate from a single stellar cluster or an accreted dwarf galaxy, but rather are the results of dynamical processes within the disk itself. A nice illustration of this is given by the study of the Arcturus group by Williams (see these proceedings). In fact, so far it is only the HR1614 moving group that has shown itself to be a single age, single abundance population, indicating that all stars come from the same cluster (De Silva *et al.* 2007).

4. Discussion and Conclusions

We find that for our enlarged, kinematically defined stellar sample, the thin and the thick disks also show distinct abundance trends. While these trends are fairly robust to statistically defined selection criteria, the distinctions are somewhat less clear than in the previous, smaller samples that we have studied. Our interpretation is that the disk is a complex entity and that clear trends probably can never be obtained when we use statistically defined selection criteria. Our criteria to define the thin disk tend to produce the most consistent abundance trends. Obviously, the thin disk is the easiest population to define both kinematically and spatially. The thick disk, represented by less than 15% the number of thin disk stars in our sample (see, e.g., Árnadóttir *et al.* 2008), has less well-defined properties at present. However, it is clear that the thick disk is more enhanced in α-elements than the thin disk at a given metallicity. It might prove fruitful to study the *in situ* thick disk stars in order to find a clean signature of the thick disk. We further exploited the thin and thick disk abundance patterns to test whether the Hercules Stream has an extragalactic origin, and its normal abundance patterns suggest that it does not.

In summary, the different Galactic stellar populations present well-defined, differentiated element abundance patterns that offer fundamental diagnostics of their origin. These data set strong constraints on Galactic chemical evolution models for the history of the Milky Way.

References

Árnadóttir, A., Feltzing, S., & Lundström, I 2008, arXiv:0807.1665
Bensby, T., Feltzing, S., & Lundström, I. 2003, *A&A*, 410, 527
Bensby, T., Feltzing, S., & Lundström, I. 2004, *A&A*, 415, 155
Bensby, T., Oey, M. S., Feltzing, S. & Gustafsson, B. 2007a, *ApJL*, 655, 89
Bensby, T., Zenn, A. R., Oey, M. S., & Feltzing, S. 2007b, *ApJL*, 663, 13
Dehnen, W. 2000, *AJ*, 119, 800
De Silva, G. M., Freeman, K. C., Asplund, M., Bland-Hawthorn, J., Bessel, M. S., & Collet, R. 2007, *AJ*, 133, 1161
Famaey, B., Jorissen, A., Luri, X., Mayor, M., Udry, S., Dejonghe, H., & Turon, C. 2005, *A&A*, 430, 165
Freeman, K. & Bland-Hawthorn, J. 2002, *ARA&A*, 40, 487
Fuhrmann, K. 2008, *MNRAS*, 384, 173
Gilmore, G. & Reid, N. 1983, *MNRAS*, 202, 1025
Gratton, R. G., Carretta, E., Desidera, S., Lucatello, S., Mazzei, P., & Barbieri, M. 2003, *A&A*, 406, 131
Nordström, B., Mayor, M., Andersen, J., Holmberg, J., Pont, F., Jørgensen, B. R., Olsen, E. H., Udry, S., & Mowlavi, N. 2004, *A&A*, 418, 989
Reddy, B. E., Lambert, D. L., & Allende Prieto, C. 2006, *MNRAS*, 367, 1329
Reddy, B. E., Tomkin, J., Lambert, D. L., & Allende Prieto, C. 2003, *MNRAS*, 340, 304
Soubiran, C. & Girard, P. 2005, *A&A*, 438, 139

Beryllium and the formation of the Thick Disk and of the Halo

Luca Pasquini[1], R. Smiljanic[1,2], P. Bonifacio[3,4,5], R. Gratton[6], D. Galli[7] and S. Randich[7]

[1]ESO
Garching bei München, Germany
email: lpasquin@eso.org

[2]Universidade de São Paolo, IAG, S. Paulo , Brazil, [3]GEPI Observatoire de Paris - Meudon, France, [4]INAF, Osservatorio di Trieste, Trieste, Italy, [5]CIFIST Marie Curie Excellence Team, [6]INAF-Osservatorio di Padova, Padova, Italy, [7]INAF- Osservatorio di Arcetri, Firenze, Italy

Abstract. We use Beryllium to investigate star formation in the early Galaxy. Be has been demonstrated to be a good indicator of time in these early epochs. By analyzing the so-far largest sample of halo and thick disk metal poor stars, we find a clear scatter in Be for a given value of [Fe/H] and [O/H]. The scatter is very pronounced for Halo stars, while it is marginal for thick disk stars. Our halo stars separate in the [α/Fe] - Be diagram, showing two main branches: one indistinguishable from the thick disk stars, and one with lower [α/Fe] ratio. The stars belonging to this branch are characterized by highly eccentric orbits and small perigalactic radius (R_{min}). Their kinematics are consistent with an accreted component.

Keywords. Stars: Abundances, Galaxy: Halo, Galaxy: Disk

1. Introduction

Beryllium has a unique nucleosynthesis: is neither a product of stellar nucleosynthesis nor created in detectable amounts by standard homogeneous primordial nucleosynthesis (Thomas et al. 1993). Its single long-lived isotope, ^{9}Be, is a pure product of cosmic-ray spallation of heavy (mostly CNO) nuclei in the interstellar medium (Reeves et al. 1970)

Early theoretical models of Be production in the Galaxy assumed the cosmic-ray composition to be similar to the composition of the interstellar medium (ISM). In this scenario, Be is produced by accelerated protons and α-particles colliding with CNO nuclei of the ISM (Meneguzzi & Reeves 1975; Vangioni-Flam et al. 1990; Prantzos et al. 1993), resulting in a quadratic dependence of the Be abundance with metallicity. However, observations of Be in metal-poor stars (Rebolo et al. 1988; Gilmore et al. 1992; Molaro et al. 1997a; Boesgaard et al. 1999) found a slope equal or close to one between log(Be/H) and [Fe/H], and just slightly higher for log(Be/H) and [O/H]†. Such slopes argue that Be behaves as a primary element and its production mechanism is independent of ISM metallicity. Thus, the dominant production mechanism is now thought to be the collision of cosmic-rays composed of accelerated CNO nuclei with protons and α-particles of the ISM (Duncan et al. 1992; Vangioni-Flam et al. 1998).

As a primary element and considering cosmic-rays to be globally transported across the Galaxy, one may expect the Be abundance to be rather homogeneous at a given time in the early Galaxy. It should have a smaller scatter than the products of stellar nucleosynthesis (Suzuki et al. 1999; Suzuki & Yoshii 2001). Thus, Be would show a good

† [A/B] = log [N(A)/N(B)]$_\star$ - log [N(A)/N(B)]$_\odot$

correlation with time and could be employed as a cosmochronometer for the early stages of the Galaxy (Beers et al. 2000; Suzuki & Yoshii 2001).

Pasquini et al. (2004, 2007) tested this suggestion deriving Be abundances in turn-off stars of the globular clusters NGC 6397 and NGC 6752. The Be ages derived from a model of the evolution of Be with time (Valle et al. 2002) are in excellent agreement with those derived from theoretical isochrones. Moreover, the Be abundances of these globular cluster stars are similar to the abundances of field stars with the same metallicity. These results strongly suggest the stellar Be abundance to be independent of the environment where the star was formed and support the use of Be as a cosmochronometer.

Pasquini et al. (2005) extended the use of Be as a measure of time to a sample of 20 halo and thick disk stars and investigated the evolution of the star formation rate in the early-Galaxy. Stars belonging to the two different kinematic components identified by Gratton et al. (2003a) seem to separate in a log(Be/H) vs. [O/Fe] diagram. Such separation is interpreted as indicating the formation of the two components to occur under different conditions and time scales.

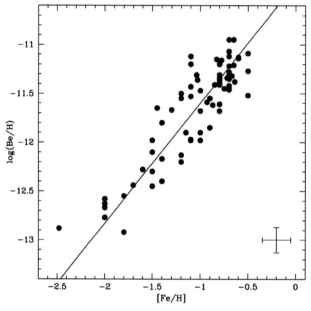

Figure 1. Be abundance vs. [Fe/H] for the sample stars. The linear relationship is clear, but a substantial scatter is present.

We present the analysis of an unprecedentedly large sample of halo and thick disk stars, and further investigate the use of Be as a cosmochronometer and its role as a discriminator of different stellar populations in the Galaxy.

2. The Sample and Analysis

The sample stars were selected from the compilation by Venn et al. (2004) of several abundance and kinematic analyses of Galactic stars available in the literature. Using the available kinematic data, Venn et al. (2004) calculated the probabilities for each star to belong to the thin disk, the thick disk, or the halo. A total of 90 stars were selected for this work; 9 of them have higher probability of being thin disk stars, 30 of being thick disk stars, 49 of being halo stars, and 2 have 50% probabilities of being halo or thick disk stars. We simply assume a star to belong to a certain kinematic group when the probability of belonging to that group is larger than the probability of being in the other two groups. Our aim being to compare stars of different populations but of similar abundances, we tried to maximize the metallicity overlap between the two sub-samples. The halo stars range from [Fe/H] = −2.48 to −0.50 and the thick disk stars from [Fe/H] = −1.70 to −0.50, although strongly concentrated in [Fe/H] ⩾ −1.00.

Spectra for all stars were obtained using UVES, (Dekker et al. 2000) fed by UT2 of the VLT. The resolving power of these spectra varies between 40,000 and 80,000 and the S/N ratio varies between 100 and 450 in the blue arm.

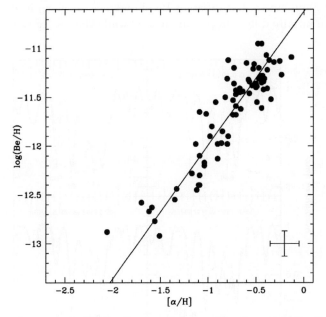

Figure 2. Be abundance vs. [α/H] for the sample stars. The linear relationship is clear, but a substantial scatter is present.

All the selected stars have previous abundance analyses. We decided to adopt the atmospheric parameters, effective temperature ($T_{\rm eff}$), surface gravity (log g), microturbulence velocity (ξ), and metallicity ([Fe/H]), as determined by the previous studies.

As Be abundances are calculated from lines of the ionized species, log g is the most relevant parameter for the analysis. Our sample stars are relatively bright and nearby, therefore we used Hipparcos parallaxes (ESA 1997) to estimate gravity. Apart from eight stars that show a significant (larger than 0.28 dex) difference between spectroscopic and astrometric gravities, the agreement is excellent.

After cleaning the sample for double line spectroscopic binaries, contaminated spectra and Be-depleted stars, we are left with 39 halo stars, 28 thick disk stars, and 9 stars classified as thin disk.

For most stars kinematic data was available, and for a few objects we have computed the orbits, following Gratton et al. (2003a).

3. Results

Figures 1 and 2 show Be abundance vs. [Fe/H] and [α/H] for the stars. The scatter, in particular in the [Fe/H] diagram, is evident. The linear relationships give:

$$Log(Be/H) = -10.37 + 1.23[Fe/H]$$
$$Log(Be/H) = -10.62 + 1.37[\alpha/H]$$

Is this scatter real or does it just reflect the abundance uncertainties? A statistical analysis shows that the scatter is significant, but the level of significance depends strongly on the assumed uncertainties. To support the hypothesis of a real scatter, we find it very convincing to directly compare the spectra of pairs of stars with similar overall stellar parameters and abundances, but different Be abundance, as given in Figure 3. Only the Be doublet substantially differs from star to star, while the rest of the spectra overlap.

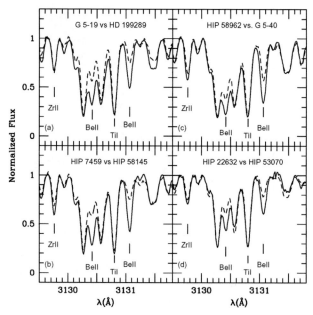

Figure 3. Comparison of spectra of couples of stars with similar parameters, but different Be content.

Once the reality of the scatter has been established, we shall consider its possible causes. In general, we can think of at least two concepts: the first is that the spread is induced by local effects. We know that this may happen, as proposed to explain the exceptionally Be-rich star HD106038 (Smiljanic et al. 2008). A second possibility is that the spread is due to the presence of a composite population. If stars with the same metallicity belong to different populations, they were formed at a different time in the Galaxy, therefore their Be abundance should be naturally different. If we divide the sample stars in thick disk and halo components, we find indeed, that while the spread among the thick disk stars is minimal, the halo component is dominating the whole scatter. This might be the signature of a composite or complex halo formation.

Figure 4 shows the Halo and Thick Disks stars in the [α/Fe] vs. Be diagram. Pasquini et al. (2005) proposed that this diagram represents star formation rate as function of time. In the same figure the models used in that work are shown. Interestingly, the halo component looks composite, with two quite distinct branches: one branch has high Be and high [α/Fe] and is indistinguishable from the thick disk stars. The other branch is well represented by the halo model and is characterized by a low [α/Fe] ratio.

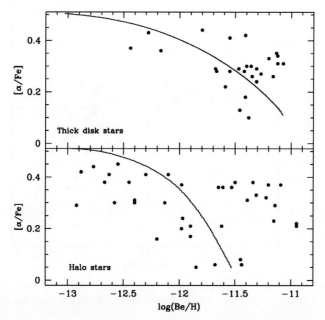

Figure 4. [α/Fe] vs. Be diagram for Halo and Thick disk stars, separately. This diagram can be interpreted as Star formation rate vs. Time. Superimposed are the models used by Pasquini et al. (2005) for the Halo and the Thick Disk. The halo seems composite, with a component at high Be and high [α/Fe] indistinguishable from thick disk stars , and a component with low [α/Fe].

If we look at the low [α/Fe] stars in more detail, they all have a similar kinematics: very low galactic rotation and orbits which reach a small perigalactic distance R_{min}. These kinematics could be compatible with an accreted component.

The hypothesis that these low [α/Fe] stars might belong to a specific population is confirmed by the fact, shown in Figure 5, that they distribute along the same line in the Fe vs. Be diagram. In conclusion, while we cannot prove that these low [α/Fe] stars are accreted, we have several indications that they separate from the other halo and thick disk stars. It is, on the other hand, also interesting that the high Be, high α stars do not show any chemical peculiarity with respect to the thick disk stars, even when a peculiar element such as Be is considered; this similarity in composition would lead us suppose a common origin.

References

Beers, T. C., Suzuki, T. K., & Yoshii, Y., 2000, in The Light Elements and their Evolution, ed. da Silva, L., de Medeiros, J. R. & Spite, M., IAU Symposium 198, 425.
Boesgaard, A. M., Deliyannis, C. P., King, J. R., Ryan, S. G., Vogt, S. S., & Beers, T. C., 1999, AJ, 117, 1549
Dekker, H., D'Odorico, S., Kaufer, A., Delabre, B., & Kotzlowski, H., 2000, SPIE, 4008, 534
Duncan, D. K., Lambert, D. L., & Lemke, M., 1992, ApJ, 401, 584
ESA, 1997, *The Hipparcos and Tycho catalogues*, ESA SP-1200
Gilmore, G., Gustafsson, B., Edvardsson, B., & Nissen, P. E., 1992, Nature, 357, 379
Gratton, R. G., Carretta, E., Desidera, S., Lucatello, S., Mazzei, P., & Barbieri, M., 2003, A&A, 404, 187

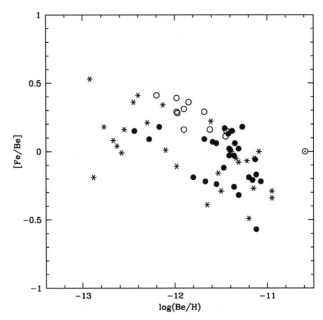

Figure 5. [Fe/Be] vs. Be for the sample stars. Open circles indicate the low [α/Fe] stars. They lie in a small range and along a sequence, suggesting a common origin.

Meneguzzi, M. & Reeves, H., 1975, A&A, 40, 99
Molaro, P., Bonifacio, P., Castelli, F., & Pasquini, L., 1997, A&A, 319, 593
Pasquini, L., Bonifacio, P., Randich, S., Galli, D., & Gratton, R. G., 2004, A&A, 426, 651
Pasquini, L., Bonifacio, P., Randich, S., Galli, D., Gratton, R. G., & Wolff, B., 2007, A&A, 464, 601
Pasquini, L., Galli, D., Gratton, R. G., Bonifacio, P., Randich, S., & Valle, G., 2005, A&A, 436, L57
Prantzos, N., Casse, M., & Vangioni-Flam, E., 1993, ApJ, 403, 630
Rebolo, R., Abia, C., Beckman, J. E., & Molaro, P., 1988, A&A, 193, 193
Reeves, H., Fowler, W. A. & Hoyle, F., 1970, Nature, 226, 727
Smiljanic, R., Pasquini, L., Primas, F., Mazzali, P. A., Galli, D., & Valle, G., 2008, MNRAS, 385, L93
Suzuki, T. K. & Yoshii, Y., 2001, ApJ, 549, 303
Suzuki, T. K., Yoshii, Y., & Kajino, T., 1999, ApJ, 522, L125
Thomas, D., Schramm, D. N., Olive, K.A., & Fields, B. D., 1993, ApJ, 406, 569
Valle, G., Ferrini, F., Galli, D., & Shore, S. N., 2002, ApJ, 566, 252
Vangioni-Flam, E., Audouze, J., Oberto, Y., & Casse, M., 1990, ApJ, 364, 568
Vangioni-Flam, E., Ramaty, R., Olive, K. A., & Cassé, M., 1998, A&A, 337, 714
Venn, K. A., Irwin, M., Shetrone, M. D., Tout, C. A., Hill, V., & Tolstoy, E., 2004, AJ, 128, 1177

The influence of star clusters on galactic disks: new insights in star-formation in galaxies

Pavel Kroupa

Argelander-Institut für Astronomie, University of Bonn, Auf dem Hügel 71,
D-53121 Bonn, Germany
email: pavel@astro.uni-bonn.de

Abstract. Stars form in embedded star clusters which play a key role in determining the properties of a galaxy's stellar population. A large fraction of newly born massive stars are shot out from dynamically unstable embedded-cluster cores spreading them to large distances before they explode. Embedded clusters blow out their gas once the feedback energy from the new stellar population overcomes its binding energy, leading to cluster expansion and in many cases dissolution into the galaxy. Galactic disks may be thickened by such processes, and some thick disks may be the result of an early epoch of vigorous star-formation. Binary stellar systems are disrupted in clusters leading to a lower fraction of binaries in the field, while long-lived clusters harden degenerate-stellar binaries such that the SNIa rate may increase by orders of magnitude in those galaxies that were able to form long-lived clusters. The stellar initial mass function of the whole galaxy must be computed by adding the IMFs in the individual clusters. The resulting integrated galactic initial mass function (IGIMF) is top-light for SFRs $< 10\, M_\odot$/yr, and its slope and, more importantly, its upper stellar mass limit depend on the star-formation rate (SFR), explaining naturally the mass–metallicity relation of galaxies. Based on the IGIMF theory, the re-calibrated Hα-luminosity–SFR relation implies dwarf irregular galaxies to have the same gas-depletion time-scale as major disk galaxies, implying a major change of our concept of dwarf-galaxy evolution. A galaxy transforms about 0.3 per cent of its neutral gas mass every 10 Myr into stars. The IGIMF-theory also naturally leads to the observed radial Hα cutoff in disk galaxies without a radial star-formation cutoff. It emerges that the thorough understanding of the physics and distribution of star clusters may be leading to a major paradigm shift in our understanding of galaxy evolution.

Keywords. stellar dynamics, gravitation, methods: n-body simulations, binaries: general, stars: formation, stars: luminosity function, mass function, Galaxy: disk, galaxies: star clusters, galaxies: kinematics and dynamics

1. Introduction

Observations of star-formation in the solar vicinity suggest that the majority of stars form in embedded clusters of which only a small fraction survive to become open clusters (Lada & Lada 2003). A similar finding has emerged for extragalactic systems (Larsen 2002a, Larsen 2002b). And in those galaxies where the star-formation rate is high such that very massive young clusters appear, Larsen(2002b) and Larsen(2004b) argue that there is also no physically meaningful distinction between a "globular-cluster" star-formation mode and a "galactic-disk" star-formation mode; a continuum of young-cluster masses is evident. Observations of the embedded cluster mass function (ECMF) show it to be a power-law, with $\beta \approx 2$ (the Salpeter index $\beta = 2.35$ is suggested by some studies, Larsen 2002a,Weidner et al. 2004), so that the physical processes related to star clusters that affect galaxies on a global scale can be calculated by integrating over the ECMF.

The problem of clustered star formation in the context of galaxies can be split into two parts: on the one hand, the physical processes of cluster birth and stellar dynamics within the environment posed by a host galaxy need to be understood, and on the other how these propagate through to galaxy scales requires illumination.

In the following a few particular problems are addressed in turn, rather than providing a comprehensive review of star clusters in galactic disks. From this compilation of problems it can be concluded that a rather remarkable amount of galactic astrophysics follows from relatively simple ideas. The notation used here follows that applied in the Cambody lectures by Kroupa(2008).

2. Cluster cores as OB-star ejection engines

It is still not quite settled whether star clusters form mass segregated, but if not then the time-scale for the massive stars ($m > 5\,M_\odot$) to sink towards their centers can be roughly estimated from the equipartition time-scale, $t_{\rm ms} \approx (m_{\rm av}/m_{\rm massive})\,t_{\rm relax}$, where $m_{\rm massive}, m_{\rm av}$ are the masses of massive and average stars, respectively, and $t_{\rm relax}$ is the two-body relaxation time. For embedded clusters $t_{\rm ms}$ can be very short, of the order of 0.1 Myr (e.g. the Orion Nebula Cluster, ONC), and so determining whether mass segregation is established by birth is a very hard observational problem.

Irrespective of whether the massive stars from in the cluster centre or not, once they are there they form a dynamically unstable cluster core which is depleted in low-mass stars that have been pushed out of the core region. The core decays by ejecting massive stars on a time-scale $t_{\rm decay} \approx N_{\rm m} \times t_{\rm core,cross}$, where $N_{\rm m}$ is the number of massive stars in the core and $t_{\rm core,cross}$ is the core-crossing time. The core-crossing time, $t_{\rm core,cross} = 2\,r_{\rm core}/\sigma_{\rm core}$, can be estimated from the core radius, $r_{\rm core}$, and the velocity dispersion of the massive stars in the core, $\sigma_{\rm core}$, whereby care must be taken to remove binary-star motions. For clusters such as the ONC, $t_{\rm decay}$ can be 0.01–0.1 Myr, which again is much shorter than its age (about 1 Myr). This suggests that the ONC may have already shot out perhaps 70 per cent of its massive stellar content (Pflamm-Altenburg & Kroupa 2006). This is consistent with the large observed fraction of runaway massive stars, which has been used by Clarke & Pringle(1992) to infer the initial dynamical configuration of massive stars as being small groups of binary-rich massive stars without the presence of many low-mass stars.

Typical ejection velocities are 5–100 km/s, and so a large fraction of massive stars explode 15 pc to 4 kpc away from their birth site, assuming the first and last SN occur 3 and 40 Myr after birth, respectively. A recent study of the distribution of massive stars is available by Schilbach & Roeser(2008), who find that 91 per cent of their sample stars can be traced to an origin in young clusters. Studying bow-shocks produced by OB stars traveling through the interstellar medium, Gvaramadze & Bomans(2008) "report the discovery of three bow shocks produced by O-type stars ejected from the open cluster NGC 6611 (M16). One of the bow shocks is associated with the O9.5Iab star HD165319, which was suggested as one of the best examples for isolated Galactic high-mass star formation".

The dynamical fact that massive stars are shot out from unstable cluster cores has barely been incorporated in galaxy evolution models, but is likely to have some important effects. In particular, the existence or non-existence of isolated O-star formation would be an important test of star-formation theories, with important implications for the existence of the $m_{\rm max} - M_{\rm ecl}$ relation that enters critically into the IGIMF theory of Section 6.

3. Binary systems

High-resolution observations of star-forming regions have been demonstrating that the fraction of binary systems is very high such that most if not all stars form as binaries (Duchene 1999, Kouwenhoven et al. 2007, Goodwin et al. 2007). Star-formation mostly in triple or quadruple systems is ruled out because their dynamical decay time is far shorter than the age of the observed stellar populations (typically older than 1 Myr): decayed higher-order multiple systems would add too many single stars to the population such that the resulting binary fraction would be too low in comparison to the observed values (Goodwin & Kroupa 2005). This sets important boundary conditions for the star-formation process, but also for star-cluster evolution models.

Numerical models of star-clusters must therefore begin with a binary fraction near unity (Kroupa, Aarseth & Hurley 2001, de la Fuente Marcos 1997, Ivanova et al. 2005) to be realistic. The evolution of a cluster then assumes a fascinating wealth of new dynamical detail as the binary population changes through the disruption of soft binaries and hardening of hard binaries (Giersz & Spurzem 2003, Davies et al. 2006, Heggie et al. 2006, Trenti et al. 2007, Portegies et al. 2007), and the cluster responds by being heated to expand (Meylan & Heggie 1997), but brief cluster cooling through early binary disruption has also been observed (Kroupa, Petr & McCaughrean 1999).

For galactic disks the main effects of initially binary-rich clusters are as follows:

- The binary fraction decreases: one rather remarkable result is that the observed high binary fraction among late-type stars in star-forming regions ($f_{\rm bin} \approx 1$) becomes nicely consistent with the observed binary fraction in the Galactic disk ($f_{\rm bin} \approx 0.5$) while at the same time the observed period- and mass-ratio distribution functions are matched as well. A unification between star-forming and disk populations has therewith been achieved (Kroupa 2008).
- The stellar ejection rate from clusters is increased significantly (Kroupa 1998, Küpper et al. 2008) leading to many more stars with velocities larger than 10 km/s. Binary-rich clusters as stellar accelerators have not been taken into account in galaxy-evolution models.
- The SNIa rate increases: As argued by Shara & Hurley (2002), the tightening of binary orbits through stellar-dynamical encounters (through sling-shot fly-bys) in long-lived clusters leads to degenerate stars becoming tight binaries which is a pathway to supernova type Ia explosions through accretion-induced or merger-induced detonation. Long lived (i.e. massive) open clusters and globular clusters therewith become SNIa-production engines, and the SNIa rate may go up by a factor of many if not by orders of magnitude in Galaxies that are able to form such clusters. This is possible in galaxies with a sufficiently high star-formation rate (SFR, Larsen 2004b): Only clusters with a life-time longer than about 5 Gyr host an environment dense enough for a sufficiently long time to allow sufficient hardening of the degenerate binaries. For example, clusters of mass $10^{4.5} M_\odot$ have such life-times in a Milky-Way type galaxy if they orbit at a distance of about 8 kpc and on circular orbits (Baumgardt et al. 2008). It follows that galaxies with $SFR > 0.1 - 1 M_\odot/{\rm yr}$ are able to produce such clusters (Weidner et al. 2004).

The implications of this type of SFR-dependence of the SNIa rate for chemical-enrichment have not been studied.

4. The expulsion of residual gas and thickened galactic disks

Because only about 30 per cent of the gas within a few pc of a star-cluster-forming cloud core ends up in stars and the remaining gas is expelled, the emerging young cluster expands. If gas-expulsion is explosive, i.e. occurs on a time-scale shorter or comparable to the crossing-time scale of the cluster, then the cluster expands with a velocity comparable to the velocity dispersion of the pre-gas-expulsion embedded cluster. A remnant cluster may survive containing a small fraction of the birth stellar population: we have a young open cluster with an associated expanding OB association (Kroupa et al. 2001). If gas expulsion occurs on a time scale of a few cluster crossing times or longer, then the exposed cluster reacts adiabatically by expanding and looses a much smaller fraction of its birth population (Lada, Margulis & Dearborn 1984, Baumgardt & Kroupa 2007).

This naturally leads to a rapid evolution of the cluster mass function (Kroupa & Boily 2002, Parmentier et al. 2008), such that the exposed-cluster mass function should appear less steep than the embedded-cluster mass function. It may thus be that the embedded-cluster mass function has a Salpeter power-law index ($\beta = 2.35$) while the exposed star-cluster mass function appears with $\beta \approx 2$. Disruptive expulsion of residual gas also implies a natural explanation for the origin of the population II stellar halo as stemming from low-mass clusters that formed in association with todays globular clusters (Baumgardt et al. 2008), given that there is no special mode of globular cluster formation but rather a continuous power-law embedded-cluster mass function, the upper mass limit of which depends on the SFR (Weidner et al. 2004).

Another implication of residual gas expulsion is that galactic disks get thickened because each star-cluster generation has associated with it the unbound young stars lost from the clusters within the first Myr. If the galaxy has a star-formation rate high enough ($SFR >$ few $M_\odot$/yr) to allow the occurrence of massive clusters, then the expanding population has a larger velocity dispersion. A galaxy may thus go through episodes of disk-thickening events if it experiences episodes of increased SFR.

A decreasing stellar velocity-dispersion–age relation towards younger stellar ages, as observed for solar-neighbourhood stars, can be accounted for quite naturally if the Milky Way has a somewhat falling SFR over the past 5–10 Gyr, whereby proper account of other disk-heating mechanisms such as scattering of stars on molecular clouds, on spiral density waves and the bar must be taken into account appropriately (Kroupa 2002).

The Milky Way (MW) thick disk may also result from this process: If the very early MW disk went into a star-burst globally such that compact clusters formed throughout the disk with masses ranging up to 10^5 $M_\odot$, and if these experienced a rapid phase of residual gas expulsion such that the "popping" clusters lost a large fraction of their stars which expanded into the young MW with a velocity dispersion comparable to the pre-gas-expulsion velocity dispersion of the embedded clusters ($30-50$ km/s), then a thick-disk component with today's thick-dick velocity dispersion can result.

This scenario has received empirical support from the observation of clumpy but straight disks in chain and spiral galaxies from redshifts $z \sim 0.5$ to 5 (Elmegreen & Elmegreen 2006). Elmegreen & Elmegreen write that these observations seem *"inconsistent with the prevailing model in which thick disks form during the violent impact heating of thin disks and by satellite debris from this mixing."* The authors discount the above "popping cluster" scenario on the basis that the observed "clusters" are massive clumps of sizes of a few hundred pc, and because the observed thickness is not likely to be carried through to todays disk galaxies because of adiabatic contraction as a result of thin-disk growth. The Elmegreen & Elmegreen high-redshift thick disks would thus become thinner as galaxies grow to the present epoch.

However, the large and massive clumps observed in chain galaxies by Elmegreen & Elmegreen are most certainly massive star-cluster complexes as observed in the Antennae galaxies, for example (Bastian et al. 2006). Within each star-cluster complex, each individual cluster would "pop" (Fellhauer & Kroupa 2005), preserving the essence of the above "disk-thickening through popping clusters" theory. If the young clusters are mass-segregated (e.g. Marks et al. 2008), then mostly low-mass stars would be lost with a velocity dispersion of 30–50 km/s. Such a population would form a thick oblate distribution about the observed young galaxies but would not be detectable. As the thin disks grow to their present masses, the spheroidal component would adiabatically contract and appear as today's thick disk.

Clearly, this scenario for the origin of thick disks does not dependent on satellite mergers and works in any disk galaxy that experienced a major star-burst throughout the disk. The "popping cluster" scenario is attractive because it relies on known physics rather than invoking for example dark-matter substructures that are merely hypothetical at the present time, but it needs more quantitative work in order to formulate predictable quantities such as the parameters of the velocity ellipsoid at different positions in the Galaxy and for different stellar age-groups. Such work is overdue, given that the data accumulated with the GAIA mission will allow very accurate testing of this scenario.

It should be noted that this "popping-cluster" scenario does not exclude some thick-disks to have originated from infalling satellite galaxies. Indeed, infalling satellites are the only viable scenario for producing counter-rotating disks.

5. The time scale for the birth of a complete star-cluster population

The stellar IMF describes the distribution of stars that form together in a cluster-forming cloud core within a time-scale of a few Myr at most. It is the statistical outcome of the physical processes that act in the core when star formation proceeds.

What about the star-cluster initial mass function, i.e. the mass function of embedded clusters (ECMF)? Star clusters form in regions of a galaxy where the molecular clouds are massive and dense enough, but it is not clear whether an ensemble of freshly formed star clusters can be defined that represent an initial population in the sense of the stellar IMF. In constructing a galaxy, it is useful though to have this tool.

In this context, it is interesting to note that Egusa et al. (2004) find a characteristic time-scale of about 5 Myr for HII regions to appear after the inter-stellar medium assembles in molecular clouds along spiral arms in disk galaxies. Thus, a disk galaxy would be churning out populations of co-eval (within a few Myr) embedded star clusters on this time-scale. Renaud et al. (2008) investigate the regions of cluster formation in interacting galaxies, and find these to be the fully-compressive tidal regions. The time-scale the inter-stellar medium (ISM) spends in these regions, emanating from them as star clusters, is 10 Myr. Thus, even in massively interacting galaxies it seems that the ISM transforms into star clusters on a time-scale of about 10 Myr when the conditions for star formation are given.

An initial star-cluster population, described by an ECMF, $\xi_{\rm ecl}(M_{\rm ecl})$, forms on some time-scale δt, and the total mass in stars thus produced is $M_{\rm tot} = \int_{M_{\rm ecl,min}}^{M_{\rm ecl,max}} M_{\rm ecl}\, \xi_{\rm ecl}(M_{\rm ecl})\, dM_{\rm ecl} = \delta t \times SFR$. Here $M_{\rm ecl}$ is the stellar mass in the embedded cluster, and the minimum and maximum values are $M_{\rm ecl,min} \approx 5\, M_\odot$ (Taurus-Auriga-type star-formation) and $1 = \int_{M_{\rm ecl,max}}^{\infty} \xi_{\rm ecl}(M_{\rm ecl})\, dM_{\rm ecl}$ since there is only one most-massive cluster assuming there to be no bound on the physically maximum cluster mass.

This set of equations relates the current SFR and $M_{\rm ecl,max}$. Using the Larsen-sample of star clusters younger than about 10 Myr in star-forming galaxies (chosen independently to the above studies and in order to minimise the effects of early cluster dissolution), Weidner et al. (2004) (WKL) fit the theoretical $M_{\rm ecl,max}$ vs SFR relation to the observational data. The best-fitting relation has a power-law ECMF with $\beta = 2.4$ (i.e. Salpeter), and $\delta t \approx 10$ Myr, independently of the SFR. The remarkable finding here is that the time-scale of forming a star-cluster population is again found to be about 10 Myr.

It therefore appears that galaxies form star-cluster populations such that on a time-scale of about $\delta t = 10$ Myr a statistically complete representation of the ECMF is given. This corresponds to the duty-cycle of molecular clouds (formation from the ISM to emerging star clusters). This constitutes an important insight, but must be challenged by further research, as one possible criticism is that the WKL result may be affected by the choice of ages of the star clusters in the sample (see also Bastian 2008).

6. The galaxy-wide IMF and new insights on galaxy evolution

Essentially all of today's understanding of how galaxies evolve and appear rests on the assumption that the stellar initial mass function (IMF) is universal, being roughly a Salpeter-power-law (with index $\alpha = 2.35$, Salpeter 1955) perhaps with a flattening below $1\,M_\odot$. The *"canonical IMF"* has $\alpha_1 = 1.3$ for $0.08 - 0.5\,M_\odot$ and $\alpha_2 = 2.3$ for $0.5 < m/M_\odot < 1$ and $\alpha_3 = \alpha_2$ for $m > 1\,M_\odot$, while the Scalo-field IMF has $\alpha_3 \approx 2.7$ (Scalo 1986).

With this assumption, the mass-metallicity relation of galaxies needs outflows to carry away metals from dwarf galaxies as otherwise the bend-down of the metal-abundances towards lower-mass galaxies cannot be understood (Kobayashi, Springel & White 2007). At the same time, the outflows must not remove the gas from late-type dwarf galaxies as these are gas-rich today, with gas-to-stellar mass ratio near 0.8 (as opposed to MW-type disk galaxies where it is about 0.2). This poses a problem for the outflow scenario. Also, the galaxy-wide SFR is commonly calculated from a measured Hα luminosity, and the widely-used SFR($L_{\rm H\alpha}$) relation (Kennicutt et al. 1994) is linear because the number of ionising massive stars scales linearly with the number of stars formed if the IMF is taken to be invariant.

Based on the neutral-gas masses measured by 21 cm observations and the SFRs measured with the Hα flux, very low star formation "efficiencies" (i.e. extremely *long* gas-consumption time-scales of many Hubble times) are deduced for dwarf galaxies, while MW-type galaxies have relatively high efficiencies, i.e. *short* gas-depletion time-scales of about 3 Gyr.

Only recently has it been fully realised that the assumption of an invariant IMF for galaxies needs revision (Kroupa & Weidner 2003): The IMF for a galaxy is given, mathematically, by the summation of all IMFs in all forming star clusters. In each star cluster the IMF is the same invariant parent distribution (the above canonical form), except that the stellar masses are bounded above by the available mass in the pre-cluster cloud core. This is a rather elementary physical constraint: for example, star-forming cloud cores of a few $M_\odot$ as in Taurus-Auriga-like stellar clusters containing a dozen stars, cannot form stars that weigh more than a few $M_\odot$. This issue that the summed IMF of many star clusters cannot be the same as the IMF had already been concluded correctly by Vanbeveren(1982).

6.1. The $m_{\rm max} - M_{\rm ecl}$ relation

The existence or non-existence of a physical maximum stellar-mass — star-cluster-mass, $m_{\rm max} - M_{\rm ecl}$, relation (Weidner & Kroupa 2006 and references therein), is of much importance for the *IGIMF theory* described below, and can be understood in terms of feedback termination of star-formation in a cloud core through the radiation and winds of the most massive stars together with a time-sequence of stellar-mass buildup such that low-mass stars form, on average, before the massive stars are able to destroy the cloud core.

Since this relation is rather fundamental, not only for our understanding of how a typical star-cluster is assembled, but also for the development of the IGIMF theory, it is worth-while to spend a few words on the $m_\mathrm{max} - M_\mathrm{ecl}$ relation: A distribution of stars always arises within a cloud core, given that the cores are turbulent and thus have a distribution of density maxima (Padoan & Nordlund 2002, Li, Klessen & Mac Low 2003). The stars formed add up to the stellar mass in the core, $\Sigma_\mathrm{stars} = M_\mathrm{ecl}$. And so the true maximum stellar mass, m_max, that can form within a cluster is limited even more strictly than $m_\mathrm{max} \leqslant M_\mathrm{ecl}$, because $m_\mathrm{max} = M_\mathrm{ecl}$ would be a cluster consisting of one star leaving no room for the range of stellar masses resulting from turbulent gas dynamics. In other words, isolated O-star formation cannot occur.

However, Maschberger & Clarke (2008) used observational $m_\mathrm{max} - M_\mathrm{ecl}$ data to suggest that the existence of a physical $m_\mathrm{max} - M_\mathrm{ecl}$ relation cannot be confirmed empirically. But, they also note that the random IMF sampling model, which admits isolated O stars (Parker & Goodwin 2007), is inconsistent with the data. Importantly, the data as used by Maschberger & Clarke (2008) cannot be interpreted in terms of a currently existing theoretical model. They can, however, be understood rather straightforwardly (Oh et al., in preparation) if the $m_\mathrm{max} - M_\mathrm{ecl}$ relation exists *and* (1) if OB stars are shot out from binary-rich and mass-segregated clusters *appearing* as isolated O stars and (2) through the rapid dissolution of intermediate-mass clusters through explosive gas-expulsion. This latter process leads to the most massive stars being surrounded by the feeble remnant of the once existing embedded cluster, and thus this $m_\mathrm{max} - M_\mathrm{ecl}$ datum would be an outlier at too low a value of observed M_ecl. Since virtually all those data that do deviate from the relation are indeed outliers at too small M_ecl values, this explanation would appear very natural.

With this in view, the data that do not lie along the proposed $m_\mathrm{max} - M_\mathrm{ecl}$ relation which are, however, used by Maschberger & Clarke, have been removed by Weidner & Kroupa (2006) because they are typically older objects. Instead, Weidner et al. (2007) used these data to estimate the damage to star clusters done by the removal of residual gas in order to constrain the star-formation efficiency in the birth clusters under the assumption that these data were originally on the relation. This demonstrates the difficulty in interpreting the existing data (Maschberger & Clarke and Parker & Goodwin *vs* Weidner & Kroupa).

In essence, the argument is concerned with the question whether a *physical* $m_\mathrm{max} - M_\mathrm{ecl}$ relation exists, or whether star-formation in a star cluster is a mere statistical affair such that stellar masses appear stochastically without any physical boundary conditions. The existence of a pronounced $m_\mathrm{max} - M_\mathrm{ecl}$ relation for star clusters would imply a self-regulatory behaviour of star-formation on cluster-forming molecular-cloud-core scales. The outline of this physical process would be that as the cloud core begins to contract the physical conditions within it are probably similar to what we observe in Taurus-Auriga. As the core contracts the density increases such that ever more massive proto-stars can condense until their feedback energy suffices to overcome gravitational collapse and the process halts, leaving an exposed and probably largely unbound stellar cluster. This would be a deterministic process in the sense that the initially available gas mass within a given region, i.e. the pressure, would determine the type of star cluster and the mass of its most massive stars, whereby the IMF remains close to the canonical value (as determined by observations). Theoretical considerations based on this line of thought do indeed yield a well-pronounced $m_{\rm max} - M_{\rm ecl}$ relation (Weidner et al. 2008).

A quantitative counter-argument against the non-existence of a physical $m_{\rm max} - M_{\rm ecl}$ relation is, finally, as follows: First of all, the observational evidence is such that in all cases of well-resolved clusters, the IMF is always found to be consistent with the canonical IMF. Now, assuming a purely random-sampling model such that stellar masses can be generated from the canonical IMF and $\Sigma_{\rm stars} = M_{\rm ecl}$, where $M_{\rm ecl}$ is a pre-defined stellar cluster mass, there are occurrences such that $m_{\rm max} \approx M_{\rm ecl}$ (Parker & Goodwin 2007). These would be the observed 4 per cent isolated O stars. However, there would also be cases where the first cluster star picked from the IMF is massive enough to destroy the cloud core through feedback energy such that no further stars can form. The resulting number of isolated O stars would therefore outnumber the observed number of isolated O stars.

Given the above and the existence of known physical processes that can explain the existence of apparently isolated O stars as being either ejected stars (§ 2) or remnants of intermediate-mass clusters that rapidly dissolved after residual gas expulsion (Weidner et al. 2007), it follows that the purely random sampling model for creating star clusters ("the non-existence model") probably needs to be rejected.

6.2. The IGIMF

By taking into account the existence of a physical $m_{\rm max} - M_{\rm ecl}$ relation, i.e. that a cluster of stellar mass $M_{\rm ecl}$ cannot have stars with $m \geqslant m_{\rm max} = {\rm fn}(M_{\rm ecl})$, and then adding up all the so-constructed IMFs for a co-eval embedded star-cluster population that is a power-law with Salpeter index $\beta = 2.35$ (see also § 1 and § 5), it follows that the resulting "integrated galactic IMF" (the IGIMF) is steeper above about $1.3\,M_\odot$ than the canonical IMF (Vanbeveren 1982, Kroupa & Weidner 2003). This immediately solves the finding that the field-star IMF is steeper than the IMF in individual clusters (Vanbeveren 1983, Vanbeveren 1984). Thus, the Scalo-field-IMF, which has $\alpha_3 \approx 3$, is unified with the canonical IMF ($\alpha_3 = 2.3$) in a straight-forward way.

Also, since the maximum cluster mass, $M_{\rm ecl,max}$, that can form within the time span δt within a galaxy depends on the SFR of the galaxy (§ 5 above), the IGIMF becomes SFR dependent, such that galaxies with a low SFR have steeper IGIMFs because of the $m_{\rm max} - M_{\rm ecl}$ relation (Weidner & Kroupa 2006, Pflamm-Altenburg et al. 2007). More importantly, the maximum stellar mass, $m_{\rm m}$, forming in a galaxy decreases with decreasing SFR.

For the MW, which has a SFR of a few $M_\odot/{\rm yr}$, the IGIMF turns out to have an index $\alpha_{\rm IGIMF} \approx 3$ above a stellar mass of $1.3\,M_\odot$ and $m_{\rm m} = 150\,M_\odot$, while for a galaxy with a SFR near $10^{-3}\,M_\odot/{\rm yr}$, $\alpha_{\rm IGIMF} \approx 3.3$ and $m_{\rm m} \approx 20\,M_\odot$. Such a variation of the IGIMF index has been reported by Hoversten & Glazebrook(2008) on the basis of analysing a hundred-thousand star-forming galaxies in the SDSS survey. Of relevance here is that the gamma-ray flux from decaying ^{26}Al yields a current SFR of about $4\,M_\odot/{\rm yr}$ for the MW if the (Scalo) IGIMF slope, $\alpha_3 = 2.7 \approx 3$, for the Milky Way field is used (Diehl et al. 2006). This constitutes an independent confirmation of the SFR-$\alpha_{\rm IGIMF}$ relation for the case of the MW. But further confirmation, for example by measuring the IGIMF in dwarf galaxies, would be essential to test this theory.

6.3. The mass-metallicity relation of galaxies

The IGIMF theory described above immediately explains the mass-metallicity relation of galaxies without the need of additional physical processes (Köppen et al. 2007). This is so because low-mass galaxies have a deficit of massive stars per low-mass star when compared to more massive galaxies – they have top-light IGIMFs. It follows that galaxy-wide metal production is compromised increasingly with decreasing galaxy mass.

This does not mean that outflows and infall do not occur, but that these processes probably play a secondary role in establishing the metal content of galaxies.

Noteworthy is that in the currently established picture outflows need to be invoked to remove the metals but such that the gas is not blown out, given that late-type dwarfs are metal-poor and very gas rich. In the IGIMF theory the metals are not produced in the first place and so unwanted gas blow-out is not an issue.

6.4. Gas consumption time scales

The IGIMF theory also implies that the Hα-luminosity–SFR calibration in general use needs to be re-calibrated. This generally used relation leads to the widely accepted result that dwarf galaxies have very low star-formation "efficiencies", i.e. gas-consumption time-scales longer than many Hubble times. Pflamm-Altenburg et al. (2007) have re-calibrated the Hα-luminosity–SFR relation based on the above IGIMF theory. Here, a dwarf galaxy with a low SFR is producing significantly fewer ionising photons due to the top-light IGIMF than a massive galaxy. For a given measured Hα luminosity the true SFR would therefore be significantly higher than hitherto thought.

Without any parameter adjustments, the immediate result is that the true SFRs of dwarf galaxies are orders of magnitude higher than thought until now, implying neutral-gas-consumption time-scales of about 3 Gyr for *all* galaxies that contain neutral gas, and that the SFR is strictly proportional to the neutral gas mass of a galaxy, $SFR = 1/(3\,\mathrm{Gyr})\,M_{\mathrm{gas}}$. Naturally, this leads to a very major revision of our understanding of galaxy evolution and of the galaxy-wide star-formation process. Note that near $M_{\mathrm{gas}} = 10^{9.5}\,M_\odot$ the Kennicutt et al. (1994) finding, that the time-scale for gas consumption is about 3 Gyr, remains valid.

The star-formation efficiency (the fraction of gas that turns into stars) over a time-scale of $\delta t = 10\,\mathrm{Myr}$ becomes $\epsilon = (1/300)\,M_{\mathrm{gas}}$, i.e. every 10 Myr a (not strongly interacting) galaxy turns 0.3 per cent of its neutral gas mass into stellar mass. According to the IGIMF theory, this would be true for all not strongly interacting galaxies with neutral gas.

6.5. The radial Hα star-formation cutoff

The IGIMF theory also immediately and naturally explains the existence of an Hα cutoff radius in disk galaxies. The current understanding, based on applying a universal IMF to galaxies, is that the radial Hα-emission cutoff comes about because beyond the Hα-cutoff radius star formation is suppressed due to dynamical processes. Basically, star-forming cloud cores cannot assemble beyond a particular radius. However, using recent GALEX observations (Boissier et al. 2007) show UV emission from young stars to continue beyond this cutoff radius, in contradiction to the theoretical work.

A *local IGIMF* description can be formulated rather straightforwardly by connecting the local star-formation rate surface density with the local neutral-gas surface density in a disk galaxy. With this star-formation rate surface density, the local star cluster density can be computed, and by summing them all up, a local IGIMF (LIGIMF) follows. Integration of the LIGIMF over radial annuli then yields the Hα flux density in dependence of the radius, and it follows quite trivially that it decays much more rapidly than the star-formation rate density (Pflamm-Altenburg & Kroupa 2008). An interesting outcome of this work is that the star-formation rate surface density is linearly proportional to the gas surface density, $\Sigma_{\mathrm{SFR}} \approx 1/(3\,\mathrm{Gyr})\,\Sigma_{\mathrm{gas}}^N$, with $N = 1$.

6.6. *Summary: IGIMF*

The implications of all of this work would be that a paradigm shift in galaxy evolution may be emerging such that the star-formation rate (density) is proportional to the neutral gas mass (density). Dwarf galaxies differ from large disk galaxies only in terms of the level of star formation because the latter have a larger neutral gas mass which supports more star-formation activity. Further implications of this are being looked into now.

7. Conclusions

The above shows that we have been missing important ingredients in our understanding of the astrophysics of disk galaxies if the physics of star clusters as the *fundamental galactic building blocks* is omitted. As important examples the following have been noted:

- Young cluster cores are stellar accelerators dispersing massive stars over large distances from their birth sites.
- The binary fraction and SN Ia rate are defined by the star-cluster population a galaxy has been able to generate over its life time.
- Galactic disks may thicken if they form ensembles of "popping" clusters.
- Star-clusters in a statistically complete ensemble, such that they represent the embedded cluster mass function, appear to form on a time-scale of about 10 Myr.
- At the same time, by realising that we must see galaxies as made up of many (mostly dissolved) star clusters we are readily led to find a new understanding of the mass-metallicity relation of galaxies, of the radial $H\alpha$ cutoff and therewith the relation of neutral gas content to the level of star formation.
- Following on from this, it emerges that dwarf irregular galaxies consume their neutral gas content on the same time scale (about 3 Gyr) as major disk galaxies, and that it is only the mass of neutral gas that drives the macroscopic evolution of these systems in terms of the buildup of stellar mass and metals.
- The last two points are a result of correctly counting massive stars in galaxies in dependence of their SFRs: massive disk galaxies have a higher SFR than low-mass irregular galaxies and consequently their IGIMF is flatter and the upper mass limit of forming stars is higher. The implications of this suggest a possibly major paradigm shift in our knowledge of how galaxies evolve.

Acknowledgments

I would like to thank the organisers for the invitation to present this material at a most memorable meeting in Copenhagen, and Jan Pflamm-Altenburg and Carsten Weidner for very important contributions. This contribution I wrote in Vienna and Canberra, and would like to thank the respective hosts (Gerhardt Hensler, Christian Theis and Helmut Jerjen) for their kind hospitality.

References

Bastian, N. 2008, *MNRAS*, in press (astro-ph:0807.4687)
Bastian, N., Emsellem, E., Kissler-Patig, M., & Maraston, C. 2006, *A&A*, 445, 471
Baumgardt, H. & Kroupa, P. 2007, *MNRAS*, 380, 1589
Baumgardt, H., de Marchi, G., & Kroupa, P. 2008, *ApJ*, in press, astro-ph/0806.0622
Baumgardt, H., Kroupa, P., & Parmentier, G. 2008, *MNRAS*, 384, 1231
Boissier, S., *et al.* 2007, *ApJS*, 173, 524
Clarke, C. J., & Pringle, J.E. 1992, *MNRAS*, 255, 423
Davies, M. B., *et al.* 2006, *New Astronomy*, 12, 201

de La Fuente Marcos, R. 1997, A&A, 322, 764
Diehl, R., et al. 2006, Nature, 439, 45
Duchêne, G. 1999, A&A, 341, 547
Egusa, F., Sofue, Y., & Nakanishi, H. 2004, PASJ, 56, L45
Elmegreen, B. G. & Elmegreen, D. M. 2006, ApJ, 650, 644
Fellhauer, M. & Kroupa, P. 2005, ApJ, 630, 879
Giersz, M. & Spurzem, R. 2003, MNRAS, 343, 781
Goodwin, S. P. & Kroupa P. 2005, A&A, 439, 565
Goodwin, S. P., Kroupa, P., Goodman, A., & Burkert A. 2007, Protostars and Planets V, 133
Gvaramadze, V. V. & Bomans, D. J. 2008, A&A, in press, astro-ph/0809.0650
Heggie, D. C., Trenti, M., & Hut P. 2006, MNRAS, 368, 677
Hoversten, E. A. & Glazebrook, K. 2008, ApJ, 675, 163
Ivanova, N., Belczynski, K., Fregeau, J. M., & Rasio, F. A. 2005, MNRAS, 358, 572
Kennicutt, R. C. Jr., Tamblyn, P., & Congdon, C. E. 1994, ApJ, 435, 22
Kobayashi, C., Springel, V., & White, S. D. M. 2007, MNRAS, 376, 1465
Köppen, J., Weidner, C., & Kroupa, P. 2007, MNRAS, 375, 673
Kouwenhoven, M. B. N., Brown, A. G. A., Portegies, Zwart S. F., & Kaper, L. 2007, A&A, 474, 77
Kroupa, P. 1998, MNRAS, 298, 231
Kroupa, P. 2002, MNRAS, 330, 707
Kroupa, P. 2008, S. Aarseth, Ch. Tout, R. Mardling (eds.), The Cambridge N-body Lectures (Lecture Notes in Physics Series, Springer Verlag), astro-ph/0803.1833
Kroupa, P., & Boily, C. M. 2002, MNRAS, 336, 1188
Kroupa, P. & Weidner, C. 2003, ApJ, 598, 1076
Kroupa, P., Aarseth, S., & Hurley, J. 2001, MNRAS, 321, 699
Kroupa, P., Petr, M. G., & McCaughrean, M. J. 1999, New Astronomy, 4, 495
Küpper, A. H. W., Kroupa, P., & Baumgardt, H. 2008, MNRAS, 389, 889
Lada, C. J. & Lada, E. A. 2003, Annual Review of A&A, 41, 57
Lada, C. J., Margulis, M., & Dearborn, D. 1984, ApJ, 285, 141
Larsen, S. S. 2002a, AJ, 124, 1393
Larsen, S. S. 2002b, in: D. Geisler, E. K. Grebel, & D. Minniti (eds.), Extragalactic Star Clusters (San Francisco: Astronomical Society of the Pacific, 2002., IAUS 207), p. 421
Larsen, S. S. 2004a, A&A, 416, 537
Larsen, S. S. 2004b in: H. J. G. L. M. Lamers, L. J. Smith & A. Nota (eds.), The Formation and Evolution of Massive Young Star Clusters (San Francisco: Astronomical Society of the Pacific, 2004., IAUS 322,), p.19
Li Y., Klessen, R. S., & Mac Low, M.-M. 2003, ApJ, 592, 975
Marks, M., Kroupa, P., & Baumgardt, H. 2008, MNRAS, 386, 2047
Maschberger, T. & Clarke, C. J. 2008, (astro-ph/:0808.4089)
Meylan, G. & Heggie, D. C. 1997, The Astronomy and Astrophysics Review, 8, 1
Padoan, P. & Nordlund, Å. 2002, ApJ, 576, 870
Parker, R. J. & Goodwin, S. P. 2007, MNRAS, 380, 1271
Parmentier, G., Goodwin, S. P., Kroupa, P., & Baumgard, H. 2008, ApJ, 678, 347
Pflamm-Altenburg, J. & Kroupa, P. 2006, MNRAS, 373, 295
Pflamm-Altenburg, J. & Kroupa, P. 2008, Nature, in press
Pflamm-Altenburg, J., Weidner, C., & Kroupa, P. 2007, ApJ, 671, 1550
Portegies Zwart, S. F., McMillan, S. L. W., & Makino, J. 2007, MNRAS, 374, 95
Renaud, F., Boily, C. M., Fleck, J., Naab, T., & Theis, C. 2008, MNRAS Letters. in press, astro-ph/0809.2927
Salpeter, E. E. 1955, ApJ, 121, 161
Scalo, J. M. 1986, Fundamentals of Cosmic Physics, 11, 1
Schilbach, E., Roeser, S. 2008, A&A, 489, 105
Shara, M. M., & Hurley, J. R. 2002, ApJ, 571, 830
Trenti, M., Heggie, D. C., & Hut, P. 2007, MNRAS, 374, 344
Vanbeveren, D. 1982, A&A, 115, 65

Vanbeveren, D. 1983, *A&A*, 124, 71
Vanbeveren, D. 1984, *A&A*, 139, 545
Weidner, C., & Kroupa, P. 2005, *ApJ*, 625, 754
Weidner, C. & Kroupa, P. 2006, *MNRAS*, 365, 1333
Weidner, C., Kroupa, P., & Larsen, S. S. 2004, *MNRAS*, 350, 1503
Weidner, C., Kroupa, P., & Goodwin, S. P. 2008, in preparation
Weidner, C., Kroupa, P., Nürnberger, D. E. A., & Sterzik, M. F. 2007, *MNRAS*, 376, 1879

Bruce Elmegreen, Jan Palouš, and Pavel Kroupa continuing the discussion after Pavel's talk.

Hans Zinnecker, Jan Palouš and Bruce Elmegreen at Carlsberg (photo: Bruce Elmegreen).

The initial luminosity and mass functions of Galactic open clusters

Hans Zinnecker[1], Anatoly E. Piskunov[2], Nina V. Kharchenko[3], Siegfried Röser[4], Elena Schilbach[4], and Ralf-Dieter Scholz[1]

[1] Astrophysical Institute Potsdam,
An der Sternwarte 16, D-14482 Potsdam, Germany
email: hzinnecker@aip.de
email: rdscholz@aip.de

[2] Institute of Astronomy of the Russian Academy of Sciences,
48 Pyatnitskaya St. 119017, Moscow, Russia
email: piskunov@inasan.rssi.ru

[3] Main Astronomical Observatory, National Academy of Sciences of Ukraine,
27 Akademika Zabolotnoho St., 03680 Kyiv, Ukraine
email: nkhar@mao.kiev.ua

[4] Ruprecht-Karls-Universität Heidelberg, Astronomisches Rechen-Institut,
Mönchhofstr. 12–14, D-69120 Heidelberg, Germany
email: roeser@ari.uni-heidelberg.de
email: elena@ari.uni-heidelberg.de

Abstract. We have derived a complete magnitude-limited sample of 440 Galactic open clusters in the solar neighborhood, with integrated V-magnitude brighter than 8 mag. This sample can be used to infer the present-day luminosity and mass functions of open clusters up to a given age; it can even be used to construct the initial mass and luminosity function (IMF, ILF) of clusters (defined as visible clusters with age 4–8 Myr). The high-mass end of the cluster IMF is a power-law with a slope of -2 or slightly shallower (-1.7) while the luminous cluster ILF has a power-slope of -1, in agreement with what is found for extragalactic clusters. Both distribution functions show a turnover, starting at $300 \, M_\odot$ and integrated magnitude -3 mag, respectively. The overall birthrate of clusters is 0.4 clusters per kpc^2 and per Myr. The average present-day cluster mass is $700 \, M_\odot$, while the average initial cluster mass is $4500 \, M_\odot$. The difference of these two average masses indicates the high infant mortality and/or weight loss of Galactic open clusters (due to dynamical evolution).

Keywords. open clusters, astrometry, photometry, luminosity function, mass function

1. Introduction

This contribution is about a new determination of the present-day and initial luminosity and mass function of open clusters in the nearby Galactic disk, based on a homogeneous cluster data base (Kharchenko *et al.* 2005a, b) inferred from the ASCC-2.5 bright star (V < 11.5 mag) catalog (Kharchenko 2001). Using astrometric and photometric information of individual stars, this uniform all-sky catalog ultimately provides cluster central positions, kinematics, ages, colors, integrated cluster magnitudes, as well as tidal cluster masses. The masses are derived from fitting 3-parameter King profiles to the radial stellar surface density distribution and thus are independent of the cluster luminosities!

We remind the reader that open clusters are typical representatives of the Galactic disk population and excellent examples to calibrate astronomical distances, reddenings, and ages. Open clusters are also great laboratories for studies of stellar evolution and of the frequency of the various stellar types, such as Cepheids, red giants, brown dwarfs, and maybe exo-planets.

The present analysis supersedes the results of an older similar study on Galactic open clusters (Piskunov et al. 2006). We stress that we are not dealing here with the IMF in clusters (e.g. Scalo 1998) but with the IMF of clusters (Piskunov et al. 2008b).

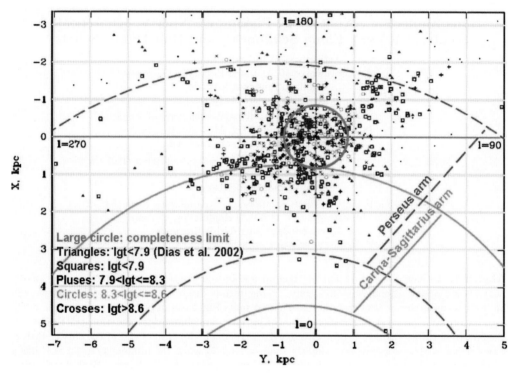

Figure 1. Distribution of open clusters in the XY-plane centered on the location of the Sun. Squares, pluses, circles, and crosses show clusters of different age groups, as indicated in the legend. Dots and triangles mark positions of clusters with known distances and ages from the Dias et al. (2002) catalogue (a somewhat heterogeneous compilation), which are not included in our sample. Long dashed and solid thick curves are the grand design spiral arms with a pitch-angle of $-6°$ fitted to young clusters in the Perseus and Carina-Sagittarius regions. The large circle around the Sun indicates the completeness limit of our sample.

2. The open cluster sample

The ASCC-2.5 catalog is a homogeneous catalog of 2.5 million stars from which a total of 650 clusters could be extracted (130 of them new, Kharchenko 2005b). Of those, 256 clusters define a volume-limited sample up to 1 kpc and a magnitude limited sample of 440 cluster, with the brightest clusters located beyond 1 kpc. Previous open cluster catalogs were published by Becker and Fenkart (1971, 216 clusters observed in UBV), by Lynga (1987, 5th Lund catalog), and by Batinelli and Capuzzo-Dolcetta (1991, a subset of 100 bright clusters inside 2 kpc). All these catalogs are heavily biased, while the ASCC-2.5 optical catalog is free from selection biases and can be used for statistical investigations.

We mention in passing that a list of 100 embedded young clusters was compiled and published in Lada and Lada (2003), while Bica et al. (2003ab) used 2MASS data to discover many new embedded clusters, both in the northern and southern hemisphere (see also Froebrich for a systematic 2MASS cluster survey). Also relevant in the context of young clusters is the catalog of 370 Galactic O-stars (GOS v1, v2 - see Maíz-Apellániz et al. 2004, and Sota et al. 2007, respectively; for an early version see Gies 1987). However, we will not make use of it here, but refer to Schilbach and Röser (2008) and their new study of the origin of O-type runaway stars in young clusters.

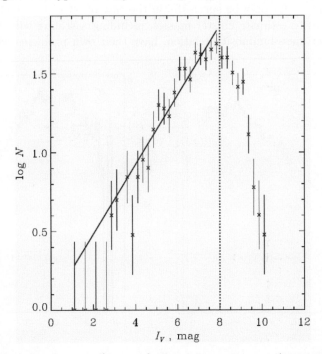

Figure 2. Number distribution of open clusters versus apparent integrated magnitude I_V (crosses). The bars are Poisson errors, the solid line is a linear fit to the distribution and the vertical dotted line is the adopted completeness limit $I_V = 8$ mag.

3. Determination of open cluster luminosities and masses

As a measure of cluster brightness needed for the construction of the luminosity function, we take the integrated V-magnitudes of each cluster. The integrated magnitudes are computed by summing up the individual fluxes of cluster members selected from the ASCC-2.5 catalogue. Doing so, we avoid contamination by field stars and can thus directly compare our results with observational results of open clusters in other galaxies. To account for the fact that the clusters reside at different distances and show Main Sequences of different lengths in a magnitude-limited survey, we introduce a small (typically ~ 0.1 mag) correction for unseen stars (7 magnitudes fainter than the bright limit of a cluster). Although small, this correction reduces all the clusters to a common and uniform system of integrated luminosities.

Our cluster masses are derived from "tidal masses" using King's (1962) formula: $M_{cluster} = 4 \, A \, (A\text{-}B) \, r_t^3 / G$, where A and B are Oort's constants of Galactic differential rotation, r_t is the cluster tidal radius, and G is the gravitational constant. Masses depend strongly on the tidal radius which is estimated from a 3-parameter King profile fit to the observed cluster structure together with the cluster core radius and central surface density. Tidal radii peak at about 10 pc and are uncertain by less than 20 %. Therefore, cluster masses should be good to about a factor of 2. For a detailed discussion of tidal radii and masses see Piskunov et al. (2007, 2008a). It has been argued that tidal masses may not be applicable to young open clusters but only to globular clusters. This concern is unfounded. As long as star clusters fill their tidal (or Roche) volume, which should be the case after their emergence from the initial embedded cluster stage, everything is fine. The alternative of deriving virial instead of tidal cluster masses (using the cluster velocity dispersion) has so far only be successful in the case of NGC 188 (Geller et al. 2008). Other alternatives to estimate cluster masses, including a scaling with the stellar IMF or using a cluster mass-luminosity relation, have their own problems which we cannot discuss here.

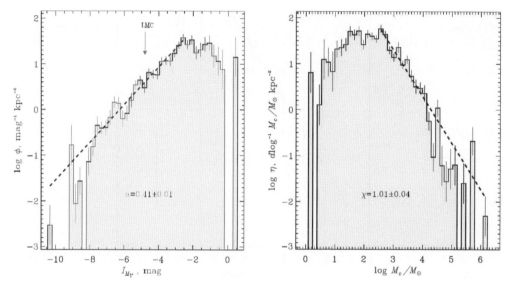

Figure 3. Luminosity (left) and mass (right) functions of Galactic open clusters based on 440 local clusters brighter than the completeness limit I_V of the sample. The bars are Poisson errors, the dashed lines show a linear fit to the brighter/massive parts of the histograms, where a and χ are the corresponding slopes. The arrow indicates the limit of integrated absolute magnitudes reached for open clusters in the LMC (Larsen 2002).

4. Cluster luminosity and mass functions

Fig. 3 shows the present-day luminosity and mass functions that we obtained (CPDLF, CPDMF): they are power-laws at the bright end and turn over towards the faint end (see abstract). Fig. 4 shows the time-resolved, integrated mass functions (dN/dlogM) including the cluster initial mass function (CIMF). The slope of the CIMF at the high-mass end (-0.7 at face value) is somewhat uncertain (fit based on just 3–4 data points) and could be consistent with the slope often quoted for extragalactic clusters (-1 per unit log mass, or -2 when expressed per unit mass). While some open cluster luminosity functions have been presented before (van den Bergh and Lafontaine 1984, Bhatt et al. 1991, Lata et al. 2002), this is the first CIMF presented for the Milky Way Galactic Disk.

5. The fraction of field stars from open clusters

We find that the fraction of nearby disk field stars that originated in Galactic open clusters is around 50 %. This result is different from previous studies in which this fraction was derived and cited to be about 10 % (Wielen 1971, Miller and Scalo 1978, Elmegreen and Clemens 1985, Janes et al. 1988, Adams and Myers 2001, Lada and Lada 2003). Even the ealier paper by Piskunov et al. (2006) confirmed this low value. At that time, the use of the present-day average open cluster mass (700 $M_\odot$) instead of the initial average mass (4500 $M_\odot$) led to a factor of 6.5 underestimate which has only been recently corrected in Piskunov et al. (2008b). Also, the birthrate of young open clusters has been revised upward, from ~ 0.2 per kpc^2 and Myr in Piskunov et al. 2006 to 0.4 per kpc^2 and Myr in Piskunov et al. (2008b). One has to multiply the birthrate with the age of the Milky Way Disk (8 – 12 Gyr) and the average initial cluster mass (4500 $M_\odot$) to obtain the surface density of field stars that originated from open clusters: this yields 18 $M_\odot$ per pc^2 as the amount of stellar mass resulting from dissolved open clusters over the age of the Galaxy (assuming a constant cluster formation rate over 10 Gyr). The local surface mass density in disk stars is 36 $M_\odot$ per pc^2 (Flynn et al. 2006), thus the ratio is 18/36 or 50 %.

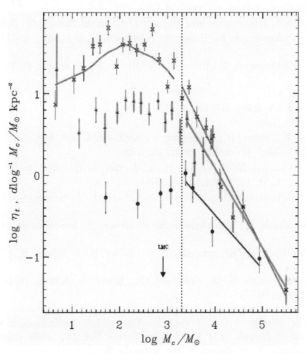

Figure 4. Evolution of the "integrated" mass function of Galactic open clusters (fixed lower age limit and variable upper age limit). Different symbols mark samples with different upper limits of cluster ages. The filled circles are for clusters with $\log t \leqslant 6.9$, the star symbols show cluster mass function for $\log t \leqslant 7.9$, and crosses indicate the CPDMF based on all 440 clusters ($\log t \leqslant 9.5$). The bars are Poisson errors. The straight lines are the corresponding fits to linear parts of the mass functions at masses greater than $\log M_c = 3.3$ indicated by the vertical dotted line. The fitted curve at $\log M_c < 3.3$ is the smoothed CPDMF. The arrow indicates the lower mass limit reached for open clusters in the LMC.

Acknowledgments

We thank the DFG for a German-Russian collaborative grant (436 RUS 113/757/0-2) to support this study. AEP gratefully acknowledges a grant from the Russian Foundation of Basic Research (RFBR 07-02-91566). HZ is grateful to the SOC for awarding IAU travel money without which this paper could not have been presented. He is also very grateful to U. Hanschur (AIP) for his help to prepare submitting this contribution.

References

Adams, F. C. & Myers, P. C. 2001, *ApJ*, 553, 744
Battinelli, P. & Capuzzo-Dolcetta, R. 1991, *MNRAS*, 249, 76
Becker, W. & Fenkart, R. 1971, *A&AS*, 4, 241
Bhatt, B. C., Pandey, A. K., & Mahra, H. S. 1991, *JApA*, 12, 179
Bica, E., Dutra, C. M., & Barbuy, B. 2003a, *A&A*, 397, 177
Bica, E., Dutra, C. M., Soares, J., & Barbuy, B. 2003b, *A&A*, 404, 223
Dias, W. S., Alessi, B. S., Moitinho, A., & Lépine, J. R. D. 2002, *A&A*, 389, 871
Elmegreen, B. G. & Clemens, C. 1985, *ApJ*, 294, 523
Flynn, C., Holmberg, J., Portinari, L., Fuchs, B., & Jahreiss, H. 2006, *MNRAS*, 372, 1149
Froebrich, D., Scholz, A., & Raftery, C. L. 2007, *MNRAS*, 374, 399
Geller, A. M., Mathieu, R. D., Harris, H. C., & McClure, R. D. 2008, *AJ*, 135, 2264
Gies, D. R. 1987, *ApJS*, 64, 545
Janes, K. A., Tilley, C., & Lynga, G. 1988, *AJ*, 95, 771
Kharchenko, N. V. 2001, *Kinematics and Physics of Celestial Bodies*, 17, 409
Kharchenko, N. V., Piskunov, A. E., Röser, S., Schilbach, E. & Scholz, R.-D. 2005a, *A&A*, 438, 1163
Kharchenko, N. V., Piskunov, A. E., Röser, S., Schilbach, E., & Scholz, R.-D. 2005b, *A&A*, 440, 403
King, I. 1962, *AJ*, 67, 471
Lada, C. J. & Lada, E.,A. 2003, *ARAA*, 41, 57
Larsen, S. S. 2002, *AJ*, 124, 1393
Lata, S., Pandey, A. K., Sagar, R., & Mohan, V. 2002, *A&A*, 388, 158
Lyngå, G. 1987, *ESO Conf. Workshop Proc.*, 28, 379
Maíz-Apellániz, J., Walborn, N. R., Galué, H. Á., & Wei, L. H. 2004, *ApJS*, 151, 103
Miller, G. E. & Scalo, J. M. 1978, *PASP*, 90, 506
Piskunov, A. E. Kharchenko, N. V., Röser, S., Schilbach, E. & Scholz, R.-D. 2006, *A&A*, 445, 545
Piskunov, A. E., Schilbach, E., Kharchenko, N. V., Röser, S., & Scholz, R.-D. 2007, *A&A*, 468, 151
Piskunov, A. E., Schilbach, E., Kharchenko, N. V., Röser, S., & Scholz, R.-D. 2008a, *A&A* 477, 165
Piskunov, A. E., Kharchenko, N. V., Schilbach, E., Röser, S., Scholz, R.-D., & Zinnecker, H. 2008b, *A&A*, accepted
Schilbach, E. & Röser, S. 2008, *A&A*, accepted
Scalo J. 1998, *The Stellar Initial Mass Function (38th Herstmonceux Conference)* eds. G. Gilmore and D. Howell. ASP Conference Series, Vol. 142, 1998, p.201
Sota, A., Maíz-Apellániz, J., Walborn, N. R., & Shida, R. Y. 2007, *astro-ph/0703005*
van den Bergh, S. & Lafontaine, A. 1984, *AJ*, 89, 1822
Wielen, R. 1971, *A&A*, 13, 309

Discussion

K. FREEMAN: Where do the other 50% of the field stars form?

H. ZINNECKER: In OB associations (like Sco-Cen) and T associations (like Taurus).

Open Clusters as tracers of the Galactic disk: the Bologna Open Clusters Chemical Evolution project

Angela Bragaglia[1], Eugenio Carretta[1], Raffaele Gratton[2] and Monica Tosi[1]

[1]INAF-Osservatorio Astronomico di Bologna
via Ranzani 1, I-40127 Bologna, Italy
email: angela.bragaglia, eugenio.carretta, monica.tosi @oabo.inaf.it

[2]INAF-Osservatorio Astronomico di Padova
vicolo dell'Osservatorio 5, I-35122 Padova, Italy
email: raffaele.gratton@oapd.inaf.it

Abstract. We present an update of the Bologna Open Clusters Chemical Evolution project (BOCCE, in short). We are conducting a photometric and spectroscopic survey of Open Clusters, to be used as tracers of the Galactic disk properties and evolution. We obtain the clusters parameters (age, distance, metallicity, and detailed abundances) in a precise and homogeneous way. We have collected data for about 40 Open Clusters and have fully analyzed the photometric data for about one half and the spectra for one quarter of them. We present here results based on these works and indicate what will come next.

Keywords. techniques: photometric, techniques: spectroscopic, stars: abundances, Hertzsprung-Russell diagram, Galaxy: abundances, Galaxy: disk, open clusters and associations: general

1. Overview

Open Clusters (OCs) are very useful tracers of the properties of the Galactic disk (e.g., Friel 1995), because they are seen over the entire disk, their ages and distances are generally measurable with better precision than those of isolated field stars up to large distances, and they cover the entire range of metallicities and ages of the disk.

In particular, OCs are very important to describe the metallicity (and detailed abundances) distribution of the disk, a key ingredient for models of chemical evolution. The metallicity distribution can be traced by other objects, like young stars, Cepheids, and Planetary Nebulae, but they are all rather young and can be used to describe the present-day situation, or the one in the near past. Instead, OCs go back to the first epochs of the disk, so we have access to the history of the disk formation and evolution and we may follow whether and how the metallicity distribution changed with time.

In the past, OCs have been used to define the metallicity distribution, with controversial results. The most commonly held view is that of a negative radial gradient (e.g., Friel *et al.* 2002), but an alternative picture, with two flat distributions at about solar and sub-solar metallicity and a discontinuity near a R_{GC} of 10 kpc, has been presented (Twarog, Ashman & Anthony-Twarog 1997). More recently, the observation of far-away OCs has permitted to reach $R_{GC} > 20$ kpc; these studies seem to indicate a negative gradient in the inner region and a flattening in the outer disk, with the transition at 10-14 kpc, see Yong, Carney & Teixera de Almeida (2005), Carraro *et al.* (2007), Sestito *et al.* (2008).

Table 1. OCs for which photometry and/or spectroscopy has been obtained. In all cases where the analysis has already been completed we give the references. Photometric parameters shown here refer to the (old) Padova tracks to give a ranking. More OCs are in our sample but the data quality has not been checked yet. In three cases (Be 19, NGC 6603, and Cr 110) telescope time has been granted but spectroscopic observations have not been carried out yet. Results obtained within a parallel collaboration (see Randich et al. 2005) are indicated within square brackets, since they are not strictly on the BOCCE spectroscopic scale; Cr 261, NGC 3960, NGC 6253 were observed in both programs.

Cluster	photom.	publ.?	$(m-M)_0$	E(B-V)	Z	age (Gyr)	high-res spec.	[Fe/H]
NGC 752	–	–					SARG@TNG	–
Be 66	TNG	(1)	13.20	1.22	0.008	3.80	no	–
NGC 1193	TNG	no					no	–
Be 17	TNG	(2)	12.20	0.62	0.008	8.50	SARG@TNG	–
Be 19	TNG	in prep					SARG@TNG?	–
Be 20	TNG	(1)	14.70	0.13	0.008	5.80	[FLAMES]	[−0.30 (6)]
Be 21	Danish	(3)	13.50	0.78	0.004	2.20	no	–
NGC 2099	CFHT	(3)	10.50	0.36	0.008	0.43	SARG@TNG	–
Be 22	NTT	(3)	13.80	0.64	0.020	2.40	no	–
NGC 2168	CFHT	(3)	9.80	0.20	0.008	0.18	SARG@TNG	–
NGC 2204	WFI	no					UVES	in prep
NGC 2243	Danish	(3)	12.74	0.08	0.004	4.80	FLAMES/SV	–
Tr 5	WFI	no					(archive)	
Cr 110	DFOSC	(3)	11.45	0.57	0.004	1.70	SARG@TNG?	–
NGC 2266	TNG	no					SARG@TNG	–
Biu 11/Be 27	SuSI2	no					no	–
Be 29	NTT	(3)	15.60	0.12	0.004	3.70	[FLAMES]	[−0.31 (6)]
Be 32	TNG	(4)	12.60	0.12	0.008	5.20	[FLAMES]	[−0.29 (7)]
Biu 13/Be 34	SuSI2	no					no	–
NGC 2323	CFHT	(3)	10.20	0.22	0.020	0.12	no	–
To 2	Danish	in prep					[FLAMES]	–
NGC 2324	WFI	no					[FLAMES]	[−0.17 (8)]
Be 36	SuSI2	no					no	–
Mel 66	–	–					[FLAMES]	[−0.33 (6)]
Mel 71	Danish	in prep					FEROS	in prep
NGC 2477	–	–					[FLAMES]	[+0.07 (8)]
NGC 2506	mosaic	(3)	12.60	0.00	0.020	1.70	FEROS	−0.20 (9)
Pi 2	Danish	(3)	12.70	1.29	0.020	1.10	no	–
NGC 2660	Dutch	(3)	12.30	0.40	0.020	0.95	[FLAMES]	[+0.04 (7)]
NGC 2849	Dutch	no					no	–
NGC 3960	DFOSC	(5)	11.60	0.29	0.020	0.90	FEROS	−0.12 (5)
Cr 261	mosaic	(3)	12.20	0.30	0.020	6.00	FEROS	−0.03 (10)
NGC 4815	FORS	no					no	–
NGC 6134	DFOSC	no					FEROS	+0.15 (9)
NGC 6253	Danish	(3)	11.00	0.23	0.050	3.00	FEROS+UVES	+0.46 (11)
IC 4651	DFOSC	no					FEROS	+0.11 (9)
NGC 6603	Danish	no					SARG@TNG?	–
IC 4756	DFOSC	no					FEROS	–
NGC 6791	CFHT	in prep					SARG@TNG	+0.47 (12)
IC 1311	TNG	no					no	–
NGC 6819	CFHT	(3)	12.20	0.12	0.020	2.00	SARG@TNG	+0.07 (13)
NGC 6939	TNG	(3)	11.30	0.34	0.020	1.30	SARG@TNG	–
King 11	TNG	(4)	11.75	1.04	0.010	4.25	no	–
NGC 7789	CFHT	no					SARG@TNG	–
NGC 7790	Loiano	(3)	12.65	0.54	0.020	0.10	SARG@TNG	–

(1) Andreuzzi, Bragaglia & Tosi (2008); (2) Bragaglia et al. (2006b); (3) Bragaglia & Tosi (2006), where references to the original papers are given; (4) Tosi et al. (2007); (5) Bragaglia et al. (2006a); (6) Sestito et al. (2008); (7) Sestito et al. (2006); (8) Bragaglia et al. (2008); (9) Carretta et al. (2004); (10) Carretta et al. (2005); (11) Carretta, Bragaglia & Gratton (2007); (12) Gratton et al. (2006); (13) Bragaglia et al. (2001)

2. The BOCCE survey

We have started a survey of OCs to derive in the most precise and homogeneous way their main parameters: age, distance, reddening, metallicity (and detailed abundances). As one of our main interest is the chemical evolution of the disk, we named our survey the "Bologna Open Clusters Chemical Evolution" project, BOCCE in short. Our goal is to build a sample large enough to be representative of the whole cluster population (in age, metallicity and position in the Galaxy). Of course, if we want to study the history of the disk we have to concentrate on old OCs: of the about 120 OCs with age larger than 1 Gyr found in the Dias et al. (2002) catalogue, we have collected data for a fair fraction, and have already studied 16 clusters in detail. We employ:

- Photometry and the Synthetic Colour-Magnitude diagrams technique to derive at the same time age, distance, reddening and a first indication of metallicity; for a review of the method and results see Bragaglia & Tosi (2006). This is the more advanced part of our survey, having already published results for more than half the original sample.
- Medium resolution spectroscopy to derive radial velocities, hence membership, for stars in crucial evolutionary phases, like the main sequence Turn-Off or the Red Giant Branch; for an application, see e.g. D'Orazi et al. (2006). This is a secondary part of our project.
- High resolution spectroscopy to measure chemical abundances, using both equivalent widths and spectrum synthesis; for a presentation of the method, see Bragaglia et al. (2001), Carretta et al. (2004). This part of the work started later and we still have to catch up with the photometric survey, but we have recently acquired spectra for many OCs. We also plan to use archive data and homogenize results obtained within a parallel program (see Randich et al. 2005).

Table 1 shows the present status of our project; in summary we have observed clusters with ages from 0.1 to 9 Gyr, R_{GC} from about 7 to 21 kpc, and metallicity from less than half solar to more than double solar.

We have a few very interesting objects in our sample, e.g., Be 17 and NGC 6791 (the oldest OCs known, with age near 9 Gyr) or Be 20, Be 29 (with R_{GC} of abour 16 and 21 kpc) and Cr 261, NGC 6253 (inside the solar circle) very important to define the metallicity gradient. Two of these clusters, NGC 6791 and NGC 6253, are also the most metal-rich OC presently known.

In the Table, to give an uniform ranking of properties, we put the values derived from photometry using the (old) Padova tracks (e.g., Bressan et al. 1993), but we always use three sets of evolutionary tracks (without overshooting and with two treatment of it). This way we are better able to quantify the systematics between results based on different stellar models. We always use the same sets of tracks even if a newer version of the same code has become available (as in the case of the Padova models) to maintain the maximum homogeneity in our determinations†

The metallicity shown in the last column of Table 1 has been obtained from the high resolution spectra. We observe a few stars per clusters, chosen if possible among confirmed members; we usually target Red Clump stars since they are the best compromise between luminosity and temperature: we can obtain high S/N spectra of stars not too cool to be a problem for model atmospheres. Figure 1 (left panel) shows an example of the quality of our data in the case of the two very metal-rich clusters NGC 6253 and NGC 6791. We

† An exception will be NGC 6791, for which we are employing new tracks at Z = 0.04 or 0.05, computed on purpose or retrieved from recent publications, because of the combination of very high metallicity and old age.

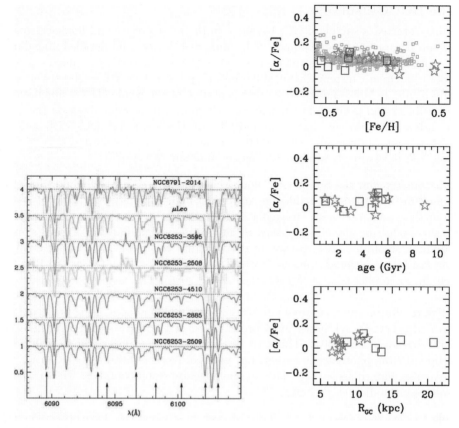

Figure 1. Left panel: Spectra obtained with FEROS and UVES for NGC 6253 (Carretta et al. 2007), with SARG for NGC 6791 (Gratton et al. 2006); also shown is the very metal-rich field giant μ Leo. The arrows indicate lines used in the abundance determination. Right panels: [α/Fe] ratio for OCs in the BOCCE sample, compared to thin disk field stars (upper plot) and shown in fuction of age (middle plot) and R_{GC} (lower plot). Red stars indicate abundances strictly on the BOCCE scale, open squares are from Sestito et al. (2006), Sestito et al. (2008), Bragaglia et al. (2008) for four OCs (from the parallel program), and Villanova et al. (2005), Carraro et al. (2004) for two OCs.

try to maintain also in this case the most homogeneous procedure, using the same line lists, gf's, model atmospheres, solar reference values, way to measure equivalent widths, synthesis, and kind of stars.

Since the photometric and spectroscopic parts of our programs are not "aligned" yet, we are not presently able to derive in a fully self-consistent way the metallicity gradient or the properties of our whole sample. However, putting together our results, literature ones, and the FLAMES survey, we confirm the presence of a radial metallicity gradient flattening in the outer regions; an indication of this was already present in Bragaglia & Tosi (2006) from the photometric metallicities, but it has recently been found on the basis of more precise metallicity determinations by Yong et al. (2005), Carraro et al. (2007), and Sestito et al. (2008).

Fig.1 (right panels) shows some results on the [α/Fe] ratios found for clusters in our sample, based on our published OCs, on work in progress, on literature, and FLAMES data. The OCs follow the same relation of [α/Fe] versus [Fe/H] of the field thin disk stars and do not seem to show any trend with age. Perhaps the most interesting feature is shown in the bottom panel, where [α/Fe] is plotted against R_{GC}: from the clusters in the BOCCE sample there is no apparent trend of increased [α/Fe] for the outer parts of the disk (see Sestito *et al.* 2008 for an extended discussion, since the two outermost clusters have abundances measured in that paper), at variance with what had been advocated by Carraro *et al.* (2004) and Yong *et al.* (2005), but in agreement with Carraro *et al.* (2007). However, we do not wish to make any strong statement until all clusters have been measured strictly on the BOCCE scale.

In summary, our effort to build a large, significant sample of open clusters with ages, distances, metallicities and detailed abundances measured with homogeneous methods is well under way and has already produced interesting results. Future efforts will be mainly directed towards increasing the number of clusters with chemical abundances measured on a common scale. Only the kind of homogeneity we are trying to achieve can guarantee that features are not created or lost because of systematic differences between analyses.

The work described here was made possible by the collaboration of many researchers, among them Gloria Andreuzzi and Luca Di Fabrizio (INAF - Fundación Galilei), Michele Cignoni (Bologna University), Jason Kalirai (UCO-Lick), and Gianni Marconi (ESO). We thank Paola Sestito and Sofia Randich for useful discussions; we make use here of results obtained in collaboration with them. We made large and fruitful use of the WEBDA (http://www.univie.ac.at/webda/). Generous allocation of observing time at Italian telescopes (TNG and Loiano), at the CFHT, and at ESO telescopes (La Silla and Paranal) is acknowledged.

References

Andreuzzi, G., Bragaglia, A., & Tosi, M. 2008, *A&A*, in press
Bragaglia, A., *et al.* 2001, *AJ*, 121, 327
Bragaglia, A. & Tosi, M. 2006, *AJ*, 131, 1544
Bragaglia, A., Tosi, M., Carretta, E., Gratton, R. G., Marconi, G., & Pompei, E. 2006a, *MNRAS*, 366, 1493
Bragaglia, A., Tosi, M., Andreuzzi, G., & Marconi, G. 2006b, *MNRAS*, 368, 1971
Bragaglia, A., Sestito, P., Villanova, S., Carretta, E., Randich, S., & Tosi, M. 2008, *A&A*, 480, 79
Bressan, A., Fagotto, F., Bertelli, G., & Chiosi, C. 1993, *A&AS*, 100, 647
Carraro, G., Bresolin, F., Villanova, S., Matteucci, F., Patat, F., Romaniello, M. 2004, *ApJ*, 128, 1676
Carraro, G., Geisler, D., Villanova, S., Frinchaboy, P. M., & Majewski, S. R. 2007, *A&A*, 476, 217
Carretta, E., Bragaglia, A., Gratton, R. G., & Tosi, M. 2004, *A&A*, 422, 951
Carretta, E., Bragaglia, A., Gratton, R. G., & Tosi, M. 2005, *A&A*, 441, 131
Carretta, E., Bragaglia, A., & Gratton, R. G. 2007, *A&A*, 473, 129
D'Orazi, V., Bragaglia, A., Tosi, M., Di Fabrizio, L., & Held, E. V. 2006, *MNRAS*, 368, 471
Dias, W. S., Alessi, B. S., Moitinho, A., & Lépine, J. R. D. 2002, *A&A*, 389, 871
Dominguez, I., Chieffi, A., Limongi, M., & Straniero, O., *ApJ*, 524, 226
Friel, E. D. 1995, *ARAA*, 33, 381
Friel, E. D., Janes, K. A., Tavarez, M., Scott, J., Katsanis, R., Lotz, J., Hong, L., & Miller, N. 2002, *AJ*, 124, 2693
Gratton, R., Bragaglia, A., Carretta, E., & Tosi, M. 2006, *ApJ*, 642, 462
Randich, S., *et al.* 2005, *ESO Messenger*, 121, 18
Sestito, P., Bragaglia, A., Randich, S., Carretta, E., Prisinzano, L., & Tosi, M. 2006, *A&A*, 458, 121
Sestito, P., Bragaglia, A., Randich, S., Pallavicini, R., Andrievski, S. M., & Korotin, S. A. 2008, *A&A*, in press

Tosi, M., Bragaglia, A., & Cignoni, M. 2007, *MNRAS*, 378, 730
Twarog, B. A., Ashman, K. M., & Anthony-Twarog, B. J. 1997, *AJ*, 114, 2556
Ventura, P., Zeppieri, A., Mazzitelli, I, & D'Antona, F. 1998, *A&A*, 334, 953
Villanova, S., Carraro, G., Bresolin, F., Patat, F. 2005, *ApJ*, 130, 652
Yong, D., Carney, B. W., & Teixera de Almeida, M. L. 2005, *AJ*, 130, 597

Hans Zinnecker and Pavel Kroupa in a joint accretion experiment at Carlsberg. Jens Hjorth in the background (photo: Bruce Elmegreen).

Origin of Star-to-Star Abundance Inhomogeneities in Star Clusters

Jan Palouš[1], Richard Wünsch[2], Guillermo Tenorio-Tagle[3], and Sergyi Silich[3]

[1] Astronomical Institute, Academy of Sciences of the Czech Republic,
Boční 1401, 140 31 Prague 4, Czech Republic
email: palous@ig.cas.cz

[2] Cardiff University, Queens Buildings, The Parade, Cardiff,
CF24 3AA, United Kingdom
email: richard@wunsch.cz

[3] Instituto Nacional de Astrofísica Optica y Electronica,
AP 51, 72000 Puebla, México
email: gtt@inaoep.mx, silich@inaoep.mx

Abstract. The mass reinserted by young stars of an emerging massive compact cluster shows a bimodal hydrodynamic behaviour. In the inner part of the cluster, it is thermally unstable, while in its outer parts it forms an out-blowing wind. The chemical homogeneity/inhomogeneity of low/high mass clusters demonstrates the relevance of this solution to the presence of single/multiple stellar populations. We show the consequences that the thermal instability of the reinserted mass has to the galactic super-winds.

Keywords. Galaxy: abundances, Galaxy: Globular clusters: general, Galaxy: open star clusters and associations: general, galaxies: star clusters, galaxies: starburst

1. Introduction

The open star clusters and stellar moving groups have internally homogeneous chemical composition. Clusters like the Hyades, Collinder 261, the Herculis stream or the moving group HR 1614 are chemically unique, distinguishable one from the other, showing no pollution from secondary star formation (De Silva *et al.*, 2008). The chemical homogeneity of open star clusters like the Hyades (De Silva *et al.*, 2006) and Collinder 261 (De Silva *et al.*, 2007) proves that they have been formed out of a well-mixed cloud and that any self-enrichment of stars did not take place there.

Young and massive stellar clusters, frequently called super star clusters, are preferentially observed in interacting galaxies. Their stellar mass amounts to several million $M_\odot$ within a region less than a few parsecs in diameter. They represent the dominant mode of star formation in starburst galaxies. Their high stellar densities resemble those of globular clusters, where several stellar populations have been observed (Piotto, 2008).

To explain the presence of multiple stellar generations in globular clusters, the slow wind emerging from a first generation of fast rotating massive stars is invoked by Decressin *et al.* (2007). (See also the review by Meynet (2008) in this volume.) The authors argue that the fast rotating massive stars function as a filter separating the H-burning products from later products of He-burning. However, it is not clear why all the massive stars rotate fast, or why the slow wind produced by stellar rotation is just retained inside the potential well of the stellar cluster.

An alternative solution, how to form the second generation of stars in massive star clusters, is proposed in models of star cluster winds described by Tenorio-Tagle *et al.*

(2007), Wünsch et al. (2007), and Wünsch et al. (2008). There, we argue that a critical mass of a cluster exists, below which the single-mode hydrodynamical solution to the cluster winds applies. Such clusters should have one stellar generation only, and show strong winds corresponding to the momentum and energy feedback of all their stars. The clusters above the critical mass should follow the bi-modal solution to their winds, where only the outer skin of the cluster participates in the wind. Their inner parts are thermally unstable, and hence being the potential places of secondary star formation.

2. The Model

The hydrodynamical behaviour of matter reinserted within a star cluster is described by Chevalier & Clegg (1985). In this adiabatic model, the authors assume that all the energy provided by stellar winds and supernovae is thermalized in random collisions of the shock waves creating a gas of temperatures $T > 10^7$ K. Chevalier & Clegg's stable solution shows almost constant density and temperature inside of the cluster. A mild outward pressure gradient drives a cluster wind with a radially increasing velocity reaching the sound speed c_{SC} at the cluster surface and approaching $2c_{SC}$ at infinity. The run of wind particle density, temperature and wind velocity in this single-mode model is

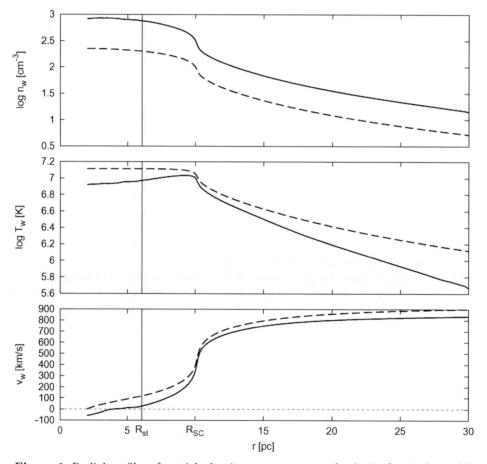

Figure 1. Radial profiles of particle density, temperature and velocity for single-modal (dashed lines) and bi-modal (solid lines) solutions.

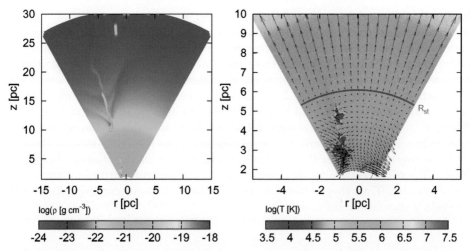

Figure 2. Bi-modal solution: density distribution (left panel), temperature distribution with velocity vectors (right panel) 0.56 Myr after the beginning of the simulation.

shown in Fig. 1. With the adiabatic model the more massive clusters produce the more powerful wind, taking away all the elements produced in stars. Thus, any inhomogeneity in stellar chemical composition reflects the abundance distribution in the parental star-forming nebula. If the original cloud is well mixed, according to the adiabatic, single-mode model of Chevalier & Clegg (1985), the star cluster will have one population of the same chemical composition.

The assumption that all the mechanical energy of winds and supernovae is thermalized need not to be completely true. The efficiency of the thermalization process η depends on the details of shock-shock collisions and different authors give different values ranging from 0.01 to 1 (Melioli, & Gouveia Dal Pino, 2004). The value of η may depend on Mach numbers of the colliding shocks and/or on the chemical composition of the colliding fronts regulating what part of the mechanical energy is directly radiated away. The value of η may be different in the case of colliding stellar winds, or colliding supernova explosions.

The density of the cluster wind n_w depends on the stellar density: for a cluster of given radius R_{SC}, n_w is proportional to the cluster mass M_{SC}. On the other hand, the radiative cooling rate is proportional to n_w^2 multiplied with a cooling function $\Lambda(T, Z)$, where Z stands for the chemical composition of the radiating plasma. Thus, the radiative cooling rate is proportional to M_{SC}^2. However, the cluster mechanical luminosity L_{SC}, or the part that is thermalized ηL_{SC}, depends linearly on cluster mass M_{SC}, which implies that a critical mass $M_{SC,crit}$ must exist, above which the energy loss due to radiative cooling and gas expansion exceeds ηL_{SC}. Its value depends on R_{SC} and η.

Less massive clusters show the single-mode behaviour. They take away at least a part of the mechanical luminosity as winds from all the cluster volume. In the adiabatic approximation, which applies to low mass clusters far below the $M_{SC,crit}$, all the mechanical energy, which is transformed to heat, is removed as winds.

For M_{SC} above the critical value, the volume inside of the cluster is split into two sectors. The inner part of the cluster, inside the stagnation radius R_{ST}, is thermally unstable. There, the instability leads to fast cooling of small parcels of gas surrounded by a hot medium. The repressurizing shocks drive the hot gas into cold gas parcels forming high density gas concentrations. The outer part of the cluster, its skin $R_{ST} < R < R_{SC}$, shows the out-blowing wind. The run of wind particle density, temperature and wind velocity in the bi-modal solution of the star cluster wind is shown in Fig. 1. The single-mode compared to the bi-modal solution shows wind of lower density and higher temperature. The velocity of the out-blowing wind is higher in the case of the single-mode solution compared to the bi-modal case; the two solutions have the same velocity at the cluster surface only. 2D hydrodynamical simulation of cluster winds are described by Wünsch et al. (2008). In Fig. 2 we show the density and temperature distribution together with velocity vectors for a star cluster with $R_{SC} = 10$ pc, $\eta = 1$, and $L_{SC} = 10^{42}$ erg s^{-1}. We can clearly distinguish the thermally unstable region within the stagnation radius $R_{ST} = 6$ pc, and the wind blowing from the outer skin of the cluster. The dense concentration, in the thermally unstable part of the cluster, shows a small velocity dispersion relative to the cluster; most of these concentrations are unable to leave the cluster. Thus they accumulate in the inner volume, becoming potential places of secondary star formation.

3. Summary

We propose a bimodal solution, where in the central part of a massive star cluster, a thermally unstable region forms (see Fig. 2). In this region, the thermal instability creates cold regions surrounded by hot medium imploding into them. The high-velocity wing of broad spectral lines Gilbert, & Graham (2007) observed in SSC may be created by imploding shock in the vicinity of thermally unstable parcels of gas.

The second generation of stars may be formed out of cold clumps produced by thermal instability. During the early evolution of a massive cluster, the first Myrs, the mechanical energy input is dominated by stellar winds (Leitherer et al. 1999). The efficiency of thermalization η may be low in this case, and a massive cluster may be above $M_{SC,crit}$, since its value is low. Later, the importance of winds fades out, and the mechanical input is dominated by supernovae. This may increase the thermalization efficiency η, increasing at the same time the value $M_{SC,crit}$. Thus the same massive cluster, which was initially in the bi-modal situation moves to single-mode situation.

The cold parcels of gas form, in the cluster central part during the early bi-modal situation, from the winds that are enriched by products of H-burning. The later He-burning products are inserted into the cluster volume when the mechanical energy input is dominated by supernovae, which may mean that the cluster is in the single-mode situation, and the wind clears its volume from He burning products. Thus, the thermal instability, which operates during a few initial Myr in the central part of the cluster, may produce a second generation of stars enriched by H-burning products. Later, the cluster moves into the single-mode situation, which means that it is able to expel the He-burning products.

The feedback of massive stars in super star clusters creates galactic winds, or super winds, reaching to large distances from the parent galaxies, transporting the products of stellar burning into intergalactic space. The bimodal solution, providing a possible explanation of multiple stellar populations in globular clusters, limits the super winds. During the initial period of cluster evolution, when the stellar winds dominate the mechanical energy input, the super wind is restricted only to the outer skin of the cluster, which means that it is rather weak. Only later, when supernovae become dominant in mechanical energy input, strong super winds blowing out from all the cluster volume may reach large distances from their parent galaxies. How effective the super winds of super star clusters can be in transporting the products of stellar evolution into intergalactic spaces should be discussed in future.

Acknowledgements

The authors gratefully acknowledge the support by the Institutional Research Plan AV0Z10030501 of the Academy of Sciences of the Czech Republic and by the project LC06014 Center for Theoretical Astrophysics of the Ministry of Education, Youth and Sports of the Czech Republic. RW acknowledges support by the Human Resources and Mobility Programme of the European Community under the contract MEIF-CT-2006-039802. This study has been supported by CONACYT-México research grant 60333 and 47534-F and AYA2004-08260-CO3-O1 from Spanish Consejo Superior de Investigaciones Científicas. The authors express their thanks to Jim Dale for careful reading of the text.

References

Chevalier, R. A. & Clegg, A. W. 1985, *Nature*, 317, 44
De Silva, G. M., Sneden, C., Paulson, D. B., Asplund, M., & Bland-Hawthorn, J. 2006, *AJ*, 131, 455
De Silva, G. M., Freeman, K. C., Asplund, M., Bland-Hawthorn, J., Bessesl, M. S., & Collet, R. 2007, *AJ*, 133, 1161
De Silva, G. M., Freeman, K, Bland-Hawthorn, J., & Asplund, M. 2008, in J. Andersen, J. Bland-Hawthorn & B. Nordstöm (eds.), *The Galaxy Disk in Cosmological Context*, Proc. IAU Symposium No. 254 (CUP), this volume
Decresssin, T., Charbonnel, C., & Meynet, G. 2007, *A&A*, 475, 859
Gilbert, A. M. & J. R. Graham. 2007, *ApJ*, 668, 168
Leitherer, C., Schaerer, D., Goldader, J. D., Gonzáles-Delgado, R. M., Robert, C., Foo Kune, D., De Mello, D. F., Devost, D., & Heckman, T. M. 1999, *ApJS*, 123, 3
Melioli, C. & de Gouveia Dal Pino, E. M. 2004, *A&A*, 424, 817
Meynet, G. 2008, in J. Andersen, J. Bland-Hawthorn & B. Nordstöm (eds.), *The Galaxy Disk in Cosmological Context*, Proc. IAU Symposium No. 254 (CUP), this volume
Piotto, G. 2008, *MemSAI*, 79, 3
Tenorio-Tagle, G., Wünsch, R., Silich, S., & Palouš, J. 2007, *ApJ*, 658, 1196
Wünsch, R., Silich, S., Palouš, J., & Tenorio-Tagle, G. 2007, *A&A*, 471, 579
Wünsch, R., Tenorio-Tagle, G., Palouš, J., & Silich, S. 2008, *ApJ*, 684, September 1, arXiv:0805.1380v1

The former Strömgren residence of honour, seen from the park.

Jan Palouš meets Adriaan Blaauw at Carlsberg, with Hans Zinnecker (left). In the background, Ole Strömgren, Ben Mottelson, and Haldor Topsøe (photo: Bruce Elmegreen).

Session 3: Accretion and the Interstellar Medium

Session chairs: Katia Cunha and Eline Tolstoy

Bengt Strömgren in 1957, 18 years after he introduced the Strömgren sphere and thus extended astrophysics to include the ISM.

Photo: NY World-Telegram and Sun.

Warm gas accretion onto the Galaxy

J. Bland-Hawthorn

School of Physics, University of Sydney, Australia, NSW 2006
email: jbh@physics.usyd.edu.au

Abstract. We present evidence that the accretion of warm gas onto the Galaxy today is at least as important as cold gas accretion. For more than a decade, the source of the bright Hα emission (up to 750 mR†) along the Magellanic Stream has remained a mystery. We present a hydrodynamical model that explains the known properties of the Hα emission and provides new insights on the lifetime of the Stream clouds. The upstream clouds are gradually disrupted due to their interaction with the hot halo gas. The clouds that follow plough into gas ablated from the upstream clouds, leading to shock ionisation at the leading edges of the downstream clouds. Since the following clouds also experience ablation, and weaker Hα (100−200 mR) is quite extensive, a disruptive cascade must be operating along much of the Stream. In order to light up much of the Stream as observed, it must have a small angle of attack ($\approx 20°$) to the halo, and this may already find support in new H I observations. Another prediction is that the Balmer ratio (Hα/Hβ) will be substantially enhanced due to the slow shock; this will soon be tested by upcoming WHAM observations in Chile. We find that the clouds are evolving on timescales of 100−200 Myr, such that the Stream must be replenished by the Magellanic Clouds at a fairly constant rate ($\gtrsim 0.1$ M$_\odot$ yr^{-1}). The ablated material falls onto the Galaxy as a warm drizzle; diffuse ionized gas at 10^4 K is an important constituent of galactic accretion. The observed Hα emission provides a new constraint on the rate of disruption of the Stream and, consequently, the infall rate of metal-poor gas onto the Galaxy. We consider the stability of H I clouds falling towards the Galactic disk and show that most of these must break down into smaller fragments that become partially ionized. The Galactic halo is expected to have huge numbers of smaller neutral and ionized fragments. When the ionized component of the infalling gas is accounted for, the rate of gas accretion is ~ 0.4 M$_\odot$ yr^{-1}, roughly twice the rate deduced from H I observations alone.

Keywords. Galaxies: interaction, Magellanic Clouds – Galaxy: evolution – ISM: individual (Smith Cloud) – shock waves – instabilities – hydrodynamics

1. Introduction

It is now well established that the observed baryons over the electromagnetic spectrum account for only a fraction of the expected baryon content in Lambda Cold Dark Matter cosmology. This is true on scales of galaxies and, in particular, within the Galaxy where easily observable phases have been studied in great detail over many years. The expected baryon fraction ($\Omega_B/\Omega_{DM} \approx 0.17$) of the dark halo mass (1.4×10^{12} M$_\odot$; Smith et al. 2007) leads to an expected baryon mass of 2.4×10^{11} M$_\odot$ but a detailed inventory reveals only a quarter of this mass (Flynn et al. 2006‡). Moreover, the build-up of stars in the Galaxy requires an accretion rate of $1-3$ M$_\odot$ yr^{-1} (Williams & McKee 1997; Binney et al. 2000), substantially larger than what can be accounted for from direct observation. We can extend the same argument to M31 where the total baryon mass is $\lesssim 10^{11}$ M$_\odot$ (Tamm et al. 2007). For the Galaxy, the predicted baryon mass may be a lower bound if the upward correction in the LMC-SMC orbit motion reflects a larger halo

---

† 1 Rayleigh (R) = $10^6/4\pi$ photons cm^{-2} s^{-1} sr^{-1}, equivalent to 5.7×10^{-18} erg cm^{-2} s^{-1} arcsec^{-2} at Hα.
‡ A decade ago, it was claimed that MACHOs may be important in the halo but these can only make up a negligible fraction by mass (Tisserand et al. 2007).

mass (Kallivayalil et al. 2006; Piatek et al. 2008; cf. Wilkinson & Evans 1999). Taken together, these statements suggest that most of the baryons on scales of galaxies have yet to be observed.

So how do galaxies accrete their gas? Is the infalling gas confined by dark matter? Does the gas arrive cold, warm or hot? Does the gas rain out of the halo onto the disk or is it forced out by the strong disk-halo interaction? These issues have never been resolved, either through observation or through numerical simulation. H I observations of the nearby universe suggest that galaxy mergers and collisions are an important aspect of this process (Hibbard & van Gorkom 1996), but tidal interactions do not guarantee that the gas settles to one or other galaxy. The most spectacular interaction phenomenon is the Magellanic H I Stream that trails from the LMC-SMC system (10:1 mass ratio) in orbit about the Galaxy. Since its discovery in the 1970s, there have been repeated attempts to explain the Stream in terms of tidal and/or viscous forces (q.v. Mastropietro et al. 2005; Connors et al. 2005). Indeed, the Stream has become a benchmark against which to judge the credibility of N-body+gas codes in explaining gas processes in galaxies. A fully consistent model of the Stream continues to elude even the most sophisticated codes.

Here, we demonstrate that Hα detections along the Stream (Fig. 1) are providing new insights on the present state and evolution of the H I gas. At a distance of $D \approx 55$ kpc, the expected Hα signal excited by the cosmic and Galactic UV backgrounds are about 3 mR and 25 mR respectively (Bland-Hawthorn & Maloney 1999, 2002), significantly lower than the mean signal of 100−200 mR, and much lower than the few bright detections in the range 400 − 750 mR (Weiner, Vogel & Williams 2002). This signal cannot have a stellar origin since repeated attempts to detect stars along the Stream have failed.

Some of the Stream clouds exhibit compression fronts and head-tail morphologies (Brüns et al. 2005) and this is suggestive of confinement by a tenuous external medium. But the cloud:halo density ratio ($\eta = \rho_c/\rho_h$) necessary for confinement can be orders of magnitude *larger* than that required to achieve shock-induced Hα emission (e.g. Quilis & Moore 2001). Indeed, the best estimates of the halo density at the distance of the Stream ($\rho_h \sim 10^{-4}$ cm^{-3}; Bregman 2007) are far too tenuous to induce strong Hα emission at a cloud face. It is therefore surprising to discover that the brightest Hα detections lie at the leading edges of H I clouds (Weiner et al. 2002) and thus appear to indicate that shock processes are somehow involved.

We summarize a model, first presented in Bland-Hawthorn et al. (2007), that goes a long way towards explaining the Hα mystery. The basic premise is that a tenuous external medium not only confines clouds, but also disrupts them with the passage of time. The growth time for Kelvin-Helmholtz (KH) instabilities is given by $\tau_{KH} \approx \lambda \eta^{0.5}/v_h$ where λ is the wavelength of the growing mode, and v_h is the apparent speed of the halo medium ($v_h \approx 350$ km s^{-1}; see §2). At the distance of the Stream, the expected timescale for KH instabilities is less than for Rayleigh-Taylor (RT) instabilities (see §3). For cloud sizes of order a few kiloparsecs and $\xi \approx 10^4$, the KH timescale can be much less than an orbital time ($\tau_{MS} \approx 2\pi D/v_h \approx 1$ Gyr). Once an upstream cloud becomes disrupted, the fragments are slowed with respect to the LMC-SMC orbital speed and are subsequently ploughed into by the following clouds.

In §2, the new hydrodynamical models are described and the results are presented; we discuss the implications of our model and suggest avenues for future research. In §3, we discuss the stability of H I clouds (high velocity clouds) moving through the corona toward the Galactic disk and briefly consider the Smith Cloud, arguably the HVC with the best observed kinematic and photometric parameters.

2. A new hydrodynamical model

There have been many attempts to understand how gas clouds interact with an ambient medium (Murray, White & Blondin 1993; Klein, McKee & Colella 1994). In order to capture the evolution of a system involving instabilities with large density gradients correctly, grid based methods (Liska & Wendroff 1999; Agertz et al. 2007) are favoured over other schemes (e.g. Smooth Particle Hydrodynamics). We have therefore investigated the dynamics of the Magellanic Stream with two independent hydrodynamics codes, *Fyris* (Sutherland 2008) and

Ramses (Teyssier 2002), that solve the equations of gas dynamics with adaptive mesh refinement. The results shown here are from the *Fyris* code because it includes non-equilibrium ionization, but we get comparable gas evolution from either code†.

The brightest emission is found along the leading edges of clouds MS II, III and IV with values as high as 750 mR for MS II. The Hα line emission is clearly resolved at $20-30$ km s^{-1} FWHM, and shares the same radial velocity as the H I emission within the measurement errors (Weiner *et al.* 2002; Madsen *et al.* 2002). This provides an important constraint on the physical processes involved in exciting the Balmer emission.

In order to explain the Hα detections along the Stream, we concentrate our efforts on the disruption of the clouds labelled MS I–IV (Brüns *et al.* 2005). The Stream is trailing the LMC-SMC system in a counter-clockwise, near-polar orbit as viewed from the Sun. The gas appears to extend from the LMC dislodged through tidal disruption although some contribution from drag must also be operating (Moore & Davis 1994). Recently, the Hubble Space Telescope has determined an orbital velocity of 378 ± 18 km s^{-1} for the LMC. While this is higher than earlier claims, the result has been confirmed by independent researchers (Piatek *et al.* 2008). Besla *et al.* (2007) conclude that the origin of the Stream may no longer be adequately explained with existing numerical models. The Stream velocity along its orbit must be comparable to the motion of the LMC; we adopt a value of $v_{MS} \approx 350$ km s^{-1}.

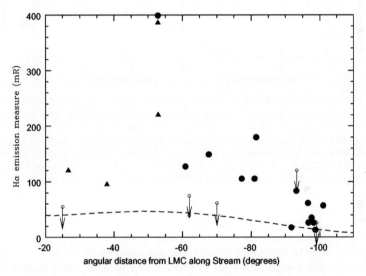

Figure 1. Hα measurements and upper limits along the Stream. The filled circles are from the WHAM survey by Madsen *et al.* (2002); the filled triangles are from the TAURUS survey by Putman *et al.* (2003). The dashed line model is the Hα emission measure induced by the ionizing intensity of the Galactic disk (Bland-Hawthorn & Maloney 1999; 2002); this fails to match the Stream's Hα surface brightness by at least a factor of 3.

2.1. *Model parameters*

Here we employ a 3D Cartesian grid with dimensions $18 \times 9 \times 9$ kpc [$(x,y,z) = (432, 216, 216)$ cells] to model a section of the Stream where x is directed along the Stream arc and the z axis points towards the observer. The grid is initially filled with two gas components. The first is a hot thin medium representing the halo corona.

Embedded in the hot halo is (initially) cold H I material with a total H I mass of 3×10^7 M$_\odot$. The cold gas has a fractal distribution and is initially confined to a cylinder with a diameter of 4 kpc and length 18 kpc (Fig. 5); the mean volume and column densities are 0.02 cm^{-3}

† Further details on the codes and comparative simulations are provided at http://www.aao.gov.au/astro/MS.

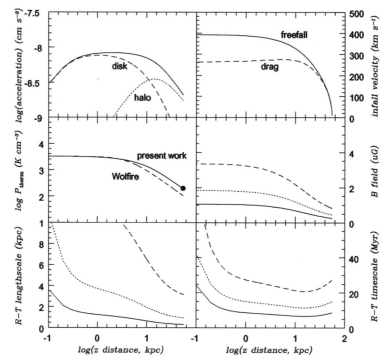

Figure 2. (a) gravitational acceleration due to the disk+halo – all plots are shown as a function of vertical height above the disk at the Solar Circle; (b) infall velocity for a point mass starting from the Magellanic Stream with halo drag ($C_D = 1$ discussed in §3; dashed curve) and without (solid); (c) coronal gas pressure for our model (solid line) compared to the Wolfire model (dashed) – the dot indicates the halo pressure used in our hydrodynamical model; (d) magnetic field strength for $\beta = 1$ (dashed), $\beta = 0.3$ (dotted), $\beta = 0.1$ (solid); (e) minimum lengthscale for RT instability (discussed in §3); (e) timescale for RT instability (discussed in §3).

and 2×10^{19} cm^{-2} respectively. The 3D spatial power spectrum ($P(k) \propto k^{-5/3}$) describes a Kolmogorov turbulent medium with a minimum wavenumber k corresponding to a spatial scale of 2.25 kpc, comparable to the size of observed clouds along the Stream.

We consider the hot corona to be an isothermal gas in hydrostatic equilibrium with the gravitational potential, $\phi(R, z)$, where R is the Galactocentric radius and z is the vertical scale height. We adopt a total potential of the form $\phi = \phi_d + \phi_h$ for the disk and halo respectively; for our calculations at the Solar Circle, we ignore the Galactic bulge. The galaxy potential is defined by

$$\phi_d(R, z) = -c_d v_{circ}^2 / (R^2 + (a_d + \sqrt{z^2 + b_d^2})^2)^{0.5} \tag{2.1}$$

$$\phi_h(R, z) = c_h v_{circ}^2 \ln((\psi - 1)/(\psi + 1)) \tag{2.2}$$

and $\psi = (1 + (a_h^2 + R^2 + z^2)/r_h^2)^{0.5}$. The scaling constants are $(a_d, b_d, c_d) = (6.5, 0.26, 8.9)$ kpc and $(a_h, r_h) = (12, 210)$ kpc with $c_h = 0.33$ (e.g. Miyamoto & Nagai 1975; Wolfire et al. 1995). The circular velocity $v_{circ} \approx 220$ km s^{-1} is now well established through wide-field stellar surveys (Smith et al. 2007).

We determine the vertical acceleration at the Solar Circle using $g = -\partial \phi(R_o, z)/\partial z$ with $R_o = 8$ kpc. The hydrostatic halo pressure follows from

$$\frac{\partial \phi}{\partial z} = -\frac{1}{\rho_h} \frac{\partial P}{\partial z} \tag{2.3}$$

After Ferrara & Field (1994), we adopt a solution of the form $P_h(z) = P_o \exp((\phi(R_o, z) - \phi(R_o, 0))/\sigma_h^2)$ where σ_h is the isothermal sound speed of the hot corona. To arrive at P_o, we adopt a coronal halo density of $n_{e,h} = 10^{-4}$ cm^{-3} at the Stream distance (55 kpc) in order to explain the Magellanic Stream Hα emission (Bland-Hawthorn et al. 2007), although this is uncertain to a factor of a few. We choose $T_h = 2 \times 10^6$ K to ensure that OVI is not seen in the diffuse corona consistent with observation (Sembach et al. 2003); this is consistent with a rigorously isothermal halo for the Galaxy. Our solution to equation (2.3) is shown in Fig. 2(c) and it compares favorably with the pressure profile derive by others (e.g. Wolfire et al. 1995; Sternberg, McKee & Wolfire 2002).

A key parameter of the models is the ratio of the cloud to halo pressure, $\xi = P_c/P_h$. If the cloud is to survive the impact of the hot halo, then $\xi \gtrsim 1$. A shocked cloud is destroyed in about the time it takes for the internal cloud shock to cross the cloud, during which time the cool material mixes and ablates into the gas streaming past. Only massive clouds with dense cores can survive the powerful shocks. An approximate lifetime† for a spherical cloud of diameter d_c is

$$\tau_c = 60 (d_c/2 \text{ kpc})(v_h/350 \text{ km s}^{-1})^{-1} (\eta/100)^{0.5} \text{ Myr}. \quad (2.4)$$

For η in the range of 100–1000, this corresponds to 60–180 Myr for individual clouds. With a view to explaining the Hα observations, we focus our simulations on the lower end of this range.

For low η, the density of the hot medium is $n_h = 2 \times 10^{-4}$ cm^{-3}. The simulations are undertaken in the frame of the cold H I clouds, so the halo gas is given an initial transverse velocity of 350 km s^{-1}. The observations reveal that the mean Hα emission has a slow trend along the Stream which requires the Stream to move through the halo at a small angle of attack (20°) in the plane of the sky (see Fig. 5). Independent evidence for this appears to come from a wake of low column clouds along the Stream (Westmeier & Koribalski 2008). Thus, the velocity of the hot gas as seen by the Stream is $(v_x, v_y) = (-330, -141)$ km s^{-1}. The adiabatic sound speed of the halo gas is 200 km s^{-1}, such that the drift velocity is mildly supersonic (transsonic), with a Mach number of 1.75.

A unique feature of the *Fyris* simulations is that they include non-equilibrium cooling through time-dependent ionisation calculations (cf. Rosen & Smith 2004). When shocks occur within the inviscid fluid, the jump shock conditions are solved across the discontinuity. This allows us to calculate the Balmer emission produced in shocks and additionally from turbulent mixing along the Stream (e.g. Slavin et al. 1993). We adopt a conservative value for the gas metallicity of [Fe/H]=-1.0 (cf. Gibson et al. 2000); a higher value accentuates the cooling and results in denser gas, and therefore stronger Hα emission along the Stream.

2.2. Results

The results of the simulations are shown in Figs. 3 – 5; we provide animations of the disrupting stream at http://www.aao.gov.au/astro/MS. In our model, the fractal Stream experiences a "hot wind" moving in the opposite direction. The sides of the Stream clouds are subject to gas ablation via KH instabilities due to the reduced pressure (Bernouilli's theorem). The ablated gas is slowed dramatically by the hot wind and is transported behind the cloud. As higher order modes grow, the fundamental mode associated with the cloud size will eventually fragment it. The ablated gas now plays the role of a "cool wind" that is swept up by the pursuing clouds leading to shock ionization and ablation of the downstream clouds. The newly ablated material continues the trend along the length of the Stream. The pursuing gas cloud transfers momentum to the ablated upstream gas and accelerates it; this results in Rayleigh-Taylor (RT) instabilities, especially at the stagnation point in the front of the cloud. We rapidly approach a nonlinear regime where the KH and RT instabilities become strongly entangled, and the internal motions become highly turbulent. The simulations track the progression of the shock fronts as they propagate into the cloudlets.

In Fig. 3, we show the predicted conversion of neutral to ionized hydrogen due largely to cascading shocks along the Stream. The drift of the peak to higher columns is due to the shocks

† Here we correct a typo in equation (1) of Bland-Hawthorn et al. (2007).

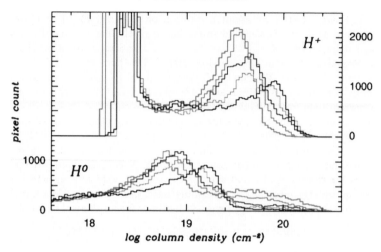

Figure 3. The dependence of the evolving fractions of H° and H$^+$ on column density as the shock cascade progresses. The timesteps are 70 (red), 120 (magenta), 170 (blue), 220 (green) and 270 Myr (black). The lowest column H I becomes progressively more compressed with time but the highest column H I is shredded in the cascade process; the fraction of ionized gas increases with time. The pile-up of electrons at low column densities arises from the x-ray halo.

eroding away the outer layers, thereby progressing into increasingly dense cloud cores. The ablated gas drives a shock into the H I material with a shock speed of v_s measured in the cloud frame. At the shock interface, once ram-pressure equilibrium is reached, we find $v_s \approx v_h \eta^{-0.5}$. In order to produce significant Hα emission, $v_s \gtrsim 35\,\mathrm{km\,s^{-1}}$ such that $\eta \lesssim 100$. In Fig. 4, we see the steady rise in Hα emission along the Stream, reaching $100-200$ mR after 120 Myr, and the most extreme observed values after 170 Myr. The power-law decline to bright emission measures is a direct consequence of the shock cascade. The shock-induced ionization rate is 1.5×10^{47} phot s^{-1} kpc^{-1}. The predicted luminosity-weighted line widths of $20\,\mathrm{km\,s^{-1}}$ FWHM (Fig. 4, inset) are consistent with the Hα kinematics. In Fig. 5, the Hα emission is superimposed onto the projected H I emission: much of it lies at the leading edges of clouds, although there are occasional cloudlets where ionized gas dominates over the neutral column. Some of the brightest emission peaks appear to be due to limb brightening, while others arise from chance alignments.

The simulations track the degree of turbulent mixing between the hot and cool media brought on by KH instabilities (e.g. Kahn 1980). The turbulent layer grows as the flow develops, mixing up hot and cool gas at a characteristic temperature of about 10^4 K. In certain situations, a sizeable Hα luminosity can be generated (e.g. Canto & Raga 1991) and the expected line widths are comparable to those observed in the Stream. Indeed, the simulations reveal that the fractal clouds develop a warm ionized skin along the entire length of the Stream. But the characteristic Hα emission (denoted by the shifting peak in Fig. 4) is comparable to the fluorescence excited by the Galactic UV field (Bland-Hawthorn & Maloney 2002). We note with interest that narrow Balmer lines can arise from pre-cursor shocks (e.g. Heng & McCray 2007), but these require conditions that are unlikely to be operating along the Stream.

2.3. Discussion

We have seen that the brightest Hα emission along the Stream can be understood in terms of shock ionization and heating in a transsonic (low Mach number) flow. For the first time, the Balmer emission (and associated emission lines) provides diagnostic information at any position along the Stream that is independent of the H I observations. Slow Balmer-dominated shocks of this kind (e.g. Chevalier & Raymond 1978) produce partially ionized media where a significant fraction of the Hα emission is due to collisional excitation. This can lead to Balmer decrements (Hα/Hβ ratio) in excess of 4, i.e. significantly enhanced over the pure recombination ratio of about 3, that will be fairly straightforward to verify in the brightest regions of the Stream.

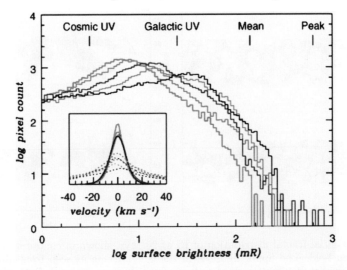

Figure 4. The evolving distribution of projected Hα emission as the shock cascade progresses. The timesteps are explained in Fig. 3. The extreme emission measures increase with time and reach the observed mean values after 120 Myr; this trend in brightness arises because denser material is ablated as the cascade evolves. The mean and peak emission measures along the Stream are indicated, along with the approximate contributions from the cosmic and Galactic UV backgrounds. **Inset:** The evolving Hα line width as the shock cascade progresses; the velocity scale is with respect to the reference frame of the initial H I gas. The solid lines are flux-weighted line profiles; the dashed lines are volume-weighted profiles that reveal more extreme kinematics at the lowest densities.

The shock models predict a range of low-ionization emission lines (e.g. OI, SII), some of which will be detectable even though suppressed by the low gas-phase metallicity. There are likely to be EUV absorption-line diagnostics through the shock interfaces revealing more extreme kinematics (Fig. 4, inset), but these detections (e.g. OVI) are only possible towards fortuitous background sources (Sembach et al. 2001; Bregman 2007). The predicted EUV/x-ray emissivity from the post-shock regions is much too low to be detected in emission.

The characteristic timescale for large changes is roughly 100−200 Myr, and so the Stream needs to be replenished by the outer disk of the LMC at a fairly constant rate (e.g. Mastropietro et al. 2005). The timescale can be extended with larger η values (equation (2.4)), but at the expense of substantially diminished Hα surface brightness. In this respect, we consider η to be fairly well bounded by observation and theory.

What happens to the gas shedded from the dense clouds? Much of the diffuse gas will become mixed with the hot halo gas suggesting a warm accretion towards the inner Galactic halo. If most of the Stream gas enters the Galaxy via this process, the derived gas accretion rate is $\sim 0.4 M_\odot$ yr^{-1}. The higher value compared to H I (e.g. Peek et al. 2008) is due to the gas already shredded, not seen by radio telescopes now. In our model, the HVCs observed today are unlikely to have been dislodged from the Stream by the process described here. These may have come from an earlier stage of the LMC-SMC interaction with the outer disk of the Galaxy.

The "shock cascade" interpretation for the Stream clears up a nagging uncertainty about the Hα distance scale for high-velocity clouds. Bland-Hawthorn et al. (1998) first showed that distance limits to HVCs can be determined from their observed Hα strength due to ionization by the Galactic radiation field, now confirmed by clouds with reliable distance brackets from the stellar absorption line technique (Putman et al. 2003; Lockman et al. 2008; Wakker et al. 2007). HVCs have smaller kinetic energies compared to the Stream clouds, and their interactions with the halo gas are not expected to produce significant shock-induced or mixing layer Hα emission, thereby supporting the use of Hα as a crude distance indicator.

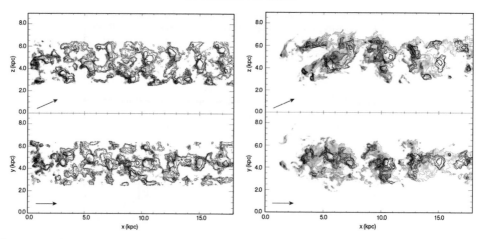

Figure 5. The initial fractal distribution of H I at 20 Myr, shown in contours, before the wind action has taken hold. The upper figure is in the $x-z$ frame as seen from above; the lower figure is the projected distribution on the plane of the sky ($x-y$ plane). Both distributions are integrated along the third axis. The logged H I contours correspond to 18.5 (dotted), 19.0, 19.5, 20.0, and 20.5 (heavy) cm^{-2}. The greyscale shows weak levels of Hα along the Stream where black corresponds to 300 mR. The predicted H I (contours) and Hα (greyscale) distributions after 120 Myr. The angle of attack in the [x,y,z] coordinate frame is indicated. The Hα emission is largely, but not exclusively, associated with dense H I gas.

Here, we have not attempted to reproduce the H I observations of the Stream in detail. This is left to a subsequent paper where we explore a larger parameter space and include a more detailed comparison with the H I and Hα power spectrum, inter alia. We introduce additional physics, in particular, the rotation of the hot halo, a range of Stream orbits through the halo gas, and so on.

If we are to arrive at a satisfactory understanding of the Stream interaction with the halo, future deep Hα surveys will be essential. It is plausible that current Hα observations are still missing a substantial amount of gas, in contrast to the deepest H I observations. We can compare the particle column density inferred from H I and Hα imaging surveys. The limiting H I column density is about $N_H \approx \langle n_H \rangle L \approx 10^{18}$ cm^{-2} where $\langle n_H \rangle$ is the mean atomic hydrogen density, and L is the depth through the slab. By comparison, the Hα surface brightness can be expressed as an equivalent emission measure, $E_m \approx \langle n_e^2 \rangle L \approx \langle n_e \rangle N_e$. Here n_e and N_e are the local and column electron density. The limiting value of E_m in Hα imaging is about 100 mR, and therefore $N_e \approx 10^{18}/\langle n_e \rangle$ cm^{-2}. Whether the ionized and neutral gas are mixed or distinct, we can hide a lot more ionized gas below the imaging threshold for a fixed L, particularly if the gas is at low density ($\langle n_e \rangle \ll 0.1$ cm^{-3}). A small or variable volume filling factor can complicate this picture but, in general, the ionized gas still wins out because of ionization of low density H I by the cosmic UV background (Maloney 1993). In summary, even within the constraints of the cosmic microwave background (see Maloney & Bland-Hawthorn 1999), a substantial fraction of the gas can be missed if it occupies a large volume in the form of a low density plasma (e.g. Rasmussen et al. 2003).

3. Direct infall of H I onto the disk

The conditions operating along the Magellanic Stream are unlikely to be representative of all H I clouds that move through the Galactic corona. A related process is the infall of individual H I clouds towards the Galactic disk. The Galactic halo is home to many HVCs of unknown origin (Wakker 2001; Lockman et al. 2008). The survival and stability of these clouds is a problem that has long been recognized (e.g. Benjamin & Danly 1997) which we now discuss.

It is likely that many or all halo clouds have experienced some deceleration during their transit through the lower halo. Using equation (2.3), we determine a freefall velocity for a cloud starting at the distance of the Stream (Fig. 1(b)) that is more than twice what is inferred for clouds at the Solar Circle (Wakker 2001), although some HVCs clearly have high space motions (e.g. Lockman *et al.* 2008). To explain this observation, Benjamin & Danly (1997) investigated a drag equation for a cloud moving through a stationary medium,

$$\mu_c \dot{v}_c = \frac{1}{2} C_D \rho_h(z) v_c^2 - \mu_c g(z) \tag{3.1}$$

where μ_c is the surface density of the cloud. Equation (3.1) only holds as long as the cloud stays together. The drag coefficient C_D is a measure of the efficiency of momentum transfer to the cloud. For the high Reynolds numbers typical of astrophysical media, incompressible objects have $C_D \approx 0.4$ (e.g. a rough sphere) which indicates that the turbulent wake behind the plunging object efficiently transfers momentum to the braking medium. The leading face of a compressible cloud may become flattened, such that the approaching medium is brought to rest in the reference frame of the cloud; in this instance, $C_D \gtrsim 1$ may be more appropriate (we adopt $C_D = 1$ here).

A solution specific to our model is shown in Fig. 2(b) where the freefall velocity is now slowed by about 35%. In practice, the cloud's projected motion can be considerably less than its 3D space velocity (e.g. Lockman *et al.* 2008). In all likelihood, infalling HVCs have experienced significant deceleration through ram pressure exerted by the corona. But even before the cloud reaches terminal velocity, the cloud is expected to break up (Murray & Lin 2004).

So how do clouds resist the destructive forces of RT and shock instabilities? In §3.1, we investigate the stabilizing influence of magnetic fields when a cloud passes through a magnetized medium. The halo magnetic field is poorly constrained at the present time (e.g. Sun *et al.* 2008). We describe the uniform magnetic field in terms of the pressure of the halo medium, or

$$\frac{B^2}{8\pi} = \beta P_h \tag{3.2}$$

such that $B \approx 1$ μG at the distance of the Stream (55 kpc) if the field is in full equipartition with the corona (see Fig. 1(d)). But there is evidence that the field is weaker than implied here ($\beta \approx 0.3$; Sun *et al.* 2008), at least within 5 kpc of the Galactic plane For the warm, denser low-latitude gas (Reynolds layer), we adopt the new parameter fits of Gaensler *et al.* (2008) from a re-analysis of pulsar data. The lower β value finds support from recent magnetohydrodynamic simulations of the Reynolds layer (Hill *et al.* 2008).

3.1. *Stability limits and growth timescales*

We consider the surface of a high velocity cloud as a boundary between two fluids. In practice, the Galactic ionizing radiation field imparts a multiphase structure to the cloud. At all galactic latitudes within the Stream distance, HVCs with column densities of order 10^{20} cm^{-2} or higher have partially ionized skins to a column depth of roughly 10^{19} cm^{-2} for sub-solar gas due to the Galactic ionizing field (see Bland-Hawthorn & Maloney 1999; Wolfire *et al.* 2003). Between the warm ionized skin and the cool inner regions is a warm neutral medium of twice the skin thickness; both outer layers have a mean particle temperature of $\lesssim 10^4$K.

The cloud is denser than the halo gas. Because of the gravitational field, RT instabilities can grow on the boundary. Furthermore, KH instabilities may also develop due to the relative motion of the cloud with respect to an external medium. Recent work has shown that buoyant bubbles in galaxy clusters are stabilized against RT and KH instabilities by viscosity and surface tension due to magnetic fields in the boundary (De Young 2003; Kaiser *et al.* 2005; Jones & De Young 2005). Here we examine whether HVC boundaries are similarly stabilized against disruption in the Galactic halo.

When there is no surface tension, no viscosity and no relative motion between the two media, the growth rate of the RT instability for a perturbation with wavenumber k is $\omega = \sqrt{gk}$, where g is the gravitational acceleration at the fluid boundary. The wavenumber is related to a perturbation length scale, $\ell = 2\pi/k$. The instability requires a few e-folding timescales to fully develop; the timescale is given by

$$t_{grow} = \omega^{-1} = \sqrt{\frac{\ell}{2\pi g}}. \qquad (3.3)$$

In the presence of a magnetic field, the transverse component (B_{tr}) provides some surface tension which can help to suppress RT instabilities below a lengthscale of

$$\ell_{min} = \frac{B_{tr}^2}{2\rho_c g} \qquad (3.4)$$

(Chandrasekhar 1961). Here ρ_c is the mass density of the denser medium, i.e. the cloud, and B_{tr} is the average value of the transverse magnetic field at the boundary.

In order to illustrate when RT instabilities become important, we assume a flat rotation curve for the Galaxy (e.g. Binney & Dehnen 1997)

$$g \approx \frac{v_{circ}^2}{R} = 1.6 \times 10^{-8} \left(\frac{v_{circ}}{220 \text{ km s}^{-1}}\right)^2 \left(\frac{R}{R_o}\right)^{-1} \text{ cm s}^{-2}. \qquad (3.5)$$

This is only a rough approximation to the form expected from equations (2.1) and (2.2). We stress that the actual behaviour discussed below, and shown in Figs. 2(e) and (f), solves for the gravitational potential correctly.

Shortly after the discovery of HVCs, it was thought that they may be self-gravitating. But this would place them at much greater distances than the Magellanic Stream (e.g. Oort 1966) which is now known not to be the case (e.g. Putman et al. 2003). Instead, we consider two cases: (i) HVCs in pressure equilibrium with the coronal gas; (ii) HVCs with parameters fixed by direct observation. In (i), because the temperature is not strongly dependent on radius, but the number density decreases rapidly with increasing radius, we expect the increased pressure to compress the clouds at lower latitudes.

We estimate the impact of RT instabilities using equations (2.3) and (3.3): for a cloud temperature of $T_c = 10^4$ K (see §2) in pressure equilibrium with the hot halo, the electron density is given by

$$n_{e,c} \approx n_{e,h} \frac{T_h}{T_c} \qquad (3.6)$$

$$= 0.02 \left(\frac{R}{55 \text{ kpc}}\right)^{-2} \left(\frac{T_h}{2 \times 10^6 \text{ K}}\right) \left(\frac{T_c}{10^4 \text{ K}}\right)^{-1} \text{ cm}^{-3}.$$

We use equations 3.4, 3.6 and 3.2 to estimate ℓ_{min} as a function of Galactocentric radius. The minimum length scale for instability is

$$\frac{\ell_{min}}{R} \sim \frac{8\pi\beta k_B T_c}{m_p v_{circ}^2} \qquad (3.7)$$

$$= 0.004 \left(\frac{T_c}{10^4 \text{ K}}\right) \left(\frac{\beta}{0.1}\right) \left(\frac{v_{circ}}{220 \text{ km s}^{-1}}\right)^{-2}$$

and its associated growth timescale using equation (3.3)

$$t_{grow} \sim \sqrt{\frac{4\beta k_B T_c}{m_p v_{circ}^2}} \, \Omega^{-1} \qquad (3.8)$$

$$= 1.1 \text{ Myr} \left(\frac{T_c}{10^4 \text{ K}}\right)^{\frac{1}{2}} \left(\frac{\beta}{0.1}\right)^{\frac{1}{2}} \left(\frac{v_{circ}}{220 \text{ km s}^{-1}}\right)^{-2} \left(\frac{R}{R_o}\right)$$

where the angular rotation rate is given by $\Omega = v_{circ}/R$.

Because we have assumed that $B^2 \propto n_{e,h}$ (equipartition) and $n_{e,c} \propto n_{e,h}$ (pressure equilibrium), neither the minimum scale length or its growth timescale depend on the halo density or temperature. They do depend on the temperature of the clouds and the ionization state. If the clouds are hotter than 10^4 K, then $n_{e,c}$ is overestimated under the assumption of pressure equilibrium. This would lead to larger minimum instability lengthscales and growth timescales. If the Galactic rotation curve drops faster than the flat profile implied by equation (2.3), we would have underestimated both the minimum instability scale length and its associated growth timescale at large radii.

Under the assumption of pressure equilibrium, the falling clouds become more compressed as they approach the disk which can hasten cooling. This effect may help to stabilize against break up, particularly if a cool shell develops (cf. Sternberg & Soker 2008).

In the absence of gravitational instability, the flow is stable against the KH instability if (Chandrasekhar 1961)

$$U^2 < \frac{B_{tr}^2(\rho_c + \rho_h)}{2\pi \rho_c \rho_h} \qquad (3.9)$$

where U is the relative velocity between the two fluids. When $\rho_c > \rho_h$, this requirement becomes

$$U < \sqrt{\frac{8\beta k_B T_h}{m_p}} \qquad (3.10)$$

$$= 115 \left(\frac{\beta}{0.1}\right)^{\frac{1}{2}} \left(\frac{T_h}{2 \times 10^6 \text{K}}\right)^{\frac{1}{2}} \text{ km s}^{-1}$$

and we have described the magnetic field in terms of the halo pressure using equation 3.2. This requirement is also independent of the halo density as we have related the magnetic field to the halo pressure, although it is dependent on the halo temperature. This requirement is nearly satisfied for HVCs if the magnetic field is near equipartition.

3.2. The Smith Cloud

Arguably, the high-latitude H I cloud that we know most about is the Smith Cloud. Lockman et al. (2008) have recently published spectacular H I data for this HVC and deduce a remarkable amount about its past and future properties. The HVC has an estimated distance of 12.4 ± 1.3 kpc, a Galactocentric radius of $R \approx 8$ kpc, a vertical height below the plane of -2.9 kpc, a mass of at least 10^6 $M_\odot$ in a volume of order 3 kpc^3 corresponding to $n_c \approx 0.014$ cm^{-3}. The cloud has a prograde orbit that is inclined 30° to the plane and appears to have come through the disk 70 Myr ago at $R \approx 13$ kpc moving from above to below the plane.

In order to have punched through the disk, the shock crossing time for the cloud must be longer than for the disk. It can be shown that

$$\frac{d_c}{z_d} > \sqrt{\frac{n_d}{n_c}} \qquad (3.11)$$

where z_d is the vertical thickness and n_d is the mean density of the H I at the crossing point. This is essentially a statement that the surface density of the cloud must be higher than the disk. If we assume the cloud punched through the Galactic hydrogen density profile determined by Kalberla & Dedes (2008), equation 3.11 indicates that the cloud was substantially thicker than the disk when it came through and somewhat more massive than what is observed now. Consistent with this picture, the observed wake may result from ablation processes induced by the impact. For cloudlets smaller than 100 pc, thermal conduction due to the halo corona (McKee & Cowie 1977) and the Galactic radiation field convert the ablated gas to a clumpy plasma.

The kinetic energy of the Smith Cloud observed today is $\sim 10^{54}$ erg – this is enough to punch through the disk if sufficiently concentrated. Impulsive shock signatures at UV to x-ray wavelengths will have largely faded away, and the H I "hole" at the crossing point will have been substantially stretched by differential shear†.

† It is sometimes claimed, this meeting notwithstanding, that outer disk H I "holes" are evidence of dark matter minihalos passing through the disk, but it can be shown that the gravitational impulse has negligible impact on the gaseous disk.

Figs. 2(e) and (f) show that a cloud of several kpc can survive RT instabilities at these latitudes, but it is difficult to see how the Smith Cloud, like several other large HVCs, could have come in from, say, the distance of the Magellanic Stream. Lockman et al. (2008) use essentially the same Galactic potential as described here to determine the cloud's orbit parameters. We conjecture that either the cloud has been dislodged from the outer disk by a passing dwarf, or the cloud has been brought in by a confining dwarf potential. A cloud metallicity of $[Fe/H]\approx-1$ is appropriate in either scenario. Interestingly, the impulse from the Galactic disk can cause the gas to become dislodged from the confining dark halo or to oscillate within it. The interloper must be on a prograde orbit which rules out some infalling dwarfs (e.g. ωCen; Bekki & Freeman 2007), but conceivably implicates disrupting dwarfs like Canis Major or Sagittarius, assuming these were still losing gas in the recent past.

3.3. Discussion

In §2, we presented evidence for a shock cascade along the Magellanic Stream arising from the disruption of upstream clouds due to their interaction with the Galactic halo. Bland-Hawthorn et al. (2007) make firm predictions that can be tested in future observations. A possible improvement is to consider the entire Magellanic System, i.e. the influence of the LMC-SMC system that lies further upstream. Mastropietro et al. (2008) present evidence for a strong interaction along the leading edge of the LMC; for their quoted model parameters, it seems plausible that this results in a stand-off bowshock ahead of the galaxy. The cross wind over the face of the LMC could be confused for a starburst-driven wind from the LMC (cf. Lehner & Howk 2007). In all likelihood, the LMC-SMC system creates a turbulent wake behind it which may impact the development of instabilities in the trailing stream.

The issue of cloud survival is highly complex. In §3, we did not consider the role of viscosity in quenching RT or KH instabilities. Simulations have shown that viscosity does lead to stabilization (Pavlovski et al. 2008) but we have not been able to estimate a lengthscale or a growth timescale appropriate for our setting. Kaiser et al. (2005) show that when the density ratio between the two media is large, KH instabilities fail to grow and the growth rate of RT instabilities depends only on the properties of higher density medium, in our case the cloud medium. However their result, taken in the limit of one density much larger than the other, $\rho_2 \gg \rho_1$, will not apply if $\nu_1 \rho_1 \gg \nu_2 \rho_2$. Here the subscripts refer to the fluids on either side of the boundary and ν is the kinematic viscosity. Because diffusivity coefficients are sensitive functions of temperature ($\propto T^{-2.5}$), they could dampen fluid instabilities. Unfortunately the expected differences in temperatures between HVCs and the halo gas (corona) suggest that $\nu_1 \rho_1 \gg \nu_2 \rho_2$ and thus we cannot apply the limit used by Kaiser et al. (2005). A proper treatment is required to cover the Galactic halo setting.

Other studies have argued that the KH instability leads to a turbulent mixing layer on the surface and so is less destructive than the RT instability (e.g. De Young 2003). At the present time, there are no relevant astrophysical codes that are capable of handling mixing in a satisfactory manner. On the issue of magnetic stability, more sophisticated treatments using MHD have been attempted, but the main conclusions appear to be contradictory (Konz et al. 2002; Gregori et al. 1999). We are not aware of MHD codes that are sufficiently capable of answering this question at the present time.

Without excessive erudition, which is inappropriate for a conference proceeding, it is difficult to mount a solid case for why hydro processes could ultimately save the day for HVCs. But the fact of the matter is that fast-moving gas clouds do survive their passage through the Galactic halo. These may be mostly shortlived entities on the road to destruction, suggesting that there is a largely hidden plasma component that we have yet to fully comprehend. This will require more extensive observations at difficult parts of the observational parameter space, matched by hydro codes that can properly treat instabilities and mixing in a multi-phase gas.

Acknowledgements

JBH is supported by a Federation Fellowship through the Australian Research Council. I thank Alice Quillen and Ralph Sutherland for their role in the work presented here. I acknowledge helpful discussions with Bob Benjamin, Chris Flynn, Bryan Gaensler, Greg Madsen and Mary Putman.

References

Agertz, O., et al. 2007, MNRAS, 380, 963
Bekki, K. & Freeman, K. C. 2003, MNRAS, 346, L11
Benjamin, R. & Danly, L. 1997, ApJ, 481, 764
Besla, G. et al. 2007, ApJ, 668, 949
Binney, J. & Dehnen, W. 1997, MNRAS
Binney, J., Dehnen, W., & Bertelli, G. 2000, MNRAS, 318, 658
Bland-Hawthorn et al. 1998, MNRAS, 299, 611
Bland-Hawthorn, J., & Maloney, P. R. 1999, ApJL, 510, L33
Bland-Hawthorn, J., & Maloney, P. R. 2002, Extragalactic Gas at Low Redshift, 254, 267
Bland-Hawthorn, J., Sutherland, R., Agertz, O., & Moore, B. 2007, ApJ, 670, L109
Bregman, J. N. 2007, ARA&A, 45, 221
Brüns, C., et al. 2005, AAp, 432, 45
Canto, J. & Raga, A. C. 1991, ApJ, 372, 646
Chandrasekhar, S. 1961, Hydrodynamic and Hydromagnetic Stability, Clarendon, Oxford
Chevalier, R. A. & Raymond, J. C. 1978, ApJL, 225, L27
Connors, T. W., Kawata, D., & Gibson, B. K. 2006, MNRAS, 371, 108
De Young, D. S. 2003, MNRAS, 343, 719
Ferrara, A. & Field, G. B. 1994, ApJ, 423, 665
Flynn, C. et al. 2006, MNRAS, 372, 1149
Gaensler, B. et al. 2008, PASA, submitted
Gibson, B. K. et al. 2000, AJ, 120, 1830
Gregori, G. et al. 1999, ApJ, 527, L113
Heng, K. & McCray, R. 2007, ApJ, 654, 923
Hibbard, J. E. & van Gorkom, J. H. 1996, AJ, 111, 655
Hill, A. et al. 2008, ApJ, 686, 363
Jones, T. W. & De Young, D. 2005, ApJ, 624, 586
Kahn, F. D. 1980, AAp, 83, 303
Kaiser, C. R., Pavlovski, G., Pope, E. D. C., & Fangohr, H. 2005, MNRAS, 359, 493
Kalberla, P. & Dedes, L. 2008, A&A, in press (0804.4831)
Kallivayalil, N., van der Marel, R. P. & Alcock, C., 2006, ApJ, 652, 1213
Keres, D., Katz, N., Weinberg, D. H. & Dave, R. 2005, MNRAS, 363, 2
Klein, R. I., McKee, C. F., & Colella, P. 1994, ApJ, 420, 213
Konz, C., Brüns, C., & Birk, G. T. 2002, A&A, 391, 713
Larson, R. B. 1969, MNRAS, 145, 405
Lehner, N. & Howk, C. 2007, MNRAS, 377, 687
Liska, R. & Wendroff, B. 1999, International Journal for Numerical Methods in Fluids, 30, 461
Lockman, F. J., Benjamin, R. A., Heroux, A. J., & Langston, G. I. 2008, ApJ, 679, L21
Madsen, G. J., Haffner, L. M., & Reynolds, R. J. 2002, ASPC, 276, 96
Maloney, P. R. & Bland-Hawthorn, J. 1999, ApJL, 522, L81
Maloney, P. 1993, ApJ, 414, 41
Mastropietro, C., Moore, B., Mayer, L., Wadsley, J., & Stadel, J. 2005, MNRAS, 363, 509
McKee, C. F. & Cowie, L. L. 1977, ApJ, 215, 213
Miyamoto, M. & Nagai, R. 1975, PASJ, 27, 533
Moore, B. & Davis, M. 1994, ApJ, 270, 209
Murray, S. D., White, S. D. M., Blondin, J. M., & Lin, D. N. C. 1993, ApJ, 407, 588
Murray, S. D. & Lin, D. C. 2004, ApJ, 615, 586
Nicastro, F., Mathur, S., & Elvis, M. 2008, Science, 319, 55
Oort, J. 1966, Bull. Astron. Inst. Neth., 18, 421
Pavlovski, G., Kaiser, C., Pope, E. C. D., & Fangohr, H. 2008, MNRAS, 384, 1377

Peek, J. E. G., Putman, M. E., & Sommer-Larsen, J. 2008, ApJ, 674, 227
Piatek, S., Pryor, C., & Olszewski, E. W. 2008, AJ, 135, 1024
Putman, M. E. et al. 2003, ApJ, 597, 948
Quilis, V. & Moore, B. 2001, ApJ, 555, L95
Rasmussen, A., Kahn, S., & Paerels, F. 2003, ASSL, 281, 109
Rosen, A. & Smith, M. D. 2004, MNRAS, 347, 1097
Ruszkowski, M., Enslin, T. A., Bruggen, M., Heinz, S., & Pfrommer, C. 2007, MNRAS, 378, 662
Savage, B. D. et al. 2003, ApJS, 146, 125
Sembach, K. R., Howk, J. C., Savage, B. D., Shull, J. M., & Oegerle, W. R. 2001, ApJ, 561, 573
Sembach, K. R., et al. 2003, ApJS, 146, 165
Sembach, K. R. et al. 2004, ApJS, 150, 387
Slavin, J. D., Shull, J. M., & Begelman, M. C. 1993, ApJ, 407, 83
Smith, M. et al. 2007, MNRAS, 379, 755
Sternberg, A., McKee, C. F., & Wolfire, M. 2002, ApJS, 143, 419
Sternberg, A. & Soker, N., 2008, MNRAS, 389, L13
Sun, X. H., Reich, W., Waelkens, A., & Enslin, T. A. 2008, A&A, 477, 573
Sutherland, R. S., 2008, ApJ, in preparation
Tamm, A., Tempel, E., & Tenjes, P. 2007, astro-ph/0707.4375
Teyssier, R. 2002, AAp, 385, 337
Thom, C., et al. 2008, astro-ph
Tisserand, P. et al. 2007, A&A, 469, 387
Tripp, T. M., et al. 2003, AJ, 125, 3122
Wakker, B. P. 2001, ApJS, 136, 463
Wakker, B. P et al. 2007, ApJ, 207, 670, L113
Weiner, B.J. & Williams, T. B. 1996, AJ, 111, 1156
Weiner, B. J., Vogel, S. N., & Williams, T. B. 2002, Extragalactic Gas at Low Redshift, 254, 256
Westmeier, T. & Koribalski, B. S. 2008, MNRAS, 388, L29
Wilkinson, M. I. & Evans, N. W. 1999, MNRAS, 310, 645
Williams, J. P. & McKee, C. F. 1997, ApJ, 476, 166
Wolfire, M. et al. 1995, ApJ, 453, 673
Wolfire, M. et al. 2003, ApJ, 587, 278

New evidence for halo gas accretion onto disk galaxies

Filippo Fraternali[1]

[1]Astronomy Department, University of Bologna, via Ranzani 1, I-40127, Bologna, Italy
email: filippo.fraternali@unibo.it

Abstract. Studies of the halo gas in the Milky Way and in nearby spiral galaxies show the presence of gas complexes that cannot be reconciled with an internal (galactic fountain) origin and are direct evidence of gas accretion. Estimating gas accretion rates from these features consistently gives values, which are one order of magnitude lower than what is needed to feed the star formation. I show that this problem can be overcome if most of the accretion is in fact "hidden" as it mixes with the galactic fountain material coming from the disk. This model not only provides an explanation for the missing gas accretion but also reproduces the peculiar kinematics of the halo gas in particular the vertical rotation gradient. In this view this gradient becomes indirect evidence for gas accretion.

1. Introduction

A large amount of fresh gas accretion onto the Milky Way has been advocated for decades since the work of Twarog (1980) showed the relative constancy of the Star Formation Rate (SFR) in the Galactic disk over the last 10 Gyrs. Roughly constant SFR requires constant accretion and this translates today into a need for gas accretion at a rate of the order $\sim 1\,M_\odot\,\mathrm{yr}^{-1}$.

Since the 1960s, the presence of a large number of neutral hydrogen (H I) clouds filling the sky around us and having, on average, negative velocities with respect to the Milky Way disk has been recognized (e.g. Hulsbosch 1968). Such clouds, named High Velocity Clouds (HVCs), given their largely anomalous radial velocities (Wakker & van Woerden 1997), were immediately regarded as possible evidence for gas accretion from intergalactic space onto our Galaxy (Oort 1970). Alternative interpretations have been proposed, for example that the clouds are produced by the cooling of material expelled from the disk via a "galactic fountain" (Bregman 1980) or that they are a much more distant Local Group population (e.g. Blitz 1999).

From the time of Oort's suggestion until very recently there were two major unknowns about the HVCs: their distances and their metallicities. It is thanks to a tenacious observational campaign carried out especially over the last few years that we now know both distances and metallicities for all the major HVCs (Wakker et al. 2007, 2008). The results leave no doubts: the HVCs are located in the Milky Way halo and have metallicities of about an order of magnitude lower than the average disk ISM metallicity. These properties make it most likely that this is accreting material falling onto the Milky Way for the first time.

The HVCs do not comprise the entire halo population, rather they are an extreme population at large heights and having particularly anomalous velocities. In the lower halo large amounts of cold gas are observed (e.g. Lockman 1984), some of which go under the name of Intermediate Velocity Clouds (IVCs) and are regarded as a likely galactic fountain component (e.g. Wakker 2001). More recently, Kalberla & Dedes (2008) showed that up to 10% of the Milky Way H I gas is in fact extra-planar and highly turbulent.

2. Halo gas in nearby spiral galaxies

In the last decade, the study of the halo (extra-planar) gas has been extended to nearby galaxies. This is a demanding task, given the low surface brightness of the halo emission, and the studies have been restricted to a relatively small number of objects. In Table 1, I summarize the results for the best studied galaxies so far. This table includes galaxies seen at different inclination angles. For edge-on galaxies the halo gas can be separated spatially from the disk gas, whilst for galaxies seen at intermediate inclinations it can be separated thanks to its peculiar kinematics. The main kinematic feature of extra-planar gas is its decreasing rotation velocity with increasing height from the plane, it is said to be "lagging" behind the disk gas. Such a velocity gradient has been estimated for a few galaxies (column 9, Table 1) and it is an important constraint for models of extra-planar gas formation (see Section 4). The presence of this gradient is also the main reason why extra-planar gas can be detected in non edge-on galaxies (Fraternali et al. 2002).

Table 1. Physical properties of extra-planar gas in spiral galaxies

Galaxy	Type	incl (°)	v_{flat} (km/s)	$M_{HI_{halo}}$ ($10^8 M_\odot$)	$M_{HI_{tot}}$ ($10^9 M_\odot$)	SFR ($M_\odot$/yr)	Accr. rate ($M_\odot$/yr)	Gradient[a] (km/s/kpc)	Ref.
Milky Way	Sb	-	220	~4	4	1 – 3	≈ 0.2[b]	−22	(1,2,3)
M 31	Sb	77	226	> 0.3	3	0.35	-	-	(4,5)
NGC 253	Sc	~75	~185	0.8	2.5	> 10	-	-	(6)
M 33	Scd	55	110	> 0.1	1	0.5	0.05[c]	-	(7,8)
NGC 2403	Scd	63	130	3	3.2	1.3	0.1	~ −12	(9)
NGC 2613	Sb	~80	~300	4.4[d]	8.7	5.1	-	-	(10)
NGC 3044	Sc	84	150	4	3	2.6[e]	-	-	(11)
NGC 4559	Scd	67	120	5.9	6.7	0.6[e]	-	~ −10	(12)
NGC 5746	Sb	86	310	~1	9.4	1.2	0.2[f]	-	(13,14)
NGC 5775	Sb	86	200	-	9.1	7.7[e]	-	−8[g]	(15)
NGC 6946	Scd	38	175	$\gtrsim 2.9$	6.7	2.2	-	-	(17)
NGC 891	Sb	90	230	12	4.1	3.8	0.2	−15	(18)
UGC 7321	Sd	88	110	$\gtrsim 0.1$	1.1	~ 0.01[h]	-	$\gtrsim$ − 25	(19)

[a] Gradient in rotation velocity with height (from the flat part of the rotation curve); [b] from complex C and other clouds with known distances in (2) without correction for the ionised fraction; [c] from the H I mass in (8) without their correction for the ionised fraction; [d] from sum of the various extra-planar clouds; [e] calculated from the FIR luminosity using the formula in Kewley et al. (2002); [f] from the counter-rotating cloud (13) using an infall time-scale of 1×10^8 yr; [g] calculated using optical lines (16); [h] SFR of only massive stars $> 5 M_\odot$. References: (1) Kalberla & Dedes (2008); (2) Wakker et al. (2007), Wakker et al. (2008); (3) Levine, Heiles & Blitz (2008); (4) Thilker et al. (2004); (5) Walterbos & Braun (1994); (6) Boomsma et al. (2005); (7) Reakes & Newton (1978); (8) Grossi et al. (2008); (9) Fraternali et al. (2002); (10) Chaves & Irwin (2001); (11) Lee & Irwin (1997); (12) Barbieri et al. (2005); (13) Rand & Benjamin (2008) (14) Pedersen et al. (2006); (15) Irwin (1994); (16) Heald et al. (2006a); (17) Boomsma et al. (2008); (18) Oosterloo, Fraternali, & Sancisi (2007); (19) Matthews & Wood (2003).

Fig. 1 shows the two types of extra-planar gas detections. On the left, the total H I map of the edge-on galaxy NGC 891 (blue + contours) is overlaid onto a DSS optical image (orange). The H I data (obtained with the Westerbork Synthesis Radio Telescope) show a massive and extended H I halo with a mass of $1.2 \times 10^9 M_\odot$ or 30% of the total H I mass (Oosterloo, Fraternali & Sancisi 2007). On the right panel of Fig. 1, the position-velocity (p-v) plot along the major axis of the galaxy NGC 2403 (inclination 60°) obtained with the VLA (Fraternali et al. 2001). A broad component of gas at rotation velocities lower

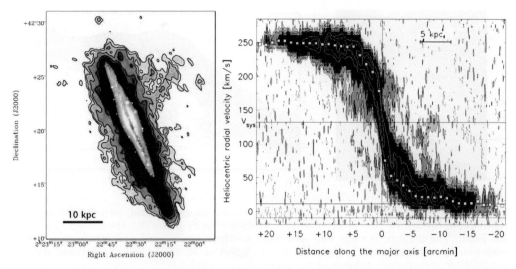

Figure 1. Two methods of detecting extra-planar gas. Left: total H I map (blue + contours) for the edge-on galaxy NGC 891, overlaid on the optical image (orange) (data from Oosterloo et al. 2007). The extra-planar gas is clearly separated on the sky and surrounds the whole galactic disk with a filament that extends up to 20 kpc. Right: position-velocity diagram along the major axis of the intermediate inclination galaxy NGC 2403 (from Fraternali et al. 2001). The extra-planar gas is kinematically separated from the disk gas. It is seen as a faint component rotating more slowly than the disk (white dots: = disk rotation curve).

than the disk gas (also called the "beard") is clearly visible at low emission levels (light grey). This beard component is the halo gas in NGC 2403.

Extra-planar gas is possibly ubiquitous as several nearby galaxies other than those reported in Table 1 show indications of its presence. It is also observed in the ionised phase. Optical studies of nearby edge-on galaxies show that roughly half of them have extended layers of diffuse ionised gas (e.g. Rossa & Dettmar 2003) and with similar kinematics as the H I layers (Heald et al. 2006b).

3. Direct evidence for gas accretion

Direct evidence of gas accretion is difficult to obtain. The strategy is to look for gas components (usually at very anomalous velocities) which are incompatible with an internal origin. The large majority of the extra-planar gas studied so far has actually a very regular kinematics that follows closely the kinematics of the disk (see for instance the p-v diagram for NGC 2403, right panel of Fig. 1). This points to a tight connection between disk and halo components.

The first features that deserve attention in the search for gas accretion are the H I filaments. Among the galaxies in Table 1 at least half show large filamentary H I structures in their halos, the most notable cases being NGC 891 and NGC 2403. NGC 891 has a long massive filament ($M_{\rm HI} \sim 1.6 \times 10^7\, M_\odot$) extending up to about 20 kpc from the plane of the disk (Fig. 1). NGC 2403 also has a filament with a similar H I mass located in projection outside the bright optical disk and clearly separated in velocity from the disk kinematics (Fig. 2). They are both very similar to Complex C in our Galaxy. We can calculate the energy needed to form these filaments assuming that they come from the disk through a galactic fountain. In the case of the NGC 891 filament it turns out that this energy should be of the order $\sim 1 \times 10^{55}$ erg. This would correspond to the explosion

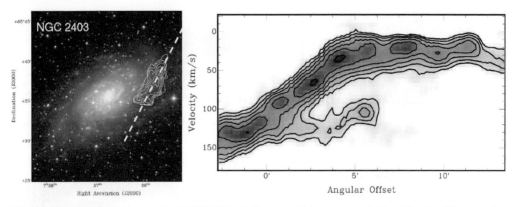

Figure 2. The H I filament in NGC 2403. Left: optical image overlaid with the H I map (in contours) of the $1 \times 10^7\,M_\odot$ filament. Right: position-velocity plot along the dashed line in the left panel. The filament is 8 kpc long and separated from the normal disk kinematics (higher contours) by about $90\,\mathrm{km\,s^{-1}}$. Note the small dispersion in velocity of the filament and its proximity to the systemic velocity of the galaxy ($v_\mathrm{sys} = 133\,\mathrm{km\,s^{-1}}$).

of about 10^5 supernovae in a specific region of space and over a time-scale shorter than a dynamical time, which is clearly a very unlikely event.

A second type of feature that has been found in these new deep surveys are clouds at very anomalous velocities that end up in the region of counter-rotation. The data of NGC 891 show two such clouds with masses of order $\sim 10^6\,M_\odot$ and counter-rotating velocities of ~ 50 and $\sim 90\,\mathrm{km\,s^{-1}}$ (Fig. 3). These clouds cannot be produced in any kind of galactic fountain and they are most likely direct evidence of gas accretion. NGC 2403 also has gas components at very anomalous velocities called "forbidden gas" (Fig. 1; Fraternali *et al.* 2002). Counter-rotating or forbidden clouds are also observed in NGC 4559, in NGC 5746 and NGC 6946.

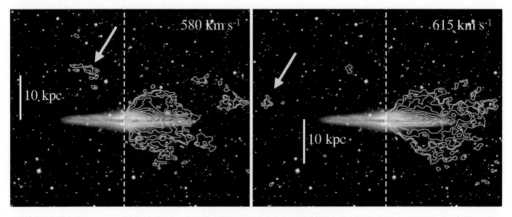

Figure 3. Counter-rotating clouds in NGC 891. These two channel maps at radial velocities 580 and 615 $\mathrm{km\,s^{-1}}$ ($v_\mathrm{sys} = 528\,\mathrm{km\,s^{-1}}$) show gas on the receding side (to the right of the vertical dashed line) of the galaxy. Two clouds are visible in isolation on the left side of the galaxy (arrows), they are counter-rotating (from Oosterloo *et al.* 2007.

If we believe that the above structures have an extragalactic origin we can estimate the rate of gas accretion by assuming typical infalling times of few $\times\,10^7 - 10^8$ yr. The resulting rates are shown in Table 1 (column 8), they are typically of the order $0.1\,M_\odot\,\mathrm{yr^{-1}}$ and generally 1 order of magnitude lower than the SFRs. These directly observed accretion rates include only H I, and they should be corrected for helium and possibly ionised gas fractions. However, it appears difficult to reconcile them with the rates of star formation (column 7, Table 1). This result is very common for nearby galaxies (Sancisi et al. 2008).

4. Indirect evidence for gas accretion

The result that the rate of gas accretion onto galaxies which is directly observed is much lower than expected implies that most of the accretion should be somewhat "hidden". I describe here possible indirect evidence of this missing gas accretion, provided by the rotation velocity gradient of the extra-planar gas. The steepness of this gradient is not reproduced by galactic fountain models (e.g. Fraternali & Binney 2006; Heald et al. 2006b) as they tend to predict shallower values (a factor half or less). Fig. 4 highlights this problem for NGC 891. The points are rotation velocities derived at heights $z = 3.9$ kpc and $z = 5.2$ kpc from the plane (Fraternali et al. 2005). Clearly the fountain clouds in the model rotate too fast (have a larger angular momentum) than the extra-planar gas in the data.

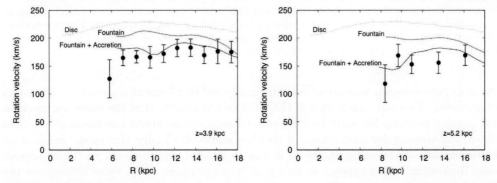

Figure 4. Indirect evidence for gas accretion. Rotation velocities (black points) of the H I extra-planar gas in NGC 891 at 3.9 kpc (left) and 5.2 kpc (right) from the plane compared to the disk rotation curve (dotted line) and predictions from two models: a pure galactic fountain model (red line above) and a model where the fountain clouds sweep up and accrete ambient gas during their passage through the halo (blue line below). The accretion rate required to produce this fit is $\sim 3 \, M_\odot \, \mathrm{yr}^{-1}$, very similar to the star formation rate of NGC 891 (from Fraternali & Binney 2008).

How can the fountain clouds loose part of their angular momentum? Fraternali & Binney (2008) consider the possibility that fountain clouds sweep up ambient gas as they travel through the halo. In this scheme ambient gas condenses onto the fountain clouds, these latter grow along their path through the halo and eventually fall down into the disk (see Fig. 5). If the ambient gas has relatively low angular momentum about the z-axis then this condensation produces a reduction in the angular momentum of the fountain gas. The only free parameter of the model is the accretion rate, which is tuned to reproduce the rotation curves of the extra-planar gas. Remarkably, the required gas accretion rate turns out to be very similar to the SFR. For NGC 891 we found a best-fit accretion rate of about $3 \, M_\odot \, \mathrm{yr}^{-1}$ (see the blue curves in Fig. 4) and for NGC 2403: $0.8 \, M_\odot \, \mathrm{yr}^{-1}$. In NGC 2403, this model is also able to reproduce the observed radial inflow of the halo gas (see Fraternali & Binney 2008).

One implication of the above fountain+accretion model is that it predicts that most of the extra-planar gas is produced by the galactic fountain and only a small fraction (about 10%) is extragalactic. This is in agreement with the metallicity of the IVCs and

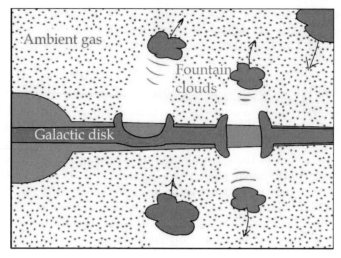

Figure 5. Schematic of the fountain gas sweeping up ambient gas. The model presented in Fraternali & Binney (2008) suggests that superbubbles blow gas into the halo which then sweeps up part of the ambient gas during its passage before falling back to the disk. This mechanism allows fountain clouds to loose part of their angular momentum as required by the data (see Fig. 4) and produces a net gas accretion at a rate similar to the SFR.

the clear links between anomalous velocity clouds and star forming regions (e.g. Boomsma *et al.* 2008). A second implication is that it does not require that the accreting gas is in any particular phase but only that its angular momentum about the z-axis is less than about half the angular momentum of the disk material. Finally, this model predicts an accretion rate of the order of the SFR and in general, proportional to the supernova rate. Interestingly, this appears to be a general requirement for galaxies throughout the Hubble time as it reconciles the observed cosmic star formation history with the gas mass in galaxies at low and high redshifts (Hopkins, McClure-Griffiths & Gaensler 2008).

References

Barbieri, C. V., Fraternali, F., Oosterloo, T., Bertin, G., Boomsma, R., & Sancisi, R. 2005, A&A, 439, 947
Blitz, L., Spergel, D. N., Teuben, P. J., Hartmann, D., & Burton, W. B. 1999, ApJ, 514, 818
Boomsma, R., Oosterloo, T. A., Fraternali, F., van der Hulst, J. M., & Sancisi, R. 2005, A&A, 431, 65
Boomsma, R., Oosterloo, T. A., Fraternali, F., van der Hulst, J. M., & Sancisi, R. 2008, A&A, in press
Bregman, J. N. 1980, ApJ, 236, 577
Chaves, T. A. & Irwin, J. A. 2001, ApJ, 557, 646
Fraternali, F., Oosterloo, T., Sancisi, R., & van Moorsel, G. 2001, ApJ, 562, L47
Fraternali, F., van Moorsel, G., Sancisi, R., & Oosterloo, T. 2002, AJ, 123, 3124
Fraternali, F., Oosterloo, T., Sancisi, R., & Swaters, R. 2005, in: ed. R. Braun, *Extra-planar Gas*, Dwingeloo, ASP Conf. Series, Vol. 331, p. 239
Fraternali, F. & Binney, J. J. 2006, MNRAS, 366, 449
Fraternali, F. & Binney, J. J. 2008, MNRAS, 386, 935
Kalberla, P. M. W., Dedes, L. 2008, A&A, in press (arXiv0804.4831)
Kewley, L. J., Geller, M. J., Jansen, R. A., & Dopita, M. A. 2002, AJ, 124, 3135
Grossi, M., Giovanardi, C., Corbelli, E., Giovanelli, R., Haynes, M. P., Martin, A. M., Saintonge, A., & Dowell, J. D. 2008, A&A, in press (arXiv0806.0412)

Heald, G. H., Rand, R. J., Benjamin, R. A., Collins, J. A., & Bland-Hawthorn, J. 2006a, ApJ, 636, 181
Heald, G. H., Rand, R. J., Benjamin, R. A., & Bershady, M. A. 2006b, ApJ, 647, 1018
Hopkins, A. M., McClure-Griffiths, N. M., & Gaensler, B. M. 2008, ApJ, 682, L13
Hulsbosch, A. N. M. 1968, BAN, 20, 33
Irwin, J. A. 1994, ApJ, 429, 618
Levine, E. S., Heiles, C., & Blitz, L. 2008, ApJ, 679, 1288
Lee, S.-W. & Irwin, J. A. 1997, ApJ, 490, 247
Lockman, F. J. 1984, ApJ, 283, 90
Matthews, L. D. & Wood, K. 2003, ApJ, 593, 721
Oosterloo, T., Fraternali, F., & Sancisi, R. 2007, AJ, 134, 1019
Oort, J. H. 1970, A&A, 7, 381
Pedersen, K., Rasmussen, J., Sommer-Larsen, J., Toft, S., Benson, A. J., & Bower, R. G. 2006, NewA, 11, 465
Rand, R. J., & Benjamin, R. A. 2008, ApJ, 676, 991
Reakes, M. L., & Newton, K. 1978, MNRAS, 185, 277
Rossa, J., & Dettmar, R.-J. 2003, A&A, 406, 505
Sancisi, R., Fraternali, F., Oosterloo, T., & van der Hulst, T. 2008, A&ARv, 15, 189
Thilker, D. A., Braun, R., Walterbos, R. A. M., Corbelli, E., Lockman, F. J., Murphy, E., & Maddalena, R. 2004, ApJ, 601, L39
Twarog, B. A. 1980, ApJ, 242, 242
Wakker, B. P & van Woerden, H. 1997, ARA&A, 35, 217
Wakker, B. P. 2001, ApJS, 136, 463
Wakker, B. P., York, D. G., Howk, C., Barentine, J. C., Wilhelm, R., Peletier, R. F., van Woerden, H., Beers, T. C., Ivezic, Z., Richter, P., & Schwarz, U. J. 2007, ApJ, 670, L113
Wakker, B. P., York, D. G., Wilhelm, R., Barentine, J. C., Richter, P., Beers, T. C., Ivezić, Z., & Howk, J. C. 2008, ApJ, 672, 298
Walterbos, R. A. M., Braun, R. 1994, ApJ, 431, 156

Joss Bland-Hawthorn urging halo gas clouds to fall onto the disk.

Adriaan Blaauw, Leo Blitz and Hans Zinnecker at Carlsberg (photo: Bruce Elmegreen).

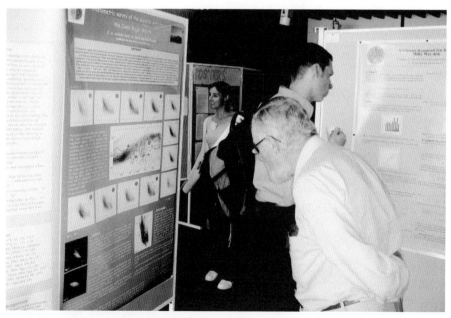
Posters were studied eagerly as well, here by Tom Lloyd Evans.

Group infall of substructures on to a Milky Way-like dark halo

Yang-Shyang Li and Amina Helmi

Kapteyn Astronomical Institute, University of Groningen,
P.O. Box 800, 9700 AV Groningen, the Netherlands
email: ysleigh@astro.rug.nl; ahelmi@astro.rug.nl

Abstract. We report the discovery that substructures/subhaloes of a galaxy-size halo tend to fall in together in groups in cosmological simulations, something that may explain the oddity of the MW satellite distribution. The original clustering at the time of infall is still discernible in the angular momenta of the subhaloes even for events which took place up to eight Gyrs ago, $z \sim 1$. This phenomenon appears to be rather common since at least 1/3 of the present-day subhaloes have fallen in groups in our simulations. Hence, this may well explain the Lynden-Bell & Lynden-Bell ghostly streams. We have also found that the probability of building up a flattened distribution similar to the MW satellites is as high as $\sim 80\%$ if the MW satellites were from only one group and $\sim 20\%$ when five groups are involved. Therefore, we conclude that the 'peculiar' distribution of satellites around the MW can be expected with the CDM structure formation theory. This non-random assignment of satellites to subhaloes implies an environmental dependence on whether these low-mass objects are able to form stars, possibly related to the nature of reionization in the early Universe.

Keywords. Methods: N-body simulations, Galaxy: formation, galaxies: dwarf, galaxies: kinematics and dynamics.

1. Introduction

The discrepancy in numbers between the substructures/subhaloes resolved in a galaxy-size cold dark matter (CDM) halo and the satellites around the Milky Way (MW) has been a long standing issue for the 'concordance' CDM structure formation theory. It implies a non-trivial mapping between the luminous satellites and the dark matter subhaloes at the (sub) galactic scales (Zentner *et al.* 2005; Libeskind *et al.* 2007) through the astrophysical processes with baryons or the theory need major modification at a fundamental level (see e.g. Kamionkowski & Liddle 2000).

In the past ten years new attention has been drawn to the dynamical properties of the MW satellites. Starting with Lynden-Bell & Lynden-Bell (1995), the existence of ghostly streams of satellites (dwarf galaxies and globular clusters) was postulated. These objects would share similar energies and angular momenta producing a strong alignment along great circles on the sky. Recently, the anisotropic distribution of satellites around the MW has been argued to be a problem for the CDM theory (Kroupa, Theis & Boily, 2005; Metz, Kroupa & Jerjen, 2007). The MW has approximately 20 satellites forming a disk-like structure while the simulated dark matter subhaloes usually distribute almost isotropically.

Here we report our findings of subhaloes falling in groups in dark matter simulations and its application to explain the oddities of the dynamical properties of MW satellites, namely, the Lynden-Bell & Lynden-Bell ghostly steams and the great MW satellites disk. Research in the past has showed that clusters of galaxies are built of galaxies coming in groups (Knebe *et al.* 2004), but it was not clear whether a similar picture also applies at

2. Substructures in a galaxy-size dark halo

2.1. The N-body Simulations

To study the dynamical properties of dark matter subhaloes, we have analysed the GAnew series of high resolution simulations of a MW-like halo in a full cosmological context ($\Omega_0 = 0.3, \Omega_\Lambda = 0.7$, $H_0 = 100h$ km s^{-1} Mpc^{-1} and $h = 0.7$). The simulations were carried out with GADGET-2 (Springel 2005) and a more detailed description on the simulation itself are reported in Stoehr (2006). In the highest mass resolution simulation (GA3new), about 10^7 particles within the virial radius resolve the MW-like halo at $z = 0$. These simulations are abundant with self-bound substructures ($\sim 4,000$ in GA3new) and the starting redshifts of the simulations are as high as $z = 37.6$, therefore rendering the simulations ideal for studies of subhalo populations and dynamics.

2.2. Group infall of dark matter subhaloes

The degree of clustering is quantified by computing the two-point 'angular correlation function', $\omega(\alpha)$, of the present-day angular momentum of the subhaloes.

$$\omega(\alpha) = \frac{N(\alpha_{ij} < \alpha)_{\text{simulation}}}{N(\alpha_{ij} < \alpha)_{\text{isotropic}}} - 1$$

The angle α is the relative orientation of the angular momenta of any two subhaloes, i.e. $\cos \alpha_{ij} = \mathbf{L}_i \cdot \mathbf{L}_j / (|\mathbf{L}_i||\mathbf{L}_j|)$. Therefore the correlation function measures the number of pairs, N, with $\alpha_{ij} < \alpha$ seen in the simulations compared to what is expected from an isotropic distribution with the same number of objects. An excess of pairs at small angular separations indicate small scale clustering in the present-day angular momentum. This clustering in the angular momentum space is still discernible even for some groups accreted about eight Gyrs ago ($z \sim 1$).

We now focus on the characteristics of groups accreted at various epochs. To identify groups we link pairs of infalling haloes whose angular momentum orientations are separated less than ten degrees, i.e., $\alpha < 10°$, and with relative distances $d < 40$ kpc at the time of accretion. We found that this combination of α and d values results in a robust set of groups, maximising their extent while minimising the number of spurious links.

We then follow the orbits of the groups identified from redshift $z \sim 4.2$ until present time. Fig. 1 shows the trajectories of some of the richest groups of subhaloes, which were accreted 2.43, 1.65 and 0.84 Gyrs ago respectively. Each dot represents the position of a subhalo colour coded from high-redshift (dark) to the present (light). The blue symbols correspond to the present-day positions while those at the time of accretion are shown in red. Fig. 1 clearly shows that groups of subhaloes follow nearly coherent orbits even long before being accreted.

Mass function of groups

We have also looked at the mass distribution of groups of subhaloes at the time of accretion. The mass function is dominated by low mass groups and can be fitted with a power-law, $dN/d\log M \propto M^n$ with $n \sim -0.5 \pm 0.15$. Note that this slope is slightly shallower compared to the mass function of the full subhalo population in a galaxy and a galaxy cluster size dark halo where $n = -0.7 - -0.9$ (e.g., Gao et al. 2004). At the limit of our simulations, we also examined the mass spectra within individual groups.

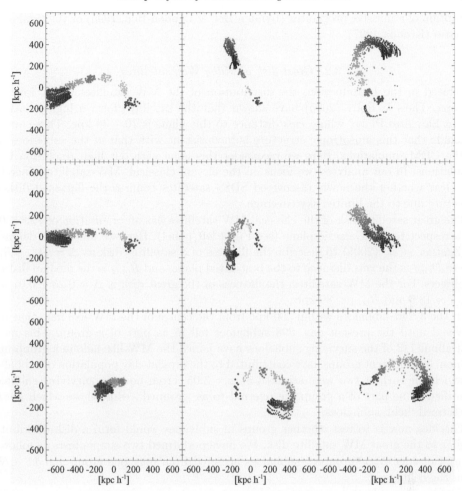

Figure 1. Three examples of trajectories of groups of subhaloes accreted at different epochs in the GA3new simulation reference frame projected onto different planes. These are some of the most abundant groups ever accreted. The colour gradients indicate the arrow of time, from dark at high redshift to light grey at the present. The positions at accretion and present time are highlighted in red and blue respectively.

The data suggests the mass distribution of subhaloes also follows a power-law trend in a group, although the signal is much noisier.

3. On the dynamical peculiarities of the MW satellites

3.1. Lynden-Bell & Lynden-Bell ghostly streams

Fig. 1 shows the group members following similar orbits and retaining the coherence for several Gyr. The clustering of subhaloes in the space is most prominent when they accrete onto the MW-like halo. But later on, the spatial coherence of groups will be destroyed due to the gravitational interactions with the host and the group members will spread along the orbit. We note that this characteristic of groups would naturally lead to structures sharing common orbital planes, analogous to Lynden-Bell & Lynden-Bell ghostly streams. Thus group infall provides an explanation clearly different from the disruption of a massive progenitor (Lynden-Bell & Lynden-Bell, 1995) or the tidal-origin scenario (Kroupa 1997).

3.2. Great disk of Milky Way satellites

As stated in the Introduction, the distribution of the MW satellites is not isotropic. Kroupa, Theis & Boily (2005) have shown that the satellites form a highly flattened 'Great Satellites Plane' whose *rms* distance to this plane is 10 − 30 kpc. These authors conclude that this anisotropic structure is inconsistent with that of the subhaloes seen in the CDM simulations. Here we reconsider this issue with our high resolution CDM simulations. In our analyses, we focus on the eleven 'classical' MW satellites since it is not clear whether the newly discovered SDSS satellites confirm the flattened disk-like structure due to the limited sky coverage.

The great satellite disk of the classical MW satellites has an orientation of $72.8 \pm 0.7°$ with respect to the Galactic plane (see Fig. 2 left panel). Here we follow the definition by Zentner et al. (2005) to describe the flatness of a satellites disk as $\Delta = D_{rms}/R_{med}$ where D_{rms} is the *rms* distance to the best-fitted plane, and R_{med} is the median distance of objects. For the MW satellites, the flatness of the great disk is $\Delta = 0.23 \pm 0.01$ with $D_{rms} \sim 18.5$ and $R_{med} \sim 80$ kpc.

Of the 3,246 subhaloes within 300 kpc from the centre of the MW-like halo that have survived until the present day, 898 subhaloes fell in as part of a group. This means that about 1/3 of the surviving subhaloes have joined the MW-like halo in a group-infall fashion. 321 different groups have contributed to the present-day population of subhaloes, of which the earliest two were accreted at $z = 3.05$. From now on, surviving subhaloes identified to be part of a group are referred to as 'grouped', while those which are not are termed 'field' subhaloes.

The idea now is to test whether groups of subhaloes would form a disk distribution similar to the great MW satellite disk. We have performed two simple tests as follows:

(a) Consider N_{sub} 'grouped' subhaloes accreted from only one group and $11 - N_{sub}$ from the field.

(b) Consider only the 'grouped' subhaloes originated in a few groups.

We generate 10^5 sets of eleven subhaloes from the present-day population which satisfy either of the above mentioned conditions and compute the flatness of the best-fit planes. The fraction of disks is then the number of sets with flatness as low as that of the great satellites disk, i.e., $N(\Delta \leqslant \Delta_{\rm GSD})$, normalised by the total number of realisations. The result shows that when considering case (a), the disk fraction increases from 4.5% to 73% as the number of selected subhaloes N_{sub} increases from 2 to 11. In comparison, 11 randomly-selected subhaloes (within 300 kpc) gives rise to flattened configurations $\sim 2.2\%$ of the time. This shows that if the Milky Way satellites fell in together, it would not be very surprising that they would be in a planar configuration at the present-day.

When considering only subhaloes originating in groups (case b), the fraction of disk-like configurations obtained in this way can be as high as $\sim 40\%$ when the subhaloes come from only two groups, and of course reaches 73% when they come from just one group. It is important to note that though the disk fraction decreases to $\sim 20\%$ when selecting from 5 different groups, it is still much higher than if one selects 11 subhaloes randomly. On the right panel of Fig. 2, we show an example of the distribution of three groups of subhaloes with 3, 4, and 4 group members respectively projected along their best-fitted plane. This distribution has $\Delta \sim 0.17$.

Given the large fraction of flattened configurations found in our simulations, we conclude that the spatial distribution of the 11 Milky Way satellites can be reproduced

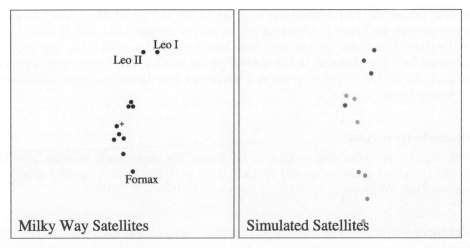

Figure 2. *Left:* Illustration of the distribution of the eleven classical Milky Way satellites where the cross sign denotes the Galactic centre. The Galactic disk would be horizontal in this orientation. *Right:* An example of the distribution of satellites in our simulations which can be traced back to come from three different groups as marked by different colours.

within ΛCDM. The requirement is that these satellites fell onto the Galactic halo in groups.

4. Conclusions and Implications

We have revisited the issue of the peculiar distribution and properties of the MW satellites and their link to the dark matter subhaloes. In particular, we have focused on the infall of substructures on to a Milky-Way like dark matter halo in a ΛCDM cosmogony utilising a series of high-resolution dark-matter simulations. We have found evidence of group infall on to the MW-like halo, which may explain the ghostly streams proposed by Lynden-Bell & Lynden-Bell (1995).

We have also explored how this planar configuration may be obtained as a result of the infall of satellites in groups. The observed correlation in the angular momentum orientation of subhaloes naturally gives rise to disk-like configurations. For example, we find that if all subhaloes are accreted from just one group, a disk-like distribution is essentially unavoidable ($\sim 80\%$ probability), while for accretion from just two groups, the likelihood of obtaining a distribution as planar as observed is 40%. Therefore the disky configuration of satellites is consistent with CDM if most satellites have their origin in a few groups. Note that in our studies, we do not need to invoke the baryon-related physics to account for the dynamical properties of the MW satellites. Thus both the 'ghostly streams' and the 'planar configuration' are manifestations of the same phenomenon: the hierarchical growth of structure down to the smallest galactic scales.

One of the possible implications of the reality of the ghostly streams is that its member galaxies formed and evolved in a similar environment before falling into the MW potential. This would have implications on the (oldest) stellar populations of these objects, such as for example, sharing a common metallicity floor (Helmi *et al.* 2006). On the other hand, this implies that there should be groups that have failed to host any luminous satellites. This would hint at a strong dependence on environment of the ability of a subhalo to retain gas (Scannapieco *et al.* 2001), or be shielded from re-ionization by nearby sources (Mashchenko, Carignan & Bouchard 2004; Weinmann *et al.* 2007).

Recent proper motion measurements of the Large and Small Magellanic clouds by Kallivayalil, van der Marel & Alcock (2006), as well as the simulations by Bekki & Chiba (2005) suggest that these systems may have become bound to each other only recently. This would be fairly plausible in the context of our results. The Clouds may well have been part of a recently accreted group and it may not even be necessary for them to ever have been a binary system.

Acknowledgements

YSL thanks the organising committee for giving the opportunity to contribute this talk and the travel grant supported by IAU. This work has been supported by a VIDI grant from the Netherlands Foundation for Scientific Research (NWO).

References

Bekki, K. & Chiba, M. 2005, *MNRAS*, 356, 680
Gao, L., White, S. D. M., Jenkins, A., Stoehr, F., & Springel, V. 2004, *MNRAS*, 355, 819
Helmi, A. *et al.* 2006, *ApJ*, 651, L121
Kallivayalil, N., van der Marel, R. P., & Alcock C. 2006, *ApJ*, 652, 1213
Kamionkowski, M. & Liddle, A. R. 2000, *Phys. Rev. Lett.*, 84, 4525
Knebe, A., Gill, S. P. D., Gibson, B. K., Lewis, G. F., Ibata, R. A., & Dopita, M. A. 2004, *ApJ*, 603, 7
Kroupa, P., Theis, C., & Boily, M. 2005, *A&A*, 431, 517
Kroupa, P. 1997, *New Astron.*, 2, 139
Li, Y.-S. & Helmi A. 2008, *MNRAS*, 385, 1365
Libeskind, N. I., Cole, S., Frenk, C. S., Okamoto, T., & Jenkins, A. 2007, *MNRAS*, 374, 16
Lynden-Bell, D. & Lynden-Bell, R. M. 1995, *MNRAS*, 275, 429
Metz, M., Kroupa, P., & Jerjen, H. 2007, *MNRAS*, 374, 1125
Mashchenko, S., Carignan, C., & Bouchard, A. 2004, *MNRAS*, 352, 168
Scannapieco, E., Thacker, R. J., & Davis, M. 2001, *ApJ*, 557, 605
Stoehr, F., White, S. D. M., Tormen, G., & Springel, V. 2002, *MNRAS*, 335, L84
Springel V. 2005, *MNRAS*, 364, 1105
Stoehr, F. 2006, *MNRAS*, 365, 147
Weinmann, S. M., Macciò, A. V., Iliev, I. T., Mellema, G., & Moore, B. 2007, *MNRAS*, 381, 367
Zentner, A. R., Kravtsov, A. V., Gnedin, O. Y., & Klypin, A. A. 2005, *ApJ*, 629, 219

Modelling the Disk (three-phase) Interstellar Medium

Gerhard Hensler

Institute of Astronomy, University of Vienna, Türkenschanzstr. 17, 1180 Vienna, Austria
email: hensler@astro.univie.ac.at

Abstract. The evolution of galactic disks from their early stages is dominated by gas dynamical effects like gas infall, galactic fountains, and galactic outflows, and more. The influence of these processes is only understandable in the framework of diverse gas phases differing in their thermal energies, dynamics, and element abundances. To trace the temporal and chemical evolution of galactic disks, it is therefore essential to model the interstellar gasdynamics combined with stellar dynamics, the interactions between gas phases, and star-gas mass and energy exchanges as detailed as possible. This article reviews the potential of state-of-the-art numerical schemes like Smooth-Particle and grid-based hydrodynamics and their ingredients, such as star-formation criteria and feedback, energy deposit and metal enrichment by stars, and the influence of gas-phase interactions on the galactic gas dynamics and chemistry.

Keywords. ISM: kinematics and dynamics, ISM: structure, Galaxy: disk, Galaxy: evolution, galaxies: evolution, galaxies: ISM

1. Introduction

Galactic disks must have formed by dissipation and cooling of the galactic gas that remained after the first star-formation (SF) episode in the spheriodal component of galaxies and was mixed with the returned and metal-enriched stellar matter. Cooling enforced the further collapse, and angular momentum kept the radially extended disk rotationally supported. Since then the disk must have been continuously fed and refreshed by infalling intergalactic gas, but also depleted by galactic outflows. As known from our Milky Way since the 70's, but now with the availability of new spectral ranges (from the FIR/submm to X-rays) and also perceived in other disk galaxies, the Interstellar Medium (ISM) consists of a multi-phase gas (Ferrière 2001), conditioning SF in the cool molecular components and driving the ISM dynamically, mainly through its very hot component.

Besides external effects, structure and evolution of galaxies are primarily determined by the energy budget of its ISM. Reasonably, also the ISM structure and its radiative and kinetic energy contents are both determined by the energy deposit of different sources on the one hand, and by the energy loss through radiative cooling on the other.

The state of the ISM is thus not an isolated and meaningless question. Since stars are born from the cool gas phase, but energize the ISM by means of their energy and mass release, the treatment of the ISM must ultimately include also the stellar component with its distribution in space, age, and metallicity. Disk formation and evolution thus depend intimately on the SF and the energy content, so that numerical simulations must account for these facts properly. These are, however, the most challenging tasks for theoretical and numerical astrophysics of the ISM and are fundamental for galaxy evolution.

Why is this so? On the one hand, the SF rate in disks obeys the well-known Schmidt-Kennicutt law (Kennicutt 1998) that relates the Hα surface brightness of disks with the gaseous surface density Σ_g; on the other hand, we know from theory and empirical results how much energy is released by stars during their different evolutionary stages. So, where is the problem? Is it not sufficient to relate the SF rate empirically to gas mass from studies of galaxies in the present local universe? Why should a detailed treatment of the ISM and its particular and temporal composition be necessary, not just a large-scale average of the state of the ISM?

In the following, let us reflect and comment on the present state of the numerical treatment of the multi-phase ISM with its dynamics, its SF and energy feedback, its necessary small-scale physical interaction processes, and the application to galaxy evolutionary models. After having explained the most important effects of the ISM on galaxy evolution and their empirical issues from observations, we will consider representative numerical developments and discuss their drawbacks and advantages without claiming completeness. Finally, we discuss the most advanced numerical efforts relative to the requirements for a fully realistic approach. Nevertheless, let me concede already in advance that the influence of magnetic fields and of Cosmic Rays are not yet included, and not even explored to first order in simulations.

2. Structural and evolutionary effects on the Interstellar Medium

2.1. Star formation and the Schmidt-Kennnicutt Law

Already in 1959 Schmidt argued that the SF rate per unit area relates to surface gas density Σ_g by a power law with exponent n. The Kennicutt (1998) relation

$$\Sigma_{H\alpha}[M_\odot/(yr \cdot pc^2)] \propto \Sigma_g[M_\odot/pc^2]^{1.4 \pm 0.15} \tag{2.1}$$

established that in equilibrium states the amount of gas determines the SF rate. Although this connection looked already pretty tight and holds over more than 4 orders of magnitude in Σ_g with a lower cut-off at almost 10 $M_\odot/pc^2$, recent advances in observation and theoretical interpretation are more confusing. Heyer et al. (2004) showed that this exponent is only valid for the molecular gas and that the correlation is tighter, while for pure H I the slope is steeper. Gao & Solomon (2004) derived the dependence of the SF rate on the dense molecular cloud cores traced by the HCN molecule.

Such a Kennicutt relation can be theoretically simply supported as

$$\Sigma_{SF} = \frac{\Sigma_g}{\tau_{SF}} \propto \frac{\Sigma_g}{\rho_g^{-0.5}} \propto \Sigma_g^{1.5} \tag{2.2}$$

if the scaleheight H is independent of Σ_g, i.e. the gas temperature is constant, so that $\Sigma_g \propto \rho_{g,0} \propto \rho_g$. This can only be understood if heating and cooling both are independent of volume gas density ρ_g or have the same dependence on it. Köppen, Theis, & Hensler (1995), however, demonstrates that the SF rate depends on ρ_g^2, if the stellar heating is compensated by collisionally-excited cooling radiation (according to Böhringer & Hensler 1989). The coefficient of this relation determines the SF timescale τ_{SF}.

But how can this be described? Elmegreen (2002) argued that the SF is determined by the free-fall time τ_{ff}. This, however, raises a conflict between the ISM conditions and observed SF rates in the sense that τ_{ff} for a typical density of 100 cm^{-3} amounts to 10^{14} sec, i.e. $3 \cdot 10^6$ yrs. For a galactic molecular cloud mass of $10^9 ... 10^{10}$ $M_\odot$ the SF rate should amount to $10^2 ... 10^3 M_\odot/yr$ – orders of magnitudes higher than observed! This means that the SF timescale must be reduced with respect to the collapse or dynamical timescale by introducing an efficiency ϵ_{SF}; its definition could be : $\tau_{SF} = \epsilon_{SF}^{-1} \cdot \tau_{ff}$. Nevertheless, Li, MacLow, & Klessen (2005), from numerical models of gravitationally unstable clouds, also obtained a Kennicutt law slightly steeper than $n = 1.4$.

Since this Schmidt-Kennicutt law is derived for gravitationally settled gas disks in rotational equilibrium, it should also reflect an equilibrium state of a self-gravitating vertically stratified ISM balanced by heating and cooling. It is therefore not surprising that the exponent is larger for starburst galaxies, i.e. if the SF is e.g. triggered by dynamical processes, than in a quiescent mode, so that the SF efficiency is increased. This also explains why high-z galaxies reveal a steeper relation with an exponent $n = 1.7 \pm 0.05$ (Bournaud, Elemegreen, & Elmegreen (2007)), i.e. at stages when the gaseous disks form preferably by gas infall.

One would reasonably expect that the Kennicutt relation can only be reproduced in self-regulated disk models, i.e. in those where energy feedback by stars determines the energy budget of the ISM. Kravtsov (2003) identified disk galaxies in his cosmological simulations and analyzed them at different high redshifts. His model parameters covered the large range observed in Σ_g, and the model values scatter around the Kennicutt slope of Σ_{SF} with Σ_g. However, surprisingly, models without energetic feedback by supernovae (SNe) did not differ from those with feedback. Similar conclusions can be drawn from the simulations by Schaye & Dalla Vecchia (2008), who neglect stellar feedback processes while varying parameters like SF efficiency, critical SF mass, etc.

It is therefore still a matter of debate whether the Kennicutt-like dependence of the surface SF rate is really a result of self-regulated SF in gas disks, or if any dependence of self-regulation processes on the physical gas state cancel out, so that the relation is really a global one.

For modelling the formation and evolution of galactic disks, however, the column density is meaningless in systems still developing dynamically. As long as the disk has not achieved its equilibrium state for the gas it is therefore necessary, and from a physical standpoint reasonable, to link the SF rate to the local state within a volume. (By the way, the same should hold for the modelling of gaseous dwarf galaxies (DGs) which are always in the state of re-arranging their gas structure because of the low gravitational potential.) This, however, means that the SF rate in a gas volume, $\partial \rho_{SF}/\partial t$, should be connected to the gas volume density ρ_g by a power-law of exponent m. But what is the value of m? Rana & Wilkinson (1986) found a wide range of $m = 1 \ldots 4$. Similar to the above derivation of the Kennicutt exponent, one can formulate

$$\frac{\partial \rho_{SF}}{\partial t} = \frac{\rho_g}{\tau_{SF}} \propto \frac{\rho_g}{\rho_g^{-0.5}} \propto \rho_g^{1.5} \qquad (2.3)$$

The proportionality again includes the SF efficiency. Such a 1.5 power-law was already applied for this reason in the first gas dynamical simulations of galaxy evolution by Larson (1969 and in the 70's) and Burkert & Hensler (1987, 1988). Under the assumption of triggered SF due to cloud-cloud collisions, a square dependence on ρ_g should apply. In a simple picture, moreover, the cloud collision rate depends not only on the cloud number density but also on the clouds' velocity dispersion, and can thereby directly affect the SF rate as a positive energetic feedback. Since dissipative collisions, however, lead to a heating of the cloudy material, and if the SF rate depends on the temperature of the cloudy ISM component, also a negative feedback is implied. Therefore, a competition between collisional heating and gas cooling will determine the sign of the SF feedback on energy and thus depends on the state and composition (metallicity, dust content, molecules, etc.) of the ISM.

Concerning the balance between direct stellar energy release to the ISM and collisionally-excited radiative cooling, Köppen, Theis, & Hensler (1995) found that for the star-forming site itself, self-regulation sets the dependence of the SF rate on ρ_g^2. In addition, the production of a SN-heated hot gas phase and its interaction with the cool gas by heat conduction leads to a limit-cycle SF rate around a characteristic temperature due to the competition between evaporation and condensation (Köppen, Theis & Hensler 1998).

Consequently, the energy release and the efficiency of its transfer into dynamics and thermalisation are of evident importance. If their description and parametrization as well as a Schmidt-dependent SF rate, respectively, are treated properly in numerical simulations, i.e. close to real conditions, a Kennicutt dependence for a gaseous galactic disk should inherently be reproduced.

2.2. *Stellar energy release and its transfer efficiency*

Because of their high power of released energy in various forms, massive stars are usually included as the only heating sources for the ISM. Moreover, a large fraction of simulations focuses the energy deposit on SN typeII explosions alone. It must be discussed, however, how effectively this energy is transferred into the ISM as turbulent and hence thermal energy, and an even more sophisticated question must be asked on the influence, transport and dissipation by Cosmic Rays and magnetic fields.

The same must be considered for the energy release of massive stars as radiation-driven and wind-blown HIIregions. Analytical estimates for purely radiative HIIregions yielded an energy transfer efficiency ϵ of the order of a few percent (Lasker 1967). Although the additional stellar wind power L_w can be easily evaluated from models and observations, the fraction that is transferred into thermal energy (ϵ_{th}) or into turbulent energy is not obvious from first principles. Transfer efficiencies for both radiative and kinetic energies remain much lower than analytically derived from detailed numerical simulations (more than one order of magnitude) and amount to only a few per mil (Hensler 2007 and references therein; Hensler *et al.* 2009, in preparation). Surprisingly, there is almost no dependence on the stellar mass, (Hensler 2007) as expected because the energy impact by Lyman continuum radiation and wind luminosity both increase non-linearly with stellar mass. Conversely, since the gas compression is stronger by a more energetic wind, the energy loss by radiation is plausibly also more efficient.

Another fact is the energization of the ISM by SNe. Although it is generally assumed that the explosive energy lies around 10^{51} ergs with an uncertainty (or intrinsic scatter) of probably one order of magnitude, SNeII do not explode as isolated and single events in the ISM. Almost without exception, massive stars remain close to the SF site and thus explode within their stellar associations. Massive stars contribute significantly to the formation of structures, like e.g. cavities, holes, and chimneys in the HI gas and superbubbles of hot gas (Recchi & Hensler 2006a). On large scales the energy release by massive stars triggers circulation of matter via galactic outflows from a gaseous disk and galactic winds. By this, also the chemical evolution is affected through the loss of metal-enriched gas from a galaxy (see e.g. Recchi & Hensler 2006b).

SN explosions as an immediate consequence of SF stir up the ISM by the expansion of hot bubbles, deposit turbulent energy into the ISM and, thereby, regulate the SF again (Hensler & Rieschick 2002). This negative energy feedback is enhanced at low gravitation, because the SN energy easily exceeds the galactic binding energy and drives a galactic wind. This low gravitation also exists in vertical direction of a rotationally supported disk. Conversely, SN and stellar wind-driven bubbles sweep up surrounding gas and can thereby excite SF self-propagation as a positive feedback mechanism (e.g. Ehlerova *et al.* 1997, Fukuda & Hanawa 2000).

Since the numerical treatment of the galactic chemo-dynamical evolution, even when it deals with two gas phases, cannot spatially resolve the ISM sufficiently, for a detailed consideration of small-scale processes the chemo-dynamical modeling has to apply parametrizations of plasma-physical processes. It should, however, be repeatedly emphasized that an appropriate chemo-dynamical description cannot vary those parameters arbitrarily, because they rely on results from theoretical, numerical, and/or empirical studies and are thereby strongly constrained. This has been described and applied to different levels of chemo-dynamical models (for an overview see Hensler 2003 and references therein, and more recently Hensler, Theis & Gallagher 2004a; Harfst, Theis & Hensler (2006)).

Investigations have been performed for ϵ of SNe (Thornton *et al.* 1998) and superbubbles (e.g. Strickland *et al.* 2004), but they are yet too simplistic for quantitative results, and the 10% efficiency derived by Thornton *et al.* is far too large. Although numerical experiments of superbubbles and galactic winds in galaxies are performed, yet they only demonstrate the destructive effect on the surrounding ISM, but lack self-consistency and complexity.

Simulations of the chemical evolution of starburst DGs by Recchi *et al.* (2006), aimed at reproducing the peculiar abundance patterns in these galaxies by different SF episodes, found that ϵ can vary widely. A superbubble which acts against the surrounding medium is cooling due to its pressure work and radiation, but compresses the swept-up shell material and imparts significant turbulent energy to the ISM. If a closely following SF episode (maybe another burst) pushes its explosive SNeII into the already formed chimney, the hot gas can easily escape without hindrance. Recchi & Hensler 2006a found that, depending on the external HI density, the chimneys do not close before a few hundred Myrs.

2.3. *Gas Infall*

Formation and evolution of galactic disks cannot be considered in isolation. As confirmed from observations, but also required in the ΛCDM framework, galaxies grow continuously during their evolution by accreting surrounding gas. Particularly during the early formation phases it still remains unclear, however, how this process works. Cold and hot accretion are proposed (see e.g. Keres *et al.* 2005, Wise & Abel 2007), both dependent on mass and providing different inherent timescales of the galaxy growth, but no clue yet exists which one dominates.

In the local Universe several disk galaxies are enveloped by huge HI halos. Refined velocity-position maps reveal cold accretion in the form of HI clouds with masses of about $10^7 M_\odot$ (Fraternali *et al.* 2002; Fraternali, this volume, and e.g. Sancisi *et al.* 2008). Since the infall rate determined by the detectable clouds fails to support the SF rate in those galaxies by almost one order of magnitude, it must be concluded that most of the infall happens by less massive clouds, perhaps like those observed in our Milky Way.

Although the element abundance properties of most Blue Compact DGs (BCDs) favour a pure young stellar population of at most 1-2 Gyr, they mostly consist of an underlying old population. Furthermore, most of these objects are also embedded into HI envelops from which at least NGC 1569 definitely suffers gas infall (Stil & Isreal 2002, Mühle *et al.* 2005). This lead Köppen & Hensler (2005) to exploit the influence of gas infall with metal-poor gas into an old galaxy with continuous SF on particular abundance patterns. Their results could match not only the observational regime of BCDs in the [12+log(O/H) -log(N/O)] space but also explain the shark-fin shape of the data distribution. Knauth *et al.* (2006) also found peculiar N/O abundance ratios in the galactic ISM, which could be explained by the infall scenario of Köppen & Hensler (2005).

The accumulated explosion of massive stars lead to the formation of a superbubble which expands out of the gaseous disk, preferentially along the steepest density gradient. In low-density environments when the thermal energy does not exert too much PdV work or lose significant radiative energy, this expansion feeds the galactic halo with hot gas. On the other hand, H I observations have long confirmed that high- und ultra-high velocity clouds of extragalactic origin fall towards the Galactic disk, demonstrating that the halo consists of a multi-phase ISM (Wolfire *et al.* 1995). Interstellar clouds with moving relative to the surrounding gas are deformed and stretched by the ram pressure. Heyer *et al.* (1996) showed the cometary deformation of clouds exposed to the outflow of hot gas in the SF region W4. Infalling clouds are not only influenced by a drag force of the hot halo gas because of its relative velocity and thereby decelerated (Benjamin & Danly 1997), the thermal interaction between the gas phases through heat conduction also affects their survival (Vieser & Hensler 2007). Due to their possible destruction and gas dispersion in hot environments, it is in general far from clear how many of the infalling high-velocity clouds really reach the gaseous disk, feed the SF, and excite turbulence, neither in the early galaxy growth phase nor during the present evolution.

2.4. *Galactic Winds*

Some disk galaxies, like e.g. NGC 253 and M82, not only supply hot halo gas but can also drive a galactic wind. This scenario can be observed in various types of galaxies and in different evolutionary stages. In the Galactic disk cavities and chimneys become visible in H I , and in external galaxies H I holes appear. In DGs, superbubbles originating from super star clusters (SSCs) are observable as closed Hα loops and in X-rays still confined to the galaxy, like in NGC 1705 (see e.g. Hensler *et al.* 1998). NGC 1569 is a most interesting object: not only does gas infall occur and probably triggers the present starburst (Mühle *et al.* 2005), but a strong galactic wind also exists. An analysis of the wind revealed a metal content only $\sim$1-2 times solar, i.e. much lower than expected from the yields by massive stars (Martin, Kobulnicki & Heckman 2002). Since this low value requires dilution with low-metallicity gas by a factor of about 10, turbulent mixing alone in the shells of the expanding superbubble with the low-Z ISM in NGC 1569 seems insufficient. More plausibly, gas clouds falling in from the enveloping H I reservoir evaporate and thereby mass-load the galactic wind.

2.5. *Structural issues*

Summarizing the above-mentioned observational facts, one must be aware that:
- the multi-phase structure of the ISM exists in any galactic region;
- the mode of SF and its positive or negative energetic feedback depend on the small-scale physics of the ISM; and
- the reliability and significance of simple, single-phase hydrodynamical models, even with feedback, are highly questionable.

3. Numerical Methods

3.1. *Star-formation prescriptions*

There are vigorous efforts to model the ISM on small scales in star-forming clouds in order to achive a deeper insight into the mechanisms controlling SF (see e.g. Ostriker, Stone, & Gammie 2001, MacLow & Klessen 2004, Li, MacLow & Klessen 2005, Krumholz & McKee 2005). Numerical simulations on intermediate scales must already prescribe the SF by some reasonable recipe. Numerous papers (see e.g. Wada & Norman 1999, de Avillez & Breitschwerdt 2004, Slyz *et al.* 2005, Tasker & Bryan 2008) have implemented SF criteria like e.g. threshold density ρ_{SF} with stars forming if $\rho_g \geqslant \rho_{SF}$, excess mass in a specified volume with respect to the Jeans mass, i.e. $M_g \geqslant M_J$, convergence of gas flows ($div \cdot \mathbf{v} < 0$), cooling timescale $t_{cool} \leqslant t_{dyn}$, temperature limits $T_g \leqslant T_{lim}$, Toomre's Q parameter, and/or temperature dependent SF rate. If at least some of these conditions are fulfilled, as a further step the gas mass which is converted into stars, i.e. the SF efficiency, has to be set. An empirical value of ϵ_{SF} is fixed from observations, i.e. $\Delta m_{SF} = \epsilon_{SF} \cdot \rho_g \cdot \Delta x^3$, where Δx^3 is the mesh volume in a grid code. If the numerical timestep Δt is smaller than the dynamical timestep τ_{ff}, Δm_{SF} must be weighted by their ratio (Tasker & Bryan 2008). Since ϵ_{SF} must inherently depend on local conditions, such that it is high in bursting SF modes as required for the Globular Cluster formation, but low (a few percent) in the self-regulated SF mode, numerical simulations often try to derive a realistic SF efficiency by comparing models of widely different ϵ_{SF} with observations (see e.g. Tasker & Bryan 2008: $\epsilon_{SF} = 0.05$ and 0.5).

While the Kennicutt law (Schaye & Dalla Vecchia 2008) is directly applied to the SF rate in many numerical treatments as the simplest prescription, another approach is a formulation based on eq. (2.3), like

$$\frac{\partial \rho_{SF}}{\partial t} = \alpha G^{1/2} \rho_g^{3/2} \qquad (3.1)$$

and to express α using the Kennicutt law again and applying an exponential disk with scaleheight Z_0 by

$$\alpha = 1.96 \cdot G^{-1/2} \frac{\Sigma_{SF}}{\Sigma_g^{3/2}} Z_0^{1/2} \qquad (3.2)$$

But, unreasonably, this is even applied to interacting galaxies, (Springel 2000) where both the application of the Kennicutt law and the usual definition of scale height is highly questionable, and in my opinion fails.

Theoretical studies by Elmegreen & Efremov (1997) derived a dependence of ϵ_{SF} on the external pressure, while Köppen, Theis, & Hensler (1995) explored a temperature dependence of the SF rate.

3.2. Smooth particle hydrodynamics

Because of its simple numerical treatment and inherent 3D representation, *Smooth Particle Hydrodynamics* (SPH) is at present the most widely applied numerical strategy for simulations of cosmological structure formation, galaxy formation, galaxy collisions and mergers, and so forth. Because of their particle character, SPH simulations easily allow simultaneous treatment of the gas and the stellar component. The SPH subdivides the gaseous component in gas packages of sizes that represent the continuum character of diffuse gas, i.e. by large spherical extents of the SPH particles, expressed by a kernel function W_{ij} so that the mass density at particle i is given through the neighbouring particles j with mass m_j by

$$\bar{\rho}_i = \sum_{j=1}^{N} m_j \cdot W_{ij}.$$

Although SPH is a quite powerful formalism for 3D problems, is has several drawbacks: 1) On the one hand, representing gas phases which change their scalelengths by orders of magnitude within a short range, i.e. adjacent hot and cool particles, is impossible. Unfortunately, many authors are misled in such applications. 2) Ad hoc assumptions on SF and feedback are necessary. As an example: over which range, i.e. over how many adjacent particles, must the SNII energy feedback be taken into account? If the mass and spatial resolution of SPH particles is insufficient, the mean thermal energy of a gas particle is forced to not increase due to SNeII, but will remain at about 10^4 K due to efficient cooling. The coarse treatment is therefore insensitive to self-regulation, and thereby cannot account for a hot gas phase. 3) Multi-fluid descriptions are almost impossible. If one decouples two kinds of particles totally, they can hardly experience the proximity of each other. Or one allows for different states but one kind of particle, so that each particle acts to its adjacent neighbour as a continuous fluid.

Because these weaknesses, which are mostly caused by the single gas-phase description and the lack of interaction processes, have been recognized by many modellers, mainly also in the field of cosmological simulations, numerous attempts to overcome them have been performed, developing various strategies for the treatment of a multi-phase ISM. Since GADGET is probably the most widely distributed and -applied publicly available SPH code (Springel 2000, Springel, Yoshida, & White 2001) let us consider specifically its application to disk galaxies and its advancements.

Dalla Vecchia & Schaye (2008) applied GADGET to study the impact of SN feedback on galactic disks in order to drive a galactic wind, and conversely the effect of the wind strength on the SF. The latter is described by the Kennicutt law combined with a density threshold. For massive galaxies with DM mass of $10^{12} M_\odot$, the mass resolution of gas particles extends down to $5 \cdot 10^4 M_\odot$. The SNII energy is deposited as kinetic wind energy, and wind particles are stochastically selected among neighbouring particles. In some models these are hydrodynamically coupled to the normal gas particles. As expected, the models show that galactic winds are stronger for larger energy release and, on the other hand, reduce the SF rate, but not in a self-consistent manner, because neither cold-gas infall nor SF trigger is included. In addition, in models with a wind the SF rates decrease by about a factor of two already within 0.5 Gyrs, which is much too short for derived SF timescales of disk galaxies.

In another recent development, Scannapieco et al. (2006) use a very sophisticated differentiation between the SPH particles into cold and hot ones, but still treat them adequately, except that particles with different thermodynamic variables (cold vs. hot) do not feel each other. The recipe to allow for SNII energy to be distributed into the cold and hot particles, but not cooled away in the cold ones, is the possibility of a so-called *promotion* of cold particles according to the state of their hot neighbors. Scannapieco et al. (2006) claim that even mass-loaded galactic winds can develop in these models. Nevertheless, SF criteria and efficiency as well as the distribution of the SN energy are freely parametrized and adapted to galaxies in the local Universe observed in equlibrium states.

In order to describe the different ISM phases by at least two particles, also differently treated, Semelin & Combes (2002) introduced four different particle species: warm gas as SPH particles, stars and DM as collisionless particles, and cold gas as sticky particles. Mass exchange between the cold and warm gas is limited by the transformation by heating from SN feedback vs. cooling. The cool particles are not intended to represent a diffuse gas distribution, but instead the clumpiness of the cloudy ISM component, and thus subject to individual cloud-cloud collisions with only partial inelasticity.

Further mass transfer between the components is allowed by SF and stellar mass loss. While the SF rate is described by a Schmidt law of the cool gas volume density with a power of 1 to 1.5, the mass fraction of formed stars is at first trapped within the cool gas particle. Not until a sufficient amount of stars exists in a specified number of neighbouring particles are these stellar masses subtracted from the gas particle and collected in a new, pure stellar particle. Semelin & Combes (2002) applied an advanced procedure to keep the particle number almost constant. Further mass exchange is achieved by the stellar mass return according to the stellar evolution and death rate. The particles are numerically treated by the so-called TREE algorithm (Barnes & Hut 1986).

Although this approach looks well justified and explored, the illustration of the best model over 2 Gyrs in Semelin & Combes (2002) shows that it obviously suffers from an unrealistically high SF rate. The gas is consumed to 20% within this short time and the SF rate has dropped from almost 100 $M_\odot$/yr to less than 5. The corresponding e-folding timescale of SF amounts to only less than 0.5 Gyrs, which stands in contradiction to the observationally derived value, e.g. in the Milky Way.

This model behaviour of a too intensive SF might have different reasons: no dynamical, energetic and matter interactions are implied into the numerical treatment. Dynamically, the drag exerted by hot gas flow against cool clouds is not sufficiently represented by only the thermal pressure gradient built up by the local SN energy release and driving the expansion of hot gas. From the energetic point of view, heat conduction transfers energy from the hot to the cool clouds and thereby regulates the SF (Köppen, Theis & Hensler 1998). In addition, heat conduction also can lead to partial evaporation of clouds and thus reduce the cloud mass reservoir available for SF.

Surprisingly, even recently the formation of exponential disks has been modelled by Bournaud, Elemegreen, & Elmegreen (2007) applying another speciality of algorithms, a particle-mesh + sticky-particle scheme. Although they dealt with the enormous number of one million particles for DM, gas and stars, their limitation to a single gas phase without any stellar feedback are much too severe to allow reliable results, in particular, with respect to the aimed issues of this exploitation. The author aimed at understanding the evolution of clump-clusters and chain galaxies by fragmentation of gaseous disks in gravitational unstable clumps. The reason for this neglect, that the feedback from SF "is not well known" and that the clump masses lie much above the influential range of SN energy, is misleading as far as it has not been tested and proven that the coupling of energetic processes between scales is really inefficient.

All the above mentioned additional processes, acting among the gas phases and, by this, coupling them even on different scales are additionally incorporated in the SPH/sticky particle treatment by Harfst, Theis & Hensler (2006). Moreover, in their model the cloudy particles show an inherent mass distribution according to their stickiness, and SF can happen in a "normal" mode, but also be induced by cloud collisions. In addition to the temperature criterion by Köppen, Theis, & Hensler (1995), the efficiency by Elmegreen & Efremov (1997) controls the SF. As a result, Harfst, Theis & Hensler (2006) demonstrated a stabilized rotating disk with moderate SF over 2 Gyrs, starting with a present-day realistic value of 3 $M_\odot$/yr for a gaseous disk and dropping off only by a factor of 4, which means a SF timescale of more than 1 Gyr. As argued by them, the reduction in the SF rate is caused by a lack of external gas that would be allowed to fall in, as it happens to the Milky Way and other disk galaxies, and maintain the SF. As a proxy, also the validity of the Kennicutt law could almost be reproduced, but with a slope of -1.5.

The model treatment allows the mass release by massive and intermediate-mass stars, the latter as Planetary Nebulae (PNe), while the energy feedback is only attributed to SNeII by 20% of the normally used SNII energy of 10^{51} ergs. Until now, none of the last-mentioned numerical treatments includes type Ia SNe. Semelin & Combes (2002) trace the metal enrichment from stellar ejecta as a simple generalized Z; Harfst, Theis & Hensler (2006) do not yet, although their prescription would allow for the differential treatment of abundance yields.

This task has been performed until now only in a further advanced development by Berczik et al. (2003), Berczik et al. (2009), which is based on a single gas-phase SPH description by Berczik (1999). The treatment is almost the same as in Harfst, Theis & Hensler (2006), except that here the cool gas phase is also treated by SPH particles. This strategy can, however, be justified by the still large minimum masses of the gas packages, 10^5 $M_\odot$, i.e. in the range of Giant Molecular Clouds (GMCs) instead of small SF clumps.

Therefore, also the cool gas can be considered as continuous. The gas phases can behave dynamically independently and are only coupled by drag and mass loading by means of heat conduction. For the SF the Jeans mass criterion is used again

The most important advancement is the chemo-dynamical treatment, which includes elements released by different progenitor stars as PNe, SNeIa, SNeII, and stellar winds and enables one to trace the galactic evolutionary epochs.

3.3. Grid codes

The advantages of particle codes, due to the easy numerical handling and the simple inspection, is paid by the drawbacks of complication with the multi-phase treatment of the ISM, the poor spatial resolution, and with necessary physical processes acting and featuring the gas on various length scales. For these purposes, numerical codes based on spatial grids are at the moment more adaptable to higher resolution and more appropriate for the physics. Nevertheless, their full 3D treatment suffers severely from low spatial resolution. Because of the numerical costs, moreover until now only a few codes exist which treat at least two separate gas phases. Another serious problem is the treatment of stars. Two possibilities exist: to incorporate the stellar component into the general fluid description, or to handle a hybrid code that switches from hydrodynamics on the grid to stellar dynamics.

Although many papers denote their published numerical applications as multi-phase gasdynamics (like e.g. Wada & Norman 1999, Wada & Norman 2001), the simulations in reality only deal with a single gas fluid and account for a range of gas temperatures and densities. Slyz et al. 2005 studied the various effects of SF, stellar feedback, self-gravity, and spatial resolution of the evolution of the ISM structure in gaseous disks. The main conclusions which can be drawn are that without SF, the structure remains more diffuse than with SF, and that with SF stellar feedback is crucial, although at high spatial resolution the SF within grid cells inherently loses its coherence. The same is still continued in the, at present, spatially best resolved (magneto-)hydrodynamical simulations of the ISM by de Avillez & Breitschwerdt 2004 (and more recent papers).

Nevertheless, also in such grid simulations, SF criteria within a grid cell like e.g. excess of a density threshold, excess of Jeans mass, convergence of flows, cooling timescale lower than dynamical one, fall below threshold temperature (see e.g. Tasker & Bryan 2008) must be applied. Furthermore, at present all these applications lack the direct interactions of different gas phases by heat conduction, dynamical drag, and dynamical instabilities through forming interfaces, but the most refined ones allow to resolve the turbulence cascade.

In addition, such mixing effects at interfaces between gas phases and due to turbulence (like e.g. those in the combined SWB/HIIcomplexes) contribute to the observations by the enhancement or, respectively, dilution of metal abundances. Hydrodynamical grid models have therefore been performed that involved the elements released from nucleosynthesis in stellar progenitors of different masses and therefore also of stellar lifetimes, so-called yields. Since also the metal-dependence of astrophysical and plasmaphysical processes should be properly included in this so-called chemo-dynamics, the combined effects of gas dynamics and chemistry can be traced.

If the gasdynamics, including SF and stellar feedback, is coupled with the chemical evolution, single-gas phase simulations in 1D are, however, misleading. The reasons are that, at first, hot metal-enriched gas is mixed with cooler gas and, secondly, the limitations to one dimension does not allow for an escape of hot gas due to 2D shape effects. As a simple exercise, one can mix 10^4 K warm gas with the ejecta of SNII explosions according to a SF efficiency of a few percent, which means that less than 1% hot gas is contributed to the specific energy and momentum of the warm gas phase. This results in two effects: the gas temperature does not exceed 10^5 K and thus drops almost instantaneously to 10^4 K again, i.e. the gas remains in its warm state; the gas dynamics of the warm gas dominates until the hot gas contributes a significant fraction to the total energy density, and the gas is already diluted by gas consumption so that the cooling is extended while the gas pressure drives an expansion. Models of that kind (Pipino, D'Ercole, & Matteucci 2008) should be considered with caution unless two independent gas phases are treated and/or this at least in 2D.

Since energetic effects affect the evolution of low-mass galaxies more efficiently than massive galaxies, their application to DGs is more spectacular and has been perfomed with a single gas-phase representation of the ISM, but in 2D (e.g. by Recchi & Hensler 2006b) in order to reproduce the peculiar element abundances in Blue Compact DGs (Recchi et al. 2006b). Since gas infall not only affects the chemistry but also SF (Hensler et al. 2004b) and dynamics of outflows, we have extended the former models by infalling clouds (Recchi & Hensler 2007) which are introduced non-self-consistently in the numerical grid as density enhancements. In these models, two kinds of cloud contributions are considered, only initially existing and continuously formed, respectively.

The issues are the following: due to dynamical processes and thermal evaporation, the clouds survive only a few tens of Myr. The internal energy of cloudy models is typically reduced by 20 – 40 % compared to models with a smooth density distribution. The clouds delay the development of large-scale outflows, helping therefore to retain a larger amount of gas inside the galaxy. However, especially in models with continuous creation of infalling clouds, their ram pressure pierce the expanding supershells so that, through these holes, freshly produced metals can more easily escape into the galactic wind. Moreover, assuming a pristine chemical composition for the clouds, their interaction with the superbubble dilutes the hot gas, reducing its metal content. The resulting final metallicity is therefore, in general, smaller (by $\sim 0.2 - 0.4$ dex) than attained by smooth models.

Starting from simple 2D HD models with metallicity inclusion (Burkert & Hensler 1987, 1988) we developed a two-gas phase chemo-dynamical prescription, which was at first applied as 1D version to the vertical formation of the Galactic disk (Burkert, Truran, & Hensler 1992). The metallicity was treated as a general Z without differentiation of particular elements between their stellar progenitors, but with various production timescales. The model could successfully reproduce the vertical stellar density and metallicity distribution while also yielding formation timescales of the thin and thick disk components.

This chemo-dynamical scheme was further extended by us to 2D and to specific elements characteristic for different stellar mass progenitors and production lifetimes. This chemo-dynamical treatment includes metal-dependent stellar yields and winds, SNeIa, SNeII, PNe, metal-dependent cooling functions and heat conduction. The gas phases are described by the Eulerian hydrodynamical equations, co-existing in every grid cell in pressure equilibrium. Thereby the cool component, as representative of interstellar clouds, is described by anisotropic velocity components derived from the Boltzmann moment equations with source and sink terms from cloud collisions. The SF is parametrized as explored in Köppen, Theis, & Hensler (1995) to guarantee self-regulation.

The application of the chemo-dynamical code to the evolution of disk galaxies (Samland & Hensler 1996) could successfully show for the Milky Way (Samland, Hensler & Theis 1997) how and on which timescale the components halo, bulge, and disk formed and how the abundances developed, could represent the temporal evolution of gas abundances, of the stellar radial abundance gradients, of the abundance ratios, and of star-to-gas ratios.

This also showed how many metals are stored in the hot gas phase and mixed with the cool one by evaporation of clouds and condensation.

Until now, the optimal grid code is considered to be the further development of the chemo-dynamical scheme (Samland, Hensler & Theis 1997) to 3D and with stellar dynamics for the stars. In addition, a cosmologically growing DM halo is included in the simulations by Samland & Gerhard (2003). These models contain all the crucial processes of SF self-regulation by stellar feedback, multi-phase ISM, and temporally resolved stellar components according to the chemo-dynamical prescription (Hensler 2003, 2007). They cannot only trace the formation and evolution of the disk galaxies' components, but also of characteristic chemical abundances, and are until now the best self-consistent evolutionary models of disk galaxies, because the disk formation is included into the global temporal galaxy evolution. Moreover, the formation and evolution of disk galaxies in their infancy were studied by these self-consistent simulations with respect to bulge formation and disk fragmentation (Immeli *et al.* 2004).

4. Discussion

While SPH codes appear plausibly to be the most appropriate numerical treatment of galaxy evolution including gaseous disks, it must be considered with caution that for the multi-component representation with interaction processes the particle number is still at present limited to a few 10,000, which means that gas packages and stellar particles in massive galaxies represent unrealistically large masses. Only for DGs does this effort lead to acceptable resolutions.

I am here not advocating against particle codes, and I am optimistic that advances in computer facilities and numerical codes will soon allow to overcome the spatial resolution problem and allow for multi-component descriptions. Nevertheless, at the moment grid strategies look much more promising because small-scale processes are more easily and reliably coupled with large-scale dynamics. In addition, physical processes can be treated more appropriately. This also allows an outlook if the self-consistent global models are further developed to include magnetic fields, radiative and Cosmic-ray transport.

A word of care should be expressed with respect to empirical relations. Agreement with the Kennicutt relation is achieved for simulations which are totally differing in the treatment of feedback processes, while the Hα surface brightness relation deviate from the Kennicutt slope of 1.4 with atomic H I gas, molecular gas, and for different modes of SF, so any representation of this relation by any modelling of galaxy evolution seems to be too weak to validate any numerical treatment by itself. Likewise, metallicity gradients and age-metallicity relations in galaxy disks also seem not to be sufficient for the validation of numerical codes. This means, on the other hand, that issues of such relations and the requirement to disentangle inherent evolutionary processes are overestimated, and that these relations are robust and insensitive to the intrinsic state of galaxies.

Furthermore, the presented models show that it seems illusory to try to correlate SF with gas densities if feedback processes are neglected.

We are still far from fully self-consistent models that allow for the inclusion of the necessary astrophysical and plasmaphysical processes on all length and time scales required. Growing computer facilities will allow soon for 3D AMR simulations, but not only refined spatial and temporal resolution is the clue; also coherence on different scales as necessary for SF, turbulence, magnetic fields, etc. require intuition, brain waves, and a large effort of hard code developing and testing.

Acknowledgements

The author is gratefully acknowledging collaborations and discussions on this topic with Peter Berczik, Dieter Breitschwerdt, Stefan Harfst, Stefan Hirche, Joachim Köppen, Simone Recchi, Andreas Rieschick, Markus Samland, Rainer Spurzem, Christian Theis, Wolfgang Vieser, and Hervé Wozniak. I also thank Simone Recchi for his careful reading of the manuscript and the organizers cordially for their invitation to this conference and their support. This work is supported by the key programme "Computational Sciences" of the University of Vienna under project no. FS538001.

References

Barnes, J. & Hut, P. *Nature*, 324, 446
Benjamin, R. & Danly, L. 1997, *ApJ*, 481, 764
Berczik P., 1999, *A&A*, 348, 371
Berczik, P., Hensler, G., Theis, C., & Spurzem, R. 2003, *Ap&SS*, 284, 865
Berczik, P., Hensler, G., Theis, C., *et al.* 2009, *A&A*, submitted
Böhringer, H. & Hensler, G. 1989, *A&A*, 215, 147
Bouche, N., Cresci, G., Davies, R., *et al.* 2007, *ApJ*, 671, 303
Bournaud, F., Elemegreen, B. G., & Elmegreen, D.M. 2007, *ApJ*, 670, 237
Burkert, A. & Hensler, G. 1987, *MNRAS*, 225, 21p
Burkert, A. & Hensler, G. 1988, *A&A*, 199, 131
Burkert, A., Truran, J., & Hensler, G. 1992, *ApJ*, 391, 651
Cox, D.P., 2005, *ARAA*, 43, 337
de Avillez, M. A. & Breitschwerdt, D. 2004, *A&A*, 425, 899
Dalla Vecchia, C. & Schaye, J. 2008, *MNRAS*, 387, 1431
Ehlerova, S., Palous, J., Theis, C., & Hensler, G. 1997, *A&A*, 328, 111
    1997, *ApJ*, 481
Elmegreen, B.G. 2002, *ApJ*, 577, 206
Elmegreen, B. G. & Efremov, Y. N. 1997, *ApJ*, 480, 235
Ferrière, K. M., 2001, *Rev. Mod. Phys.*, 73, 1031
Fraternali, P., van Moorsel, G., Sancisi, R., & Oosterloo, T. *AJ*, 123, 3124
Freyer, T., Hensler, G., & Yorke, H. W., 2003, *ApJ*, 594, 888
Freyer, T., Hensler, G., & Yorke, H. W., 2006, *ApJ*, 638, 262
Fukuda, N. & Hanawa, T. 2000, *ApJ*, 533, 911
Gao, Y. & Solomon, P. M., 2004, *A&A*, 606, 271
Harfst, S., Theis, C., & Hensler, G. 2006, *A&A*, 499, 509
Hensler, G. 2003, *ASP Conf. Ser. Vol.*, 304, eds. C. Charbonnel *et al.*, p. 371
Hensler, G. 2007, *EAS Publ. Ser. Vol.*, 24, eds. E. Ensellem *et al.*, p. 113
Hensler, G., Dickow, R., Junkes, N., & Gallagher, J.S., III. 1998, *ApJ*, 502, L17
Hensler, G., Kppen, J., Pflamm, J., & Rieschick, A. 2004, IAU Symp. Ser. 217, eds. P.-A. Duc, J. Braine, E. Brinks, p. 178
Hensler, G. & Rieschick, A. 2002, *ASP-CS*, 285, 341
Hensler, G., Theis, Ch., & Gallagher, J. S., III. 2004, *A&A*, 426, 25
Heyer, M. H., Brunt, C., Snell, R. L., *et al.* 1996, *ApJ*, 464, L175
Heyer, M. H., Corbell, E., Schneider, S. E., & Young, J.S., 2004, *ApJ*, 602, 723
Immeli, A., Samland, M., Gerhard, O., & Westera, P. 2004, *A&A*, 413, 547
Kennicutt, R. J. 1998, *ApJ*, 498, 541
Keres, D., Katz, N., Weiberg, D. H., & Dave, 2005, *MNRAS*, 363, 2
Knauth, D. C., Meyer, D. M., & Lauroesch, J. T. 2006, *ApJ*, 647, L115
Köppen, J. & Hensler, G. 2005, *A&A*, 434, 531
Köppen, J., Theis, C., & Hensler, G. 1995, *A&A*, 296, 99
Köppen, J., Theis, C., & Hensler, G. 1998, *A&A*, 328, 121
Krumholz, M. R. & McKee, C. F. 2005 *ApJ*, 630, 250

Kravtsov, A. V. 2003, *ApJ*, 590, L1
Larson, R. B. 1969, *MNRAS*, 145, 405
Larson, R. B. 1974, *MNRAS*, 169, 229
Larson, R. B. 1975, *MNRAS*, 173, 671
Larson, R. B. 1976, *MNRAS*, 176, 31
Lasker, B. M. 1967, *ApJ*, 149, 23
Li, Y., MacLow, M.-M., & Klessen, R. S. 2005, *ApJ*, 626, 823
MacLow, M.-M. & Klessen, R. S. 2004, *Rev.Mod.Phys.*, 76, 125
Martin, C. L., Kobulnicki, H. A., & Heckman, T. M. 2002, *ApJ*, 574, 663
Mühle, S., Klein, U.,Wilcots, E. M., & Hüttermeister, S. 2005, *AJ*, 130, 524
Ostriker, E., Stone, J. M., & Gammie, C. F. 2001, *ApJ*, 546, 980
Pipino, A., DErcole, A., & Matteucci, F. 2008, *A&A*, 484, 679
Rana, N. C. & Wilkinson, D. A. 1986, *MNRAS*, 218, 497
Recchi, S. & Hensler, G. 2006a, *A&A*, 445, L39
Recchi, S. & Hensler, G. 2006b, *Rev. Mod. Astronomy*, 18, 164
Recchi, S. & Hensler, G. 2007, *A&A*, 476, 841
Recchi, S., Hensler, G., Angeretti, L., & Matteucci, F. 2006, *A&A*, 445, 875
Rieschick, A. & Hensler, G. 2000, *ASP-CS*, 215, 130
Samland, M. & Hensler, G. 1996, *Rev. Mod. Astron.*, 9, 277
Samland, M., Hensler, G., & Theis, C. 1997, *ApJ*, 476, 544
Samland, M. & Gerhard, O. 2003, *A&A*, 399, 961
Sancisi, R., Fraternali, F., Oosterloo, T., & van der Hulst, T. 2008, *A&A Rev*, 15, 189
Scannapieco, C., Tissera, P. B., White, S. D. M, & Springel, V. 2006, *MNRAS*, 371, 1125
Schaye, J. & Dalla Vecchia, C. 2008, *MNRAS*, 383, 1210
Schmidt, M. 1959, *ApJ*, 129, 243
Semelin, B. & Combes, F. 2002, *A&A*, 388, 826
Slyz, A. D., Devriendt, J. E. G., Bryan, G., & Silk, J. 2005, *MNRAS*, 356, 737
Springel, V. 2000, *MNRAS*, 312, 859
Springel, V., Yoshida, N., & White, S. D. M. 2001, *New Astron.*, 6, 79
Stil, J. M. & Israel, F. P. 2002, *A&A*, 392, 473
Strickland, D. K., Heckman, T. M., Colbert, E. J. M., et al. 2004, *ApJ*, 606, 829
Tasker, E. J. & Bryan, G. L. 2008, *ApJ*, 673, 810
Thornton, K., Gaudlitz, M., Janka, H.-Th., & Steinmetz, M. 1998, *ApJ*, 500, 95
Vieser, W. & Hensler, G. 2007, *A&A*, 472, 141
Wada, K. & Norman, C. A. 1999, *ApJ*, 516, L13
Wada, K. & Norman, C. A. 2001, *ApJ*, 547, 172
Wada, K. & Norman, C. A. 2007, *ApJ*, 660, 276
Wise, J. H. & Abel, T. 2007, *ApJ*, 665, 899
Wolfire, M. G., McKee, C. F., Hollenbach, D., & Tielens, A. G. G. M. 1995, *ApJ*, 453, 673

Measuring Outer Disk Warps with Optical Spectroscopy

Daniel Christlein[1] and Joss Bland-Hawthorn[2]

[1] Max-Planck-Institut für Astrophysik,
Karl-Schwarzschild-Str. 1, 85748 Garching, Germany
email: dchristl@mpa-garching.mpg.de

[2] Institute of Astronomy, School of Physics
University of Sydney, NSW 2006, Australia
email: jbh@physics.usyd.edu.au

Abstract. Warps in the outer gaseous disks of galaxies are a ubiquitous phenomenon, but it is still unclear what generates them. One theory is that warps are generated internally through spontaneous bending instabilities. Other theories suggest that they result from the interaction of the outer disk with accreting extragalactic material. In this case, we expect to find cases where the circular velocity of the warp gas is poorly correlated with the rotational velocity of the galaxy disk at the same radius. Optical spectroscopy presents itself as an interesting alternative to 21-cm observations for testing this prediction, because (i) separating the kinematics of the warp from those of the disk requires a spatial resolution that is higher than what is achieved at 21 cm at low HI column density; (ii) optical spectroscopy also provides important information on star formation rates, gas excitation, and chemical abundances, which provide clues to the origin of the gas in warps. We present here preliminary results of a study of the kinematics of gas in the outer-disk warps of seven edge-on galaxies, using multi-hour VLT/FORS2 spectroscopy.

Keywords. galaxies: evolution, galaxies: formation, galaxies: kinematics and dynamics, galaxies: structure

---

1. Introduction

Warps in the outer disks of galaxies are a ubiquitous phenomenon. They are seen both in the distribution of stars (Sanchez-Saavedra *et al.* 1990; Cox *et al.* 1996) and neutral hydrogen (e.g., Sancisi 1976; Bosma 1981), and surveys estimate that possibly 50% or more of all galaxies show evidence for warps beyond the isophotal radius R_{25} (Briggs 1990), which typically contains $\sim 90\%$ of the total light. Usually, these manifest themselves in the form of the outer disk bending away from the plane defined by the inner disk, defining either the shape of a bowl, or, more commonly, an integral sign as seen edge-on. This suggests that specific angular momentum of material in the warps is not aligned with that of the inner disk.

Numerous suggestions have been made over the years for the responsible mechanisms. Among the earliest such proposals were internal bending modes in the disk (Lynden-Bell 1965), but such modes were soon recognized to be persistent only in a disk with an unrealistically sharp mass truncation (Hunter & Toomre 1969). Revaz & Pfenniger (2004) have revived this discussion by identifying short-lived bending instabilities as a possible cause. Other proposed mechanisms focus on the interaction of the baryonic disk with its environment: A bending of the outer baryonic disk may be introduced by a misalignment between the angular momentum of the inner baryonic disk and the hypothetical non-spherical, dark matter halo that it is embedded in (Toomre 1983; Dekel & Shlosman 1983; Kuijken 1991; Sparke & Casertano 1998), creating a gravitational torque on the disk. However, it has been argued that the inner halo would realign with the baronic disk over time, and the warp would dissipate (Nelson & Tremaine 1999; Binney, Jiang & Dutta 1998; Dubinski & Kuijken 1995; New et al. 1998). Ostriker & Binney (1989), Jiang & Binney (1999), Shen & Sellwood (2006) consider the impact of ongoing accretion onto an outer dark matter halo and argue that, since the angular momentum of infalling material will in general not be aligned with the present galaxy disk, this will create an ongoing torque on the outer galaxy disk. Binney (1992) has furthermore suggested that, if infalling material loses angular momentum to the halo, it might penetrate as far as the outer edge of the baryonic disk itself.

It is this hypothesis that we wish to test with the present work. How galaxies acquire gas is one of the key questions in our understanding of how they evolve, and determining whether warps are indeed signatures of such processes therefore is of great importance.

How can the direct accretion hypothesis be tested? The specific angular momentum vector of accreting material will, in general, neither be aligned exactly with that of the inner disk, nor have the same size. If such infalling material is indeed in direct contact and exchanging angular momentum with gas in the outermost baryonic disk, then it will introduce kinematic anomalies, i.e., a deviation from disk-like rotation, such as a lag or excess in the circular velocity. Measuring the circular velocity of gas in the warps therefore becomes an important observational discriminator.

Our project has measured line-of-sight velocities of gas in the outer disks of seven galaxies, the majority of which display clear signs of optical warps. Our measurements were obtained via optical spectroscopy of the Hα line. Although the traditional way of observing gas in the outer disk is via the 21-cm line of neutral hydrogen, optical spectroscopy has proven a surprisingly successful alternative for studying the outer disk (Bland-Hawthorn, Freeman & Quinn 1997; Christlein & Zaritsky 2008), for a number of reasons: 1) Most importantly, the spatial resolution — an order of magnitude better than even the highest-quality interferometric HI maps — allows us to clearly separate gas in the warps from gas in the plane of the galaxy. This, in turn, allows us to access galaxies with smaller angular diameters at larger redshifts, greatly increasing the number of suitable targets. 2) Sporadic local star formation or illumination of gas in the warps by escaping UV flux from the inner star-forming disk guarantee low-level Hα flux from the outer disk far beyond what is usually perceived as the star formation threshold. It has been demonstrated (Bland-Hawthorn, Freeman & Quinn 1997; Christlein & Zaritsky 2008) that multi-hour optical spectroscopy of such low-level emission can, in some cases, probe outer galaxy disks to similar extents as 21-cm. 3) Optical spectroscopy yields a plethora of ancillary data, particularly stellar continuum and metal lines. In determining the origin of gas in the outer disks and warps, metallicity indicators may provide valuable additional insight (although an analysis exceeds the scope of these proceedings). Our aim is to a) measure the rotational velocity of gas in the warps and b) determine whether there are kinematic anomalies in this rotational velocity, i.e., sudden breaks or upturns that are not consistent with an extrapolation of the rotation curve from the inner disk, which may be associated with the onset of the warp.

2. Observations

Our sample consists of seven galaxies, of which six were taken from the catalog of optically warped galaxies by Sanchez-Saavedra et al. (2004); one object was targetted blindly without prior knowledge of a warp. All are nearly edge-on, with redshifts in the range of several thousand km/s, and their angular size is well-matched to the field of view.

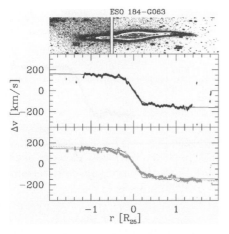 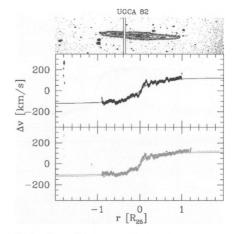

Figure 1. ESO 184-G063. For explanation, see text. Hα is more extended in the offset spectrum than along the major axis, but no kinematic disruption is associated with the warp.

Figure 2. UGCA 23. For explanation, see text. Hα is more extended in the offset spectrum than along the major axis, but no kinematic disruption is associated with the warp.

Our observations were carried out over three nights in September 2007, using FORS2 on the VLT-UT1. The typical observing strategy was to observe for a total of one hour with a long slit of 0.5″ width aligned along the major axis (to obtain a reference for the rotational velocity and angular extent gas in the plane of the galaxy), then for two hours with a position angle offset from the major axis PA by a few degrees. The purpose of the latter observation was to assure that Hα emission emanating from the warps would be observed, rather than from the plane of the galaxy. Sense and size of this offset were determined on a case-by-case basis after inspecting plates from the Digitzed Sky Survey as well as our own acquisition images, and typically chosen so that the slit would pass within or just beyond the outermost contours of stellar continuum light at the end of the warp.

3. Results

- **ESO 184-G063** (Fig. 1) is a small ($M = -18.5$) Sb-type galaxy at cz = 3207 km/s, and has a strong integral-sign warp, which is distinctly stronger on one side than the other.

This figure, as well as all subsequent ones, shows the rotation curve along the major axis (top panel) and in the offset position (bottom panel; major axis rotation curve plotted with small dots/dashed lines for reference). The radial extent is 2 R_{25} on both sides of the nucleus; the acquisition image is plotted to the same scale.

Along the major axis, our rotation curve extends to 1.4 R_{25}, which in itself is remarkable, given that normal Hα rotation curves with conventional exposure times and smaller instruments typically reach $0.7R_{25}$, and rarely as far as R_{25}. Our off-axis spectrum extends out to 1.6 R_{25}, clearly intersecting the tip of the optical warp. Both the major axis and the offset rotation curves are consistent, and there is no sign of a sudden break in the rotational velocity coincident with the onset of the warp.

- **UGCA 23** (Fig. 2) is an $M = -19$ Sd-type galaxy at cz=3864 km/s. Acquisition images show lumpy emission and slightly bent isophotes on one side. The major axis rotation curve extends to $\sim R_{25}$ in this case. The offset spectrum samples the warp feature

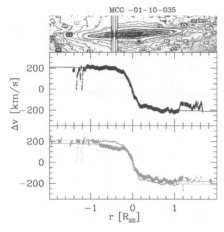 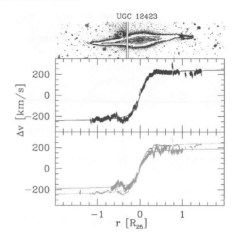

Figure 3. MCG -01-10-035. Hα is more extended in the offset spectrum than along the major axis. A break in the rotation curve is found along the major axis, coinciding with the onset of a tidal-tail-like warp feature.

Figure 4. UGC 12423. See text for explanation.

and shows emission as far as 1.2 R$_{25}$. The two rotation curves are indistinguishable; the kinematics of the extraplanar feature are disk-like.

- **MCG -01-10-035** (Fig. 3) is a small ($M = -18.3$) Sc-type galaxy at cz = 3788 km/s. Our acquisition images reveal a very extended tail of low surface brightness extending as far as $2R_{25}$ and then curving back. Its appearance is very suggestive of a tidal feature. While the other side of the galaxy does not exhibit an equivalently striking feature, the outer isophotes appear bent on both sides.

The major axis spectrum samples far into this low-surface brightness tail and extends to 1.6 R$_{25}$ on that side, and $1.4R_{25}$ on the other. Remarkably, at the onset of this feature, around R$_{25}$, we see a sudden break in the rotation curve *along the major axis* in the sense of a jump to lower rotational velocities by $\sim 60 - 70$ km/s. On the other side, no such strong break is discernible, but there is Hα emission at velocities distinctly below the extrapolated rotation velocity, going as far down as the systemic velocity of the galaxy itself.

Our offset spectrum intercepts the tail at its largest extent. While we do not see a continuous Hα signal beyond R$_{25}$, there are individual Hα-bright regions extending as far as $2R_{25}$ on both sides of the galaxy. The rotation curve fitted to this spectrum shows no breaks in the velocity; however, at large radii, its line-of-sight velocity is consistent with the lower-velocity gas in the tidal feature.

- **UGC 12423** (Fig. 4) is an Sc-type galaxy at cz = 4839 km/s with an absolute magnitude $M = -19.4$. It shows a strong and asymmetrical optical warp. The major-axis rotation curve extends to 1.4 R$_{25}$ on the side with the stronger warp, which is even farther than the extent of the warped disk. At the outermost point, the rotational velocity is still consistent with the extrapolated rotation curve. A possible dip in the rotation velocity at the position of the warp feature does not appear significant.

The offset spectrum is, in this case, much less extended; it reaches R$_{25}$ on the side of the stronger warp, and only 0.7 R$_{25}$ on the other side, with the exception of one small Hα-emitting region at 1.4 R$_{25}$. There are no signs for any kinematic anomalies in this rotation curve.

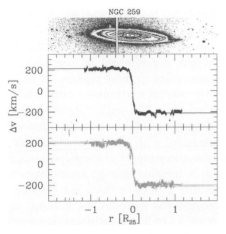

Figure 5. NGC 259. See text for explanation.

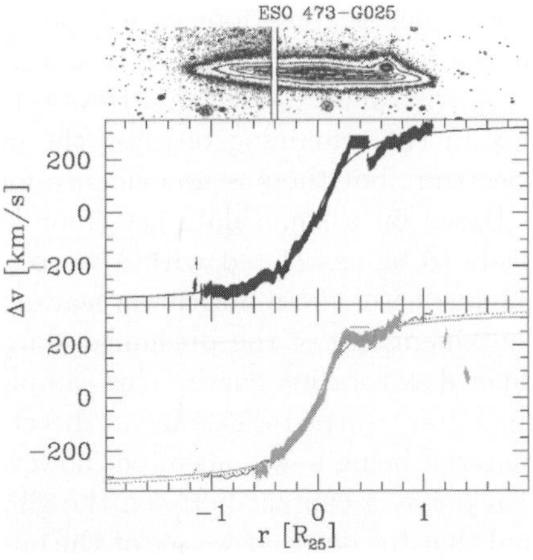

Figure 6. ESO 473-G025. text for explanation.

- **NGC 259** (Fig. 5) is among the brighter galaxies in this sample ($M = -20$), an Sbc-type at cz=4045 km/s. It is also the least inclined among our targets. The acquisition images show that the supposed warp feature is in fact more suggestive of a continuation of a spiral arm. The major axis spectrum samples its kinematics out to 1.2 R_{25}, but shows no kinematic disturbance. The offset spectrum only extends to $\sim R_{25}$.

- **ESO 473-G025** (Fig. 6), the brightest object in our sample ($M = -20.3$), is an Sb-type galaxy at cz=7110 km/s. Acquisition images show what is at most a very subtle warp. The major axis spectrum extends a little beyond R_{25}. The kink visible in the rotation curve in Fig.6 corresponds to a region where two components of the Hα line are visible at the same position in projection along the slit. However, there are no signs of kinematic anomalies towards the edge of the disk. The offset rotation curve is much less extended. A single emission region around 1.4 R_{25} coincides with what appears to be a small satellite galaxy on the acquisition image.

- **ESO 340-G026** (not shown due to space constraints) is an Sc-type galaxy at cz=5483 km/s with $M = -19.5$. There are low-surface-brightness tails at the edge of the disk, but they show no significant warp. We exposed in two different offset positions, but obtained sufficient data for an analysis in only one case. The major axis spectrum, in any case, extends to 1.3 R_{25} and thus far enough to sample the kinematics of the outlying features. The rotation curve, again, shows no signs of kinematic anomalies.

4. Conclusions

In an effort to measure the kinematics of gas in the warps in the outer disks of galaxies with high spatial resolution, we have targetted seven edge-on galaxies, most of them with known optical warps, using multi-hour long slit spectroscopy.

- In at least three of seven cases, we have observed more extended Hα emission if the slit was aligned along the warp rather than along the major axis. This shows that we have been successful at detecting gas in the warps, and that we have thus attained our observational goal. Counting the single emission region detected in UGC 12423, this ratio is four out of seven.

- In three of these four cases, there are no signs of kinematic anomalies in the rotation curve associated with the onset of the warp.

- In one of these four cases, there is a sharp break in the rotational velocity roughly at the onset of the warp. In this case, the optical appearance of the "warp" is highly suggestive of a tidal feature. However, the break occurs in the major axis rotation curve.
- In the remaining objects, the outer-disk features are sampled by the major-axis spectrum, but there is no evidence for kinematic anomalies in these regions.

Based on a small data set (four kinematic detections of extraplanar outer disk gas likely to be associated with a warp), we find evidence for kinematic anomalies only in the one case whose optical appearance strongly suggests a tidal feature. In other cases, the kinematics of the presumed warp gas are consistent with the extrapolation of the inner disk rotation curve. This sample is therefore not in support of the hypothesis that the outer, warped disks are in direct contact and exchanging angular momentum with material being newly accreted; however, the possibility cannot be ruled out yet that the real point of contact between the galaxy disk and accreted material lies at larger radii, and that the gas that we see at the onset of the optical warp has already settled into disk-like kinematics. We propose to explore three avenues for further investigation: (i) Gas at larger radii, substantially beyond the optical warps, should be targetted; ideally, targets should thus be selected on the basis of HI maps where available. (ii) The sample should be increased in size from the present three clear detections. (iii) Additional indicators that might constrain the origin of the warp gas, in particular, metallicity, should be included in the analysis.

Based on observations made with ESO Telescopes at the Paranal Observatories under programme ID <079.B-0426>.

References

Binney, J. 1992, *ARA&A*),30, 51
Binney, J., Jiang, I.-G., & Dutta, S. 1998, *MNRAS* 297, 1237
Bland-Hawthorn, J., Freeman, K. C., & Quinn, P. J. 1997, *ApJ*), 490, 143
Bosma, A. 1981, *AJ*, 86, 1791
Briggs, F. H. 1990, *ApJ*, 352, 15
Christlein, D. & Zaritsky, D. 2008, *ApJ*), 680, 1053
Cox, A. L., Sparke, L. S., van Moorsel, G., & Shaw, M. 1996, *AJ*, 111, 1505
Dekel, A. & Shlosman, I. 1983, *IAUS*, 100, 187
Dubinski, J. & Kuijken, K. 1995, *ApJ*, 442, 492
Hunter, C. & Toomre, A. 1969, *ApJ*, 155, 747
Jiang, I.-G. & Binney, J. 1999, *MNRAS*), 303, 7
Kuijken, K. 1991, *ApJ*, 376, 467
Lynden-Bell, D. 1965, *MNRAS*, 129, 299
Nelson, R. W. & Tremaine, S. 1999, *MNRAS*, 306, 1
New, K. C. B., Tohline, J. e., Frank, J., & Vaeth, H. M. 1998, *ApJ*, 503, 632
Ostriker, E. C. & Binney, J. 1989, *MNRAS*, 237, 785
Revaz, Y. & Pfenniger, D. 2004, *A&A*, 425, 67
Sancisi, R. 1976, *A&A*, 53, 159
Sanchez-Saavedra, M. L., Battaner, E., & Florido, E. 1990, *MNRAS*, 246, 458
Sanchez-Saavedra, M. L., Battaner, E., Guijarro, A., López-Correidora, M., & Castro-Rodríguez, N 2003, *A&A*, 399, 457
Shen, J. & Sellwood, J. A. 2006, *MNRAS*), 370, 2
Sparke, L. S. & Casertano, S. 1988, *MNRAS*, 234, 873
Toomre, A. 1983, *IAUS*, 100, 177

Star Formation in Disks: Spiral Arms, Turbulence, and Triggering Mechanisms

Bruce G. Elmegreen[1]

[1]IBM Research Division, T.J. Watson Research Center, 1101 Kitchawan Road,
Yorktown Hts. 10598, USA
email: bge@us.ibm.com

Abstract. Star formation is enhanced in spiral arms because of a combination of orbit crowding, cloud collisions, and gravitational instabilities. The characteristic mass for the instability is 10^7 $M_\odot$ in gas and 10^5 $M_\odot$ in stars, and the morphology is the familiar beads on a string with 1-2 kpc separation. Similar instabilities occur in resonance rings and tidal tails. Sequential triggering from stellar pressure occurs in two ways. For short times and near distances, it occurs in the bright rims and dense knots that lag behind during cloud dispersal. For long times, it occurs in swept-up shells and along the periphery of cleared regions. The first case should be common but difficult to disentangle from independent star formation in the same cloud. The second case has a causality condition and a collapse condition and is often easy to recognize. Turbulent triggering produces a hierarchy of dense cloudy structure and an associated hierarchy of young star positions. There should also be a correlation between the duration of star formation and the size of the region that is analogous to the size-linewidth relation in the gas. The cosmological context is provided by observations of star formation in high redshift galaxies. Sequential and turbulent triggering is not yet observable, but gravitational instabilities are, and they show a scale up from local instabilities by a factor of ~ 3 in size and ~ 100 in mass. This is most easily explained as the result of an increase in the ISM turbulent speed by a factor of ~ 5. In the clumpiest galaxies at high redshift, the clumps are so large that they should interact with each other and merge in the center, where they form or contribute to the bulge.

Keywords. stars: formation, ISM: evolution, galaxies: formation, galaxies: high-redshift

1. Star Formation in Spiral Arms

Star formation in local disk galaxies, including the Milky Way, is highly concentrated in spiral arms. This is partly because the gas is concentrated in the arms as a result of density-wave streaming. The stellar spiral arm potential pulls out the interarm gas nearly radially, causing it to traverse the interarm region quickly, and it shocks and pulls back the gas when it enters the arm, causing it to deflect and move nearly parallel to the arm. Because the time spent at each position relative to an arm is inversely proportional to the perpendicular component of the velocity, and because the local density is also inversely proportional to this velocity, the gas lingers in an arm for a time proportional to its density. Thus most star formation occurs where the arm is dense because most of the time and most of the gas occurs there. If this were all that happened, then the star formation rate should be directly proportional to the total gas density in an azimuthal profile around the disk. There are no recent studies of the azimuthal dependence of star formation, although more on this topic was done over a decade ago (e.g., Garcia-Burillo et al. 1993). In addition to this purely kinematical effect, there may also be a dynamical effect in the sense that the star formation rate per unit gas mass increases in the arms relative to the interarm regions. In this case, the star formation rate should increase with the total gas density to a power greater than unity for azimuthal profiles. In either case, the sizes of the star-forming regions are much larger in the arms than between the arms (e.g., Lundgren et al. 2004).

Consideration of dynamical processes and the time available suggest that spiral arms could promote local gravitational instabilities that trigger star formation in giant complexes (e.g., Tosaki et al. 2002). In the low-shear environment of an arm (e.g., Luna et al. 2006), such instabilities produce 10^7 $M_\odot$ clumps rather than spiral wavelets. Magnetic fields help by removing gas angular momentum during subsequent collapse (Kim & Ostriker 2002). The morphology of star formation then consists of HI clouds with molecular cores lining the stellar spiral arms (e.g., Engargiola et al. 2004). Feathers and spurs extend into the interarm region as the arm clouds feel an increased shear (e.g., La Vigne et al. 2006). Deep in the interarm regions, star formation lingers in the long-lived envelopes and diffuse debris of spiral arm clouds (Elmegreen 2007).

In galaxies without strong stellar spirals, star formation occurs throughout the disk in flocculent arms that are probably the result of local, swing-amplified instabilities (e.g., Fuchs et al. 2005). Thus, star-forming clouds form or grow in (1) spiral-wave triggered gravitational instabilities and cloud collisions in dense dust lanes, and (2) random gravitational instabilities everywhere if there are no stellar spirals, or if the stellar spirals are weak. In either case, star formation follows cloud formation in a few cloud dynamical times. It lingers and gets triggered at a low level in the molecular cloud debris for a much longer time.

2. Triggering

Triggering of star formation is well known in places like the pillars of M16 (Hester et al. 1996), and other bright rims (e.g., Sugitani et al. 1989; Reach et al. 2004). The large-scale structure of Ophiuchus also suggests that the main star formation was triggered in the head of a cometary cloud shaped by pressures from the Sco-Cen OB association (de Geus 1992). Other triggering takes the form of shells, with old stars in the center and young stars along the periphery. Such shells have been observed in local star-forming regions (e.g., Zavagno et al. 2006) and in other galaxies (e.g., Brinks & Bajaja 1986). Some triggering shells can be very large (Wilcots & Miller 1998; Walter & Brinks 1999; Yamaguchi et al. 2001a) and some contain triggered pillars also (e.g., Yamaguchi et al. 2001b; Oey et al. 2005).

Generally, the spatial scale for triggering is the shock speed in the ambient gas multiplied by the timescale for the pressure source, $L \sim (P/\rho_0)^{1/2} \times T$. If the spatial scale is smaller than the cloud in which the first generation of star formation occurs, then pillars and bright rims form by the push-back of interclump gas. Star formation triggering can be fast in this case, prompted by the direct squeezing of pre-existing dense gas. Also in this case, the relative velocity of the triggered stars will be small because they are in the dense gas that is left behind. If the triggering scale is larger than the size of the cloud, then shells form by the push-back of all the gas, both in the cloud and in the intercloud medium. This is a slow process because new clumps have to form, and the time scale for this is $\sim (G\rho_{shell})^{-1/2}$. The velocity of triggered stars in this second case can be large, comparable to the shock speed, $\sim (P/\rho_0)^{1/2}$. There is also a causality condition, where the triggering distance between generations equals the time difference multiplied by the velocity difference. Criteria for gravitational collapse in expanding shells were derived by Elmegreen (1994), Elmegreen, Palous & Ehlerova (2002), and Whitworth et al. (1994).

There are many types and sources of energy in the ISM and many possible causes for triggering. Energy *types* include thermal, magnetic, turbulent, cosmic ray, and rotational. The first four are all comparable and equal to several tenths of an eV cm^{-3}. Rotational energy is much denser, several hundred eV cm^{-3}. Of the first four, only supersonic turbulence has been associated with star formation triggering because such turbulence can compress the gas a lot. Thermal instabilities compress the gas too, but usually in small regions where self-gravity is not important. Magnetic fields cause slight compression when the gas rearranges itself on the field lines following a Parker instability, but this rearrangement alone is not enough to trigger star formation – that requires self-gravity too. Cosmic rays interact with the ISM primarily through heating and the generation of small scale MHD turbulence. If this turbulence compresses the gas significantly, and on sufficiently large scales, then cosmic rays could trigger star formation (there has been no theoretical work on this mechanism as far as I know). Rotational energy compresses the gas and triggers star formation more than any of the others when something gets in the way of the uniform circular motion. Then a galactic-scale shock forms and star formation results in the dense gas. Spiral density waves in the stars and stellar bars can do this, as mentioned in the previous section.

Among the various *sources* of ISM energy, we include supernovae, HII regions and stellar winds as stellar sources, and galaxy rotation as an energy source for the magnetic field through the dynamo. The magnetic field drives turbulence and convection into the halo by the Parker instability (e.g., Kosiński, & Hanasz 2006; Lee & Hong 2007), and it drives turbulence in the midplane by amplifying epicyclic oscillations in the radial direction, which is the Magneto-Rotational instability (e.g., Kim, Ostriker, & Stone 2003). Galactic rotation also generates turbulence at spiral waves (Bonnell, *et al.* 2006; Kim, Kim, & Ostriker 2006; Dobbs & Bonnell 2007). In addition, there is star-light energy, which is comparable in density to turbulence, cosmic rays, and other energies mentioned above, but is not so easily coupled to the gas. There is also ISM self-gravity, which may be considered as a source of energy for motions if it is replenished by cloud disruption, in which case the real energy source is that of the disruption (i.e., young stars; self-gravity only stores the other energies in potential form.)

Of these sources, most do not trigger star formation because they do not interact with gas in the right way. The most important energy sources are those that produce high pressures for long times – long enough for gravity to act in the compressed regions. Single supernovae, for example, are short-lived, with radiative lifetimes only a percent of the local dynamical time of the gas they are in (Dekel & Silk 1986). Magnetic energy, cosmic rays, and starlight do not compress the gas much. HII regions and O-star winds compress only the most local gas, but they do this for a long time and often trigger star formation in adjacent molecular clouds. Combined pressures from HII regions, winds and supernovae can cause major triggering: they act in OB associations for a relatively long time, ~ 5 Myr or more, and can trigger star formation on a scale of 10 to 100 pc or more in the surrounding gas, which includes remnants of the molecular clouds that formed these stars, neighboring molecular clouds, and intercloud gas.

In summary, star formation either follows cloud formation, or it is stimulated in existing clouds by external processes. Gaseous self-gravity alone triggers cloud formation through: (1) dust-lane fragmentation in stellar density waves, (2) ring fragmentation in Lindblad resonance rings, (3) tidal clump or dwarf galaxy formation in the tidal arms of interacting galaxies, and (4) swing amplified clumps if there are no imposed stellar structures. Self-gravity also causes existing clouds to collapse and fragment into denser pieces where stars form. The morphologies of these processes are relatively easy to recognize in ideal cases: they appear as "beads on a string" of star formation in spiral arms, resonance rings, and tidal tails, or they are sheared spiral-like clumps from local instabilities in the gas. All of the clouds or cores formed by self-gravity have masses exceeding the relevant Jeans mass. In the case of instabilities in disks, filaments and shocked layers, the elongated regions are always unstable to condensation into spheroids. If the resulting spheroids are massive enough, then they can collapse further into clusters or single stars. If the disk is rotating, or the layer is expanding, then the largest scales are stabilized and there is a column density threshold that has to be exceeded so that regions large enough to exceed the Jeans length (where gravity exceeds pressure) are also smaller than the stabilization length (where rotation or expansion exceed gravity).

Inside the clouds formed by these processes, star formation is also triggered by locally high pressures from HII regions, winds, and multiple SNe, and from cloud collisions, or in supersonically turbulent media, by compressions from converging flows. All of these compressions tend to enhance magnetic diffusion and decrease the dynamical time. The morphology of this triggering is also fairly easy to recognize in ideal cases because the compressed regions are shells, comets, and moving layers, all adjacent to high pressures, and all with the causality constraint mentioned above.

3. Turbulence Triggering

The compressed regions in a supersonically turbulent cloud act like seeds for gravitational attraction and can lead to the local accretions necessary to make stars (see reviews in Mac Low & Klessen 2004; Bonnell et al. 2007). Simulations by several groups have shown how turbulence in a cloud core can produce a star cluster (e.g., Li et al. 2004; Bate & Bonnell 2005; Jappsen et al. 2005; Padoan et al. 2005; Nakamura & Li 2005; Martel et al. 2006; Tilley & Pudritz 2007).

There are several signatures of turbulence triggering. First, the cloudy structure in a turbulent medium has a power law power spectrum, which means there is no characteristic scale except the largest scale (e.g., Stützke et al. 1998). In fact, the morphology tends to be hierarchical, with large clouds containing small subclouds over many levels. This hierarchy in gas produces a similar hierarchy in young stars, which is evident as substructure in embedded clusters (e.g., Testi et al. 2000; Dahm & Simon 2005; Gutermuth et al. 2005), and as nested super-structures on larger scales (e.g., several subgroups are collected into each OB association, and several OB associations are collected into each star complex; see review in Elmegreen 2008a). Each galactic cluster seems to be the inner mixed region of the hierarchy, where the efficiency of star formation is automatically high (Elmegreen 2008b).

Second, the hierarchy of structures produces a mass spectrum for clusters, and nearly the same mass spectrum for stars, that is $dN/dM \propto M^{-2}$ (the Salpeter IMF would have a power -2.35). This is because each layer in the hierarchy contains the same total mass, just divided up in different ways. Because the hierarchy is a sequence in the log of the mass (each level divides up the mass of the previous higher level), we have the mass conserving requirement that $MdN/d\log M = $ constant, that is, the total mass in each $\log M$ interval is constant. This converts to $M^2 dN/dM = $ constant, as above. Another way to view this is to consider a hierarchy of levels where the final product of star formation, a cluster, for example, or an OB association, comes from some cloud at one of many possible levels in the hierarchy. That cloud is contained, along with other clouds, in the next higher cloud, and also is subdivided into several sub-clouds. Then it turns out that the same mass function follows by randomly sampling all clouds in the hierarchical tree. That is, if every cloud (and all of its subclouds) is equally likely to produce a stellar object as every other cloud, then $M^2 dN/dM =$ constant again. This may be seen with a simple example. Imagine clouds subdivided by twos: one cloud of mass 32 units divided into 2 clouds of mass 16 units, which are each divided into 2 more clouds of 8 units, and so on until the smallest level, which has 32 clouds of 1 unit mass each. The total number of clouds, counting everything, is $32+16+8+4+2+1 = 63$. The probability that a cloud of mass 8, say, is selected, is the number of clouds with mass 8, namely 4 clouds, divided by the total number of clouds, 63. The probability that a cloud of mass 4 is chosen is similarly $8/63$. In general, the probability that a cloud of mass M is chosen is $\propto 1/M$. Since we have intervals of $\log M$ again, and we are assuming the number of clouds is proportional to their probability, we get $dN/d\log M \propto 1/M$, or $dN/dM \propto 1/M^2$, as above. Thus turbulence, and the hierarchical structure it always produces, is the likely cause of the M^{-2} mass functions for clusters and OB associations (Elmegreen & Efremov 1997), and maybe even stars.

The mass function for clouds is measured differently. Clouds are usually defined by the resolution of a survey. That is, most cloud mass functions have clouds with sizes within a factor of 10 of the survey resolution. The total span of mass is therefore only a factor of 100 (since mass is proportional to size-squared). Larger clouds are not counted because they can always be subdivided into their smaller pieces. This is a problem with defining clouds as regions inside of closed contours, for example, and of disallowing any multiple-counting of mass. Turbulent media are not really composed of separate clouds within an intercloud medium. Contours do not represent the power law structure correctly and mass spectra obtained from contours do not represent the true distribution of mass into all of its parts. For stellar structures, we can measure objects arbitrarily large (unlike the case for contoured clouds) because the stellar objects are not defined by their structure (e.g., contours) but by their stellar content: IR-excess stars, for example, define a young cluster, OB stars define an OB association, and red supergiants define a star complex (Efremov 1995). This connection between stellar types and structure nomenclature is a selection effect, resulting from the third aspect of turbulence triggering, discussed next. Also, stellar aggregates tend to be defined by friends-of-friends algorithms, which is intrinsically hierarchical, unlike contouring. Better algorithms for counting clouds have recently been devised (Rosolowsky *et al.* 2008).

Third, turbulence triggering tends to occur on the timescale for the turbulent motions to move through a region, i.e., the crossing time. Because the turbulent speed scales with a fractional power of the size, the crossing time (size divided by speed) also increases with a fractional power of the size. This means that larger regions form stars longer. This correlation has been observed in the LMC clusters and cepheids (Efremov & Elmegreen 1998). As a result, regions defined by stars with a certain age range, such as IR-excess stars, O-type stars, etc., tend to have a certain size and mass. They are not distinct objects, however. OB associations are not intrinsically different from T associations or embedded clusters or star complexes, aside from their difference in selected age range. Recall that turbulence is scale free. It is only the selection of an age that corresponds to the selection of a certain scale or mass of star formation. This is true up to the largest scale for star formation, which is a flocculent spiral arm or one of the beads-on-a-string in a stellar arm (Elmegreen & Efremov 1996; Odekon 2008).

Turbulence triggering seems to work along side sequential triggering on scales smaller than the ambient Jeans length, which is L_J defined below. This is also about the galactic scale height. Thus a simplified model for all of these processes would be that gravitational instabilities produce clouds and drive turbulence on the largest scales (L_J), while turbulence and gravitational collapse trigger the first generation of stars inside these primary clouds. Sequential triggering prolongs star formation in the debris during the process of cloud disruption. The fractions of stars triggered by turbulence and by the various sequential processes would be interesting to observe.

4. The Cosmological Connection

Deep surveys in the optical and infrared have produced images of thousands of young galaxies, some of which are half, or even one-tenth, the age of the current universe. It is interesting to ask whether star formation in these galaxies has the same cause and morphology as local star formation.

To investigate this, we have measured regions of star formation in all of the large (> 10 pixels) galaxies in the Hubble Space Telescope Ultra Deep Field (UDF, Beckwith 2006). There are about 1000 of them (Elmegreen et al. 2005). Galaxies look different at high redshift. Irregular structures dominate, and interactions are relatively common (Abraham et al. 1996a,b; Conselice, Blackburne, & Papovich 2005). The one irregularity that seems to be most common is the presence of giant blue clumps from star formation (Elmegreen & Elmegreen 2005). These clumps are usually observed in the restframe ultraviolet because of the high redshifts involved, but even there they have absolute magnitudes that can be brighter than -18 mag (Elmegreen & Elmegreen 2006b; Elmegreen et al. 2007a, 2008c). Population models suggest that clump stellar masses are in the range $10^7 - 10^8$ $M_\odot$, with some clumps larger than 10^9 $M_\odot$ (Elmegreen et al. 2007a, 2008c). Clump diameters are ~ 1.5 kpc, and stellar ages in the clumps are typically within a factor of 3 of 10^8 yrs (Elmegreen & Elmegreen 2005; Elmegreen et al. 2008c).

There are no obvious shells or comets of triggered star formation at high redshift, but none are expected at the available resolution of ~ 200 pc out to $z \sim 5$. In the most clumpy galaxies, which are the *clump clusters* and *chains* (Elmegreen et al. 2005), there are no spirals either, even though spiral structure would be seen at ~ 5 if it was bright enough in the uv. Other galaxies clearly have spirals and bulges. At $z > 1$, clump clusters and chains in the UDF outnumber spirals 2:1.

In a recent survey of bulge properties using H-band NICMOS observations in the UDF (Elmegreen et al. 2008c), we found that $\sim 50\%$ of the clump clusters and $\sim 30\%$ of the chains have bright, often central, red clumps indicative of bulges. Others may have only a red and smooth underlying disk. We also found that when there are bulges in the most clumpy galaxies, these bulges are more similar to the clumps with respect to age and mass than the bulges in spiral galaxies are to their clumps. Thus clump clusters and chains either have no bulges or they have young bulges.

Figure 1 shows a selection of UDF clump cluster galaxies without obvious bulges (from Elmegreen et al. 2008c, which contains a color version of this figure). Each galaxy is shown as a pair of images, with the color Skywalker image on the left and the NICMOS H-band image on the right. There are bright clumps at NICMOS-H, but nothing centralized and nothing obviously connected with a disk population; most are associated with a star formation feature, and some could be independent galaxies. Either these clump clusters will turn into late-type spirals with small bulges, or they have not yet formed their bulges. Conceivably, one of the prominent clumps could migrate to the center to make a bulge, or several could collide and make a bulge. This clump-migration model for bulge formation was proposed by Noguchi (1999) and Immeli et al. (2004a,b). More details and a match to observations is in recent papers by Bournaud et al. (2007, 2008) and Elmegreen et al. (2008a,c).

The recent batch of models starts with a disk that is half gas and half stars, in addition to a live halo. The stellar part is Toomre-stable but the gas+stellar disk is unstable. Almost immediately, the gas and some of the stars collapse into four or five giant clumps with 10^8 $M_\odot$, like the observed clumps in clump-cluster galaxies. When the disk is slightly off-center from the bulge, the clumps get flung around to large radii, as observed in UDF 6462 (Bournaud et al. 2008), contributing to the general appearance of irregularity. The basic process of clump formation is a gravitational instability. The collapse is rapid and it forms clumps rather than spirals because the self-gravitational forces dominate the background galactic forces. This result seems reasonable for a young galaxy primed with fresh gas from cosmological accretion. The instability has two important differences compared with that in local galaxies: first, the high gas fraction makes the clumps round rather than spiral-like, and second, a high turbulent speed makes the clumps massive. Recall that the bulk Jeans length is $L_J = 2\sigma^2/G\Sigma$ for turbulent speed σ and disk column density Σ, and the Jeans mass is $M_J = \sigma^4/G^2\Sigma$. The key to a large unstable mass is a large velocity dispersion.

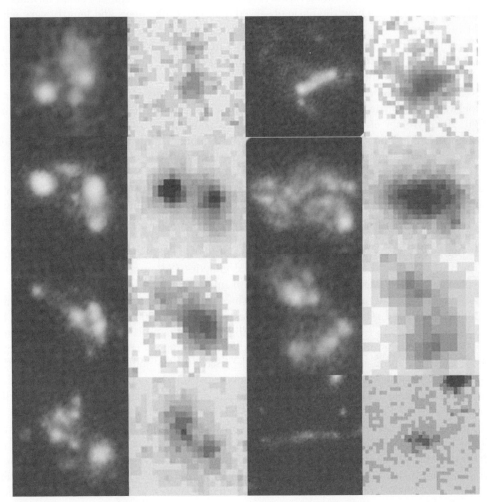

Figure 1. Eight clumpy galaxies in the UDF with no obvious central red object that might be a bulge. Each galaxy is shown twice, on the left using the high resolution of the ACS from the color Skywalker image, and on the right using the lower resolution NICMOS image in H band. Red emission is either extended in a seemingly old population of stars, or it is associated with bright star-forming regions.

The basic scales L_J and M_J come from the dispersion relation for gravitational instabilities in an infinitely thin and extended disk, which is $\omega^2 = \sigma^2 k^2 - 2\pi k G \Sigma$; $i\omega$ is the growth rate and k is the wavenumber. The wavenumber at fastest growth is obtained by setting $d\omega/dk = 0$, which gives $k_{fast} = \pi G \Sigma / \sigma^2$. The Jeans length is $L_J = 2\pi/k_{fast}$ and the Jeans mass is $M_J \sim (L_J/2)^2 \Sigma = \sigma^4/G^2 \Sigma$, as written above. This mass could be written in a variety of ways, such as $\pi (L_J/2)^2 \Sigma$ or $(L_J)^2 \Sigma$, depending on assumptions about what constitutes a cloud; i.e., what fraction of the unstable mass gets into the cloud. The preferred form is a compromise, and selected partly to match local observations of giant cloud masses. Additional things modify the dispersion relation, such as magnetic fields, spiral arms, finite disk thickness, turbulent motions, a non-isothermal equation of state, molecule formation, and so on. Simulations reproduce many of these effects in a way that simple expressions cannot. A detailed model for star formation in a galaxy disk, such as that by Robertson & Kravtsov (2008) discussed at this conference, should do a better job of defining a characteristic scale or outer-scale for cloud formation. Still, it is useful to see how the basic observed quantities, like mass and length, vary with ISM properties.

The mass and size of star formation clumps in high redshift galaxies exceed those in local galaxies by factors of ~ 100 and ~ 3, respectively. Thus σ^2 has to increase by $100/3 \sim 30$, which means that σ has to increase by $\times 5.5$, making it 30 or 40 km s^{-1} instead of the local 6 or 7 km s^{-1}. This requirement is satisfied by the observation of high velocity dispersions in high redshift galaxies (Förster Schreiber, et al. 2006; Genzel et al. 2006, 2008; Weiner et al. 2006). Similarly, Σ has to be larger by a factor of $100/3^2 \sim 10$ at high redshift. The gaseous disks of local spiral galaxies have $\Sigma \sim 10\ M_\odot$ pc^{-2}, which makes $M_J \sim 2 \times 10^7\ M_\odot$, comparable to the observed local gas clump mass, for $\sigma = 6$ km s^{-1}. At high redshift, the unstable column density in the disk has to be $\sim 100\ M_\odot$ pc^{-2}, which is comparable to the total mass column density of the inner regions of today's spirals. Considering that high redshift spiral galaxies are slightly smaller than today's spirals (Elmegreen et al. 2007b), the clump clusters and chains could be forming the inner thick disks and bulges of today's galaxies.

A big uncertainty for star formation studies of high redshift galaxies is the neutral gas abundance. CO has been observed in several galaxies (Solomon & Vanden Bout 2005; Tacconi et al. 2008) but with little resolution into clouds. HI has not been observed in emission yet and the absorption of HI, in the form of damped Lyman α lines, has an unknown geometry relative to stellar galaxies (Wolfe et al. 2008). CO is also present in DLA gas (Srianand et al. 2008). Quite possibly, the gas mass is comparable to or larger than the observed stellar mass, and the gas is as irregular as the stars in these clumpy disks. Our prediction is that the velocity dispersion of the neutral gas should be high, something like 40 km s^{-1} or more, which is observed for the ionized gas. The dispersion inside the clumps should be high also, although perhaps not quite as high if only the cores are observed in molecular transitions.

There are two other pieces of evidence that star formation is prompted by gravitational instabilities in high redshift disks. First there is a linear alignment of clumps in chain galaxies, which are presumably edge-on clumpy disks. Clump positions are aligned along the midplane of the chain to within a fraction of a pixel (Elmegreen & Elmegreen 2006a) or ~ 100 pc, on average (i.e., for 112 chain galaxies in the UDF). This implies that most clumps are not extragalactic objects in the process of coalescence; they formed in a pre-existing disk. Second, a large fraction of interacting galaxies with tidal features and rings have regularly spaced clumps in those features (Elmegreen et al. 2007a). They therefore had to form there, and their separation should be comparable to L_J. This gives the same requirement on velocity dispersion as the clumps in non-interacting disks.

5. Conclusions

Giant cloud formation is often triggered by gravitational and associated instabilities in gas disks. The clump scale is ~ 600 pc for local galaxies, and what forms is a "star-complex," composed of OB associations and dense clusters with $10^5 - 10^6\ M_\odot$ of stars. The cloud mass at the beginning of the process is $\sim 10^7\ M_\odot$. Most of this gas is in the form of low-density HI except in the dense inner regions of galaxies, where it can be largely molecular because of the higher ambient pressure.

In high-redshift disks, the clump scale is larger, ~ 1500 pc, and the stellar mass is larger, $\sim 10^7 - 10^8\ M_\odot$. The associated gas mass in a clump is unknown but may be $10^8 - 10^9\ M_\odot$ with much of that molecular. This scale-up of star formation at high z seems to be the result of a high turbulent speed, which, combined with a higher gas column density, makes the gravitational length and characteristic mass larger by factors of ~ 3 and ~ 100, respectively.

In local galaxies, star formation begins quickly in cloud cores once they become self-gravitating. There is no reason to think otherwise for high redshifts. Locally, the cloud cores are cold, dense, molecular, magnetic, and turbulent – all necessary attributes contributing to star formation in one way or another (including the requirement of angular momentum loss during star formation). The same should be true at high redshift too. Outside the local dense cores, star formation lingers in isolated cloud debris and it can be triggered for a long time in the dispersing cloud envelopes. The analogous final stages of star formation at high redshift are unknown. Cloud dispersal by star formation feedback should be more difficult with a higher velocity dispersion (Elmegreen et al. 2008b). If this difficulty increases the efficiency of star formation, then comparatively little gas could remain in the cloud envelopes for prolonged triggering. Generally, cloud envelopes are more stable than their cores because of the envelope exposure to background starlight and the resulting longer magnetic diffusion time. This is why triggering can be important in the final stages of cloud disruption: compression enhances the diffusion rate while it shortens the dynamical time, causing low density regions to form stars where they otherwise would not. There is no evidence for sequentially-triggered star formation at high redshift, however, but the resolution is not good enough yet to see it.

Other differences between low and high redshift star formation concern the fate of the stellar clumps. At low z, where the disk gas fraction is low and the clump formation and evolution times are comparable to the shear time, star-forming clumps dissolve slowly into star streams and add to the thin disk without changing their galactocentric radii. At high z, clump formation appears to be much more violent and rapid compared to background galactic processes. This is observed directly in clump clusters and chain galaxies, and it is true to a certain degree also in spirals, where the clumps are more massive than they are locally. These differences follow from the observed relatively high turbulent speed and the expected high gas mass fraction. Simulations suggest that the clumps in the most clumpy disks interact with each other, stir the preexisting stars to make a thicker disk, shed half of their own stars during these interactions to add to this thick disk, and then spiral in the remaining half to make or add to a bulge.

The general theme of this conference is to understand the formation and structure of the Milky Way in the context of various models and observations of galaxies on cosmological scales. In terms of star formation, the youngest resolvable disk galaxies look both strange, in terms of their increased clumpiness, and familiar, in terms of the likely processes involved. Simulations can reproduce the basic structures without difficulty if they start with ideal initial conditions. At some point, however, the galaxy formation process has to be important, and this involves the gas accretion rate and geometry, and the galaxy interaction rate. It is more difficult to understand these aspects of young galaxies without the local analogues that are so revealing for star formation.

References

Abraham, R., Tanvir, N., Santiago, B., Ellis, R., Glazebrook, K., & van den Bergh, S. 1996a, MNRAS, 279, L47
Abraham, R., van den Bergh, S., Glazebrook, K., Ellis, R., Santiago, B., Surma, P., & Griffiths, R. 1996b, ApJS, 107, 1
Bate, M. R. & Bonnell, I. A. 2005, MNRAS, 356, 1201
Beckwith, S. V. W., et al. 2006, AJ, 132, 1729
Bonnell, I. A., Dobbs, C. L., Robitaille, T. P., & Pringle, J. E. 2006, MNRAS, 365, 37
Bonnell, I. A., Larson, R. B., & Zinnecker, H. 2007, in: B. Reipurth, D. Jewitt, & K. Keil (eds), *Protostars and Planets VI* (Tucson, Univ of Arizona), p. 149
Bournaud, F., Elmegreen, B. G., & Elmegreen, D. M. 2007a, ApJ, 670, 237
Bournaud, F., Daddi, E., Elmegreen, B. G., Elmegreen, D. M., & Elbaz, D. 2008, A&A in press, astroph/0803.3831
Brinks, E. & Bajaja, E. 1986, A&A, 169, 14
Conselice, C. J., Blackburne, J. A., & Papovich, C. 2005a, ApJ, 620, 564
Dahm, S. E. & Simon, T. 2005, AJ, 129, 829
de Geus, E. J. 1992, A&A, 262, 258
Dekel, A. & Silk, J. 1986, ApJ, 303, 39
Dobbs, C. L. & Bonnell, I. A. 2007, MNRAS, 374, 1115
Efremov, Y. N. 1995, AJ, 110, 2757
Efremov, Yu. N. & Elmegreen, B. G. 1998, MNRAS, 299, 588
Elmegreen, B. G. 1994, ApJ, 427, 384
Elmegreen, B. G. 2007, ApJ, 668, 1064
Elmegreen, B. G. 2008a, in: A. de Koter, L. J. Smith, & L. B. F. M. Waters (eds), *Mass Loss from Stars and the Evolution of Stellar Clusters* (San Francisco: Astronomical Society of the Pacific), p. 249
Elmegreen, B. G. 2008b, ApJ, 672, 1006
Elmegreen, B. G. & Efremov, Y. N. 1996, ApJ, 466, 802
Elmegreen, B. G. & Efremov, Y. N. 1997, ApJ 480, 235
Elmegreen, B. G., Palous, J., & Ehlerova, S. 2002, MNRAS, 334, 693
Elmegreen, B. G. & Elmegreen, D. M. 2005, ApJ, 627, 632
Elmegreen, D. M., Elmegreen, B. G., Rubin, D. S., & Schaffer, M. A. 2005, ApJ, 631, 85
Elmegreen, B. G. & Elmegreen, D. M. 2006a, ApJ, 650, 644 thick disks
Elmegreen, D. M. & Elmegreen, B. G. 2006b, ApJ, 651, 676 rings bend chains
Elmegreen, D. M., Elmegreen, B. G., Ferguson, T., & Mullan, B. 2007a, ApJ, 663, 734 tidal tails
Elmegreen, D. M., Elmegreen, B. G., Ravindranath, S., & Coe, D. A. 2007b, ApJ, 658, 763
Elmegreen, B. G., Bournaud, F., & Elmegreen, D. M. 2008a, ApJ, 684, in press
Elmegreen, B. G., Bournaud, F., & Elmegreen, D. M. 2008b, ApJ, submitted
Elmegreen, D. M., Elmegreen, B. G., Fernandez, M. X., & Lemonias, J. J. 2008c, ApJ, submitted
Engargiola, G., Plambeck, R. L., Rosolowsky, E., & Blitz, L. 2003, ApJS, 149, 343
Förster Schreiber, N. M., et al. 2006, ApJ, 645, 1062
Fuchs, B., Dettbarn, C., & Tsuchiya, T. 2005, A&A, 444, 1
Garcia-Burillo, S., Guelin, M., & Cernicharo, J. 1993, A&A, 274, 123
Genzel, R., et al. 2006, Nature, 442, 786
Genzel, R., et al. 2008, ApJ, in press, aarXiv:0807.1184
Gutermuth, R. A., Megeath, S. T., Pipher, J. L., Williams, J. P., Allen, L. E., Myers, P. C., & Raines, S. N. 2005, ApJ, 632, 397
Hester, J. et al. 1996, AJ, 111, 2349
Immeli, A., Samland, M., Gerhard, O., & Westera, P. 2004a, A&A, 413, 547
Immeli, A., Samland, M., Westera, P., & Gerhard, O. 2004b, ApJ, 611, 20
Jappsen, A.-K., Klessen, R. S., Larson, R. B., Li, Y., & Mac Low, M.-M. 2005, A&A, 435, 611
Kim, Chang-Goo; Kim, Woong-Tae; Ostriker, & Eve C. 2006, ApJ, 649, L13
Kim, W.-T. & Ostriker, E. C. 2002, ApJ, 570, 132
Kim, W.-T., Ostriker, E. C., & Stone, J. M. 2003, ApJ, 599, 1157

Kosiński, R. & Hanasz, M. 2006, *MNRAS*, 368, 759
La Vigne, M. A., Vogel, S. N., & Ostriker, E. C. 2006, *ApJ*, 650, 818
Lee, S. M. & Hong, S. S. 2007, *ApJS*, 169, 269
Li, P. S., Norman, M. L., Mac Low, M.-M., & Heitsch, F. 2004, *ApJ*, 605, 800
Luna, A., Bronfman, L., Carrasco, L., & May, J. 2006, *ApJ*, 641, 938
Lundgren, A. A., Olofsson, H., Wiklind, T., & Rydbeck, G. 2004, *A&A*, 422, 865
Mac Low, M.-M. & Klessen, R. S. 2004, *RvMP*, 76, 125
Martel, H., Evans, N. J., II, & Shapiro, P. R. 2006, *ApJS*, 163, 122
Nakamura, F. & Li, Z.-Y. 2005, *ApJ*, 631, 411
Noguchi, M. 1999, *ApJ*, 514, 77
Odekon, M. C. 2008, *ApJ*, 681, 1248
Oey, M. S., Watson, A. M., Kern, K., & Walth, G. L. 2005, *AJ*, 129, 393
Padoan, P., Kritsuk, A., Norman, M. L., & Nordlund, A. 2005, *Memorie della Societa Astronomica Italiana*, 76, 187
Reach, W. T., et al. 2004, *ApJS*, 154, 385
Robertson, B. E. & Kravtsov, A. V. 2008, *ApJ*, 680, 1083
Rosolowsky, E. W., Pineda, J. E., Kauffmann, J., & Goodman, A. A. 2008, *ApJ*, 679, 1138
Solomon, P. & Vanden Bout, P. 2005, *ARA&A*, 43, 677
Srianand, R. Noterdaeme, P., Ledoux, C., & Petitjean, P. 2008, *A&A*, 482, L39
Stützki, J., Bensch, F., Heithausen, A., Ossenkopf, V., & Zielinsky, M. 1998, *A&A*, 336, 697
Sugitani, K., Fukui, Y., Mizuni, A., & Ohashi, N. 1989, *ApJ*, 342, L87
Tacconi, L. J. et al. 2008, *ApJ*, 680, 246
Testi, L., Sargent, A. I., Olmi, L., & Onello, J. S. 2000, *ApJ*, 540, L53
Tilley, D. A. & Pudritz, R. E. 2007, *MNRAS*, 382, 73
Tosaki, T., Hasegawa, T., Shioya, Y., Kuno, N., & Matsushita, S. 2002, *PASJ*, 54, 209
Yamaguchi, R., Mizuno, N., Onishi, T., Mizuno, A., & Fukui, Y. 2001a, *PASJ*, 53, 959
Yamaguchi, R., Mizuno, N., Onishi, T., Mizuno, A., & Fukui, Y. 2001b, *PASJ*, 553, L185
Walter, F. & Brinks, E. 1999, *AJ*, 118, 273
Weiner, B. J., et al. 2006, *ApJ*, 653, 1027
Whitworth, A. P., Bhattal, A. S., Chapman, S. J., Disney, M. J., & Turner, J. A. 1994, *A&A*, 290, 421
Wilcots, E. M. & Miller, B. W. 1998, *AJ*, 116, 2363
Wolfe, A. M., Prochaska, J. X., Jorgenson, R. A., & Rafelski, M. 2008, arXiv:0802.3914
Zavagno, A., Deharveng, L., Comerón, F., Brand, J., Massi, F., Caplan, J., & Russeil, D. 2006, *A&A*, 446, 171

Coffee break. Camilla Juul Hansen and Anna Árnadottir outside the lecture hall.

From the Saturday excursion to Hven. The group tries, with mixed success, to emulate the posture of the statue of Tycho Brahe. Photo: Bruce Elmegreen (in front, at centre).

H I in Galactic Disks

Elias Brinks[1], Frank Bigiel[2], Adam Leroy[2], Fabian Walter[2], W. J. G. de Blok[3], Ioannis Bagetakos[1], Antonio Usero[1], and Robert C. Kennicutt, Jr.[4]

[1] Centre for Astrophysics Research, University of Hertfordshire, Hatfield AL10 9AB, UK
[2] Max–Plank–Institut für Astronomie, Königstuhl 17, 69117, Heidelberg, Germany
[3] Univ. of Cape Town, Dept. of Astronomy, Private Bag X3, Rondebosch 7701, South Africa
[4] Institute of Astronomy, University of Cambridge, Madingley Road, Cambridge CB3 0HA, UK

Abstract. Studies of the atomic phase of the interstellar medium, via the 21–cm spectral line of neutral hydrogen (H I), play a key rôle in our attempts to understand the structure and evolution of disk galaxies. We present here results from The H I Nearby Galaxy Survey (THINGS) and focus on the mass distribution as derived from the observed kinematics, and on the link between gas and star formation rate surface density, i.e., the Schmidt–Kennicutt law. Also, we briefly dwell on the wealth and wide variety of structures, often outlining what seem to be expanding shells surrounding sites of recent, massive star formation.

Keywords. galaxies: structure, galaxies: spiral, galaxies: ISM

1. Introduction

Results coming from The H I Nearby Galaxy Survey (THINGS) are beginning to shed new light on the structure and properties of galaxy disks. The THINGS team have observed 34 galaxies with the NRAO† Very Large Array in its B–, C–, and D–configuration, resulting in maps at a spatial resolution of $6''$. The velocity resolution is $5.2\,\mathrm{km\,s}^{-1}$ or better and the observations reach typical 1σ column density sensitivities of $4 \times 10^{19}\,\mathrm{cm}^{-2}$ at $30''$ resolution. Full details regarding the observations and data reduction can be found in Walter *et al.* (2008). THINGS covers a wide range of Hubble types, star formation rates, absolute luminosities, and metallicities and is being used to address key science questions regarding the kinematics and matter distribution in galaxies (de Blok *et al.* 2008; Oh *et al.* 2008; Trachternach *et al.* 2008), their star formation properties (Bigiel *et al.* 2008; Leroy *et al.* 2008a) and the structure of the ISM (Bagetakos *et al.* 2008; Usero *et al.* 2008). Results from a study assessing the extent of the neutral gas disks and what causes them to drop sharply in surface brightness were presented at this meeting as a poster by Portas (this volume) whereas some of the results derived by Leroy *et al.* (2008a) were highlighted in the talk by Blitz (this volume).

Most of the galaxies in THINGS were drawn from the *Spitzer* Infrared Nearby Galaxies Survey (SINGS; Kennicutt *et al.* 2003), a multi-wavelength project designed to study the properties of the dusty ISM in nearby galaxies. This ensures that multi–wavelength observations for each galaxy are available for further analysis. To this aim we also collaborate with the team running the *GALEX* (Galaxy Evolution Explorer) Nearby Galaxies Survey (Gil de Paz *et al.* 2007). Our final resolution of $6''$ is well matched to that of the *Spitzer* Space Telescope (e.g., resolution at $24\,\mu\mathrm{m}$: $6''$) and *GALEX* (resolution in the Near–Ultraviolet: $5''$). A further crucial ingredient in several of the papers which

† The National Radio Astronomy Observatory is a facility of the National Science Foundation operated under cooperative agreement by Associated Universities, Inc.

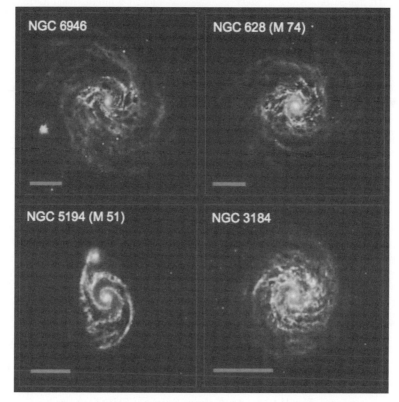

Figure 1. Some examples of spiral galaxies observed as part of THINGS. The H I emission is colour–coded in blue, the older stellar population is assigned an orange hue. Purple is a measure of the recent star formation activity and is a linear combination of the FUV flux measured with *GALEX*, and *Spitzer* 24 μm emission. The green bar measures 10 kpc.

are currently being prepared by us for publication is access to high quality CO observations. This is being provided by a parallel survey (HERACLES) with the HERA array of the IRAM 30–m telescope of 18 THINGS galaxies which we observed in the CO(2-1) transition (Leroy *et al.* 2008b).

Fig. 1 shows a mosaic of false colour composites for four spiral galaxies (see figure caption for details). The figure illustrates the quality achieved by THINGS and gives an indication of what can be achieved by combining THINGS with SINGS and *GALEX* data. In the sections which follow, we will highlight a few of the aspects of the analysis which is currently in progress.

2. Kinematics and Mass models

In the paper by de Blok *et al.* (2008) we present a rotation curve analyis of 19 galaxies. These are the highest quality H I rotation curves available to date for a large sample of nearby galaxies, spanning a wide range of H I masses and luminosities. The high quality of the data allows us to systematically derive the geometrical and dynamical parameters using H I data alone, for a much larger sample than has hitherto been possible. Fig. 2 illustrates how the analysis is performed.

We do not find any declining rotation curves unambiguously associated with a cut–off in the mass distribution out to the last measured point. The rotation curves are combined with 3.6 μm data from SINGS to construct mass models. Our best–fit dynamical disk

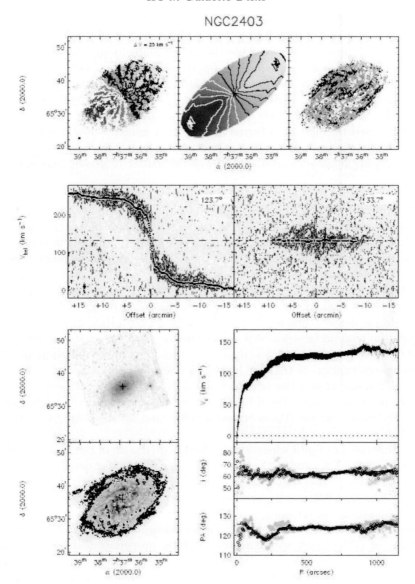

Figure 2. Tilted–ring analysis of the galaxy NGC 2403. The top row shows the observed velocity field (left), the tilted–ring model (centre), and residuals (right). The systemic velocity is indicated by the thick contour. The approaching side can be identified by the light gray–scales and the dark contours. The adopted dynamical center is indicated with a cross. The beam size is indicated by the ellipse in the bottom–left corner. The centre row presents a position–velocity cut along the major (left) and minor axis (right). The systemic velocity and position of the center are indicated by dashed lines. Over–plotted is the rotation curve projected onto the average major axis using the derived radial variations of PA and i. The bottom left panel shows at the top a *Spitzer* 3.6 μm emission map and at the bottom the H I surface brightness map. The bottom right panel is made up of the derived rotational velocity (top), inclination (middle) and position angle of the receding kinematical major axis (bottom) as a function of radius (figure taken from de Blok *et al.* 2008).

masses, derived from the rotation curves, are in good agreement with photometric disk masses derived from the 3.6 μm images in combination with stellar population synthesis arguments. This implies that we can use the 3.6 μm maps to infer the stellar mass rather than having to assume a maximum disk for the stellar component, and thus removing considerable uncertainty in the determination of the mass profile of the non–baryonic matter. The stellar mass–modelling is described in detail in the papers by de Blok et al. (2008) and Leroy et al. (2008a). We also confirm that lower mass (dwarf) systems are dark matter dominated throughout: whereas in high mass galaxies the baryonic mass within the radius where the peak disk rotational velocity is found, can account for the velocity observed, in lower mass systems about 50% of the matter is already non–baryonic.

3. Star Formation Law

Following the pioneering work of Schmidt (1959), it is common to relate gas density to the SFR density using a power law. Although originally formulated in terms of volume densities, for a constant scale height disk the Schmidt–law can be written as $\Sigma_{SFR} \sim (\Sigma_{gas})^N$. The value of N has been found in the literature to cover a range of 0.9 to 3.5 (see Bigiel et al. 2008, for a summary). Kennicutt (1998), extending his earlier work (Kennicutt 1989), studied the relation between the disk–averaged SFR surface density and gas surface density, finding $N = 1.4 \pm 0.15$. Using THINGS we are now able to investigate the relation between SFR surface density and gas surface density on a pixel by pixel basis, the pixels being chosen to be at a common linear resolution of 750 pc. We use a linear combination of GALEX FUV and Spitzer 24 μm emission to determine the SFR. A full account of how this is done is given in the appendix of Leroy et al. (2008a). Moreover, we can investigate the Schmidt–Kennicutt law for just the neutral atomic gas (H I), for the molecular gas (H$_2$ as traced by CO), and for the total gas surface density (H I+H$_2$).

This is illustrated in Fig. 3 where we plot $\log \Sigma_{SFR}$ (the log of the SFR surface density) against $\log \Sigma_{H\,I}$, $\log \Sigma_{H_2}$, and $\log \Sigma_{H\,I+H_2}$ (see figure caption for details). In the paper by Bigiel et al. (2008) we find that the Σ_{SFR} and Σ_{H_2} obey a tight Schmidt–type power law with index $N = 1.0 \pm 0.2$, i.e., H$_2$ forms stars at a constant efficiency in the disks of spiral galaxies. The average molecular gas depletion time is $\sim 2 \times 10^9$ years. We interpret the linear relation and constant depletion time as evidence that stars in the disks of spirals are forming in GMCs with approximately uniform properties and that Σ_{H_2} may be a measure of the filling fraction of GMCs rather than real variations in surface density. The relationship between total gas surface density $\Sigma_{H\,I+H_2}$ and Σ_{SFR} varies dramatically among and within spiral galaxies which is due to the fact that most galaxies show little or no correlation between $\Sigma_{H\,I}$ and Σ_{SFR}. In addition to a molecular Schmidt law, the other general feature of our sample is a sharp saturation of H I surface densities at $\Sigma_{H\,I} = 9\,M_\odot\,pc^{-2}$ (or $12\,M_\odot\,pc^{-2}$ when including He) above which we observe gas to be molecular.

4. H I Supershells

The superior resolution and sensitivity of THINGS reveals a wealth of structure in the ISM of gas–rich systems and is exploited in a paper by Bagetakos et al. (2008). Fig. 4, adapted from their paper, shows as example the H I surface brightness map of NGC 6946. The H I closely traces the flocculent spiral arms. In addition, a large number of elliptical features can be seen, both in the H I surface brightness maps as well as in the data cubes. They have in the past been described as H I shells or supershells (depending on their size), H I superbubbles, or H I holes. For the remainder of this paper we refer to them as H I holes as this is how they appear in maps and avoids prejudicing their interpretation.

The most widely accepted explanation for the origin of H I holes is that they are due to the almost instantaneous deposition of kinetic energy as the result of Type II supernovae from rapidly evolving, massive stars which formed in super star clusters (SSCs) or OB associations (Oey & Clarke 1997). We have detected more than 1000 holes in a total of 20 galaxies. This is the first time that the same detection technique is applied using such a large data set of uniformly high quality. The sizes of the H I holes range from about 100 pc (our resolution limit) to 2000 pc. Their expansion velocities vary from 5 to 35 km s^{-1}. We estimate their ages at 6 – 150 Myr and their energy requirements, based on a simple single blast approximation (Chevalier 1974), at $10^{50} - 10^{53}$ erg. In most galaxies, H I holes are found all the way out to the edge of the H I disk. Assuming that holes are the result of massive star formation we estimate the star formation rate and find that it correlates with values obtained by other SF tracers, lending support to the idea that H I holes and SF are linked. We also calculated the 2– and 3–dimensional porosity of the ISM in the galaxies studied and find that it tends to increase from early–type spirals to dwarf

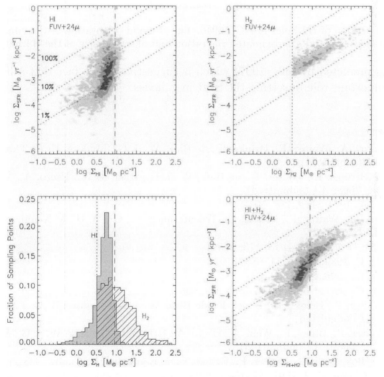

Figure 3. Plots of $\log \Sigma_{\rm SFR}$ versus $\log \Sigma_{\rm H\,I}$ (top left), $\log \Sigma_{\rm H_2}$ (top right), and $\log \Sigma_{\rm H\,I+H_2}$ (bottom right); the results for 7 spiral galaxies are plotted together in these diagrams. Diagonal dotted lines show lines of constant SFE, indicating the level of SFR needed to consume 1%, 10% and 100% of the gas reservoir (including helium) in 10^8 years. Thus, the lines also correspond to constant gas depletion times of, from top to bottom, $10^8, 10^9$, and 10^{10} yr. The bottom left panel shows the normalized distribution of H I and H$_2$ surface densities in the sample. The dashed vertical line is drawn at $9\,{\rm M}_\odot\,{\rm pc}^{-2}$ which is where we observe gas in excess of this surface density to be in the molecular phase; the dotted vertical line corresponds to the sensitivity of the CO observations used to infer the H$_2$ column density and lies at $3\,{\rm M}_\odot\,{\rm pc}^{-2}$ (figure taken and adapted from Bigiel et al. 2008).

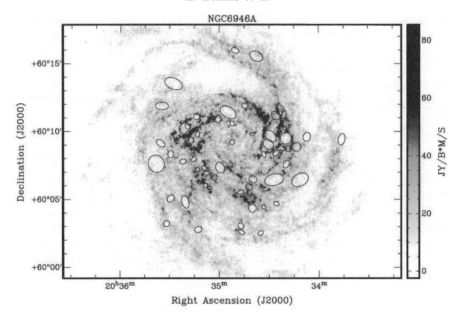

Figure 4. Location of H I holes in NGC 6946. The size and orientation of the ellipses indicate the location and approximate orientation and observed size of the H I holes.

galaxies, as predicted by Silk (1997) which is suggestive of the fact that self–regulation plays an important rôle in setting the SFR in galaxies.

References

Bigiel, F., Leroy, A., Walter, F., Brinks, E., de Blok, W. J. G., Madore, B., & Thornley, M. D. 2008, *AJ* (submitted)
Bagetakos, I., Brinks, E., Walter, F., de Blok, W. J. G., Rich, J. W., Usero, A., & Kennicutt, R. C., Jr. 2008, *AJ* (submitted)
Chevalier, R. A. 1974, *ApJ*, 188, 501
de Blok, W. J. G., Walter, F., Brinks, E., Trachternach, C., Oh, S.-H., & Kennicutt, R. C., Jr. 2008, *AJ* (accepted)
Gil de Paz, A., *et al.* 2007, *ApJS*, 173, 185
Kennicutt, R. C. 1989, *ApJ*, 344, 685
Kennicutt, R. C. 1998, *ApJ*, 498, 541
Kennicutt, R. C., Jr., *et al.* 2003, *PASP*, 115, 928
Leroy, A., Walter, F., Brinks, E., Bigiel, F., de Blok, W. J. G., Madore, B., & Thornley, M. D. 2008a, *AJ* (submitted)
Leroy, A., Walter, F., Bigiel, F., Weiß, A., Usero, A., Brinks, E., de Blok, W. J. G., & Kennicutt, R. C., Jr. 2008b, *AJ* (submitted)
Oh, S.-H., de Blok, W. J. G., Walter, F., Brinks, E., & Kennicutt, R. C., Jr. 2008, *AJ* (accepted)
Oey, M. S. & Clarke, C. J. 1997, *MNRAS*, 289, 570
Schmidt, M. 1959, *ApJ*, 129, 243
Silk, J. 1997, *ApJ*, 481, 703
Trachternach, C., de Blok, W. J. G., Walter, F., Brinks, E., & Kennicutt, R. C., Jr. 2008, *AJ* (submitted)
Usero, A., Brinks, E., Walter, F., de Blok, W. J. G., & Kennicutt, R. C., Jr. 2008, *AJ* (submitted)
Walter, F., Brinks, E., de Blok, W. J. G., Bigiel, F., Kennicutt, R. C., Jr., Thornley, M. D., & Leroy, A. 2008, *AJ* (accepted)

The Molecular Gas Component of Galaxy Disks

Leo Blitz

University of California, Berkeley
Radio Astronomy Lab., 601 Campbell Hall
Berkeley, CA 94720-3411, USA
email: blitz@astro.berkeley.edu

Abstract. The molecular gas in galaxy disks shows much more galaxy to galaxy variation than does the atomic gas. Detailed studies show that this variation can be attributed to differences in hydrostatic pressure in the disks due largely to variations in the stellar surface density and the total gas surface density. One prediction of pressure modulated H_2 formation is that the location where HI and H_2 have equal surface densities occurs at a constant value of the stellar surface density in the disk. Observations confirm this constancy to 40%.

Keywords. star formation, galaxies, galaxy disks

1. Introduction

Among the relatively few statements that can be made with confidence about star formation in galaxies, one stands out: stars form from clouds of molecular gas, now and always. We know this empirically at the present and earlier epochs from observations that show that the youngest stars are always found embedded in molecular gas. From theory, astronomers generally believe that stars form as the result of a Jeans instability. In the form of the Jeans density, we can write:

$$\rho_{Jeans} = \left(\frac{kT}{\mu m_H G}\right)^3 \frac{\pi^5}{(6M_J)^2} \qquad (1)$$

Since T is typically ~ 10 K and can, in any event, never be less than 2.7 K, to form a one solar mass star requires a number density of $\sim 10^6$ cm^{-3}, densities found only in molecular clouds. Thus, when one considers star formation in galaxies from the ISM , one need only consider the molecular gas; the atomic gas is irrelevant except as a reservoir from which molecular gas can associate.

Galaxy disks exhibit relatively little variation in HI surface density with radius and from galaxy to galaxy (for late types) except in dwarfs. Typically, the surface density peaks at about 10 $M_\odot$ and remains relatively constant throughout the stellar disk; regions with higher gas surface density are almost invariably primarily molecular. The variation in molecular gas content and surface density is however much greater than for HI. This can be seen in Figures 1 and 2. Figure 1 shows the CO emission from 20 strong CO emitters from the BIMA SONG Survey (Helfer *et al.* 2003). Figure 2 shows the CO emission from the remaining 24 galaxies in the survey displayed on the same scale. To be detectable at all in these images, the CO must have an H_2 surface density of ~ 6 $M_\odot$. The range of H_2 surface densities shown in the two figures is more than two orders of magnitude. The Milky Way most closely resembles NGC 3351 in its CO content (second row and third column of Figure 2). NGC 4535 in the third row and first column of

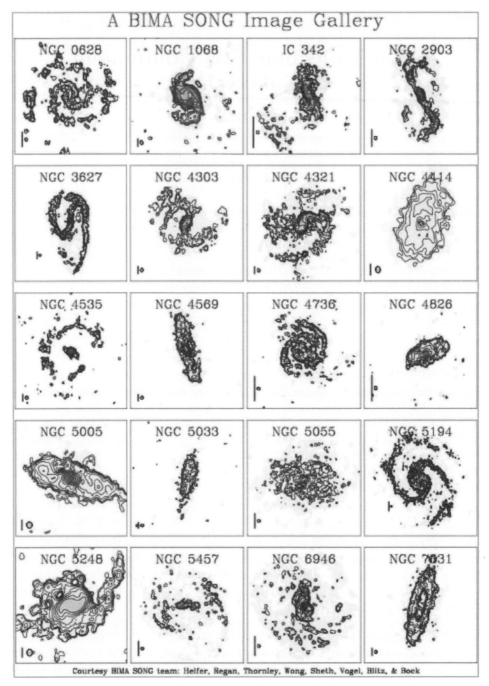

Figure 1. Maps of CO emission from 20 of the strongest emitters from the 44 galaxy BIMA SONG survey. Note the wide range in CO morphologies in the maps. REF

Figure 1 is morphologically similar, and is also a multi-armed spiral, but has a ring of emission with a higher surface density than that of the Galaxy.

What causes the differences from galaxy to galaxy? Following the suggestion of Wong and Blitz (2002), Blitz and Rosolowsky (2004, 2006) argued that the ratio of $N(H_2)/N(HI)$

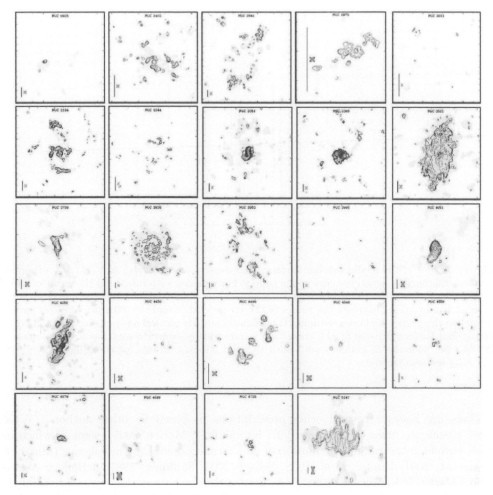

Figure 2. Maps of the CO emission from the remaining 24 galaxies in the survey which are not as frequently displayed as those in Figure 1. Some galaxies are not detected at all, some only in the nucleus.

on a pixel by pixel basis is primarily due to the hydrostatic pressure in the galactic disk and they empirically derived the following relation:

$$R_{mol} = \left[\frac{P_{ext}/k}{(3.5 \pm 0.6) \times 10^4}\right]^{0.92 \pm 0.07},$$

where $R_{mol} = 2N(H_2)/N(HI)$.

Because galaxy disks exhibit little variation in the velocity dispersion of the gas and the scale height of the stars both within and between galaxies, one of the surprising consequences of the R_{mol}-pressure relation is that $R_{mol} = 1$ should occur at constant *stellar* surface density in every galaxy. Blitz and Rosolowsky (2004) showed that for the 22 galaxies they analyzed, the constancy in stellar surface density is good to 40% even though the radius where the equivalence in hydrogen surface density varies by more than an order of magnitude (see Figure 3). From the R_{mol}-pressure relation, Blitz and Rosolowsky (2006) developed a star formation prescription that they argued was an improvement over the Kennicutt (1998a) star formation prescription.

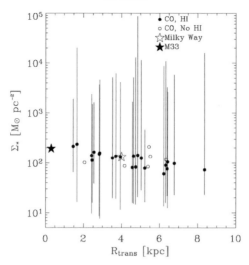

Figure 3. Plot of the median stellar surface mass density where $N(H_2) \approx N(HI)$ as a function of where this surface density occurs in the galaxy. For galaxies with measured HI densities (filled circles, 22 galaxies), the *range* of stellar surface densities is plotted as a vertical line (this is *not* an error bar), running between the values of the 1st percentile and the 99th percentile of surface density in the galaxy. Galaxies without HI measurements are plotted as open circles (6 galaxies). The stellar surface density at the transition is remarkably constant and has a mean value of 120 $M_\odot$ pc^{-2} for the 22 galaxies with both CO and HI data. Points for the Milky Way (MW) and M33 are also plotted.

There are however sevaral other prescriptions proposed by other authors, (Wyse, 1986; Kennicutt 1998a; Elmegreen 1997; Krumholz & McKee, 2005), some of which include cutoffs in the radial distribution of star formation effiiency in galaxies (Martin & Kennicutt, 2001; Hunter *et al.* 1998; Schaye 2004; Skillman, 1987; deBlok & Walter 2006). Martin & Kennicutt (2001), e.g., argued that star formation cuts off beyond the radius where the disk becomes Toomre stable. Clearly, star formation prescriptions that include atomic gas must have either a cutoff in the star formation efficiency, or some sort of falloff at the edge of the stellar disk because the HI disks tend to be much more extended, often with no change in surface density at the edge of the stellar disk (Bigiel 2008).

The most extensive investigation of various star formation prescriptions and cutoffs has been done in a recent paper by Leroy *et al.* (2008) using the THINGS HI Survey (Walter *et al.* 2008), and the BIMA SONG CO Survey (Helfer *et al.* 2003) to get the neutral hydrogen content of a sizable number of disks. Leroy *et al.* (2008) used the GALEX Survey (Gil de Paz 2007) and Spitzer SINGS observations (Kennicutt *et al.* 2003) to measure the star formation in the disks of the same galaxies, supplemented by additional observations of their own (Leroy *et al.* 2009)

This vast work has come to a number of important conclusions, some of which are listed below:

1) The star formation efficiency of H_2, (the rate at which a given surface density of H_2 turns into stars) is constant independent of environment or any other variable they investigated. the value is $5.25 \pm 2.5 \times 10^{-10}$ yr^{-1}. The inverse of this quantity is

the molecular gas depletion time, the time to use up all of the molecular gas to star formation, and is equal to 1.9×10^9 y.

2) In spiral galaxies, the transition between a mostly HI and a mostly H_2 ISM is a well defined function of local conditions. It occurs at a characteristic radius $r = 0.43 \pm 0.18$ r_{25}, $\Sigma_* = 81 \pm 25$ $M_\odot$ pc^{-2}, $\Sigma_{gas} = 14 \pm 6$ $M_\odot$ pc^{-2}, $P_{hydro} = 2.3 \pm 1.5 \times 10^4$ k_B cm^{-3} K, and $\tau_{orb} = 1.8 \pm 0.4 \times 10^8$ yr.

3) R_{mol} appears to be a continuous function of environment from the HI dominated ($R_{mol} \sim 0.1$) to H_2 dominated ($R_{mol} \sim 10$) regime. The variation in R_{mol} is too strong to be reproduced only by varying τ_{orb} or τ_{ff}. Physics other than these timescales must also play an important role in cloud formation

4) Thresholds for large scale stability do not offer an obvious way to predict R_{mol}. There is no clear relationship (continuous or stepfunction) between the star formation efficiency and Q_{gas}, $Q_{stars+gas}$, or to a shear threshold. Disks appear to be stable or marginally stable throughout once the effects of stars are included.

5) The power law relationship between R_{mol} and hydrostatic pressure found by Leroy et al. (2008) is roughly consistent with expectations by Elmegreen (1993), observations by Wong & Blitz (2002) and Blitz & Rosolowsky (2006), and simulations by Robertson & Kravtsov (2008). In its simplest form, this is a variation on the classical Schmidt law, i.e., R_{mol} is set by gas *volume* density.

6) Power law fits of R_{mol} to P_h, radius, τ_{orb}, and Σ_* reproduce the observed star formation efficiency reasonably well in spiral galaxies but yield large scatter or higher than expected star formation efficiencies in the outer parts of dwarf galaxies, offering indirect evidence that differences between the dwarf and normal spiral subsample such as metallicity (dust), radiation field, and strong spiral shocks play a role in setting these relations.

References

Bigiel, F., Leroy, A., Walter, F., Brinks, E., de Blok, W. J. G., Madore, B., & Thornley, M. D. 2008, arXiv:0810.2541
Blitz, L. & Rosolowsky, E. 2004, ApJL, 612, L29
Blitz, L. & Rosolowsky, E. 2006, ApJ, 650, 933
de Blok, W. J. G. & Walter, F. 2006, AJ, 131, 363
Elmegreen, B. G. 1993, ApJ, 411, 170
Elmegreen, B. G. 1997, Revista Mexicana de Astronomia y Astrofisica Conference Series, 6, 165
Gil de Paz, A., et al. 2007, ApJS, 173, 185
Helfer, T. T., Thornley, M. D., Regan, M. W., Wong, T., Sheth, K., Vogel, S. N., Blitz, L., & Bock, D. C.-J. 2003, ApJS, 145, 259
Hunter, D. A., Elmegreen, B. G., & Baker, A. L. 1998, ApJ, 493, 595
Kennicutt, R. C., Jr. 1998, ARAA, 36, 189
Kennicutt, R. C., Jr. 1998, ApJ, 498, 541
Kennicutt, R. C., Jr., et al. 2003, PASP, 115, 928
Krumholz, M. R. & McKee, C. F. 2005, ApJ, 630, 250
Leroy, A. K., Walter, F., Brinks, E., Bigiel, F., de Blok, W. J. G., Madore, B., & Thornley, M. D. 2008, arXiv:0810.2556
Leroy et al. 2009, AJ submitted
Martin, C. L. & Kennicutt, R. C., Jr. 2001, ApJ, 555, 301
Robertson, B. E. & Kravtsov, A. V. 2008, ApJ, 680, 1083

Schaye, J. 2004, ApJ, 609, 667
Skillman, E. D. 1987, NASA Conference Publication, 2466, 263
Walter, F., Brinks, E., de Blok, W. J. G., Bigiel, F., Kennicutt, R. C., Jr., Thornley, M. D., & Leroy, A. K. 2008, arXiv:0810.2125
Wong, T. & Blitz, L. 2002, ApJ, 569, 157
Wyse, R. F. G. 1986, ApJL, 311, L41

Linda Sparke, Gerry Gilmore and Ken Freeman discussing at the welcome reception.

Disk Stability and Turbulence Generation: Effects of the Stellar Component

Woong-Tae Kim

Department of Physics and Astronomy, FPRD, Seoul National University, Seoul 151-742, Republic of Korea
email: wkim@astro.snu.ac.kr

Abstract. Galactic disks consist of both stars and gas. The stars gravitationally influence the gas either in disks at large or within spiral arms, leading to the formation of giant clouds and turbulence driving in the gas. In featureless disks as in flocculent galaxies, swing amplification operating in a combined star-gas disk is efficient to form bound condensations and feed a significant level of random gas motions. This occurs when the gaseous Toomre parameter is less than 1.4 for the stellar parameters similar to the solar neighbourhood conditions. In disks with spiral features, on the other hand, spiral-arm spurs and associated giant clouds develop as a consequence of magneto-Jeans instability in which magnetic tension counterbalances the stabilizing Coriolis force. Spiral shocks are inherently unstable when the vertical dimension is taken into account, exhibiting flapping motions of the shock front. This naturally converts the kinetic energy in galaxy rotation into random kinetic energy of the gas. The resulting turbulent motions are supersonic and persist despite strong shock dissipation. Thermal instability occurring in gas flows across spiral arms prompts phases transitions that produce a significant fraction of thermally-unstable, intermediate-temperature gas in the postshock expansion zones.

Keywords. galaxies: ISM, instabilities, ISM: kinematics and dynamics, ISM: magnetic fields, method: numerical, stars: formation

1. Introduction

Understanding the physical processes that regulate star formation in disk galaxies is of paramount importance for galaxy formation and evolution, which is one of the main themes of this symposium. The fact that most star formation in the Milky Way and external disk galaxies takes place in cold, giant molecular clouds (GMCs) (e.g., Williams, Blitz, & McKee 2000) implies that the compression of diffuse gas (or small clouds) into GMCs is the first step in initiating galactic star formation. The fact that the largest GMCs, including atomic envelopes, typically have mass $\sim 10^7 M_\odot$ and spacing ~ 1 kpc (e.g., Elmegreen 1995a and references therein), similar to the characteristic Jeans mass and wavelength in galactic gaseous disks at large, strongly supports the idea that GMCs originate from large-scale gravitational instabilities (e.g., Elmegreen & Elmegreen 1983). Although Parker instability has also been proposed as a candidate for GMC formation (Blitz & Shu 1980), it is stabilized by magnetic tension at the non-linear stage and is thus unable to produce high-density structures like GMCs (e.g., Santillán et al. 2000; Kim & Ostriker 2006).

Turbulence appears to dominate the internal dynamics responsible for fragmentation, compression/dispersal, and hierarchical structures of GMCs and plays an important role in controlling star formation within GMCs (e.g., Elmegreen 2007; McKee & Ostriker 2007). Various mechanisms have been proposed for interstellar turbulence, including ones based on stellar energies such as expansion of H II regions, outflows from young stellar objects, supernova explosions, etc., and ones based on non-stellar sources such as magnetorotational instability, gravitational instability, and galactic spiral shocks, etc. (see Mac Low & Klessen 2004; Elmegreen & Scalo 2004; Ostriker 2006 for recent reviews). Stellar sources have often been favoured since supernovae at the Galactic rate are estimated to produce the observed level of the ISM turbulence. However, radio observations of external galaxies indicate that turbulent H I velocity dispersions in non-star-forming outer disks are comparable to those in star-forming inner disks (van Zee & Bryant 1999) and they are not correlated with spiral arms, as well (Petric & Rupen 2007). These invoke the possibility of turbulence driving by non-stellar sources, many of which simply transport kinetic energy in disk rotation into random gas motions.

In this article we review recent numerical studies on GMC formation and turbulence generation via gravitational instability and other dynamical processes in disk galaxies, concentrating on the role played by stellar gravity. In addition to galactic differential rotation, thermal/turbulent pressure, self-gravity, and magnetic fields, gas in disk galaxies is also influenced by gravity from a dynamically-active stellar disk or by stellar spiral potential perturbations. For more detailed, quantitative results, the reader is refereed to Kim et al. (2006, 2008) and Kim & Ostriker (2006, 2007).

2. Gravitational Instabilities in Featureless Disks

2.1. Axisymmetric Instability

Disk galaxies contain stellar and gaseous disks with finite vertical thickness. Despite large velocity dispersions, the large mass fraction ($\sim 75\% - 90\%$) of the stellar disk makes its contribution to the growth of self-gravitating modes in the combined system almost comparable to that of the gaseous part. Most previous works on the effects of the live stellar component approximated stars as an isothermal "fluid" (e.g., Jog & Solomon 1984, Elmegreen 1995b, Jog 1996). Taking a true kinetic treatment of the stellar component, Rafikov (2001) derived the dispersion relation for axisymmetric modes in razor-thin, two-component disks. Taking allowance for the dilution of self-gravity due to finite vertical thickness, the axisymmetric dispersion relation becomes

$$\frac{2\pi Gk\Sigma_g}{(\kappa^2 + k^2 c_g^2 - \omega^2)(1 + kH_g)} + \frac{2\pi Gk\Sigma_s \mathcal{F}(\omega/\kappa, k^2\sigma_R^2/\kappa^2)}{(\kappa^2 - \omega^2)(1 + kH_s)} = 1, \qquad (2.1)$$

where ω and k denote the frequency and wavenumber of perturbations, respectively, H_g and H_s are the vertical scale heights of the gaseous and stellar disks, respectively, and $\mathcal{F}$ is the stellar reduction factor (Kim & Ostriker 2007). Other symbols have their usual meanings. For razor-thin disks ($H_g = H_s = 0$), equation (2.1) yields the sufficient conditions $Q_g \equiv \kappa c_g/\pi(G\Sigma_g) > 1$ for gas-only disks and $Q_s \equiv \kappa \sigma_R/(3.36 G\Sigma_s) > 1$ for star-only disks.

Figure 1a plots the marginal stability curves ($\omega^2 = 0$) on the Q_g–Q_s plane for some values of $R \equiv c_g/\sigma_R$. The results of Rafikov (2001) for razor-thin disks with are plotted as thin lines, while thick lines are for disks with finite thickness with $H_g = 0.87 c_g/\kappa$ and $H_s = 0.4\sigma_R/\kappa$. Note that a two-component disk marked by a filled circle at $(Q_g, Q_s) = (1.4, 2.1)$ and $R = 0.3$ corresponding to the solar neighborhood conditions, is highly stable to axisymmetric perturbations when finite disk thickness is considered, while it would be marginally stable if treated as razor-thin. Figure 1b plots the critical Q_g values as functions of the gaseous scale height. Clearly, the presence of the stellar component makes the gaseous disk more unstable, increasing $Q_{g,c}$ from unity to 1.27 for razor-thin disks. On the other hand, finite disk thickness decreases this value to $Q_{g,c} \simeq 0.67$ when

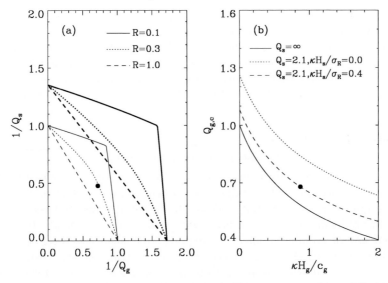

Figure 1. (a) Marginal stability curves for axisymmetric gravitational instability in two-components disks with $R \equiv c_g/\sigma_R$; the region below each curve represents stable configurations. Thin lines correspond to razor-thin disks, while thick lines are for disks with $H_g = 0.87c_g/\kappa$ and $H_s = 0.4\sigma_R/\kappa$. (b) The critical Q_g values for gas-only disks (solid line), for combined disks with a razor-thin, stellar component (dotted line), and combined disks with a thick, stellar component (dashed line). In both panels, the dots represent the solar neighborhood conditions.

$H_g = 0.87c_g/\kappa$ and $H_s = 0.4\sigma_R/\kappa$, indicating that the stabilizing effect of disk thickness overwhelms the destabilizing effect of stellar gravity. Given that $Q_{g,c} \simeq 0.67$ is well below observed critical values at $Q_g \sim 1.4$ (Martin & Kennicutt 2001), star formation thresholds are unlikely due to axisymmetric gravitational instability in two-component disks.

2.2. Swing Amplification

Even if disks are stable to axisymmetric perturbations, nonaxisymmetric modes in shearing disks are allowed to grow as they swing from leading to trailing configurations. This occurs due to conspiracy among rotational shear, self-gravity, and epicycle shaking (Toomre 1981). While the amplitude of density growth in swing is a continuous function of Q_g in the linear theory, nonlinear simulations show that disks exhibit Q_g threshold behavior such that swing-amplified disturbances in disks with Q_g small enough become nonlinear and subsequently gravitationally unstable. Kim & Ostriker (2001) found that the critical values for nonlinear threshold are $Q_{g,c} \sim 1.2 - 1.4$ for razor-thin, gas-only disks. By solving the orbits of stars using a particle-mesh method (instead of treating them as a fluid) and evolving the gaseous disk based on a hydrodynamic method, Kim & Ostriker (2007) found $Q_{g,c} \sim 1.4$ for thick, two-component disks with the stellar parameters similar to the solar neighborhood values. Similar critical values were obtained by Li et al. 2006 for global galaxy models. Disks with Q_g less than the critical value develop high-density filaments that eventually undergo fragmentation and produce gravitationally bound condensations of mass $6 \times 10^7 M_\odot$ each.

In addition to forming bound condensations, swing amplification in two-component disks can also be vigorous enough to tap rotational and gravitational energy to feed random motions in the gaseous component. When $Q_g \simeq Q_{g,c}$, the gaseous velocity dispersions induced in two-component disks amount to $\sim 0.6c_g$, which is $\sim 3-4$ times larger than the values obtained from gas-only disks (Kim & Ostriker 2007). This level of turbulence driven by swing amplification in marginally-stable disks is comparable to that by magnetorotational instability in non-self-gravitating, gas-only disks (Kim et al. 2003). The turbulence driven by swing is anisotropic. The ratio of total power in the shearing to compressive parts is about 1.5, which is $\sim 3-7$ times lower than found in hypersonic turbulence. This indicates that the turbulence in two-component models is subsonic and that the pumping by self-gravity is preferentially via compressional motions.

3. Galactic Spiral Shocks

3.1. *Magneto-Jeans Instability*

The stellar component has a dramatic effect on the evolution of the gaseous component through spiral density waves. Being dynamically much more responsive, the gas passing through the gravitational potential wells of stellar spiral arms is strongly compressed and shocked, producing narrow dust lanes with strong magnetic fields. In addition, the conservation law of angular momentum requires that shear be reversed inside spiral arms, making the overall shear experienced by gas traversing the arm regions very small. These imply that spiral arms are prone to magneto-Jeans instability (Lynden-Bell 1966; Elmegreen 1994; Kim & Ostriker 2002) in which magnetic tension forces negate the stabilizing effects of galaxy rotation. When finite disk thickness is considered, the most unstable disturbances have wavelengths about 10 times the Jeans length at the arm peaks (Kim & Ostriker 2006). They first grow in the direction perpendicular to the arms before trailing away in the interarm regions, similar to dust filaments or spurs (feathers) found in recent *Spitzer* and *HST* images of M51. In the nonlinear stage, spurs experience fragmentation to produce bound clumps of mass $\sim 10^7 M_\odot$ each, similar to the largest observed GMCs, which may evolve to H II regions in both arm and interarm regions.

3.2. *Turbulence Driving*

Spiral shocks act as one of the important mechanisms that transform a fraction of kinetic energy in galaxy rotation into gaseous turbulence. One-dimensional spiral shocks with fluid quantities varying only along the direction perpendicular to the shocks are known to readily stationary, indicating that they represent stable equilibria (Roberts 1969 Woodward 1975; Kim & Ostriker 2002). When the vertical degrees of freedom are allowed, however, spiral shocks never achieve a steady state, moving back and forth relative to the mean positions (Kim & Ostriker 2006). These shock flapping motions arise mainly because the vertical oscillation period of the gas is incommensurable with the arm-to-arm crossing time, so that the gas streamlines are not closed. The shock flapping is able to feed random gas motions on the scale of disk scale height that persist despite shock dissipation. The induced velocity dispersions amount to $\sim 7-10$ km s^{-1} in the plane and $\sim 3-4$ km s^{-1} in the vertical direction, and are larger inside spiral arms than in interarm regions by about a factor of 2 (Kim & Ostriker 2006).

3.3. *Effect of Thermal Instability*

One of the main characteristics of galactic spiral shocks is that strongly compressed gas at the shock front becomes less dense as it expands in the postshock regions. This implies that the gas may undergo thermal phase transitions when the shocked gas is subject to thermal instability. Recently, Kim et al. (2008) explored nonlinear evolution of galactic gas flows across spiral arms that are explicitly exposed to interstellar heating, cooling and thermal conduction. Figure 2 plots an exemplary density profile, when a quasi-steady state is reached, of a spiral shock in a one-dimensional model with the initial number density $n_0 = 2$ cm^{-3} and the arm strength $F = 5\%$. In the interarm regions ($x/L_x < 0$

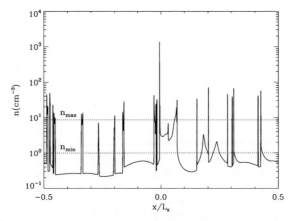

Figure 2. Density profile of a quasi-stationary galactic spiral shock with thermal instability in a model with the initial number density of $n_0 = 2$ cm^{-3}. Dotted lines marked by $n_{\max} = 8.6$ cm^{-3} and $n_{\min} = 1.0$ cm^{-3} divide the cold ($n > n_{\max}$), intermediate-temperature ($n_{\max} > n > n_{\min}$), and warm ($n < n_{\min}$) phases.

or $x/L_x > 0.2$) cold clouds with $n > n_{\max} = 8.6$ cm^{-3} are roughly in pressure balance with the surrounding warm intercloud medium with $n < n_{\min} = 1$ cm^{-3}.

The warm/cold gas in the interarm regions is shocked and immediately turns to denser cold gas in the arm region ($0 < x/L_x < 0.03$). Due to the post-shock expansion, this subsequently enters the unstable transition zone ($0.03 < x/L_x < 0.2$) and undergoes thermal instability to evolve back to the warm/cold interarm phases. For the model shown in Figure 2, the gas stays at the arm, transition, and interarm zones approximately for 14%, 22%, and 64% of the arm-to-arm crossing time, respectively. The gas at intermediate temperature (with $n_{\min} < n < n_{\max}$) is found to be $\sim 0.25 - 0.30$ of the total by mass. Of this, about 70% is found in the transition zone, while the remaining 30% corresponds to the boundaries of the cold interarm clouds, suggesting that the postshock expanding flows are an important source for the intermediate-temperature gas. Kim *et al.* (2008) also found that thermal instability in association with one-dimensional spiral shocks can drive random gas motions at ~ 1.5 km s^{-1} in the interarm and transition zones, which is $\sim 5 - 7$ times larger than the values from pure thermal instability (e.g., Kritsuk & Norman 2002; Piontek & Ostriker 2004). Given that one-dimensional isothermal spiral shocks are stable, these results suggest that flapping motions with thermal instability in multi-dimensional spiral shocks can be even more efficient in driving turbulence in the arm and interarm regions.

4. Summary

Recent numerical simulations have shown that for the stellar parameters similar to those in the solar neighborhood, a two-component (stars plus gas) disk without spiral features is subject to gravitational runaway provided the gaseous disk has a Toomre stability parameter less than 1.4. In this case, swing amplification is able to form bound clumps and drive a significant level of turbulent energy in the gas. In disks with spiral arms, the gas is easily compressed to form a spiral shock where magneto-Jeans instability is efficient to produce arm substructures such as spurs and giant clouds. Spiral shocks in two or higher dimensions are inherently unstable to generate random gas motions that remain supersonic even in the presence of shock dissipation. In short, the effect of the stellar component on disk instability and turbulence driving is by no means negligible.

Acknowledgements

Much of the material contained in this article is based on the collaborative work with E. Ostriker and C.-G. Kim. The financial support from the organizers of this symposium is gratefully acknowledged. This work has been supported in part by Korea Research Foundation Grant funded by the Korean Government (MOEHRD) (KRF-2007-313-C00328).

References

Blitz, L. & Shu, F. H. 1980, *ApJ*, 238, 148
Elmegreen, B. G. 2007, *ApJ*, 668, 1064
Elmegreen, B. G. 1995a, in *The 7th Guo Shoujing Summer School on Astrophysics: Molecular Clouds and Star Formation*, eds. C. Yuan & Hunhan You (Singapore:World Scientific), 149
Elmegreen, B. G. 1995b, *MNRAS*, 275, 944
Elmegreen, B. G. & Elmegreen, D. M. 1983, *MNRAS*, 203, 31
Elmegreen, B. G. & Scalo, J. 2004, *ARAA*, 42, 211
Elmegreen, B. G. 1994, *ApJ*, 433, 39
Jog, C. J. 1996, *MNRAS*, 278, 209
Jog, C. J. & Solomon, P. M. 1984, *ApJ*, 276, 127
Kim, C.-G., Kim, W.-T., & Ostriker, E. C. 2006, *ApJ*, 649, L13
Kim, C.-G., Kim, W.-T., & Ostriker, E. C. 2008, *ApJ*, in press; arXiv:0804.0139
Kim, W.-T. & Ostriker, E. C. 2001, *ApJ*, 559, 70
Kim, W.-T. & Ostriker, E. C. 2002, *ApJ*, 570, 132
Kim, W.-T. & Ostriker, E. C. 2006, *ApJ*, 646, 213
Kim, W.-T. & Ostriker, E. C. 2007, *ApJ*, 660, 1232
Kim, W.-T. Ostriker, E. C, & Stone, J. M. 2003, *ApJ*, 599, 1151
Kritsuk, A. G. & Norman, M. L. 2002, *ApJ*, 569, L127
Li, Y., Mac Low, M.-M., & Klessen, R. S. 2006, *ApJ*, 626, 823
Lynden-Bell, D. 1966, *Observatory*, 86, 57
Mac Low, M.-M. & Klessen, R. S. 2004, *Rev. Mod. Phys.*, 76, 125
Martin, C. L. & Kennicutt, R. C. 2001, *ApJ*, 555, 301
McKee, C. F. & Ostriker, E. C. 2007, *ARAA*, 45, 567
Ostriker, E. C. 2006, in IAU Symp. 237, *Triggered Star Formation in a Turbulent ISM*, ed. B. G. Elmegreen & J. Palous (Cambridge: Cambridge Univ. Press), 70
Petric, A. O. & Rupen, M. P. 2007, *AJ*, 134, 1952
Piontek, R. A. & Ostriker, E. C. 2004, *ApJ*, 601, 905
Rafikov, R. R. 2001, *MNRAS*, 323, 445
Roberts, W. W. 1969, *ApJ*, 158, 123
Santillán, A., Kim, J., Franco, J., Martos, M., Hong, S.S., & Ryu, D. 2000, *ApJ*, 545, 353
Toomre, A. 1981, in *Structure and Evolution of Normal Galaxies*, eds. S. M. Fall & D. Lynden-Bell (Cambridge:Cambridge Univ. Press), 111
van Zee, L., & Bryant, J. 1999, *AJ*, 118, 2172
Williams, J. P., Blitz, L., & McKee, C. F. 2000, in *Protostars and Planets IV*, eds. Mannings, Boss, & Russell (Tuscon:Univ. of Arizona press), 97
Woodward, P. R. 1975, *ApJ*, 195, 61

Spiral Arm Tangencies in the Milky Way

Robert A. Benjamin[1]

[1]Department of Physics, University of Wisconsin-Whitewater,
800 West Main St., Whitewater, WI, USA
email: benjamir@www.uww.edu

Abstract. The historical directions of spiral arm tangencies in the Milky Way are presented and compared to results of mid-infrared star counts using the Spitzer Space Telescope. While the Scutum and Centaurus tangency directions show a 20-30% excess of star counts, all other expected tangency directions show no similar increases. These two tangencies are probably associated with a density wave arm that comes off the near side of the bar of the Galaxy while the other arms whose tangencies are not detected may be compression in the gas, but not in the old stellar disk.

Keywords. Galaxy: structure – Galaxy: disk – Galaxy: fundamental parameters – Galaxy: kinematics and dynamics – Galaxy: evolution

1. Galactic Spiral Structure in a Cosmological Context

Whatever happened to the study of Galactic spiral structure? IAU Symposium No. 1: *Coordination of Galactic Research*, which gathered 27 participants near Groningen in June 1953, marked the beginning of an era focused on Galactic structure in general and spiral structure in particular. Thirty years later, IAU Symposium No 106: *The Milky Way Galaxy*, which reassembled 150 participants in Groningen in June 1983, marked the end of spiral structure. The review of the subject in this second volume (Liszt 1985) collects over a half dozen maps, all strikingly different, many based on the same HI data, concludes that "there is really no adequate method available for solving this problem", and further warns that "the newly-begun process of deriving galactic structure in CO seems to be recapitulating the history laid down by HI observers." And at this IAU Symposium No. 254, twenty five years later almost to the day, the Milky Way has become a *disk* galaxy, as opposed to a *spiral* galaxy, and we debate its cosmological context. We seem to have dodged the question posed so innocently back in 1953.

All of which leads one to wonder, does Galactic spiral structure even *matter* in a cosmological context?

It might be that obtaining a complete understanding of the details of the spiral pattern is not necessary to construct adequate models for the global evolution and star formation history of disk galaxies. A study of spiral galaxies by Elmegreen & Elmegreen (1986), for example, showed little difference between the star formation rate of grand design and flocculent spiral galaxies, leading them to argue that well-organized spiral arms contribute less than 50% to the overall star formation rate of a disk galaxy. And in this volume, Bruce Elmegreen shows that for younger galaxies in the process of assembly, the bulk of star formation might have been much more irregular and episodic. Recent work shows evolution in the fraction of disk galaxies that show bars (Sheth *et al.* 1986), but not much is known about the cosmological evolution of spiral structure. Perhaps spiral arms only show up after disks have quieted down: an interesting, but irrelevant, regularity in the structure of disk galaxies.

But if we truly understand galactic disks, particularly the interrelation between the gaseous, stellar, and star formation components, I would think that we would also have a secure model of spiral structure: Spiral arms provide a test that any robust disk model must be able to pass. Unfortunately, although the Milky Way is the one galaxy in the Universe where we can separate out all the different stellar populations, measure kinematics, and study the ISM on a cloud-by-cloud basis, it is also the one spiral galaxy for which we do not have a satisfactory map of spiral structure.

2. Spiral Arm Tangencies for the Milky Way

Hunt through *Galactic Astronomy* (Binney & Merrifield 1998), and you will find three not-very-compelling maps of kinematic distances to molecular clouds, selected HII regions, and the distribution of Cepheids. From the latter two, the authors describe "a picture of the distribution of young stars that is woefully out of focus so that a measure of imagination will be required to make out features that may in reality be well-defined." The pitfalls of mapping spiral structure using kinematic distances to HI and CO emission are well known (Burton et al. 1992), the chief difficulty being that arms will be marked by deviations from circular rotation, while gas is mapped by *assuming* circular rotation. Another complicating factor is the fact that maps of spiral structure could depend on the tracer one uses. Binney & Merrifield (1998) make a distinction between gas arms, mass arms, and star formation arms. These features may not have similar amplitudes, or even be spatially coincident.

However, one aspect of Galactic spiral structure *ought* to be robustly established: the directions of spiral arm tangencies. In an ideal grand-design spiral galaxy, mapping spiral structure would be as simple as (1) identifying tangencies in direction and distance in the first quadrant ($l = 0 - 90°$), (2) identifying the corresponding tangencies in the fourth quadrant ($l = 270 - 360°$), and (3) connecting these points with a logarithmic spiral. A check of the method would be that the fourth quadrant tangencies should be a larger angle from the Galactic center than the corresponding first quadrant tangencies, since we expect the Milky Way to be a trailing spiral.

Figure 1 shows two different methods for locating the spiral arm tracers. The pairs of tangencies come from integrating the CO intensity of Dame et al. (2001) over a ± 15 km s^{-1} range around the tangent point velocity to identify the directions with the greatest tangent point emission (Dame, priv. comm). The historical range of tangencies, taken from a selective compilation of the literature by Englmaier & Gerhard (1999), is shown as well. A reassessment of the direction of spiral arm tangencies, using modern high angular resolutions surveys in CO and HI, would be extremely desirable. For arms interior to the Sun, a rational naming system would identify the arms by their first and fourth quadrant tangency locations. As this figure shows, the system is far from rational. The *Sagittarius-Carina* arm, for example, should really be the *Sagitta-Vela* arm, and the *Scutum-Crux* arm should probably be called the *Aquila-Centaurus* arm!

All of the previous work of spiral structure has focused on the distribution of gas and star formation, but GLIMPSE results (Benjamin et al. 2005) have shown that the high resolution, low extinction mid-infrared view of the stellar disk can yield surprises. GLIMPSE (Galactic Legacy Infrared MidPlane Survey Extraordinaire) and GLIMPSE 2 are *Spitzer Space Telescope Legacy Programs* to survey the inner Galaxy ($|l| \leqslant 65°$ and $|b| \lesssim 1°$) at 3.6, 4.5, 5.8, and 8.0 μm using the Infrared Array Camera (Benjamin et al. 2003). Fitting the average number of sources per square degree in the outer Galaxy with the model expectations for an exponential disk yielded an exponential stellar disk scale length, $H_* = 3.9 \pm 0.6$ kpc. This disk has been divided out in Figure 1. We found an extended Galactic bar, characterized by an enhancement of red clump giants at ∼12th magnitude with a brightness decreasing with decreasing longitude, yielding a bar angle, $\phi = 44° \pm 10°$ and half-length, $R_{bar} = 4.4 \pm 0.5$ kpc. This is different from the $20 - 25°$ seen for the shorter COBE/DIRBE bar (Gerhard 2002), which appears to be a distinct structure. This maximum longitude of the Long Bar is also the expected direction for the Scutum spiral arm tangency, suggesting that the spiral arm joins onto the bar at this point. We also detected a 25% excess of sources corresponding to the Centaurus spiral arm tangency, confirming the claim of Drimmel & Spergel (2001) based on the K-band light distribution as studied with COBE/DIRBE.

3. The "Missing" Spiral Arms

The clear detection of the Centaurus stellar disk tangency in Figure 1, and the excess of stars associated with the Bar/Scutum tangency seems to indicate that this feature of the Galaxy is characterized by an overdensity in stars as well as gas and star formation. But then, what is one to make of the fact that all the other expected tangencies are not evident in the GLIMPSE data? The "missing" Sagittarius arm tangency at $l \approx 50°$ was also noted by Drimmel (2000) and Drimmel & Spergel (2001). We agree with these

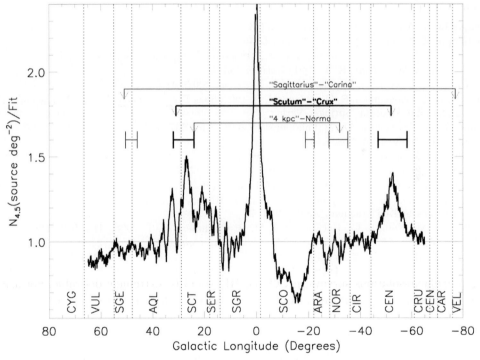

Figure 1. GLIMPSE 4.5 μm star counts between 6.5 and 12.5 magnitude normalized to a Bessel function fit to the data to take out the contribution of the exponential disk (Benjamin et al. 2005). Much of the jaggedness of the curve is attributable to extinction. The *Long Bar* ($l = 30°$ to $l \approx -15°$) can be seen clearly. Tangencies locations and spiral arm names based on CO studies (Dame, 2008, priv comm) are given with the horizontal bars with arrows. Note that in most cases, the name of the arm does not match with the constellation labels shown at the bottom of the graph. The "historical" tangency directions compiled by Englmaier & Gerhardt (1999) are shown for the *Sagittarius, Scutum, 3 kpc Norma* and *Centaurus* tangency. Note that for all these tangency directions, the only two which seem to show an excess in star counts over an exponential stellar disk are the Scutum and Centaurus dirccctions.

authors that the most likely explanation is that there is a qualitative difference between the different spiral arms of the galaxy: Observations of other galaxies show that it is not unusual for galaxies to have optically visible arms, without underlying enhancements in the old stellar disk (Block & Wainscoat 1991). Models, both old (Shu, Milione, & Roberts 1973) and new (Martos et al. 2004), show how that it is possible to form arms of compressed gas without increasing the stellar surface density.

If the model of a principally two-armed spiral for the Galaxy is correct, the Centaurus tangency provides an ideal testing ground for models of spiral density wave theory. Certainly, the $l = 302 - 313^o$ direction is known for several distinct anomalies, including large deviations in the HI velocity field (McClure-Griffiths & Dickey 2007) and a clear magnetic field reversal (Brown et al. 2007). In addition, the CS detection rate of MSX-selected dark clouds drops from about 80% to 20% in this direction (Jackson et al. 2008), suggesting that the densest molecular gas in the inner Galaxy lies principally in this Scutum-Centaurus arm which, we argue, is the region of the deepest gravitational potential.

The field of Galactic spiral structure has been moribund for more than a decade, but thanks to new surveys and tracers, I believe that a resurgence is at hand.

References

Benjamin, R. A., et al., 2003, PASP, 115, 953
Benjamin, R. A., et al. 2005, ApJL, 630, L149
Binney, J. & Merrifield, M. 1998, Galactic Astronomy, Princeton Univ. Press, N.J.
Block, D. L. & Wainscoat, R. J. 1991, Nature, 353, 48
Brown, J. C., Haverkorn, M., Gaensler, B. M., et al. 2007, ApJ, 663, 258
Burton, W. B., Elmegreen, B. G., & Genzel, R. 1992, Saas-Fee Advanced Course 21: The Galactic Interstellar Medium, Springer-Verlag
Dame, T. M., Hartmann, D., & Thaddeus, P. 2001, ApJ, 547, 792
Drimmel, R. 2000, A&A, 358, L13
Drimmel, R. & Spergel, D. N. 2001, ApJ, 556, 181
Elmegreen, B. G. & Elmegreen, D. M. 1986, ApJ, 311, 554
Englmaier, P. & Gerhard, O. 1999, MNRAS, 304, 512
Gerhard, O. 2002, in The Dynamics, Structure & History of Galaxies: A Workshop in Honour of Professor Ken Freeman, ASPC, 273, 73
Jackson, J. M., Finn, S. C., Rathborne, J. M., & Chambers, E. T. 2008, ApJ, 680, 349
Liszt, H. S. 1985, IAU Symp No 106: The Milky Way Galaxy, Kluwer, p. 283
Martos, M., Hernandez, X., Yáñez, M., Moreno, E., & Pichardo, B. 2004, MNRAS, 350, L47
McClure-Griffiths, N. M., & Dickey, J. M. 2007, ApJ, 671, 427
Sheth, K. et al. 2008, ApJ, 675, 1141
Shu, F. H., Milione, V., & Roberts, W. W., Jr. 1973, ApJ, 183, 819

Session 4: Stars as Drivers and Tracers of Chemical Evolution

Session chairs: Linda Sparke, Franziska Piontek and David Latham

Ole Strömgren bringing greetings from the family in his parents' home at Carlsberg.

Evolution and chemical and dynamical effects of high-mass stars

Georges Meynet[1], Cristina Chiappini[1,2], Cyril Georgy[1], Marco Pignatari[3,4], Raphael Hirschi[3], Sylvia Ekström[1] and André Maeder[1]

[1] Geneva University, Geneva Observatory
CH-1290 Versoix, Switzerland
email: georges.meynet@obs.unige.ch

[2] Osservatorio Astronomico di Trieste
Via G. B. Tiepolo 11, I - 34131 Trieste, Italia
email: Cristina.Chiappini@obs.unige.ch

[3] Keele University, Keele
Staffordshire ST5 5BG, United Kingdom
email: mpignatari@gmail.com

[4] Joint Institute for Nuclear Astrophysics
University of Notre Dame, Notre Dame, IN 46556, United States

Abstract. We review general characteristics of massive stars, present the main observable constraints that stellar models should reproduce. We discuss the impact of massive star nucleosynthesis on the early phases of the chemical evolution of the Milky Way (MW). We show that rotating models can account for the important primary nitrogen production needed at low metallicity. Interestingly such rotating models can also better account for other features as the variation with the metallicity of the C/O ratio. Damped Lyman Alpha (DLA) systems present similar characteristics as the halo of the MW for what concern the N/O and C/O ratios. Although in DLAs, the star formation history might be quite different from that of the halo, in these systems also, rotating stars (both massive and intermediate) probably play an important role for explaining these features. The production of primary nitrogen is accompanied by an overproduction of other elements as ^{13}C, ^{22}Ne and s-process elements. We show also how the observed variation with the metallicity of the number ratio of type Ibc to type II supernovae may be a consequence of the metallicity dependence of the line-driven stellar winds.

Keywords. stars: early-type, evolution, Wolf-Rayet, supernovae; Galaxy: halo; nucleosynthesis

1. Massive stars as "cosmic engines"

Massive stars are cosmic engines (see the recent proceedings entitled "Massive stars as Cosmic Engines" of the IAU Symp. 250) and thus represent a key ingredient in the evolution of the galaxies. At first sight this might be very surprising since these objects are very rare. Indeed one counts 3 stars with initial masses above 8 $M_\odot$ for 1000 stars with masses between 0.1 and 120 $M_\odot$ (estimate based on Salpeter's IMF). They contain about 14% of the stellar mass.

Thus both in number and in mass, massive stars represent small fractions. What makes them nevertheless very important objects is that massive stars are very "generous" objects, injecting in the interstellar medium, in *short timescales* (between 3 and 30 million years) great amounts of radiation, mass and mechanical energy:

- **Radiation:** using the mass-luminosity relation, a 100 $M_\odot$ has a luminosity about 1 million times higher than the Sun. The high luminosity of massive stars allows them to be observed as individual objects well beyond the Local Group. Of course when the star explodes as a core collapse supernova, individual events can be seen at still much greater distance. Long soft GRB are believed to be associated to core collapse supernova events. One of the farthest event of that kind presently known is at a redshift of 6.26 (Cusumano et al. 2006)!

Even in galaxies so far away that individual stars cannot be seen in them, it is still possible to detect in the spectrum of their integrated light signature of the presence of very massive stars undergoing strong stellar winds (Wolf-Rayet stars). The broad emission lines, which are formed in the fast expanding envelope of these stars, are sufficiently important to emerge above the continuum produced by the background stellar populations (see Brinchmann et al. 2008 for a recent survey of galaxies with Wolf-Rayet signatures in the low-redshift Universe). The analysis of such features allow to infer information about the number of such stars. From narrow emission lines as Hβ, it is possible to infer the number of O-type stars and thus to obtain a number fraction of WR to O-type stars in the starburst regions of distant galaxies. The WR/O ratio contains information on the star formation history, the strength of the starburst and its age. Of course the quality of the information deduced in that way depends on the quality of the stellar models used. The collective effect of massive stars is of first importance to understand the photometric evolution of galaxies. Typically about 2/3 of the visible light of galaxies arises from the massive star populations. Their high ionizing power gives birth to HII regions which trace the regions of recent star formation. The strong UV luminosity of massive stars has also been (and is still) used to deduce the history of star formation in the Universe (see e.g. Hopkins & Beacom 2006 for a discussion of the cosmic star formation history). In dusty galaxies, part of the UV light heats the dust and makes the galaxies to glow in IR (see e.g. the discussion of Pérez-González et al. 2006 on M81). Interestingly the ionizing front which expands with time around massive stars can trigger star formation in their vicinity. Examples of such behavior is seen in the Triffid nebula (see Hester et al. 2005). The ionising flux of the Pop III stars played a key role in reionizing the early Universe (Barkana 2006).

- **Mass:** either through stellar winds or at the time of the supernova explosion great amounts of mass are injected back into the interstellar medium. As seen above, about 14% of the mass of stars formed in a stellar generation, called M_* in the following, is in the form of massive stars. Nearly all this mass is ejected back into the interstellar medium by massive stars (13% of M_*), only a very small amount (about 1% of M_*) remains locked into compact remnants (neutron stars or black holes). If during the formation of a black hole, all the mass is swallowed by the black hole, smaller amounts of mass are returned. Typically, if all stars more massive than 30 $M_\odot$ follow such a scenario, then only a little more than 7% of M_* are returned. Part of the material returned, between 3.5 and 4.5% of M_*, is under the form of new synthesized elements and thus contributes to the chemical evolution of the galaxies and of the Universe as a whole.

- **Mechanical energy:** during its lifetime a star with an initial mass of 60 $M_\odot$ will lose through stellar winds more than 3 fourths of its initial mass through stellar winds (at solar metallicity!). Assuming a mass loss rate of 4×10^{-5} $M_\odot$ per year, a velocity of the wind of 3000 km s^{-1}, one obtains a mechanical luminosity equal to 30000 solar luminosities or to about 10% the radiation luminosity of the massive star. Integrating over the WR lifetime (about half a million years), one obtains that the mechanical energy injected into the surrounding amounts to 2×10^{51} erg, *i.e.* a quantity of the same order of magnitude as the energy injected by a supernova explosion. Note that only a small fraction of this energy (and of the ionising radiation) is still in the circumstellar gas at the end of the stellar lifetime. For instance, Freyer et al. (2003), using a two-dimensional radiation hydrodynamics code, find that 0.4% percent of the energy emitted by a 60 $M_\odot$ under the form of ionising radiation and of wind kinetic energy is in the circumstellar

Table 1. Overview of current grids of stellar models with rotation and mass loss.

Z	Mass $M_\odot$	$v_{\rm ini}$ km s^{-1}	Magn. Field	Reference
0.0	9,15,25,40,60,85,200	0 & 800	No	Ekström et al. (2008b)
	3,9,20,60	40 - 1400	No	Ekström et al. (2008a)
10^{-8}	9,20,40,60,85	0 & 500 - 800	No	Hirschi (2007)
10^{-5}	2,3,5,7,9,15,20,40,60	0 & 300	No	Meynet & Maeder (2002)
	20,30,40,50,60	230 - 600	Yes	Yoon & Langer (2005)
	12,16,20,25,40,60	0 - 940	Yes	Yoon et al. (2006)
	3,9,20,60	40 - 1000	No	Ekström et al. (2008a)
0.0005	20,40,60,120,200	0, 600, 800	No	Decressin et al. (2007)
0.001	20,40,60	230 -600	Yes	Yoon & Langer (2005)
	12,16,20,25,30,40,60	0 - 750	Yes	Yoon et al. (2006)
0.002	3,9,20,60	30 - 880	No	Ekström et al. (2008a)
	12,16,20,25,30,40,60	0 - 650	Yes	Yoon et al. (2006)
0.004	9,12,15,20,25,40,60	0 & 300	No	Maeder & Meynet (2001)
	30,40,60,120	300	No	Meynet & Maeder (2005)
	12,16,20,25,30,40,60	0 - 500	Yes	Yoon et al. (2006)
0.008	30,40,60,120	300	No	Meynet & Maeder (2005)
0.020	8,10,12,15,20,25	0 - 470	No	Heger & Langer (2000)
	8,10,12,15,20,25	200	No	Heger et al. (2000)
	9,12,15,20,25,40,60,120	0 & 300	No	Meynet & Maeder (2000)
	9,12,15,20,25,40,60,85,120	0 & 300	No	Meynet & Maeder (2003)
	12,15,20,25,40,60	0 & 300	No	Hirschi et al. (2004)
	12,15,20,25,35	200	Yes & No	Heger et al. (2005)
	16, 30, 40	210 - 560	Yes	Yoon et al. (2006)
	3,9,20,60	30 -730	No	Ekström et al. (2008a)
0.040	20,25,40,60,85,120	0 & 300	No	Meynet & Maeder (2005)

gas at the end of the stellar lifetime (the supernova injection energy is not accounted for here)†. From this fraction 32.5% is kinetic energy of bulk motion, 45% is thermal energy and the remaining 22.5% is ionization energy of hydrogen. Freyer et al. (2006) performed a similar computation for a 35 $M_\odot$. They obtain that, at the end of the stellar lifetime, 1% of the energy released as Lyman continuum radiation and stellar wind has been transferred to the circumstellar gas. From that fraction 10% is kinetic energy of bulk motion, 36% is thermal energy and the remaining 54% is ionization energy of hydrogen. Freyer et al. (2006) conclude that it is necessary to consider both ionizing radiation and stellar winds for describing the interaction of OB stars with their circumstellar environment.

Collective effects of stellar winds and supernova explosion may trigger in certain circumstances galactic superwinds (see the review by Veilleux et al. 2005). These galactic winds are also loaded in new chemical species and participate to the enrichment of the intergalactic medium or intra galactic medium of clusters of galaxies.

2. Observed constraints for massive star models at various metallicities

To have a complete view of the impact of the evolution of stars on the evolution of a galaxy, models spanning the whole range of metallicities from $Z = 0.0$ (Pop III stellar models) up to the highest metallicities observed in the considered galaxy have to be

† The circumstellar gas encompasses the shocked wind and photoionized HII regions, i.e. the mass inside a sphere with a radius of about 50 pc

used. Different grids of stellar models are available in the literature accounting for various effects as overshooting, mass loss, rotation, magnetic fields, binary interactions... A subsample of recent grids of models for single stars accounting for the effects of mass loss, rotation and overshooting are presented in Table 1‡. To use results of stellar models with some confidence a necessary prerequisite is to check that they are able to reproduce well observable constraints. Among the most important constraints, let us cite the following ones:

- **Surface chemical composition of massive stars:** Many observations show that the surface chemical abundances of massive stars present variations during evolutionary stages and in mass domain where such changes are not predicted by standard models† (see for instance the recent review by Meynet *et al.* 2008, which summarizes the results of recent large massive star surveys in the Milky Way and the Magellanic Clouds). Understanding such changes are not only interesting for improving stellar physics, they are a key element to obtain reliable predictions for the evolutionary tracks and the chemical yields. These surface abundances are the sign that the chemical abundances in the stellar interiors are different from those expected from standard models. Since what governs the evolution of a star is its internal changes of chemical composition, one can easily understand that the process responsible for these changes of the surface abundances has strong impact also on all the outputs of stellar models. Explanations for these changes of the surface abundances have been proposed in the literature: rotational mixing (Heger & Langer 2000; Meynet & Maeder 2000) and/or mass transfer in close binary systems (Langer *et al.* 2008) are the two most important invoked processes. According to Hunter *et al.* (2008) about 60% of the observed B-type stars present observed characteristics compatible with the rotational mixing theory.

- **Populations of Wolf-Rayet stars:** as is well known Wolf-Rayet (WR) stars are evaporating stars losing during their lifetime a large fraction of their initial stellar mass under the form of strong stellar winds (see the recent review by Crowther 2007). Observation shows that the number fraction of WR to O-type stars increases with the metallicity. The number ratio of WN (WR with He and N emission lines) to WC (WR with C emission lines) star also increases with the metallicity. Such features have to be explained by stellar models. They depend on both the physics of mixing in stellar interiors and the effects of mass loss. Reasonable agreement between models and observations can be obtained when the effects of rotational mixing and metallicity dependence of the stellar winds are accounted for (Meynet & Maeder 2005). Models accounting for the effects of mass transfer in close binaries may also reproduce the observed trend (Eldridge *et al.* 2008). According to Foellmi *et al.* (2003ab) the proportion of detected binaries among WR stars in the Magellanic Clouds is between 30-40%. Taken at face, these numbers indicate that binarity might not be the dominant WR channel mechanism in these systems, especially when one considers the fact that not all stars in binary systems undergo a Roche Lobe Overflow event (e.g. those binary systems in which the components are too far away from each other).

- **Populations of supergiants:** it is a well known fact that the number ratio of blue to red supergiant increases when the metallicity increases (Meylan & Maeder 1982; Eggenberger *et al.* 2002). Standard models predict exactly the inverse behavior. As discussed in Langer & Maeder (1995) this feature points probably toward a missing mixing process in massive stars. It has been showed that in the SMC the observed blue to red

‡ Some of them include magnetic field effects according to the dynamo theory proposed by Spruit (2002).
† Standard models are considered here to be models in which chemical mixing occur only in convective regions.

supergiant ratio can be reproduced by rotating models (Maeder & Meynet 2001). The blue to red supergiant ratio is also affected by mass loss.
- **The frequency of different core-collapse supernovae:** the main types of core collapse supernovae are type II (H lines present in the spectrum), type Ib (no H lines detected but He lines detected) and type Ic (no H, no He-lines detected). Many different subtypes are defined into these three categories (see e.g. Cappellaro et al. 2001). The frequency of the type Ibc supernovae normalized to the type II supernovae depends on the minimum initial mass required for stars to end their lifetime with no H-rich envelope. This is correct as long as the star formation rate can be considered constant in the last 50 million years and provided the evolution of single stars is mainly responsible for the value of this ratio. Prantzos & Boissier (2003) show that the above ratio increases with the metallicity. This has recently been confirmed by new data collected by Prieto et al. (2008). Models should explain such trend. We shall come back on that topic in Sect. 4 below.

Many other observational features might be added to the list above like the number ratios of Be stars (fast rotating stars presenting an expanding equatorial disk) to B-type stars which increases when the metallicity decreases (Maeder et al. 1999; Wisniewski & Bjorkman 2006), the rotation rate of young pulsars (see the discussion in Heger et al. 2005), the shape of fast rotating stars as deduced from interferometry (see e.g. Carciofi et al. 2008), the variation with the latitude of the effective temperature at the surface of fast rotating stars (see e.g. Monnier et al. 2007), the wind anisotropies observed for LBV stars and for Be stars (see Weigelt et al. 2007; Meilland et al. 2007ab), the ratio of the angular velocity of the core to that of the surface and the size of the convective core obtained through asteroseismology (see e.g. Aerts 2008)... Ideal stellar models should be able to reproduce satisfactorily all these important constraints. This is the price to pay to obtain reliable chemical yields, quantities of energy and momentum injected by massive stars. In the rest of this paper, we shall discuss two topics: the effects of rotation on the yields of CNO elements at low metallicity and the effect of rotation and mass loss on the properties of supernova progenitors.

3. Early chemical evolution of the Milky Way

Very metal poor halo stars have formed (at least in part) from matter ejected by very metal poor massive stars (see e.g. Chiappini et al. 2005, 2008). Their surface compositions thus reflect the nucleosynthesis occurring in the first generations of massive stars (provided of course that no other processes as accretion or in-situ mixing mechanism has changed their surface composition).

Interestingly, many observations of these stars show puzzling features. Among them let us cite the two following ones shown in Fig. 1: 1.-spectroscopic observations (e.g. Spite et al. 2005) indicate a primary production of nitrogen over a large metallicity range; 2.-halo stars with $\log(O/H)+12$ inferior to about 6.5 present higher C/O ratios than halo stars with $\log(O/H)+12$ between 6.5 and 8.2 (Akerman et al. 2004; Spite et al. 2005).

Fast rotating massive stars are very interesting candidates for producing primary nitrogen at low metallicity. Their short lifetimes, together with their ability, when rotating sufficiently fast, to be important sources of primary nitrogen, allow them to account for the high observed N/O ratio at very low O/H values. Moreover these stars can also reproduce the observed C/O upturn mentioned just above. This is illustrated in Fig. 1, where predictions for the evolution of N/O and C/O of chemical evolution models using different sets of yields are compared (Chiappini et al. 2006a†). We see that the observed N/O ratio is much higher than what is predicted by a chemical evolution model using the yields of the slow-rotating $Z = 10^{-5}$ models from Meynet & Maeder (2002) down to $Z = 0$. When adding the yields of the fast-rotating $Z = 10^{-8}$ models from Hirschi (2007) the fit is much improved. The same improvement is found for the C/O ratio, which presents an upturn at low metallicity. Thus these comparisons support fast rotating massive stars as the sources of primary nitrogen in the galactic halo.

High N/O and the C/O upturn of the low-metallicity stars are also observed in low-metallicity DLAs (Pettini et al. 2007, see the crosses in Fig. 1). We note that the observed points are below the points for the halo stars in the N/O versus O/H plane. This may be attributed to two causes: either the observed N/O ratios observed in halo stars are somewhat overestimated or the difference is real and might be due to different star formation histories in the halo and in DLAs. Let us just discuss these two possibilities.

Measures of nitrogen abundances at the surface of very metal poor stars is quite challenging, much more than the measure of nitrogen in the interstellar medium as is done for the DLAs, therefore one expects that the data for DLAs suffer much smaller uncertainties than those for halo stars. In that respect the observed N/O ratios in DLAs give more accurate abundances than halo stars. Most probably the star formation history in DLAs is not the same as in the halo. While, as recalled above, in the halo we see the result of a strong and rapid star formation episode, in DLAs one might see the result of much slower and weaker star formation episodes. In that case, both massive stars and intermediate mass stars contributed to the build up of the chemical abundances and the chemical evolution models presented in Fig. 1 no longer apply to these systems (see Chiappini et al. 2003; Dessauges-Zavadsky 2007 for chemical evolution models of DLAs). It will be very interesting to study the results of chemical evolution models adapted to this situation and accounting for stellar yields from both rotating massive and intermediate stars. Let us just mention at this stage that primary nitrogen production in metal poor intermediate mass stars is also strongly favored when rotational mixing is accounted for (Meynet & Maeder 2002). Thus also in that case, rotation may play a key role.

The primary nitrogen production is accompanied by other interesting features such as the production of primary ^{13}C (see Chiappini et al. 2008), and of primary ^{22}Ne. Primary ^{22}Ne is produced by diffusion of primary nitrogen from the H-burning shell to the core He-burning zone, or by the engulfment of part of the H-burning shell by the growing He-burning core. These processes occur in rotating massive star models (Meynet & Maeder 2002; Hirschi 2007). In the He-burning zone, ^{14}N is transformed into ^{22}Ne through the classical reaction chain ^{14}N$(\alpha,\gamma)^{18}$F$(\beta^+ \nu)^{18}$O$(\alpha,\gamma)^{22}$Ne.

In the He-burning zones (either in the core at the end of the core He-burning phase or in the He-burning shell during the core C-burning phase and in the following convective C-burning shell), neutrons are released through the reaction ^{22}Ne$(\alpha,n)^{25}$Mg. These neutrons then can either be captured by iron seeds and produce s-process elements or be captured by light neutron poisons and thus be removed from the flux of neutrons which is useful for s-process element nucleosynthesis. The final outputs of s-process elements will depend on at least three factors: the amounts of 1.- ^{22}Ne, 2.- neutron poisons and 3.- iron seeds. In standard models (without rotation), when the metallicity decreases, the amount of ^{22}Ne decreases (less neutrons produced), the strength of primary neutron poisons becomes relatively more important in particular for [Fe/H]$\leqslant$-2 with respect to solar, and the

---

† The details of the chemical evolution models can be found in Chiappini et al. (2006b), where they show that such a model reproduces nicely the metallicity distribution of the Galactic halo. This means that the timescale for the enrichment of the medium is well fitted.

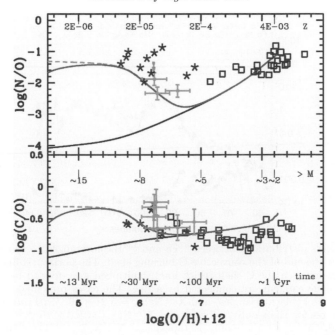

Figure 1. Variation of the N/O and C/O ratios as a function the O/H ratios. Data points for halo stars are from Israelian et al. (2004, open squares) and of Spite et al. (2005, stars). The points with error bars are for DLA systems from Pettini et al. (2008). The lower continuous curve is the chemical evolution model obtained with the stellar yields of slow rotating $Z = 10^{-5}$ models from Meynet & Maeder (2002) and Hirschi et al. (2004). The dashed line includes the yields of fast rotating $Z = 10^{-8}$ models from Hirschi (2007) at very low metallicity. The intermediate curve is obtained using the yields of the $Z = 0$ models presented in Ekström et al. (2008) up to $Z = 10^{-10}$. The chemical evolution models are from Chiappini et al. (2006a), see Table 1 for the initial velocities of the stellar models.

amount of iron seeds also decreases (e.g., Raiteri et al. 1992). Thus very small quantities of s-process elements are expected (see the triangles in Fig. 2). When primary nitrogen and therefore primary ^{22}Ne is present in quantities as given by rotating models which can reproduce the observed trends for the N/O and C/O ratios in the halo stars, then a very different output is obtained. The abundances of several s-process elements are increased by many orders of magnitudes. In particular, the elements are produced in the greatest quantities in the atomic mass region between strontium and barium, and no long in the atomic mass region between iron and strontium as in the case of standard models.

These first results need to be extended for other masses, rotation and metallicities. However they already show that some heavy s-process elements, not produced in standard models (without rotation), might be produced in significant quantities in metal poor rotating stellar models. It will be very interesting in the future to find some non ambiguous signature of the occurrence of this process in the abundance pattern of very metal poor halo stars.

4. Type Ib and Ic supernovae

Core collapse supernovae of type Ib and Ic are very interesting events for many reasons. One of them is that in four cases, the typical spectrum of a type Ic supernova has been observed together with a long soft Gamma Ray Burst (GRB) event (Woosley & Bloom 2006). Also, recent observations (Prieto et al. 2008) present new values for the variation

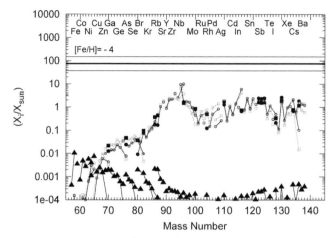

Figure 2. s-process distributions between ^{57}Fe and ^{138}Ba normalized to solar for the 25 M$_\odot$ and [Fe/H]=-4 at the end of the convective C-burning shell. The horizontal line corresponds to the ^{16}O overabundance in the C shell (thick line), multiplied and divided by two (thin lines). Isotopes of the same element are connected by a line. The cases presented are the following: *i)* non-rotating model (black triangles) and X(^{22}Ne)$_{\rm ini}$ = 5.21 × 10^{-5}; *ii)* rotating model (open squares, full squares for the s-only isotopes) and X(^{22}Ne)$_{\rm ini}$ = 5.0 × 10^{-3}; *iii)* rotating model (open circles, full circles for the s-only isotopes) and X(^{22}Ne)$_{\rm ini}$ = 1.0 × 10^{-2}. Figure from Pignatari et al. (2008).

with the metallicity of the number ratio (SN Ib+SN Ic)/SN II to which theoretical predictions can be compared. Finally, according at least to single star scenarios, these supernovae arise from the most massive stars. They offer thus a unique opportunity to study the final stages of these objects which have a deep impact on the photometric and spectroscopic evolution of galaxies and also contribute to its chemical evolution .

We shall now discuss the predictions of single star models for the type Ib/Ic supernovae frequency. Since these supernovae do not show any H-lines in their spectrum, they should have as progenitors stars having removed *at least* their H-rich envelope by stellar winds, *i.e.* their progenitors should be WR stars of the WNE type (stars with no H at their surface and presenting He and N lines) or of the WC/WO type.

In Fig. 3, the duration of the different WR subphases is plotted as a function of the initial mass for various metallicities. Only the results from models with rotation are plotted. The greatest part of the WR lifetime is spent in the WNL phase (WR with still H in the envelope). Rotation increases the duration of this phase by allowing stars to enter this phase at an earlier stage of its evolution. For identical initial rotational velocities, the duration of the WNL phase is greater at higher metallicity. At higher metallicity, the higher mass loss rates by stellar winds enable the star to enter the WR phase at an earlier stage. The WNE phase is also longer at higher metallicity (as is also the case for non–rotating models). The WC phase keeps more or less the same duration for all the metallicities in the higher mass star range. In the lower mass star range the WC phase is longer at higher metallicity as a result of the shift toward a lower value of the minimum initial mass of single stars needed to become a Wolf–Rayet star.

To link these results with the type of the supernova event we adopted the following rules: as long as some hydrogen is present in the ejecta, a type II supernova will occur. We considered that all supernovae ejecting less than about 0.6 M$_\odot$ of helium are of type Ic. All progenitors satisfying neither the criterion for becoming a type II (no H present in the ejecta), neither the one for becoming a type Ib (more than ∼0.6 M$_\odot$ of He) are

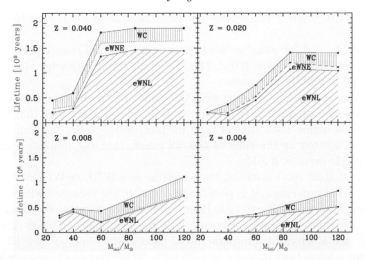

Figure 3. Variation of the durations of the WR subphases as a function of the initial mass at various metallicities. All the models begin their evolution with $v_{\rm ini} = 300$ km s^{-1}. Figure taken from Meynet & Maeder (2005). The small "e" in front of the different WR subtype indicates that criteria based on evolutionary computations have been used to classify the stars. In the text we have dropped the "e" and e.g. WNL=eWNL.

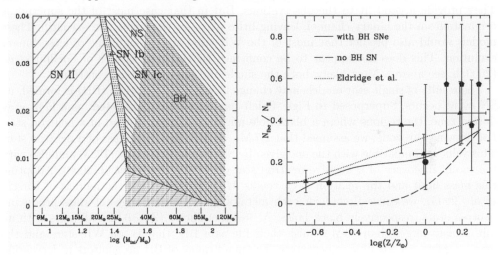

Figure 4. *Left panel :* Ranges of masses of various types of SNe. The area where formation of BH is expected is indicated as a superposed gray region. *Right panel :* Rate of SN Ibc / SN II if all models produce a SN (solid line) or if models producing a black hole do not explode in a SN (dashed line). Pentagons are observational data from Prieto *et al.* (2008), and triangles are data from Prantzos & Boissier (2003). The dotted line represents the binary models of Eldridge *et al.* (2008).

considered to give rise to type Ib SN events. Fig. 4 shows the different supernova types expected for various initial masses and metallicities. Looking at that figure, we can make the following remarks:

• As expected the lower mass limit for having a type Ibc SN decreases when the metallicity increases. We see that the dependence on the metallicity of this limit is much stronger at low Z than at high Z;

• The mass range of stars ending their life in a type Ib supernova event is relatively narrow. Below this mass range, the stars do not succeed in entering the WR phase, above, mass loss rates are efficient enough for allowing the star to further evolve into the WC or WO phase.

- Above 30 $M_\odot$ and for metallicities higher than 0.008, all stars end their life as type Ic SNe. For metallicities below 0.008, the minimum initial mass for stars ending their lifetime as type Ic rapidly increases.

The mass range for type Ic supernovae is thus much larger than the mass range for type Ib SNe. Even when weighted with a Salpeter IMF, one expects thus that the frequency of type Ic SNe is higher than that of type Ib. Interestingly also, one can note from the mass of helium ejected at the time of the SN event, that the minimum amount of He ejected by type Ic event is 0.3 $M_\odot$.

Considering that all models ending their lifetime as a WNE or WC/WO phase will explode as a type Ibc supernova, it is possible to compute the variation with the metallicity of the number ratio of type Ibc to type II supernovae. The result is shown in Fig. 4 (right panel). One sees that this ratio increases with the metallicity. This is due to the fact that at higher metallicity, the minimum initial mass of stars ending their life as WNE, or WC/WO stars is lower than at lower metallicities. Single star models can reasonably well reproduce the observed trend with the metallicity. They however give slightly too small values with respect to the observations, which may indicate that a portion of the type Ibc supernovae may originate from close binary evolutions. Models accounting for single and binary channel (but without rotation) are shown as a dotted line (Eldridge et al. 2008). They provide a good fit to the observations. But in that case most of the supernovae originate from the binary channel, leaving little place for the single star scenario. These models would also predict that most of the WR stars are the outcome of close binary evolution. This does not appear to be confirmed by the observations of Foellmi et al. (2003ab, see above). Most likely, both the single and binary channel contribute.

The results of single star models may change if, when a Black Hole (BH) is formed, no SN event occurs. Superposed to Figs. 4 (left panel), we have indicated, as a light gray dotted zone, the regions where a black hole might be formed instead of a neutron star. For drawing this zone, we assumed that 2.7 $M_\odot$ as the maximum mass of a neutron star. This value is compatible with the one given by Shapiro & Teukolsky (1983) and also with the recent discovery of a massive neutron star of 2.1 $M_\odot$ (Freire et al. 2008). Adopting this mass limit and the relation M_{NS} versus M_{CO} deduced from the models of Hirschi et al. (2005), we can estimate for each metallicity the mass ranges of stars producing neutron stars, respectively black holes. At low metallicity ($Z \leqslant 0.01$), all stars with an initial mass larger than 30 $M_\odot$ finish their life as a BH, that is, all WR stars and the most massive red– or blue–supergiants. The mass range of BH progenitors is decreasing when more and more metal rich environments are considered. For $Z \geqslant \sim 0.040$, no BH is expected from single star scenario. (see Fig. 4). This is of course an effect due to increased mass loss at higher metallicities. The stars lose too much mass for forming BHs.

We note that in the frame of our hypothesis, all the type Ib SNe give birth to a neutron star. For $Z \geqslant \sim 0.03$, most of the type Ic supernovae also produce a neutron star. Below a metallicity of 0.008, all WR progenitors produce a black hole. If, when a BH is formed no supernova event is obtained, then one expects no type Ibc supernovae (from single stars) below a metallicity of about 0.008.

We computed new (SN Ib + SN Ic)/ SN II ratios with the assumption that all models massive enough to form a black hole do not produce a SN. Comparing with the observed rates in Fig. 4 (see dashed line in right panel) we see that in the case no supernova event occurs when a BH is formed, single star models might still account for a significant fraction of the type Ibc supernovae for $Z > 0.02$. At $Z = 0.004$ all type Ibc should arise from other evolutionary scenarios, probably involving close binary evolution with mass transfer. If we take a smaller maximal mass for neutron stars, this situation is still more extreme: almost all the WR stars produce a black hole, thus the rate of SN Ibc / SN II is null or very small at each metallicity.

Probably, the hypothesis according to which no supernova event is associated when a BH is formed, is too restrictive. For instance, the collapsar scenario for Gamma Ray Bursts (Woosley 1993) needs the formation of a black holes (Dessart, private communication) and this formation is at least accompanied in some cases by a type Ic supernova event. Also the observation of the binary system GRO J1655-40 containing a black hole (Israelian et al. 1999) suggest that a few stellar masses have been ejected and therefore a SN event occurred when the BH formed. This is deduced from the important chemical anomalies observed at the surface of the visible companion, chemical anomalies whose origin is attributed to the fact that the (now visible) companion accreted part of the SN ejecta. This gives some support to the view that, at least in some cases, the collapse to a BH does not prevent a supernova event to occur.

To conclude, we see that still major improvements of massive star stellar models are still needed in order to improve our understanding of their impacts in galaxies. Probably a key step will be made forward when we shall also have a better understanding on how massive stars form. This is crucial to link stellar physics to the physics of galaxies.

References

Akerman, C. J., Carigi, L., Nissen, P. E., Pettini, M., & Asplund, M. 2004, A&A, 414, 931
Aerts, C. 2008, in Massive Stars as Cosmic Engines, Proceedings of the International Astronomical Union, IAU Symposium, Volume 250, p. 237-244
Barkana, R. 2006, Science, 313, 931
Brinchmann, J., Kunth, D., & Durret, F. 2008, A&A, 485, 657
Cappellaro, E. & Turatto, M. 2001, The influence of binaries on stellar population studies, Dordrecht: Kluwer Academic Publishers, 2001, xix, 582 p. Astrophysics and space science library (ASSL), Vol. 264. ISBN 0792371046, p.199
Carciofi, A. C., Domiciano de Souza, A., Magalhes, A. M., Bjorkman, J. E. & Vakili, F. 2008, ApJL, 676, 41
Chiappini, C., Matteucci, F., & Meynet, G. 2003, A&A, 410, 257
Chiappini, C., Matteucci, F., & Ballero, S.K. 2005, A&A, 437, 429
Chiappini, C., Hirschi, R., Meynet, G., Ekström, S., Maeder, A., & Matteucci, F. 2006a, A&A Letters, 449, 27
Chiappini, C., Hirschi, R., Matteucci, F., Meynet, G., Ekström, S., & Maeder, A. 2006b, in "Nuclei in the Cosmos IX", Proceedings of Science, 9 pages (arXiv:astro-ph/0609410)
Chiappini, C., Ekström, S., Meynet, G., Hirschi, R., Maeder, A., & Charbonnel, C. 2008, A&A, 479, L9
Dessauges-Zavadsky, M., Calura, F., Prochaska, J. X., D'Odorico, S., & Matteucci, F. 2007, A&A, 470, 431
Cusumano, G., Mangano, V., Chincarini, G., et al. 2006, Nature, 440, 164
Crowther, P. A. 2007, ARAA, 45, 177
Decressin, T., Meynet, G., Charbonnel, C., Prantzos, N., & Ekström, S. 2007, A&A, 464, 1029
Eggenberger, P., Meynet, G., & Maeder, A. 2002, A&A, 386, 756
Ekström, S., Meynet, G., Maeder, A., & Barblan, F. 2008a, A&A, 478, 467
Ekström, S., Meynet, G., Chiappini, C., Hirschi, R., & Maeder, A. 2008b, A&A, in press (arXiv:0807.0573)
Eldridge, J. J., Izzard, R. G., & Tout, C. A. 2008, MNRAS, 384, 1109
Foellmi, C., Moffat, A. F. J., & Guerrero, M.A. 2003a, MNRAS, 338, 360
Foellmi, C., Moffat, A. F. J., & Guerrero, M.A. 2003b, MNRAS, 338, 1025
Freire, P. C. C., Wolszczan, A., van den Berg, M., & Hessels, J. W. T. 2008, ApJ, 679, 1433
Freyer, T., Hensler, G, & Yorke, H. W. 2003, ApJ, 594, 888

Freyer, T., Hensler, G., & Yorke, H. W. 2006, ApJ, 638, 262
Heger, A. & Langer, N. 2000, *ApJ*, 544, 1016
Heger, A., Langer, N., & Woosley, S. E. 2000, ApJ, 528, 368
Heger, A., Woosley, S. E., & Spruit, H. C. 2005, ApJ, 626, 350
Hester, J. J. & Desch, S. J. 2005, in Chondrites and the Protoplanetary Disk, ASP Conf. Ser., 341, eds. A. N. Krot, E. R. D. Scott, B. Reipurth, p. 107
Hirschi, R. 2007, A&A, 461, 571
Hirschi, R., Meynet, G., & Maeder, A. 2004, A&A, 425, 649
Hirschi, R., Meynet, G., & Maeder, A. 2005, A&A, 443, 581
Hopkins, A. M. & Beacom, J. F. 2006, ApJ, 651, 142
Hunter, I., Brott, I., Lennon, D. J. *et al.* 2008b, *A&A* submitted, astro-ph0711.2267v1
Israelian, G., Ecuvillon, A., Rebolo, R., García-López, R., Bonifacio, P., & Molaro, P. 2004, *A&A*, 421, 649
Israelian, G. *et al.* 1999, Nature, 401, 142
Langer, N. & Maeder, A. 1995, A&A, 295, 685
Langer, N., Cantiello, M., Yoon, S.-C., Hunter, I., Brott, I., Lennon, D., de Mink, S., & Verheijdt, M. 2008, Massive Stars as Cosmic Engines, Proceedings of the International Astronomical Union, IAU Symposium, Volume 250, p. 167-178
Maeder, A. & Meynet, G. 2001, *A&A*, 373, 555
Maeder, A., Grebel, E. K., & Mermilliod, J.-C. 1999, *A&A*, 346, 459
Meilland, A., Millour, F., & Stee, P. 2007a, *A&A*, 464, 73
Meilland, A., Stee, P., & Vannier, M. 2007b, *A&A*, 464, 59
Meylan, G. & Maeder, A. 1982, A&A, 386, 576
Meynet, G. & Maeder, A. 2000, *A&A*, 361, 101
Meynet, G. & Maeder, A., 2002, A&A, 390, 561
Meynet, G. & Maeder, A. 2003, A&A, 404, 975
Meynet, G. & Maeder, A. 2005, *A&A*, 429, 581
Meynet, G., Ekström, S., Maeder, A., Hirschi, R., Georgy, C. & Beffa, C. 2008, Massive Stars as Cosmic Engines, Proceedings of the International Astronomical Union, IAU Symposium, Volume 250, p. 147-160
Monnier, J. D., Zhao, M., & Pedretti, E., 2007, Science, 317, pp. 342
Pignatari, M., Gallino, R., Meynet, G., Hirschi, R., Herwig, F., & Wiescher, M. 2008, submitted to ApJ Letter
Prantzos, N. & Boissier, S. 2003, *A&A*, 406, 259
Pérez-Gonzàlez, P. G., Kennicutt, R. C., & Gordon, K. D. 2006, ApJ, 648, 987
Pettini, M., Zych, B. J., Steidel, C. C., & Chaffee, F. H. 2008, MNRAS, 385, 2011
Prieto, J. L., Stanek, K. Z., & Beacom, J. F. 2008, ApJ, 673, 999
Raiteri, C. M., Gallino, R., & Busso, M. 1992, ApJ, 387, 263
Shapiro, S. L. & Teukolsky, S. A. 1983, in Black holes, white dwarfs, and neutron stars: The physics of compact objects, Wiley-Interscience
Spite, M., Cayrel, R., & Plez, B. 2005, *A&A*, 430, 655
Spruit, H. C. 2002, A&A, 381, 923
Veilleux, S., Cecil, G., & Bland-Hawthorn, J. 2005, ARAA, 43, 769
Weigelt, G., Kraus, S., Driebe, T. 2007, *A&A*, 464, 87
Wisniewski, J. P. & Bjorkman, K. S. 2006, *ApJ*, 652, 458
Woosley, S. E. 1993, ApJ, 405, 273
Woosley, S. E. & Bloom, J. S. 2006, ARAA, 44, 507
Yoon, S.-C. & Langer, N. 2005, A&A, 443, 643
Yoon, S.-C., Langer, N., & Norman, C. 2006, A&A, 460, 199

The First Galaxies

Volker Bromm[1]

[1]Department of Astronomy, University of Texas, Austin, TX 78712, U.S.A.

Abstract. An important open frontier in astrophysics is to understand how the first sources of light, the first stars and galaxies, ended the cosmic dark ages at redshifts $z \simeq 15 - 20$. Their formation signaled the transition from the simple initial state of the universe to one of ever increasing complexity. We here review recent progress in understanding the assembly process of the first galaxies with numerical simulations, starting with cosmological initial conditions and modelling the detailed physics of star formation. The key drivers in building up the primordial galaxies are the feedback effects from the first stars, due to their input of radiation and of heavy chemical elements in the wake of supernova explosions. In addition, the conditions inside the first galaxies are governed by the gravitationally-driven turbulence generated during the virialization of the dark matter host halo. Our theoretical predictions will be tested with upcoming near-infrared observatories, such as the *James Webb Space Telecope*, in the decade ahead.

Keywords. cosmology: theory, early universe — galaxies: formation — stars: formation, supernovae

1. Introduction

After the parameters of the cosmological background model have now been successfully determined to high precision, the next frontier of cosmology is to understand structure formation. Galaxy formation, however, is very complex, and we still lack a first-principle based theoretical picture. Within the hierarchical ΛCDM model, the key to galaxy formation might lie in its initial stages, when the first, low-mass, building blocks, the first galaxies, emerged. With the formation of the first stars, the so-called Population III (Pop III), the universe was rapidly transformed into an increasingly complex, hierarchical system, due to the energy and heavy element input from the first stars and accreting black holes (Barkana & Loeb 2001; Bromm & Larson 2004; Ciardi & Ferrara 2005; Miralda-Escudé 2003). Currently, we can directly probe the state of the universe roughly a million years after the Big Bang by detecting the temperature anisotropies in the cosmic microwave background (CMB), thus providing us with the initial conditions for subsequent structure formation. Complementary to the CMB observations, we can probe cosmic history all the way from the present-day universe to roughly a billion years after the Big Bang, using the best available ground- and space-based telescopes. In between lies the remaining frontier, and the first stars and galaxies are the sign-posts of this early, formative epoch.

To simulate the build-up of the first stellar systems, we have to address the feedback from the very first stars on the surrounding intergalactic medium (IGM), and the formation of the second generation of stars out of material that was influenced by this feedback. There are a number of reasons why addressing the feedback from the first stars and understanding second-generation star formation is crucial:
(i) The first steps in the hierarchical build-up of structure provide us with a simplified laboratory for studying galaxy formation, which is one of the main outstanding problems in cosmology.
(ii) The initial burst of Pop III star formation may have been rather brief due to the strong negative feedback effects that likely acted to self-limit this formation mode

(Greif & Bromm 2006; Yoshida et al. 2004). Second-generation star formation, therefore, might well have been cosmologically dominant compared to Pop III stars.

(iii) A subset of second-generation stars, those with masses below $\simeq 1\ M_\odot$, would have survived to the present day. Surveys of extremely metal-poor Galactic halo stars therefore provide an indirect window into the Pop III era by scrutinizing their chemical abundance patterns, which reflect the enrichment from a single, or at most a small multiple of, Pop III SNe (Beers & Christlieb 2005; Frebel et al. 2007; Karlsson et al. 2008). Stellar archaeology thus provides unique empirical constraints for numerical simulations, from which one can derive theoretical abundance patterns to be compared with the data.

Existing and planned observatories, such as HST, Keck, VLT, and the *James Webb Space Telescope (JWST)*, planned for launch around 2013, yield data on stars and quasars less than a billion years after the Big Bang. The ongoing *Swift* gamma-ray burst (GRB) mission provides us with a possible window into massive star formation at the highest redshifts (Lamb & Reichart 2000; Bromm & Loeb 2002, 2006). Measurements of the near-IR cosmic background radiation, both in terms of the spectral energy distribution and the angular fluctuations provide additional constraints on the overall energy production due to the first stars (Santos et al. 2002; Magliocchetti et al. 2003; Dwek et al. 2005; Fernandez & Komatsu 2006; Kashlinsky et al. 2005). Understanding the formation of the first galaxies is thus of great interest to observational studies conducted both at high redshifts and in our local Galactic neighborhood.

2. Primordial Star Formation

The first stars in the universe formed a few 100 Myr after the Big Bang, when the primordial gas was first able to cool and collapse into dark matter (DM) minihalos with masses of the order of $10^6\ M_\odot$ (Abel et al. 2002; Bromm et al. 2002; Yoshida et al. 2006). These stars are believed to have been very massive, with masses of the order of $100 M_\odot$, owing to the limited cooling ability of primordial gas in minihalos via the radiation from H_2 molecules. While the initial conditions for the formation of the very first stars are known from precision measurements of cosmological parameters (Spergel et al. 2007), the situation for the subsequent generations of stars is much more compex. It has become evident that Pop III star formation might actually consist of two distinct modes: one where the primordial gas collapses into a DM minihalo, and one where the metal-free gas becomes significantly ionized prior to the onset of gravitational runaway collapse (Johnson & Bromm 2006). To clearly indicate that both modes pertain to *metal-free* star formation, we here follow the new classification scheme suggested by Chris McKee (see McKee & Tan 2008; Johnson et al. 2008). Within this scheme, the minihalo Pop III mode is termed Pop III.1, whereas the second mode is called Pop III.2.

While the formation of the very first, Pop III.1, stars in minihalos relied on H_2 cooling, the HD molecule can play an important role in the cooling of primordial gas in several situations, allowing the temperature to drop well below 200 K (Abel et al. 2002; Bromm et al. 2002). In turn, this efficient cooling may lead to the formation of primordial stars with masses of the order of $10\ M_\odot$, the so-called Pop III.2 (Johnson & Bromm 2006). In general, the formation of HD, and the concomitant cooling that it provides, is found to occur efficiently in primordial gas which is strongly ionized, owing ultimately to the high abundance of electrons which serve as catalyst for molecule formation in the early universe (Shapiro & Kang 1987). Efficient cooling by HD can be triggered within the relic H II regions that surround Pop III.1 stars at the end of their brief lifetimes (Alvarez et al. 2006), owing to the high electron fraction that persists in the gas as it cools and recombines (Johnson et al. 2007; Yoshida et al. 2007b). The efficient formation of HD

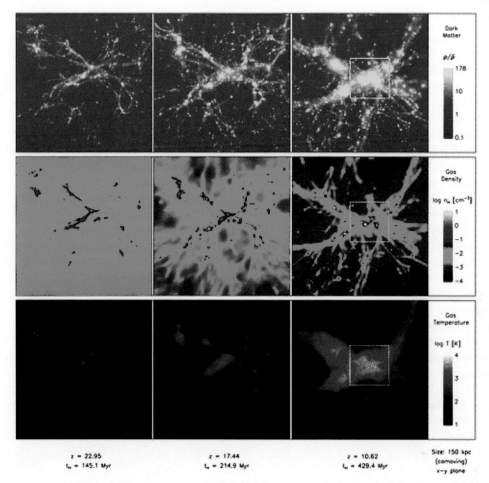

Figure 1. Assembly of the first galaxy (from Greif *et al.* 2008). Gravity is assembling the dark matter (*top row*) and pristine gas (*middle row*) into a primordial galaxy. Shown is the evolution at three different stages (*from left to right*). White crosses indicate individual Pop III stars, which form prior to the first galaxy. The assembly process is accompanied by powerful virialization shocks that heat up the infalling gas to $> 10^4$ K (*bottom row*).

can also take place when the primordial gas is collisionally ionized, such as behind the shocks driven by the first SNe or in the virialization of massive DM halos (Machida *et al.* 2005; Johnson & Bromm 2006; Shchekinov & Vasiliev 2006).

There might thus be a progression of characteristic masses of the various stellar populations that form in the early universe. In the wake of Pop III.1 stars (typically with $M_* \sim 100 M_\odot$) formed in DM minihalos, Pop III.2 star formation (with $M_* \sim 10 M_\odot$) ensues in regions which have been previously ionized, typically associated with relic H II regions left over from massive Pop III.1 stars collapsing to black holes, while even later, when the primordial gas is locally enriched with metals, Pop II (with $M_* \sim 1 M_\odot$) stars begin to form (Bromm & Loeb 2003; Greif & Bromm 2006). Recent simulations confirm this picture, as Pop III.2 star formation ensues in relic H II regions in well under a Hubble time, while the formation of Pop II stars after the first SN explosions is delayed by more than a Hubble time (Greif *et al.* 2007; Yoshida *et al.* 2007a,b; but see Whalen *et al.* 2008).

3. The First Galaxies and the Onset of Turbulence

How massive were the first galaxies, and when did they emerge? Theory predicts that DM halos containing a mass of $\sim 10^8 M_\odot$ and collapsing at $z \sim 10$ were the hosts for the first bona fide galaxies. These dwarf systems are special in that their associated virial temperature exceeds the threshold, $\sim 10^4$ K, for cooling due to atomic hydrogen (Oh & Haiman 2002). These so-called 'atomic-cooling halos' did not rely on the presence of molecular hydrogen to enable cooling of the primordial gas. In addition, their potential wells were sufficiently deep to retain photoheated gas, in contrast to the shallow potential wells of minihalos (Dijkstra et al. 2004). These are arguably minimum requirements to set up a self-regulated process of star formation that comprises more than one generation of stars, and is embedded in a multi-phase interstellar medium.

One of the important consequences of atomic cooling is the softening of the equation of state below the virial radius, allowing a fraction of the potential energy to be converted into kinetic energy (Wise & Abel 2007; Greif et al. 2008). This implies that perturbations in the gravitational potential can generate turbulent motions on galactic scales, which are then transported to the centre of the galaxy. In this context the distinction between two fundamentally different modes of accretion becomes important (Greif et al. 2008). Gas accreted directly from the IGM is heated to the virial temperature and comprises the sole channel of inflow until cooling in filaments becomes important. This mode is termed hot accretion, and dominates in low-mass haloes at high redshift. The formation of the virial shock and the concomitant heating in an atomic cooling halo are visible in Fig. 1. The second mode, termed cold accretion, becomes important as soon as filaments are massive enough to enable molecule reformation, which allows the gas to cool and flow into the central regions of the nascent galaxy with high velocities. Although the assembly of the first galaxies provides us with an idealized laboratory for galaxy formation in general, the degree of complexity that is exhibited in the corresponding merger tree (see Fig. 2) is already considerable. Still, current supercomputer simulations have just reached the capacity to address the first galaxy formation process in an a-priori fashion, one star at a time.

Cold accretion is a viable agent for driving turbulence, due to the large amount of kinetic energy it brings to the center of the galaxy. Two physically distinct mechanisms are responsible for creating shocks (Greif et al. 2008). The virial shock forms where the ratio of infall velocity to local sound speed approaches unity, while a multitude of unorganized shocks forms near the center of the galaxy and is mostly caused by accretion of cold, high-velocity gas from filaments. These are more pronounced than the virial shock and have a significantly higher angular component. They create transitory density perturbations that could in principle become Jeans-unstable and trigger the gravitational collapse of individual clumps. In concert with metal enrichment by previous star formation in minihaloes, metal mixing in the first galaxies will likely be highly efficient and could lead to the formation of the first low-mass star clusters (Clark et al. 2008), in extreme cases possibly even to metal-poor globular clusters (Bromm & Clarke 2002). Some of the extremely iron-deficient, but carbon and oxygen-enhanced stars observed in the halo of the Milky Way may thus have formed as early as redshift $z \simeq 10$ (Karlsson et al. 2008).

4. Outlook

Understanding the formation of the first galaxies marks the frontier of high-redshift structure formation. It is crucial to predict their properties in order to develop the optimal

Figure 2. Hierarchical assembly of the first galaxy, expressed in the corresponding merger tree (from Greif *et al.* 2008). Each line represents an individual progenitor halo and is color-coded according to its mass. The target halo hosting the galaxy is represented by the rightmost path, which ultimately attains $\sim 10^8 M_\odot$, thus fulfilling the criterion for the onset of atomic hydrogen cooling (*red oval*). Star symbols denote the formation of Pop III.1 stars in minihalos. Here, ~ 10 such stars form prior to the assembly of the first galaxy. Their feedback will set the initial conditions for the formation of the second generation of stars inside the galaxy.

search and survey strategies for the *JWST*. Whereas *ab-initio* simulations of the very first stars can be carried out from first principles, and with virtually no free parameters, one faces a much more daunting challenge with the first galaxies. Now, the previous history of star formation has to be considered, leading to enhanced complexity in the assembly of the first galaxies. One by one, all the complex astrophysical processes that play a role in more recent galaxy formation appear back on the scene. Among them are external radiation fields, comprising UV and X-ray photons, and possibly cosmic rays produced in the wake of the first SNe. There will be metal-enriched pockets of gas which could be pervaded by dynamically non-negligible magnetic fields, together with turbulent velocity fields built up during the virialization process. However, the goal of making useful predictions for the first galaxies is now clearly drawing within reach, and the pace of progress is likely to be rapid.

Acknowledgements

I would like to thank the organizers for their kind hospitality during a memorable week of stimulating talks and discussions. Support from NSF grant AST-0708795 is gratefully acknowledged. The simulations presented here were carried out at the Texas Advanced Computing Center (TACC).

References

Abel, T., Bryan, G. L., & Norman, M. L. 2002, Science, 295, 93
Alvarez, M. A., Bromm, V., & Shapiro, P. R. 2006, ApJ, 639, 621
Barkana, R. & Loeb, A. 2001, Phys. Rep., 349, 125
Beers, T. C. & Christlieb, N. 2005, ARA&A, 43, 531
Bromm, V. & Clarke, C. J. 2002, ApJ, 566, L1
Bromm, V., Coppi, P. S., & Larson, R. B. 2002, ApJ, 564, 23
Bromm, V. & Larson, R. B. 2004, ARA&A, 42, 79
Bromm, V. & Loeb, A. 2002, ApJ, 575, 111
—. 2003, Nature, 425, 812
—. 2006, ApJ, 642, 382
Ciardi, B. & Ferrara, A. 2005, Space Science Reviews, 116, 625
Clark, P. C., Glover, S. C. O., & Klessen, R. S. 2008, ApJ, 672, 757
Dijkstra, M., Haiman, Z., Rees, M. J., & Weinberg, D. H. 2004, ApJ, 601, 666
Dwek, E., Arendt, R. G., & Krennrich, F. 2005, ApJ, 635, 784
Fernandez, E. R. & Komatsu, E. 2006, ApJ, 646, 703
Frebel, A., Johnson, J. L., & Bromm, V. 2007, MNRAS, 380, L40
Greif, T. H. & Bromm, V. 2006, MNRAS, 373, 128
Greif, T. H., Johnson, J. L., Bromm, V., & Klessen, R. S. 2007, ApJ, 670, 1
Greif, T. H., Johnson, J. L., Klessen, R. S., & Bromm, V., 2008, MNRAS, 387, 1021
Johnson, J. L. & Bromm, V. 2006, MNRAS, 366, 247
Johnson, J. L., Greif, T. H., & Bromm, V. 2007, ApJ, 665, 85
Johnson, J. L., Greif, T. H., & Bromm, V. 2008, MNRAS, 388, 26
Karlsson, T., Johnson, J. L., & Bromm, V. 2008, ApJ, 679, 6
Kashlinsky, A., Arendt, R. G., Mather, J., & Moseley, S. H. 2005, Nature, 438, 45
Lamb, D. Q. & Reichart, D. E. 2000, ApJ, 536, 1
Machida, M. N., Tomisaka, K., Nakamura, F., & Fujimoto, M. Y. 2005, ApJ, 622, 39
Magliocchetti, M., Salvaterra, R., & Ferrara, A. 2003, MNRAS, 342, L25
McKee C. F. & Tan, J. C. 2008, ApJ, 681, 771
Miralda-Escudé, J. 2003, Science, 300, 1904
Oh, S. P. & Haiman, Z. 2002, ApJ, 569, 558
Santos, M. R., Bromm, V., & Kamionkowski, M. 2002, MNRAS, 336, 1082
Shapiro, P. R. & Kang, H. 1987, ApJ, 318, 32
Shchekinov, Y. A. & Vasiliev, E. O. 2006, MNRAS, 368, 454
Spergel, D. N. et al. 2007, ApJS, 170, 377
Whalen, D., van Veelen, B., O'Shea, B. W., & Norman, M. L. 2008, ApJ, 682, 49
Wise, J. H. & Abel, T. 2007, ApJ, 665, 899
Yoshida, N., Bromm, V., & Hernquist, L. 2004, ApJ, 605, 579
Yoshida, N., Oh, S. P., Kitayama, T., & Hernquist, L. 2007a, ApJ, 663, 687
Yoshida, N., Omukai, K., & Hernquist, L. 2007b, ApJ, 667, L117
Yoshida, N., Omukai, K., Hernquist, L., & Abel, T. 2006, ApJ, 652, 6

Chemical enrichment in the early Galaxy

Torgny Karlsson[1,2]

[1]NORDITA, AlbaNova University Center, SE-106 91, Stockholm, Sweden
[2]School of Physics, The University of Sydney, 2006 NSW, Australia
email: torgny@physics.usyd.edu.au

Abstract. The chemical enrichment by the first sources of light in the universe ultimately set the stage for the subsequent evolution of the Milky Way system. The oldest and, usually, the most-metal poor stars are our 'near-field' link to this ancient epoch as they, apart from tracing the chemical enrichment itself, also indirectly hold information on, e.g., the conditions for star formation and feed-back effects in the early universe. In particular, I will discuss the possible origins of the relatively large number of carbon enhanced metal-poor stars in the Galactic halo. Furthermore, I will argue that the apparent absence of the chemical signature of so-called pair-instability supernovae (PISNe), which are a natural consequence of current theoretical models for primordial star formation at the highest masses, may arise from a subtle observational selection effect. Whereas most surveys traditionally focus on the most metal-poor stars, early PISN enrichment is predicted to 'overshoot', reaching enrichment levels of [Ca/H] ~ -2.5 that would be missed by current searches.

Keywords. stars: abundances, stars: formation, stars: Population II, Galaxy: evolution, Galaxy: formation, Galaxy: halo

1. Introduction

Extremely metal-poor (EMP) stars are defined to have [Fe/H] < -3† (Beers & Christlieb 2005) and they populate the metal-poor tail of the Galactic halo metallicity distribution function (MDF). Curiously, EMP stars appear to be the only objects in the known universe found to have metallicities significantly below [Fe/H] $= -3$ and as such, they are commonly believed to be, possibly, our closest link (i.e., both spatially and relation-wise) to the era of the first stars.

There is an ongoing debate on which types of systems that were the original building blocks of the Galactic halo. In accordance with the ΛCDM paradigm, it has been argued that systems similar to those of the dwarf spheroidal (dSph) satellite galaxies were a dominant formation site of Galactic halo stars. However, this view appear to be incompatible with the low levels of α-enhancement observed in individual dSph stars (e.g., Venn *et al.* 2004) and rules out present-day dSphs, like Fornax and Sculptor, as the formation site of Galactic halo stars in general, at least the high-α stars (see contribution by P. E. Nissen, this volume). It also turns out that the observed lack of EMP stars in the more luminous dSphs (Helmi *et al.* 2006) appear to rule out these objects as the formation site of the Galactic halo EMP stars in particular. These dSphs probably formed relatively late, in lower-density peaks already enriched to a level of [Fe/H] $\simeq -3$ (Helmi *et al.* 2006; Salvadori *et al.* 2008).

At present, it is not clear exactly in which environment the EMP stars were formed originally. Since they are believed to be second generation stars, born out of gas enriched

† [A/B] $= \log_{10}(N_A/N_B)_\star - \log_{10}(N_A/N_B)_\odot$, where N_X is the number density of element X.

only by the first supernovae (SNe), the majority of these stars must have formed relatively early on, presumably in so-called 'atomic cooling halos', which are small dark matter halos on the order of 10^8 $\mathcal{M}_\odot$ (see, e.g., Greif et al. 2008). These halos were probably the first objects able to form Pop II stars (e.g., Karlsson et al. 2008). Interestingly, the newly discovered ultra-faint dSphs (e.g., Willman et al. 2005; Belokurov et al. 2007), which may be relic galaxies formed prior to reionization (Bovill & Ricotti 2008), appear to contain a fair fraction of EMP stars (Kirby et al. 2008). If so, these objects may turn out to be surviving members of a population of 'first galaxies' that would give us a unique possibility to directly study the birth sites of second generation EMP stars.

2. Star-to-star scatters in chemical abundance ratios for EMP stars

Recent observations by a number of authors (e.g., Cayrel et al. 2004; Arnone et al. 2005; Barklem et al. 2005) have revealed surprisingly small star-to-star scatters in the α and iron-peak elements for very metal-poor and EMP stars, scatters which are consistent with observational uncertainties alone. This has been interpreted as evidence of a chemically well-mixed interstellar medium (ISM), even at very early epochs, in which the mixing time-scale was extremely short.

To circumvent unphysically short mixing time-scales, Arnone et al. (2005) speculated that the longer cooling time-scales in metal-poor gas may prevent subsequent star formation until the ejecta from whole generations of SNe were efficiently mixed. As a result, a small star-to-star scatter would be achieved. However, very long cooling time-scales would wipe out any observed trend with metallicity, such as a decreasing [Zn/Fe] with increasing [Fe/H] observed for EMP giants (Cayrel et al. 2004), unless the SN yields, integrated over the initial mass function (IMF), show a significant metallicity dependence. At least in the case of [Zn/Fe], theoretical yield calculations do not seem to predict such a dependence (e.g., Chieffi & Limongi 2004; Nomoto et al. 2006). Another interesting example is the decreasing [C/O] trend with increasing [O/H] observed for EMP stars (Akerman et al. 2004). Also worth noting is that, even though all stars may form in clusters, at least in the Galaxy today, the majority ($\sim 50 - 85\%$) of the massive stars exploding as SNe, will do so in isolation and not in groups. Hence, the extra averaging of SN ejecta due to clustered star formation will probably not be very significant.

In fact, the presence of trends in the EMP Galactic halo may instead suggest that the ISM at this point was fairly unmixed. In a poorly mixed, extremely metal-poor ISM, two low-mass stars can form out of gas enriched by two SNe of different masses producing different amounts of heavy elements. In such a scenario, trends may simply result from the different mass-dependences of the SN yields (Karlsson & Gustafsson 2005; Nomoto et al. 2006). However, in contrast to what is assumed in homogeneous chemical evolution models, where the most metal-poor stars are enriched by the stars with the shortest lifetimes, i.e., the highest-mass stars, the most metal-poor stars will instead, in most cases, be enriched by the least massive SNe as they eject the least amount of metals.

A number of other elements instead show clear evidence of inhomogeneous chemical enrichment. Large and real scatters are found in the n-capture elements (e.g., François et al. 2007) and probably also in elements like Na, Al, and Si (Cayrel et al. 2004) as well as in N (Spite et al. 2005). Although the scatter in the n-capture elements may result from the possibility that these elements only are produced in a small SN mass range as compared to light and intermediate-mass elements (e.g., Cescutti 2008), this would not explain the observed scatter in those other elements. As regards nitrogen, Chiappini et al. (2006) speculate that the large observed scatter may originate from variations in the initial rotational velocity of extremely metal-poor massive stars, which are shown to produce large variations in the N-yield (Hirschi 2007).

The issue of the star-to-star scatter dichotomy, where some elements show evidence of an unmixed ISM while others do not, needs to be further addressed. However, worth keeping in mind is that *1)* incomplete and/or biased samples of stars may not capture the true cosmic star-to-star scatter, I will briefly come back to this in Sect. 5, and *2)* the effects of inhomogeneous chemical enrichment are only expected to be observed below [Fe/H] ~ -3 (Karlsson 2005), a metallicity regime which still is quite poorly sampled.

3. Carbon-enhanced metal-poor stars

One of the major discoveries found by the HK and the Hamburg/ESO surveys of metal-poor stars is the large fraction of metal-poor stars having [C/Fe] > 1. These stars are called carbon-enhanced, metal-poor (CEMP) stars. The fraction of CEMP stars below [Fe/H] = -2 is determined to $\sim 10-20\%$ (e.g., Cohen *et al.* 2005; Lucatello *et al.* 2006), a fraction which tend to increase with decreasing metallicity. About 80% of the CEMP stars have been shown to be enhanced in s-process elements as well. This subgroup is named CEMP-s, and there are now convincing evidence that the vast majority of the CEMP-s stars are members of binary systems and that they, like the more metal-rich CH and Ba stars, likely obtained their peculiar abundance pattern by mass-transfer from an evolved primary (e.g., Lucatello *et al.* 2005).

What about the remaining 20% of the CEMP stars, the so-called CEMP-no stars, "no" standing for no s-process enrichment? Here, the CEMP-r stars, which are enhanced in r-process (but no s-process) elements, are included in this group. Interestingly, the three ultra metal-poor stars (see Beers & Christlieb 2005) found below [Fe/H] = -4.5, HE 0107 − 5240 (Christlieb *et al.* 2002), HE 1327 − 2326 (Frebel *et al.* 2005), and HE 0557 − 4840 (Norris *et al.* 2007) are all CEMP-no stars and none of them has a detected binary companion. A number of different scenarios have been put forward to explain the anomalous abundance patterns, in particular the high CNO abundances, observed in these ultra metal-poor stars. These scenarios include pre-enrichment of the primordial cloud by a faint SN (e.g., Umeda & Nomoto 2003), enrichment by massive star winds (Meynet *et al.* 2006), mass-transfer from a binary companion (e.g., Suda *et al.* 2004; Tumlinson 2007), and atmospheric dust-gas separation (Venn & Lambert 2008).

Instead of treating the ultra metal-poor stars individually, as unique objects, it could, alternatively, be worthwhile to explore what one may learn if they were treated as 'normal' stars in the general context of chemical evolution. If the gas out of which the EMP stars were formed, experienced a period of low or delayed star formation, e.g., due to negative feedback effects from the first generations of stars, a small population of very C-enhanced, ultra metal-poor stars is to expect (Karlsson 2006). Figure 1 shows the predicted distribution of EMP stars (shaded areas) in the [C/Fe] − [Fe/H] plane. Evidently, the predicted fraction of CEMP stars increases with decreasing [Fe/H], which partly is due to the fact that low-mass star formation may be inhibited in regions with low contents of C (and O, see, e.g., Frebel *et al.* 2007). The predicted fraction of CEMP-no stars below [Fe/H] = -2 is roughly $1-7\%$, depending on which set of stellar yields is used. This is consistent with the observed fraction of $2-5\%$.

As a result of the delayed star formation, the current model is also able to explain the apparent deficit of stars in the metallicity range $-5 \lesssim$ [Fe/H] $\lesssim -4$. Note that the shortfall of stars in this region, should, in this scenario, rather be interpreted as an extension or a stretching of the Galactic halo MDF below [Fe/H] $\simeq -4$. Finally, a small

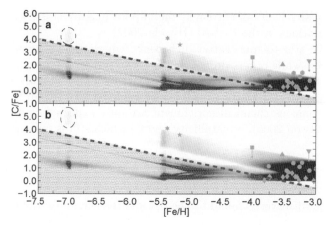

Figure 1. The predicted distribution of stars in the [C/Fe] − [Fe/H] plane. *Upper panel:* The thick, dashed line indicates a carbon abundance of [C/H] = −3.5. The shaded area below this limit should contain very few low-mass stars. Symbols denote various observations of stars in the Milky way's halo (see Karlsson 2006). The predicted group of mega metal-poor stars is encircled. *Lower panel:* Same as above but with a carbon yield increased by a factor of 4 for stars in the mass range $30 \leqslant m/\mathcal{M}_\odot \leqslant 60$. Observed abundance ratios are not corrected for 3D effects.

population of mega metal-poor stars ([Fe/H] < −6) is predicted to exist in this scenario as a result of an early enrichment by O-Ne-Mg SNe (see Figure 1; Karlsson 2006).

4. The missing second stellar generation

Current simulations of primordial star formation tend to agree on the fact that the initial conditions in the early universe should favor the formation of massive and very-massive stars, in excess of $100\mathcal{M}_\odot$ (e.g., Bromm *et al.* 1999,2002; Abel *et al.* 2002). This is partly due to the inefficient cooling of the gas which result in initially high Jeans masses (e.g., Larson & Starrfield 1971) and partly due to the apparent inability of primordial collapsing gas clouds to fragment at later stages (Yoshida *et al.* 2006). Interestingly, Eta Carinae and the Pistol Star are two examples of very massive stars in our own galaxy and recently, an extremely powerful SN explosion in the nearby galaxy NGC 1260 was suggested to be the final death of a very massive star in excess of $100\mathcal{M}_\odot$ (Smith *et al.* 2007).

Assuming that primordial very massive stars were able to form and that at least some of them where able to explode as so-called pair-instability SNe (PISNe), in what metallicity regime should we expect to find their chemical signature? Due to their very short life-times, PISNe must have been the first stars to enrich the ISM in metals. In the classical picture of chemical evolution, an initial enrichment by primordial PISNe would generate a metallicity floor out of which the second generation, low-mass stars were able to from. Naturally, these stars would be the most metal-poor stars to be found in the Galaxy and would show a unique PISN signature, characterized by, by not limited to, a pronounced odd-even effect and a lack of r- and s-process elements. Such a signature is, however, not observed, neither in the EMP stars, nor in the ultra metal-poor stars. These stars all show chemical signatures more resembling those of normal core collapse SNe (see, however, Sect. 3 for a further discussion). This has been taken as an indication that very massive stars were exceedingly few in the early universe (e.g., Tumlinson *et al.* 2004; Ballero *et al.* 2006), if not altogether absent, in contrast to the predictions.

A rather different conclusion may, however, be drawn if the instantaneous mixing approximation assumed in homogeneous chemical evolution modelling is relaxed and

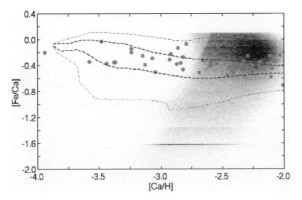

Figure 2. Predicted distribution (shaded area) of low-mass stars in the [Fe/Ca] − [Ca/H] plane. Only the partial distribution of simulated stars for which > 90% of the total atmospheric Ca abundance is synthesized in PISNe is displayed. Symbols denote various observations of Galactic halo stars. The dashed, thick and thin lines indicate, respectively, the 1σ (innermost 68.3%) and 3σ (innermost 99.7%) cosmic scatter in the simulation (for details, see Karlsson et al. 2008).

the enrichment by SNe is instead allowed to occur locally, where the ejecta are mixed relatively slowly with their surroundings by turbulent diffusion. In order to simulate the stochastic enrichment by primordial PISNe in a cosmological context, we followed the initial star formation and chemical enrichment of collapsing atomic cooling halos (Karlsson et al. 2008). We allowed for a primordial star formation mode in which only massive and very massive stars were able to form. This primordial mode ceased as the gas became metal-enriched and turned into a normal Pop II star formation mode in which only low- and high mass stars were allowed to form. The larger explosion energies of the PISNe was also taken into account as they may sweep up larger amounts of gas before merging with the ISM.

The result is shown in Figure 2. The predicted distribution of stars with a dominant contribution from primordial PISNe, here shaded in gray, is located around [Ca/H] $\sim$ -2.5, significantly above the most metal-poor stars in the simulation (indicated by the thick and thin dashed lines). These stars have, in fact, such high Ca abundances that a fair fraction of them may risk to remain undetected in current surveys of metal-poor stars. Furthermore, for realistic estimates of the fraction of primordial very massive stars exploding as PISNe (say $\sim 10\%$ by mass in the primordial stellar population, Greif & Bromm 2006), the fraction of low-mass stars below [Ca/H] $= -2$ with a dominant PISN signature is predicted to be very low, only $\sim 3 \times 10^{-4}$, which may well explain the fact that such a star has not yet been found. Current observational data are able to constrain the fraction of metal-free, very massive stars to $\lesssim 40\%$, by mass (Karlsson et al. 2008).

5. Concluding remarks

Although the EMP stars are crucial for the understanding of the star formation process, feedback effects and the overall chemical enrichment in the early Galaxy, they do seem to tell us little about the primordial PISNe. Instead, it may be the "not-so-extremely-metal-poor" stars that hold the clue to the first sources of light in the universe, i.e., if our understanding of primordial star formation is correct. Furthermore, as argued in Sect. 3, CEMP-no stars belong to a population of stars which was born with a high [C/Fe]. If so, these stars should *not* be excluded from the general picture of chemical evolution. In particular, they should be considered when computing the cosmic star-to-star scatter.

Acknowledgment. The author acknowledges support from the IAU and Danish sponsors.

References

Abel, T., Bryan, G. L., & Norman, M. L. 2002, *Science*, 295, 93
Akerman, C. J., Carigi, L., Nissen, P. E., Pettini, M., & Asplund, M. 2004, *A&A*, 414, 931
Arnone, E., Ryan, S. G., Argast, D., Norris, J. E., & Beers, T. C. 2005, *A&A*, 430, 507
Ballero, S. K., Matteucci, F., & Chiappini, C. 2006, *New Astron.*, 11, 306
Barklem, P. S., et al. 2005, *A&A*, 439, 129
Beers, T. C. & Christlieb, N. 2005, *ARAA*, 43, 531
Belokurov, V., et al. 2007, *ApJ*, 654, 897
Bovill, M. S. & Ricotti, M. 2008, *ApJ*, submitted *(astro-ph/0806.2340)*
Bromm, V., Coppi, P. S., & Larson, R. B. 1999, *ApJ* (Letters), 527, L5
—. 2002, *ApJ*, 564, 23
Cayrel, R., et al. 2004, *A&A*, 416, 1117
Cescutti, G. 2008, *A&A*, 481, 691
Chiappini, C., Hirschi, R., Meynet, G., Ekström, S., Maeder, A., & Matteucci, F. 2006, *A&A* (Letters), 449, L27
Chieffi, A. & Limongi, M. 2004, *ApJ*, 608, 405
Christlieb, N., et al. 2002, *Nature*, 419, 904
Cohen, J. G., et al. 2005, *ApJ* (Letters), 633, L109
François, P., et al. 2007, *A&A*, 476, 935
Frebel, A., Johnson, J. L., & Bromm, V. 2007, *MNRAS* (Letters), 380, L40
Frebel, A., et al. 2005, *Nature*, 434, 871
Greif, T. H. & Bromm, V. 2006, *MNRAS*, 373, 128
Greif, T. H., Johnson, J. L., Klessen, R. S., & Bromm, V. 2008, *MNRAS*, submitted *(astro-ph/0803.2237)*
Helmi, A., et al. 2006, *ApJ* (Letters), 651, L121
Hirschi, R. 2007, *A&A*, 461, 571
Karlsson, T. 2005, *A&A*, 439, 93
—. 2006, *ApJ* (Letters), 641, L41
Karlsson, T. & Gustafsson, B. 2005, *A&A*, 436, 879
Karlsson, T., Johnson, J. L., & Bromm, V. 2008, *ApJ*, 679, 6
Kirby, E. N., Simon, J. D., Geha, M., Guhathakurta, P., & Frebel, A. 2008, *ApJ* (Letters), submitted *(astro-ph/0807.1925)*
Larson, R. B. & Starrfield, S. 1971, *A&A*, 13, 190
Lucatello, S., Beers, T. C., Christlieb, N., Barklem, P. S., Rossi, S., Marsteller, B., Sivarani, T., & Lee, Y. S. 2006, *ApJ* (Letters), 652, L37
Lucatello, S., Tsangarides, S., Beers, T. C., Carretta, E., Gratton, R. G., & Ryan, S. G. 2005, *ApJ*, 625, 825
Meynet, G., Ekström, S., & Maeder, A. 2006, *A&A*, 447, 623
Nomoto, K., Tominaga, N., Umeda, H., Kobayashi, C., & Maeda, K. 2006, *Nucl. Phys. A*, 777, 424
Norris, J. E., et al., N. 2007, *ApJ*, 670, 774
Salvadori, S., Ferrara, A., & Schneider, R. 2008, *MNRAS*, 386, 348
Smith, N., et al. 2007, *ApJ*, 666, 1116
Spite, M., et al. 2005, *A&A*, 430, 655
Suda, T., Aikawa, M., Machida, M. N., Fujimoto, M. Y., & Iben, I., Jr. 2004, *ApJ*, 611, 476
Tumlinson, J. 2007, *ApJ*, 665, 1361
Tumlinson, J., Venkatesan, A., & Shull, J. M. 2004, *ApJ*, 612, 602
Umeda, H. & Nomoto, K. 2003, *Nature*, 422, 871
Venn, K. A., Irwin, M., Shetrone, M. D., Tout, C. A., Hill, V., & Tolstoy, E. 2004, *AJ*, 128, 1177
Venn, K. A. & Lambert, D. L. 2008, *ApJ*, 677, 572
Willman, B., et al. 2005, *AJ*, 129, 2692
Yoshida, N., Omukai, K., Hernquist, L., & Abel, T. 2006, *ApJ*, 652, 6

Halo chemistry and first stars.
The chemical composition of the matter in the early Galaxy, from C to Mg †

M. Spite[1], P. Bonifacio[1,2,3], R. Cayrel[1], F. Spite[1], P. Francois[1],
H. G. Ludwig[1,2], E. Caffau[1], S. Andrievsky[4], B. Barbuy[5], B. Plez[6],
P. Molaro[3], J. Andersen[7], T. Beers[8], E. Depagne[9], B. Nordström[7],
F. Primas[10]

[1] GEPI - Observatoire de Paris-Meudon F92195 Meudon Cedex, France
email: monique.spite@obspm.fr

[2] CIFIST Marie Curie Excellence Team

[3] INA - Osservatorio Astronomico di Trieste, Via Tiepolo 11, I-34143 Trieste, Italy

[4] Dept of Astronomy, Odessa Nat. University, Shevchenko Park, 65014 Odessa, Ukraine

[5] Univ. de Sao Paulo, Depto. de Astronomia, Rua do Matao 1226, Sao Paulo 05508-900, Brazil

[6] GRAAL, Univ. de Montpellier II, F-34095 Montpellier Cedex 05, France

[7] The Niels Bohr Institute, Astronomy group, Juliane Maries Vej 30,
DK-2100 Copenhagen, Denmark

[8] Michigan State Univ., East Lansing, MI 48824, USA

[9] Las Cumbres Observatory, Goleta, CA 93117, USA

[10] ESO, Karl Scwarzschild-Str. 2, D 85749, Garching bei München, Germany

Abstract. From NLTE computations of the magnesium abundance in a sample of extremely metal-poor giants we derive [Mg/Fe]=+0.7, leading to [Al/Mg]=–0.80 and [Na/Mg]=–0.85 in the early Galaxy. The ratio [O/Mg] should be near to the solar value. Measurements of nitrogen abundances derived from the analysis of the NH band in eight more stars confirm the large scatter of the ratios [N/Fe] and [N/O] in the early Galaxy.

Keywords. Stars: abundances, stars: atmospheres, stars: Populations II, Galaxy:abundances, Galaxy:evolution, nucleosynthesis

1. Introduction

The "Extremely metal-poor" stars formed less than 1 Gyr after the formation of the Galaxy. They are extremely metal-poor because, at that time, the matter had been enriched by a very small number of supernovae. Moreover, these first supernovae could only be massive type II supernovae which have a very short lifetime.

In contrast, the chemical composition in the Galactic disk is the result of enrichment not only by SNII, but also by AGB stars and SNI.

Seven years ago, to take advantage of the high-resolution spectrograph UVES at the VLT, and of the existence of a large sample of EMP stars (H&K survey: Beers, Preston & Schectman 1992), Roger Cayrel stimulated the formation of a "First stars team" to apply for an ESO Large Program dedicated to the study of the chemical composition of the Galactic matter in the early times. Other teams in the world have chosen to work on similar subjects, in particular at the Keck and SUBARU telescopes in Hawaii or at

† Based on observations made with the ESO Very Large Telescope at Paranal Observatory, (Large Programme "First Stars", ID 165.N-0276; P.I.: R. Cayrel, and Programme 078.B-0238; P.I.: M. Spite).

2. Observations and Spectrum analysis

In summary, 52 very metal-poor stars (34 with [Fe/H] < −2.9) were observed in the "First Stars" programme (S/N ≈ 200, R ≈ 45000, 3350 < λ < 10000). All these stars were "normal" metal-poor stars, not carbon-enriched, but some carbon-rich stars were observed for comparison. Later in a complementary run focused on the problem of the nitrogen abundance, ten more giant stars in the lower part of the HR diagram were observed in the same conditions.

In a first step, a classical LTE analysis was performed using MARCS theoretical 1D plane-parallel model atmospheres (Gustafsson et al. 2008). The stellar effective temperature was determined from photometry (or the wings of the Hα line, for dwarfs) and the gravity from the ionization equilibrium of iron.

More details can be found in Cayrel et al. (2004) and Bonifacio et al. (2007, 2008).

3. [O/Mg], [Na/Mg] and [Al/Mg] in the Galaxy, Influence of NLTE effects.

In extremely metal-poor stars, the lines are often very weak and the abundance of several elements like sodium or aluminum can be determined only from the resonance lines. However, it is well known that these lines are very sensitive to NLTE effects. As a first approximation we applied a systematic non-LTE correction actually estimated for dwarfs to all the stars (Cayrel et al. 2004). But in both cases the scatter in the abundance ratios was very large; for Na there was strong disagreement between dwarfs and giants. After a complete NLTE computation of the Na and Al resonance lines, the agreement between the Na and Al abundances in giants and in dwarfs became very good and the scatter much smaller (Andrievsky et al. 2007, 2008).

But as noted by Gehren et al. (2006) non-LTE effects are also not negligible when computing Mg abundances in metal-poor stars. Therefore we decided to compute also NLTE profiles for Mg lines. For giants we found a mean NLTE correction of about +0.4 dex and [Mg/Fe] ≈ +0.7 (computations for dwarfs are in progress). In Cayrel et al. (2004) we measured [O/Fe] ≈ +0.7 (without 3D correction). This measurement of the oxygen abundance is not sensitive to NLTE effects (forbidden line). The new value of the magnesium abundance in giants would imply [O/Mg] ≈ +0.0 in the early Galaxy.

In figures 1 and 2 we show the variation of the ratios [Al/Mg] and [Na/Mg] in the Galaxy. For the disk stars [Al/Mg] is larger than –0.1dex (Gehren, 2006), in the halo this ratio decreases, the minimum is reached at about [Fe/H]=−3.0 with [Al/Mg] = −0.8, then at lower metallicity it seems to increase slightly.

At very low metallicity the scatter of [Na/Mg] is larger than for [Al/Mg], but all the "Na-rich" stars are "mixed" stars. These stars have a high N abundance coupled with a low C abundance and a low value of the $^{12}C/^{13}C$, ratio, indicating a mixing with the H burning layer. If this mixing is particularly deep, it can bring products of the O, Ne, Na cycle to the surface (Denissenkov & Pinsonneault 2008). The relation between [Na/Mg] and [Fe/H] is a continuous enrichment through all three galactic populations, halo, thick disk, thin disk. The minimum is reached at about [Fe/H]=−3.0 with [Na/Mg] ≈ −0.85, then at lower metallicity it is possible that [Na/Mg] increases slightly.

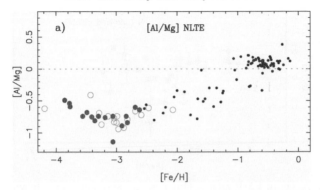

Figure 1. Evolution of the [Al/Mg] ratio in the Galaxy as a function of [Fe/H]. The small black dots are the data by Gehren et al. (2004, 2006) for the disk and the halo. The large blue dots represent our sample of "unmixed" giants, the open circles the "mixed" giants.

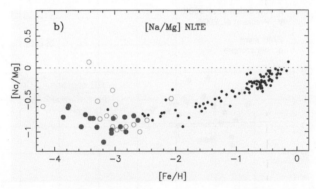

Figure 2. Same as Fig. 1, but for [Na/Fe]. At very low metallicity the scatter is larger than for [Al/Mg] but all the "Na-rich" stars are mixed stars, and the enrichment is probably due to mixing with the deep H burning layer.

4. C and N abundances in the early Galaxy

In extremely metal-poor stars, carbon and nitrogen abundances are derived from the G band of CH (430 nm) and the violet band of NH (336 nm). In general, the NH (and CN) bands are not visible in the spectra of EMP turnoff stars (the stars are too hot), so the N abundance can be determined only in giants. On the other hand, the abundances of C and N in the pristine gas can be deduced only from stars located below the bump in the HR diagram. At higher luminosity, the atmospheres of the stars are significantly altered by mixing and thus are not a good diagnostic of the initial chemical composition (Spite et al., 2005, 2006).

In Spite et al. (2005), we showed from our sample of unmixed stars that the N abundance in the early Galaxy showed large scatter. But since there were only few unmixed giants, we could not decide between a simple scatter, two different levels of abundance, or even an anti-correlation between [Fe/H] and the [N/Fe] or [N/Mg] ratios.

We have recently observed the region of the NH band in eight more "unmixed" EMP giants, selected from the sample of Barklem et al. (2005). The N abundances in these stars have been computed from synthetic spectra. The results are shown in Fig. 3. These new measurements support the existence of significant scatter in [N/Fe] in EMP stars. This scatter could reflect different rotational velocities in the massive first stars (e.g. Meynet et al., 2008). The scatter seems to be even smaller at very lowest metallicities.

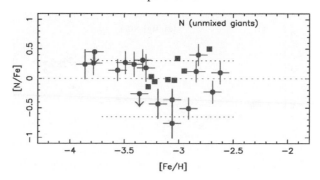

Figure 3. [N/Fe] in EMP giants. Our new measurements are shown as squares.

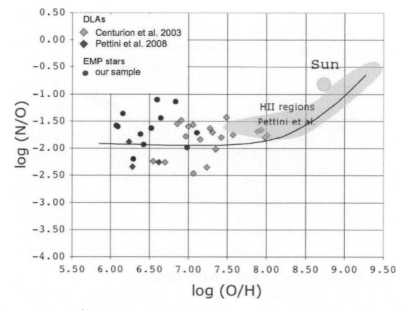

Figure 4. log N/O for H2 regions, DLAs (diamonds) and EMP stars (blue dots)

However, for [Fe/H] < −3.4 the NH band is only measurable if [N/Fe] > 0, so the smaller scatter in this domain could be a spurious effect.

In three of the newly observed stars it was possible to measure the oxygen abundance from the forbidden O I line (see Cayrel *et al.* 2004). These measurements are in good agreement with the previous measures.

For all the EMP stars with measured N and O abundances, the ratio N/O can be compared to the values obtained in DLAs by Centurion *et al.* 2003, or Pettini *et al.* 2008 (Fig. 4). At low metallicity the scatter of the ratio N/O is large in both stars and DLAs. There is no indication of a decrease of N/O with O/H as it would be expected if the formation of N would depend on the metallicity of the progenitor. In fact, the mean N/O ratio remains almost constant. Massive stars may contribute significantly to the nucleosynthesis of O, but also of C and N (Meynet *et al.*, 2008).

5. Conclusions

Precise computations of the abundances of the light metals Na, Al and Mg in extremely metal-poor stars have shown that it is important to take NLTE effects into account. From our preliminary NLTE computations of the Mg abundance in EMP giant stars, we found that: [Mg/Fe] $\approx$ +0.7dex, [Al/Mg] $\approx$ −0.80dex and [Na/Mg] $\approx$ −0.85dex.

We note that all the ratios [X/Mg] in Cayrel *et al.* (2004) would have to be corrected in the same way, and in particular, the mean value of [O/Mg] at very low metallicity could be close to the solar value.

New observations of a sample of "unmixed" metal-poor giants confirm the large scatter of the ratios N/O and N/Fe. The mean value of N/O is almost constant in the interval $6 < \log(O/H) < 7.5$.

References

Andrievsky, S. M., Spite M., Korotin, S. A., Spite, F., Bonifacio P., Cayrel, R., Hill, V., & François, P., 2007, *A&A 464, 1081*

Andrievsky, S. M., Spite, M., Korotin, S. A., Spite, F., Bonifacio, P., Cayrel, R., Hill, V., & François, P., 2008, *A&A 481, 481*

Barklem, P. S., Christlieb, N., Beers, T. C., Hill, V., Bessell, M. S., Holmberg, J., Marsteller, B., Rossi, S., Zickgraf2, F.-J., & Reimers, D., 2005, *A&A 439, 129*

Beers, T. C., Preston, G. W., & Schectman, S. A., 1992, *AJ 103, 1987*

Bonifacio, P., Molaro, P., Sivarani, T., Cayrel, R., Spite, M., Spite, F., Plez, B., Andersen, J., Barbuy, B., Beers, T. C., Depagne, E., Hill, V., François, P., Nordström, B., & Primas, F., 2007, *A&A 462, 851*, (**"First Stars VII"**)

Bonifacio, P., Spite, M., Cayrel, R., Hill, V., Spite, F., François, P., Plez, B., Ludwig, H-G., Caffau, E., Molaro, P., Depagne, E., Andersen, J., Barbuy, B., Beers, T. C., Nordström, B., & Primas, F., 2004, *A&A submitted*, (**"First Stars XII"**)

Cayrel, R., Depagne, E., Spite, M., Hill, V., Spite, F., François, P., Plez, B., Beers, T. C., Primas, F., Andersen, J., Barbuy, B., Bonifacio, P., Molaro, P., & Nordström, B., 2004, *A&A 416, 1117*, (**"First Stars V"**)

Centurion, M., Molaro, P., Vladilo, G., Péroux, C., Levshakov, S. A., & D' Odorico, V., 2003, *A&A 403, 55*

Denissenkov, P. A. & Pinsonneault, M., 2008, *ApJ 679, 1541*

Gehren, T., Liang, Y. C., Shi, J. R., Zhang, H. W., & Zhao, G., 2004, *A&A 413, 1045*

Gehren, T., Shi, J. R., Zhang, H. W., Zhao, G., & Korn, A. J., 2006, *A&A 451, 1065*

Gustafsson, B., Edvardsson, B., Eriksson, K., Jørgensen, U. G., Nordlund, A., & Plez, B., 2008, *arXiv0805.0554v1*

Meynet, G., Ekstrom, S., Georgy, C., Maeder, A., & Hirschi, R., 2008, *arXiv0806.4063M*

Pettini, M., Zych, B. J., Steidel, C. C., & Chaffee, F. H. 2008 *Mont. Not. R., Astron. Soc. 385, 2011*

Spite, M., Cayrel, R., Plez, B., Hill, V., Spite, F., & Depagne, E., François, P. Bonifacio. P. Barbuy, B. Beers, T. Andersen, J. Molaro, P., Nordström, B., Primas, F., 2005, *A&A 430, 655*, (**"First Stars VI"**)

Spite, M., Cayrel, R., Hill, V., Spite, F., François, P., Plez, B., Bonifacio, P., Molaro, P., Depagne, E., Andersen, J., Barbuy, B., Beers, T. C., Nordström, B., & Primas, F., 2006, *A&A 455, 291*, (**"First Stars IX"**)

A stellar abundance gang relaxing outside the lecture hall: Andreas Korn, Sofia Feltzing, Frank Grundahl, Bengt Gustafsson, and Anna Önehag.

Preben Grosbøl, Carme Jordi, and Claus Fabricius at Carlsberg.

Chemical Yields from Supernovae and Hypernovae

Ken'ichi Nomoto[1,2], Shinya Wanajo[1,2], Yasuomi Kamiya[1,2], Nozomu Tominaga[3], and Hideyuki Umeda[2]

[1] Institute for the Physics and Mathematics of the Universe, University of Tokyo, Kashiwa, Chiba 277-85668, Japan
email: nomoto@astron.s.u-tokyo.ac.jp
[2] Department of Astronomy, University of Tokyo, Bunkyo-ku, Tokyo 113-0033, Japan
[3] National Astronomical Observatory, Mitaka, Tokyo 113-0033, Japan

Abstract. We review the final stages of stellar evolution, supernova properties, and chemical yields as a function of the progenitor's mass. (1) 8 - 10 $M_\odot$ stars are super-AGB stars when the O+Ne+Mg core collapses due to electron capture. These AGB-supernovae may constitute an SN 2008S-like sub-class of Type IIn supernovae. These stars produce little α-elements and Fe-peak elements, but are important sources of Zn and light p-nuclei. (2) 10 - 90 $M_\odot$ stars undergo Fe-core collapse. Nucleosynthesis in aspherical explosions is important, as it can well reproduce the abundance patterns observed in extremely metal-poor stars. (3) 90 - 140 $M_\odot$ stars undergo pulsational nuclear instabilities at various nuclear burning stages, including O and Si-burning. (4) Very massive stars with $M \gtrsim 140 M_\odot$ either become pair-instability SNe, or undergo core-collapse to form intermediate mass black holes if the mass loss is small enough.

Keywords. Galaxy: halo — gamma rays: bursts — nuclear reactions, nucleosynthesis, abundances — stars: abundances — stars: AGB — supernovae: general

1. Core-Collapse Supernovae and Progenitor Masses

The final stages of massive star evolution, supernova properties, and their chemical yields depend on the progenitor's masses M as follows (e.g., Arnett 1996):

(1) *8 - 10 $M_\odot$ stars*: These stars are on the AGB phase when the O+Ne+Mg core collapses due to electron captures. The exact mass range depends on the mass loss during the AGB phase. They undergo weak explosions being induced by neutrino heating. These stars produce little α-elements and Fe-peak elements, but are important sources of Zn and light p-nuclei. These AGB supernovae may constitute an SN 2008S-like sub-class of Type IIn supernovae.

(2) *10 - 90 $M_\odot$ stars*: These stars undergo Fe-core collapse to form either a neutron star (NS) or a black hole (BH), and produce a large amount of heavy elements from α-elements and Fe-peak elements. Observations have shown that the explosions of these Fe-core collapse supernovae are quite aspherical. In the extreme case, the supernova energy is higher than 10^{52} erg s^{-1}, i.e. a Hypernova. Nucleosynthesis in these jet-induced explosions is in good agreement with the abundance patterns observed in extremely metal-poor stars.

(3) *90 - 140 $M_\odot$ stars*: These massive stars undergo nuclear instabilities and associated pulsations (ϵ-mechanism) at various nuclear burning stages depending on the mass loss and thus metallicity. In particular, if the mass loss is negligible, pulsations of O-cores and/or Si-cores due to O, Si-burning could produce dynamical mass ejection. Eventually, these stars undergo Fe-core collapse.

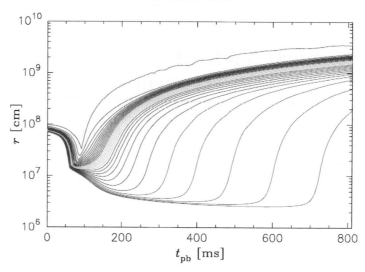

Figure 1. The change in the radius as a function of post-bounce time for material ejected from the collapsing O-Ne-Mg core (Kitaura & Janka 2006).

(4) *140 - 300 $M_\odot$ stars*: If these very massive stars (VMS) do not lose much mass, they become pair-instability supernovae (PISN). The star is completely disrupted without forming a BH and thus ejects a large amount of heavy elements, especially Fe.

(5) *Stars with $M \gtrsim 300 M_\odot$*: These VMSs are too massive to be disrupted by PISN but undergo core collapse (CVMS), forming an intermediate-mass black hole (IMBH). Some mass ejection could be possible, associated with the possible jet-induced explosion.

Here we summarize the properties of the above supernovae and chemical yields in some detail.

2. 8 - 10 $M_\odot$ AGB Stars undergoing Electron Capture Supernovae

2.1. *Collapse of O+Ne+Mg Core induced by Electron Capture*

For 8 - 10 $M_\odot$ stars, electrons become degenerate already in a C+O core. In a semi-degenerate C+O core, neutrino cooling leads to off-center ignition of carbon when the C+O core mass exceeds the critical mass of 1.06 $M_\odot$. The off-center carbon burning shell moves inward all the way to the center due to heat conduction (Nomoto 1984; Timmes & Woosley 1992; Garcia-Berro et al. 1997).

After exhaustion of carbon in the central region, an O+Ne+Mg core forms. The core mass does not exceed the critical mass of 1.37 $M_\odot$ for neon ignition and, hence, neon burning is never ignited (Nomoto 1984). Then the O+Ne+Mg core becomes strongly degenerate. The envelope becomes similar to the asymptotic giant-branch (super-AGB) stars (Hashimoto, Iwamoto, & Nomoto 1993; Poelarends et al. 2008) with a thin He burning shell that undergoes thermal pulses and s-process nucleosynthesis.

The final fate depends on the competition between the mass loss that reduces the envelope mass and the increase in the core mass through the H-He shell burning. If the mass loss is fast, an O+Ne+Mg white dwarf is formed, which could be the case for 8 $M_\odot$ - $M_{\rm up}$ stars, where $M_{\rm up} \sim 9 \pm 0.5 M_\odot$ being smaller for smaller metallicity (Poelarends et al. 2008). For $M_{\rm up}$ - 10 $M_\odot$ stars, the core mass grows to 1.38 $M_\odot$ and the central density reaches 4×10^9 g cm^{-3}. The electron Fermi energy exceeds the threshold for electron captures ^{24}Mg(e$^-$, ν) ^{24}Na (e$^-$, ν) ^{20}Ne and ^{20}Ne (e$^-$, ν) ^{20}F (e$^-$, ν) ^{20}O. The resultant decrease in Y_e triggers collapse (Nomoto 1987).

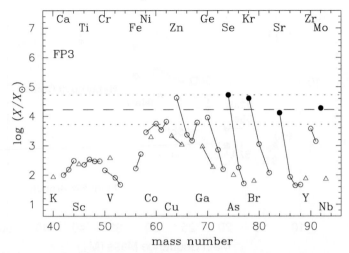

Figure 2. Mass fractions of isotopes (after decay) in the ejecta of model FP3 (Wanajo et al. 2008) relative to their solar values (Lodders 2003) as a function of the mass number. The even-Z and odd-Z isotopes are denoted by open circles and triangles, respectively. The p-nuclei are represented by filled symbols. The dotted horizontal lines indicate a "normalization band" between the largest production factor and a factor of ten smaller than that, along with the median value (*dashed line*).

The hydrodynamical behavior of collapse and bounce is somewhat different from the iron core collapse of more massive stars (Fig. 1: Kitaura & Janka 2006). The explosion energy is as low as $E \sim 10^{50}$ erg. The explosion is also suggested be close to spherical, thus producing little pulsar kick. The existence of pulsars in globular clusters might be explained by the electron capture supernovae (Kalgera et al. 2008).

2.2. *Nucleosynthesis in Electron Capture Supernovae*

Nucleosynthesis in the supernova explosion of a 9 $M_\odot$ star has been investigated (Hofman et al. 2008; Wanajo et al. 2008) using thermodynamic trajectories taken from the explosion model (Kitaura & Janka 2006). Here we summarize the results by Wanajo et al. (2008).

1. The unmodified model produces small amounts of α-elements and iron, but large amounts of ^{64}Zn, ^{70}Ge, and in particular, ^{90}Zr, with some light p-nuclei (e.g., ^{92}Mo; Fig. 2). This is due to the ejection of a large amount of neutron-rich matter ($Y_e = 0.46 - 0.49$), and might put severe constraints on the frequency of occurrence of this type of supernovae (Hofman et al. 2008). However, the production of ^{90}Zr does not serve as a strong constraint, because it is easily affected by a small variation of Y_e (see below).

2. The overproduction of ^{90}Zr becomes more moderate if the minimum Y_e is only $1 - 2\%$ larger than that in the unmodified model. Such a change in the initial Y_e profile might be caused by convection that is not considered in the 1-D simulation (Kitaura & Janka 2006). In this case (model FP3: Fig. 2), the largest overproduction, which is shared by ^{64}Zn, ^{70}Se, and ^{78}Kr, falls to one-tenth that of the unmodified model. The ^{64}Zn production provides an upper limit to the occurrence of exploding O-Ne-Mg cores at about 20% of all core-collapse supernovae.

3. The ejecta mass of ^{56}Ni is $0.002 - 0.004\,M_\odot$, much smaller than the $\sim 0.1\,M_\odot$ in more massive progenitors. Convective motions near the mass cut may also affect the Y_e-distribution and thus the ^{56}Ni mass. See Wanajo et al. (2008) for a recent comparison between the electron capture supernova yields and abundances in the Crab Nebula.

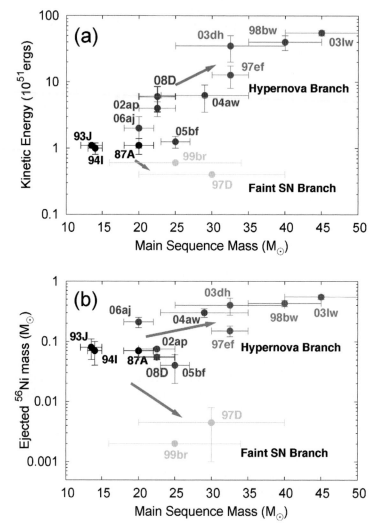

Figure 3. *Upper*: The kinetic explosion energy E as a function of the main sequence mass M of the progenitors for several supernovae/hypernovae. Hypernovae are the SNe with $E_{51} > 10$. *Lower*: Same as the upper panel, but for the ejected mass of ^{56}Ni.

2.3. Connection to Faint Supernovae

The expected small amount of ^{56}Ni as well as the low explosion energy of electron capture supernovae have been proposed as an explanation of the observed properties of low-luminosity SNe IIP, such as SN 1997D (Chugai & Utrobin 2000; Kitaura & Janka 2006) and of the low luminosity of SN 2008S-like transients (Prieto et al. 2008; Thompson et al. 2008). The estimated ^{56}Ni masses of $\sim 0.002 - 0.008\, M_\odot$ for observed low-luminosity SNe II-P (Zampieri et al. 2003; Hendry et al. 2005) are in reasonable agreement with the presented results from O-Ne-Mg core explosions. An alternative possibility of such supernovae is that more massive stars ($\gtrsim 20\, M_\odot$) with low explosion energies suffer from fallback of freshly synthesized ^{56}Ni (Turatto et al. 1998; Nomoto et al. 2003; Zampieri et al. 2003). A recent analysis of the progenitors of SNe IIP by Smartt et al. (2008) favors low mass progenitors. A lack of α-elements such as O and Mg in the case of collapsing O-Ne-Mg cores will be a key to spectroscopically distinguish between these two scenarios.

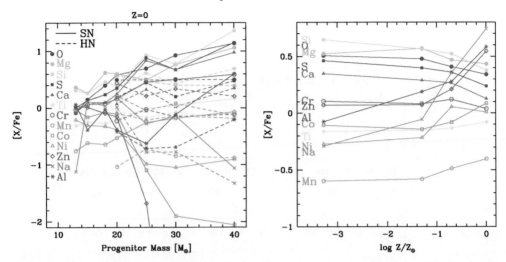

Figure 4. (*Left:*) Relative abundance ratios as a function of progenitor mass with $Z = 0$. The solid and dashed lines show normal SNe II with $E_{51} = 1$ and HNe. (*Right:*) The IMF weighted abundance ratios as a function of metallicity of progenitors, where the HN fraction $\epsilon_{\rm HN} = 0.5$ is adopted. Results for $Z = 0$ are plotted at $\log Z/Z_\odot = -4$ (Nomoto et al. 2006; Kobayashi et al. 2006).

Recently, the progenitor of SN 2008S was discovered in the infrared, and it has been suggested that it was an AGB star (Prieto et al. 2008). If so, SN 2008S could belong to the electron capture supernovae.

The envelope of the AGB star is carbon-enhanced (Nomoto 1987). Then dust could easily be formed to induce mass loss. This may result in a deeply dust-enshrouded object such as the progenitor of SN 2008S (Prieto et al. 2008; Thompson et al. 2008). This might also imply that the mass range of the stars that end their lives as O-Ne-Mg supernovae is $\sim 9.5 - 10\,M_\odot$; their frequency, $\sim 7 - 8\%$ of all the core-collapse events, satisfies the constraint from our nucleosynthesis results ($< 30\%$).

3. 10 - 90 $M_\odot$ Stars undergoing Aspherical Explosions

These stars undergo Fe-core collapse and become Type II-P (plateau) supernovae (SNe II-P) if the red-supergiant-size H-rich envelope remains, and Type Ibc supernovae if the H-rich envelope has been stripped off by a stellar wind or Roche-lobe overflow. These SNe are the major sources of heavy elements from C to the Fe-peak.

Their yields depend on the progenitor's mass M, metallicity, and the explosion energy E. From the comparison between the observed and calculated spectra and light curves of supernovae, we can estimate M and E as shown in Figure 3 (Nomoto et al. 2006).

Three SNe (SNe 1998bw, 2003dh, and 2003lw) are associated with long Gamma-Ray Bursts (GRBs) (e.g., Woosley & Bloom 2006). The progenitors of these GRB-SNe tend to be more massive than $\sim 30 M_\odot$. Also GRB-SNe are all very energetic with the kinetic energy E exceeding 10^{52} erg, more than 10 times the kinetic energy of normal core-collapse SNe. Here we use the term 'Hypernova (HN)' to describe such hyper-energetic supernovae with $E_{51} = E/10^{51}$ erg $\gtrsim 10$ (Fig.3; Nomoto et al. 2004, 2006; Kobayashi et al. 2006).

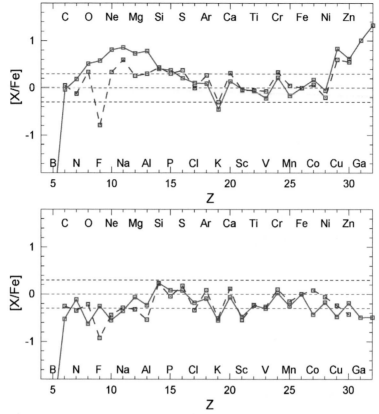

Figure 5. (*Upper*): Comparison of yields for $M = 15 M_\odot$, $Z = 0.02$, and $E = 1 \times 10^{51}$ erg (red-solid: Nomoto et al. 2006; blue-dashed: Limongi et al. 2000). (*Lower*): Same as upper, but for $M = 25 M_\odot$.

3.1. Supernova and Hypernova Yields

Theoretical models of stellar evolution depend on the treatment of complicated physical processes, such as mixing due to convection and rotation, convective overshooting, mass loss, etc. Thus SN yields obtained by various groups are not necessarily in agreement. Figure 5 compares the yields of the models with $E = 1 \times 10^{51}$ erg and $Z = 0.02$ for $M = 15 M_\odot$ (upper) and $M = 25 M_\odot$ (lower) (Nomoto et al. 2006; Limongi et al. 2000). These models include mass loss but not rotation. It is seen that two yields are in good agreement.

Figure 6 (upper) compares the yields of models with $E = 1 \times 10^{51}$ erg and $Z = 0.00$ for $M = 20 M_\odot$ between the three groups (Nomoto et al. 2006; Limongi et al. 2000; Heger & Woosley 2008). These three yields are in good agreement. The smooth pattern in Heger & Woosley (2008) is due to the mixing-fallback effect (Umeda & Nomoto 2002) being taken into account in their model. This implies that the difference in the treatment of such mixing-fallback during the explosion is larger than other differences in presupernova models among the three groups.

Figure 6 (upper) also compares the observed averaged abundance pattern of the EMP stars (Cayrel et al. 2004) with the three theoretical models. We note that theoretical predictions of Zn, Co, Ti/Fe are much smaller than the observed ratios. The underproduction of these elements relative to Fe is much improved in the hypernova models

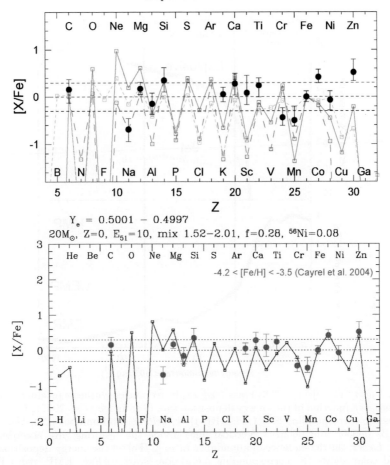

Figure 6. Averaged elemental abundances of stars with [Fe/H] = −3.7 (Cayrel *et al.* 2004) compared with yields for $M = 20 M_\odot$ and $Z = 0.0$ (*upper*: Nomoto *et al.* 2006; Limongi *et al.* 2000; Heger & Woosley 2008) and the hypernova yield (*lower*: 20 $M_\odot$, $E_{51} = 10$).

(lower); this suggests that hypernovae play an important role in the chemical enrichment during early galactic evolution.

In the following section, therefore, we focus on nucleosynthesis in the high energy jet-induced explosions (Tominaga *et al.* 2007; Tominaga 2008). In the jet-like explosion, fallback can occur even for $E > 10^{52}$ erg.

3.2. *Nucleosynthesis in Jet-Induced Explosions and GRB-SN Connection*

The observed late-time spectra indicate that the explosions of these Fe-core collapse supernovae are quite aspherical (Maeda *et al.* 2008; Modjaz *et al.* 2008). The extreme case is the hyper-aspherical explosions induced by relativistic jet(s) as seen in the GRB-SNe.

Recent studies of nucleosynthesis in jet-induced explosions have revealed the connection between GRBs and EMP stars as summarized in Figure 7 (Tominaga *et al.* 2007; Tominaga 2008). In this model for the $40 M_\odot$ star, the jets are injected at a radius $R_0 \sim 900$ km with energy deposition rates in the range $\dot{E}_{\mathrm{dep},51} \equiv \dot{E}_{\mathrm{dep}}/10^{51}\,\mathrm{ergs\,s^{-1}} = 0.3 - 1500$. The diversity of $\dot{E}_{\mathrm{dep}}$ is consistent with the wide range of the observed isotropic equivalent γ-ray energies and timescales of GRBs (Amati *et al.* 2007 and references therein).

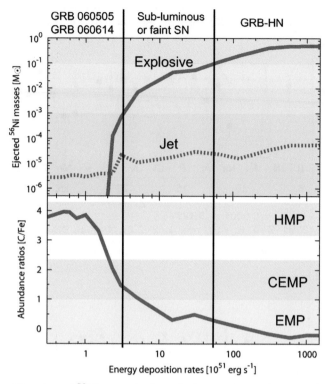

Figure 7. *Upper:* The ejected ^{56}Ni mass (*red*: explosive nucleosynthesis products, *blue*: the jet contribution) as a function of the energy deposition rate (Tominaga et al. 2007). The background color shows the corresponding SNe (*red*: GRB-HNe, *yellow*: sub-luminous SNe, *blue*: faint SNe, *green*: GRBs 060505 and 060614). Vertical lines divide the resulting SNe according to their brightness. *Lower:* the dependence of abundance ratio [C/Fe] on the energy deposition rate. The background color shows the corresponding metal-poor stars (*yellow*: EMP, *red*: CEMP, *blue*: HMP stars).

Variations of activities of the central engines, possibly corresponding to different rotational velocities or magnetic fields, may well produce the variation of $\dot{E}_{\rm dep}$.

The ejected $M(^{56}{\rm Ni})$ depends on $\dot{E}_{\rm dep}$ as follows (Fig. 7). Generally, higher $\dot{E}_{\rm dep}$ leads to the synthesis of larger $M(^{56}{\rm Ni})$ in explosive nucleosynthesis because of higher post-shock densities and temperatures (e.g., Maeda & Nomoto 2003; Nagataki et al. 2006). If $\dot{E}_{\rm dep,51} \gtrsim 60$, we obtain $M(^{56}{\rm Ni}) \gtrsim 0.1 M_\odot$, which is consistent with the brightness of GRB-HNe. Some C+O core material is ejected along the jet direction, but a large amount of material along the equatorial plane falls back.

For $\dot{E}_{\rm dep,51} \gtrsim 60$, the remnant mass is initially $M_{\rm rem}^{\rm start} \sim 1.5 M_\odot$ and grows as materials is accreted from the equatorial plane. The final BH mass is generally larger for smaller $\dot{E}_{\rm dep}$. The final BH masses range from $M_{\rm BH} = 10.8 M_\odot$ for $\dot{E}_{\rm dep,51} = 60$ to $M_{\rm BH} = 5.5 M_\odot$ for $\dot{E}_{\rm dep,51} = 1500$, which are consistent with the observed masses of stellar-mass BHs (Bailyn et al. 1998). The model with $\dot{E}_{\rm dep,51} = 300$ synthesizes $M(^{56}{\rm Ni}) \sim 0.4 M_\odot$, and the final mass of BH left after the explosion is $M_{\rm BH} = 6.4 M_\odot$.

For low energy deposition rates ($\dot{E}_{\rm dep,51} < 3$), in contrast, the ejected ^{56}Ni masses ($M(^{56}{\rm Ni}) < 10^{-3} M_\odot$) are smaller than the upper limits for GRBs 060505 and 060614 (Della Valle et al. 2006; Fynbo et al. 2006; Gal-Yam et al. 2006).

If the explosion is viewed from the jet direction, we would observe the GRB without SN re-brightening. This may be the situation for GRBs 060505 and 060614. In particular,

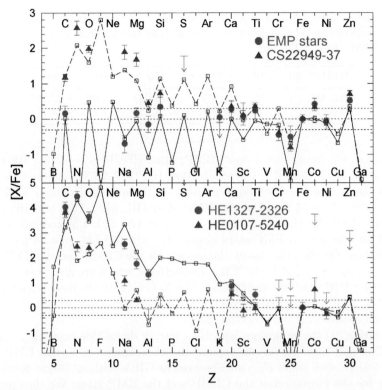

Figure 8. A comparison of the abundance patterns between metal-poor stars and models (Tominaga et al. 2007). *Upper*: typical EMP stars (*red dots*, Cayrel et al. 2004) and CEMP (*blue triangles*, CS 22949–37, Depagne et al. 2002) and models with $\dot{E}_{\mathrm{dep},51} = 120$ (*solid line*) and $= 3.0$ (*dashed line*). *Lower*: HMP stars: HE 1327–2326, (*red dots*, e.g., Frebel et al. 2005), and HE 0107–5240, (*blue triangles*, Christlieb et al. 2002, Bessell & Christlieb 2005) and models with $\dot{E}_{\mathrm{dep},51} = 1.5$ (*solid line*) and $= 0.5$ (*dashed line*).

for $\dot{E}_{\mathrm{dep},51} < 1.5$, ^{56}Ni cannot be synthesized explosively, and the jet component of the Fe-peak elements dominates the total yields (Fig. 8). The models eject very little $M(^{56}\mathrm{Ni})$ ($\sim 10^{-6} M_\odot$).

For intermediate energy deposition rates ($3 \lesssim \dot{E}_{\mathrm{dep},51} < 60$), the explosions eject $10^{-3} M_\odot \lesssim M(^{56}\mathrm{Ni}) < 0.1 M_\odot$, and the final BH masses are $10.8 M_\odot \lesssim M_{\mathrm{BH}} < 15.1 M_\odot$. The resulting SN is faint ($M(^{56}\mathrm{Ni}) < 0.01 M_\odot$) or sub-luminous ($0.01 M_\odot \lesssim M(^{56}\mathrm{Ni}) < 0.1 M_\odot$).

3.3. Abundance Patterns of Extremely Metal-Poor Stars

The abundance ratio [C/Fe] depends on $\dot{E}_{\mathrm{dep}}$ as follows. Lower $\dot{E}_{\mathrm{dep}}$ yields larger M_{BH} and thus larger [C/Fe], because the infall reduces the amount of inner core material (Fe) relative to that of outer material (C) (see also Maeda & Nomoto 2003). As in the case of $M(^{56}\mathrm{Ni})$, [C/Fe] changes dramatically at $\dot{E}_{\mathrm{dep},51} \sim 3$.

The observed abundance patterns of EMP stars are good indicators of SN nucleosynthesis because the Galaxy was effectively unmixed at [Fe/H] < -3 (e.g., Tumlinson 2006). They are classified into three groups according to [C/Fe]:

(1) [C/Fe] ~ 0, normal EMP stars ($-4 <$ [Fe/H] < -3, e.g., Cayrel et al. 2004);

(2) [C/Fe] $\gtrsim +1$, Carbon-enhanced EMP (CEMP) stars ($-4 <$ [Fe/H] < -3, e.g., CS 22949–37, Depagne et al. 2002);

Table 1. Stability of Pop III and Pop I massive stars: ○ and × indicate that the star is stable or unstable, respectively. The e-folding time for the fundamental mode is shown after × in units of 10^4 yr (Nomoto et al. 2003).

$M(M_\odot)$	80	100	120	150	180	300
Pop III	○	○	○	× (9.03)	× (4.83)	× (2.15)
Pop I	○	× (7.02)	× (2.35)	× (1.43)	× (1.21)	× (1.71)

(3) [C/Fe] ∼ +4, hyper metal-poor (HMP) stars ([Fe/H] < −5, e.g., HE 0107–5240, Christlieb et al. 2002; Bessell & Christlieb 2005; HE 1327–2326, Frebel et al. 2005).

Figure 8 shows that the abundance patterns of the averaged normal EMP stars, the CEMP star CS 22949–37, and the two HMP stars (HE 0107–5240 and HE 1327–2326) are well reproduced by the models with $\dot{E}_{\rm dep,51}$ = 120, 3.0, 1.5, and 0.5, respectively. The model for the normal EMP stars ejects $M(^{56}\rm Ni) \sim 0.2 M_\odot$, i.e., a factor of 2 less than SN 1998bw. On the other hand, the models for the CEMP and the HMP stars eject $M(^{56}\rm Ni) \sim 8 \times 10^{-4} M_\odot$ and $4 \times 10^{-6} M_\odot$, respectively, which are always smaller than the upper limits for GRBs 060505 and 060614. The N/C ratio in the models for CS 22949–37 and HE 1327–2326 is enhanced by partial mixing between the He and H layers during presupernova evolution (Iwamoto et al. 2005).

To summarize, (1) the explosions with large energy deposition rate, $\dot{E}_{\rm dep}$, are observed as GRB-HNe, and their yields can explain the abundances of normal EMP stars, and (2) the explosions with small $\dot{E}_{\rm dep}$ are observed as GRBs without bright SNe and can be responsible for the formation of the CEMP and the HMP stars. We thus propose that GRB-HNe and GRBs without bright SNe belong to a continuous series of BH-forming massive stellar deaths with relativistic jets of different $\dot{E}_{\rm dep}$.

4. $90 - 140 M_\odot$ Stars undergoing Pulsational Nuclear Instabilites

These massive stars undergo nuclear instabilities and associated pulsations (ϵ-mechanism) at various nuclear burning stages. Because of the large contribution of radiation pressure in these stars, dynamical stability is very close to neutral. Even a slight contribution of electron-positron pair creation affects the stability. Thus the pulsation behavior is sensitive to their mass, and thus to the mass loss rate and metallicity.

To determine the above upper mass limit, the non-adiabatic stability of massive Pop III ($Z = 0$) stars has been analyzed (Ibrahim et al. 1981; Baraffe et al. 2001; Nomoto et al. 2003). As summarized in Table 1 (Nomoto et al. 2003), the critical mass of a Pop III star is $128 M_\odot$, while that of a Pop I star is $94 M_\odot$. This difference comes from the very compact structure (with high central temperature) of Pop III stars. Stars more massive than the critical mass will undergo pulsation and mass loss. We note that the e-folding time of instability is much longer for Pop III stars than Pop I stars with the same mass, and thus the mass loss rate is much lower. Thus, massive Pop III stars could survive the instabilities without losing much mass.

Massive Pop III stars are formed through mass accretion, starting from a tiny core through collapse (e.g., Yoshida et al. 2008). Such an evolution with mass accretion starting from $M \sim 1 M_\odot$ has recently been studied by Ohkubo et al. (2006, 2008).

Figure 9 shows the evolutionary tracks of the central density and temperature in the later phases. For the models with mass accretion (M-2, YII), the central entropy in the early stage is low, corresponding to the small stellar mass. During the main-sequence

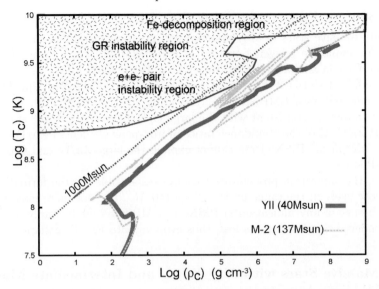

Figure 9. Evolutionary tracks of the central temperature and central density of very massive stars (Ohkubo et al. 2008). The numbers in brackets are the final masses for models YII and M-2. The $1000M_\odot$ stars (Ohkubo et al. 2006) are also shown.

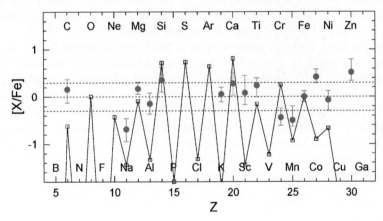

Figure 10. Comparison of the abundance patterns in the pair-instability supernova model ($M = 200M_\odot$; Umeda & Nomoto 2002) and extremely metal-poor stars (Cayrel et al. 2004).

phase, the stellar mass increases to $\sim 40M_\odot$ (YII) and $\sim 137M_\odot$ (M-2), and the density-temperature track shifts toward corresponding higher entropy through hydrogen burning.

The star YII ($\sim 40M_\odot$) ends its life in Fe-core collapse to form a black hole. The star M-2, whose final mass is $137M_\odot$, undergoes nuclear instability due to oxygen and silicon burning and pulsates, as seen in Figure 9. Such pulsations are seen in stars with masses $90M_\odot \lesssim M \lesssim 140M_\odot$ (Nomoto et al. 2005; Woosley et al. 2007; Umeda & Nomoto 2008; Ohkubo et al. 2008). In the extreme case, the pulsation could induce dynamical mass ejection and optical brightening as might be observed in the brightest SN 2006gy (Woosley et al. 2007). The nuclear energy released is not sufficient to explode the whole star. After several oscillations, the star finally collapses to form a black hole.

5. 140 - 300 $M_\odot$ Stars undergoing Pair-Instability Supernovae

These very massive stars (VMS) undergo pair-creation instability and are disrupted completely by explosive oxygen burning, as pair-instability supernovae (PISNe) (e.g., Arnett 1996; Umeda & Nomoto 2002; Heger & Woosley 2002).

The abundance patterns of the ejected material for the 200 $M_\odot$ star (Umeda & Nomoto 2002) are compared with EMP stars (Cayrel *et al.* 2004) in Figure 10. It is clear that PISN ejecta cannot be consistent with the large C/Fe observed in HMP stars and other C-rich EMP stars. Also, the abundance ratios of iron-peak elements ([Zn/Fe] < -0.8 and [Co/Fe] < -0.2) in the PISN ejecta cannot explain the large Zn/Fe and Co/Fe ratios in typical EMP stars.

Therefore, the supernova progenitors that are responsible for the formation of EMP stars are most likely in the range of $M \sim 20 - 140~M_\odot$, but not more massive than 140 $M_\odot$. The absence of any indication of PISNe in EMP stars might imply that 140 - 300 $M_\odot$ stars undergo significant mass loss, thus evolving into Fe core-collapse.

6. Very Massive Stars with $M > 300 M_\odot$ and Intermediate Mass Black Holes

It is possible that the First Stars were even more massive than $\sim 300 M_\odot$ (Ohkubo *et al.* 2006, 2008) if rapid mass accretion continues during the whole main-sequence phase of Pop III stars. [Another possible scenario for any metallicity is that VMSs are formed by merging of less massive stars in the environment of very dense star clusters (e.g., Ebisuzaki *et al.* 2001; Portezies Zwart & van den Heuvel 2007)].

Such massive stars undergo core-collapse (CVMS: core-collapse VMS) as seen from the 1000 $M_\odot$ star track in Figure 9. If such stars formed rapidly rotating black holes, jet-like mass ejection could produce processed material (Ohkubo *et al.* 2006). In fact, for moderately aspherical explosions, the patterns of nucleosynthesis match the observational data of both intracluster medium and M82 (Ohkubo *et al.* 2006). This result suggests that explosions of CVMS contribute significantly to the chemical evolution of gases in clusters of galaxies. For Galactic halo stars, predicted [O/Fe] ratios are smaller than the observational abundances. This result may support the view that Pop III CVMS could be responsible for the origin of intermediate mass black holes (IMBH).

References

Amati, L., Della Valle, M., Frontera, F., *et al.* 2007, *A&A* 463, 913
Arnett, W. D. 1996, *Supernovae and Nucleosynthesis* (Princeton: Princeton Univ. Press)
Bailyn, C. D., Jain, R. K., Coppi, P., & Orosz, J. A. 1998, *ApJ* 499, 367
Baraffe, I., Heger, A., & Woosley, S. E. 2001, *ApJ* 550, 890
Bessell, M. S. & Christlieb, N. 2005, in V. Hill *et al.* (eds.), *From Lithium to Uranium*, Proc. IAU Symposium No. 228 (Cambridge: Cambridge Univ. Press), 237
Cayrel, R., *et al.* 2004, *A&A* 416, 1117
Christlieb, N., *et al.* 2002, *Nature* 419, 904
Chugai, N. N. & Utrobin, V.P. 2000, *A&A* 354, 557
Della Valle, M., *et al.* 2006, *Nature* 444, 1050
Depagne, E., *et al.* 2002, *A&A* 390, 187
Ebisuzaki, T., *et al.* 2001, *ApJ* 562, L19
Fynbo, J. P. U., *et al.* 2006, *Nature* 444, 1047
Frebel, A., *et al.* 2005, *Nature* 434, 871
Garcia-Berro, E., Ritossa, C., & Iben, I., Jr. 1997, *ApJ* 485, 765
Gal-Yam, A., *et al.* 2006, *Nature* 444, 1053

Hashimoto, M., Iwamoto, K., & Nomoto, K. 1993, *ApJ* 322, L206
Heger, A. & Woosley, S. E. 2002, *ApJ* 567, 532
Heger, A. & Woosley, S. E. 2008, arXiv:0803.3161
Hendry, M. A., et al. 2005, *MNRAS* 359, 906
Hofman et al. 2008, *ApJ* 395, L672
Ibrahim, A., Boury, A., & Noels, A. 1981, *A&A* 103, 390
Iwamoto, K., Mazzali, P. A., Nomoto, K., et al. 1998, *Nature* 395, 672
Iwamoto, N., Umeda, H., Tominaga, N., Nomoto, K., & Maeda, K. 2005, *Science* 309, 451
Kalgero, J., et al. 2008, *ApJ* 670, 774
Kitaura & Janka, T. 2006, *A&A* 133, 175
Kobayashi, C., Umeda, H., Nomoto, K., Tominaga, N., & Ohkubo, T. 2006, *ApJ* 653, 1145
Limongi, M., Straniero, & Chieffi, A. 2000, *ApJS* 129, 625
Lodders, K. 2003, *ApJ* 591, 1220
Maeda, K. & Nomoto, K. 2003, *ApJ* 598, 1163
Maeda, K., et al. 2008, Science 319, 1220
Modjaz, M., et al. 2008, *ApJ* in press (arXiv:0801.0221)
Nagataki, S., Mizuta, A., & Sato, K. 2006, *ApJ* 647, 1255
Nomoto, K. 1984, *ApJ* 277, 791
Nomoto, K. 1987, *ApJ* 322, 206
Nomoto, K., Maeda, K., Umeda, H., Ohkubo, T., Deng, J., & Mazzali, P. 2003, in *IAU Symp. 212, A Massive Star Odyssey*, ed. V. D. Hucht, et al. (San Fransisco: ASP), 395
Nomoto, K., et al. 2004, in C. L. Fryer (ed.), *Stellar Collapse* (Astrophysics and Space Science: Kluwer), p. 277 (astro-ph/0308136)
Nomoto, K., et al. 2005, in *The Fate of Most Massive Stars*, ed. R. Humphreys & K. Stanek (ASP Ser. 332), 374 (astro-ph/0506597)
Nomoto, K., et al. 2006, *Nuclear Phys A* 777, 424 (astro-ph/0605725)
Ohkubo, T., Umeda, H., Maeda, K., Nomoto, K., Suzuki, T., Tsuruta, S., & Rees, M. J. 2006, *ApJ* 645, 1352
Ohkubo, T., Nomoto, K., Umeda, H., Yoshida, N., & Tsuruta, S. 2008, *ApJ* submitted
Poelarends, A. J. T., Herwig, F., Langer, N., & Heger, A. 2008, *ApJ* 675, 614
Portezies Zwart, S. F., & van den Heuvel, E. P. J. 2007, *Nature* 450, 388
Prieto, J. L., et al. 2008, *ApJ* 681, L9
Smartt, S. J., et al. 2008, *MNRAS* submitted (arXiv:0809.0403)
Thompson, T. A., et al. 2008, *ApJ* submitted (arXiv:0809.0510)
Timmes, F. X., & Woosley, S. E., 1992, *ApJ* 396, 649
Tominaga, N., Maeda, K., Umeda, H., Nomoto, K., Tanaka, et al. 2007, *ApJ* 657, L77
Tominaga, N. 2008, *ApJ* in press (arXiv:0711.4815)
Tumlinson, J. 2006, *ApJ* 641, 1
Turatto, M., et al. 1998, *ApJ* 498, L129
Umeda, H. & Nomoto, K. 2002, *ApJ* 565, 385
Umeda, H. & Nomoto, K. 2008, *ApJ* 673, 1014
Wanajo, S., Nomoto, K., Janka, H.-T., Kitaura, F. S., & Muller, B. 2008, *ApJ* submitted
Woosley, S. E. & Bloom, J. S. 2006, *ARA&A* 44, 507
Woosley, S. E., Blinnikov, S., & Heger, A. 2007, *Nature* 450, 390
Yoshida, N., Omukai, K., & Hernquist, L. 2008, *Science* 321, 669
Zampieri, L., et al. 2003, *MNRAS* 338, 711

Eileen Friel, Michael Chabin and Birgitta Nordström meet again at the welcome reception.

Chiaki Kobayashi and Laura Portinari relaxing offstage at Carlsberg. Photo: Bruce Elmegreen.

Effects of Supernova Feedback on the Formation of Galaxies

Cecilia Scannapieco[1], Patricia B. Tissera[2], Simon D. M. White[1] and Volker Springel[1]

[1] Max-Planck Institute for Astrophysics
Karl-Schwarzschild Str. 1, D85748, Garching, Germany
email: cecilia@mpa-garching.mpg.de

[2] Instituto de Astronomía y Física del Espacio
Casilla de Correos 67, Suc. 28, 1428, Buenos Aires, Argentina

Abstract. We study the effects of Supernova (SN) feedback on the formation of galaxies using hydrodynamical simulations in a ΛCDM cosmology. We use an extended version of the code GADGET-2 which includes chemical enrichment and energy feedback by Type II and Type Ia SN, metal-dependent cooling and a multiphase model for the gas component. We focus on the effects of SN feedback on the star formation process, galaxy morphology, evolution of the specific angular momentum and chemical properties. We find that SN feedback plays a fundamental role in galaxy evolution, producing a self-regulated cycle for star formation, preventing the early consumption of gas and allowing disks to form at late times. The SN feedback model is able to reproduce the expected dependence on virial mass, with less massive systems being more strongly affected.

Keywords. galaxies: formation, galaxies: evolution, methods: n-body simulations

1. Introduction

Supernova explosions play a fundamental role in galaxy formation and evolution. On one side, they are the main source of heavy elements in the Universe and the presence of such elements substantially enhances the cooling of gas (White & Frenk 1991). On the other hand, SNe eject a significant amount of energy into the interstellar medium. It is believed that SN explosions are responsible of generating a self-regulated cycle for star formation through the heating and disruption of cold gas clouds, as well as of triggering important galactic winds such as those observed (e.g. Martin 2004). Smaller systems are more strongly affected by SN feedback, because their shallower potential wells are less efficient in retaining baryons (e.g. White & Frenk 1991).

Numerical simulations have become an important tool to study galaxy formation, since they can track the joint evolution of dark matter and baryons in the context of a cosmological model. However, this has shown to be an extremely complex task, because of the need to cover a large dynamical range and describe, at the same time, large-scale processes such as tidal interactions and mergers and small-scale processes related to stellar evolution.

One of the main problems that galaxy formation simulations have repeatedly found is the inability to reproduce the morphologies of disk galaxies observed in the Universe. This is generally refered to as the angular momentum problem that arises when baryons transfer most of their angular momentum to the dark matter components during interactions and mergers (Navarro & Benz 1991; Navarro & White 1994). As a result, disks are too small and concentrated with respect to real spirals. More recent simulations which include prescriptions for SN feedback have been able to produce more realistic disks (e.g. Abadi et al. 2003; Robertson et al. 2004; Governato et al. 2007). These works have pointed out the importance of SN feedback as a key process to prevent the loss of angular momentum, regulate the star formation activity and produce extended, young disk-like components.

In this work, we investigate the effects of SN feedback on the formation of galaxies, focusing on the formation of disks. For this purpose, we have run simulations of a Milky-Way type galaxy using an extended version of the code GADGET-2 which includes chemical enrichment and energy feedback by SN. A summary of the simulation code and the initial conditions is given in Section 2. In Section 3 we investigate the effects of SN feedback on galaxy morphology, star formation rates, evolution of specific angular momentum and chemical properties. We also investigate the dependence of the results on virial mass. Finally, in Section 4 we give our conclusions.

2. Simulations

We use the simulation code described in Scannapieco et al. (2005, 2006). This is an extended version of the Tree-PM SPH code GADGET-2 (Springel & Hernquist 2002; Springel 2005), which includes chemical enrichment and energy feedback by SN, metal-dependent cooling and a multiphase model for the gas component. Note that our star formation and feedback model is substantially different from that of Springel & Hernquist (2003), but we do include their treatment of UV background.

We focus on the study of a disk galaxy similar to the Milky Way in its cosmological context. For this purpose we simulate a system with $z = 0$ halo mass of $\sim 10^{12}\ h^{-1}$ $M_\odot$ and spin parameter of $\lambda \sim 0.03$, extracted from a large cosmological simulation and resimulated with improved resolution. It was selected to have no major mergers since $z = 1$ in order to give time for a disk to form. The simulations adopt a ΛCDM Universe with the following cosmological parameters: $\Omega_\Lambda = 0.7$, $\Omega_m = 0.3$, $\Omega_b = 0.04$, a normalization of the power spectrum of $\sigma_8 = 0.9$ and $H_0 = 100\ h$ km s^{-1} Mpc^{-1} with $h = 0.7$. The particle mass is 1.6×10^7 for dark matter and $2.4 \times 10^6\ h^{-1}\ M_\odot$ for baryonic particles, and we use a maximum gravitational softening of $0.8\ h^{-1}$ kpc for gas, dark matter and star particles. At $z = 0$ the halo of our galaxy contains $\sim 1.2 \times 10^5$ dark matter and $\sim 1.5 \times 10^5$ baryonic particles within the virial radius.

In order to investigate the effects of SN feedback on the formation of galaxies, we compare two simulations which only differ in the inclusion of the SN energy feedback model. These simulations are part of the series analysed in Scannapieco et al. (2008), where an extensive investigation of the effects of SN feedback on galaxies and a parameter study is performed. In this work, we use the no-feedback run NF (run without including the SN energy feedback model) and the feedback run E-0.7. We refer the interested reader to Scannapieco et al. (2008) for details in the characteristics of these simulations.

3. Results

In Fig. 1 we show stellar surface density maps at $z = 0$ for the NF and E-0.7 runs. Clearly, SN feedback has an important effect on the final morphology of the galaxy. If SN feedback is not included, as we have done in run NF, the stars define a spheroidal component with no disk. On the contrary, the inclusion of SN energy feedback allows the formation of an extended disk component.

The generation of a disk component is closely related to the star formation process. In the left-hand panel of Fig. 2 we show the star formation rates (SFR) for our simulations. In the no-feedback case (NF), the gas cools down and concentrates at the centre of the

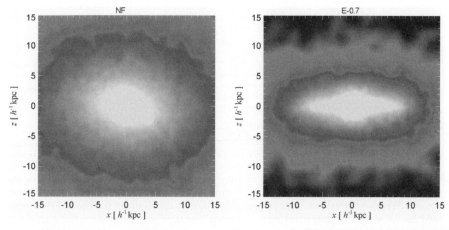

Figure 1. Edge-on stellar surface density maps for the no-feedback (NF, left-hand panel) and feedback (E-0.7, right-hand panel) simulations at $z = 0$. The colors span 4 orders of magnitude in projected density, with brighter colors representing higher densities.

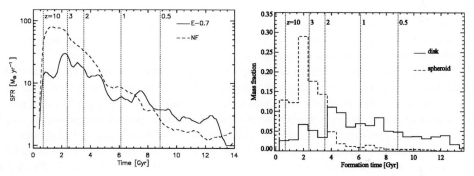

Figure 2. Left: Star formation rates for the no-feedback (NF) and feedback (E-0.7) runs. Right: Mass fraction as a function of formation time for stars of the disk and spheroidal components in simulation E-0.7.

potential well very early, producing a strong starburst which feeds the galaxy spheroid. As a result of the early consumption of gas to form stars, the SFR is low at later times. On the contrary, the SFR obtained for the feedback case is lower at early times, indicating that SN feedback has contributed to self-regulate the star formation process. This is the result of the heating of gas and the generation of galactic winds. In this case, the amount of gas available for star formation is larger at recent times and consequently the SFR is higher. In the right-hand panel of Fig. 2 we show the mass fraction as a function of formation time for stars of the disk and spheroidal components in our feedback simulation (see Scannapieco et al. 2008 for the method used to segregate stars into disk and spheroid). ¿From this plot it is clear that star formation at recent times ($z \lesssim 1$) significantly contributes to the formation of the disk component, while stars formed at recent times contribute mainly to the spheroid. In this simulation, ~ 50 per cent of the mass of the disk forms since $z = 1$. Note that in the no-feedback case, only a few per cent of the final stellar mass of the galaxy is formed since $z = 1$.

Our simulation E-0.7 has produced a galaxy with an extended disk component. By using the segregation of stars into disk and spheroid mentioned above, we can calculate the masses of the different components, as well as characteristic scales. The disk of the

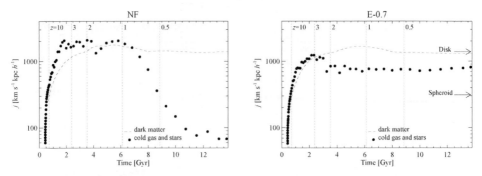

Figure 3. Dashed lines show the specific angular momentum as a function of time for the dark matter that, at $z = 0$, lies within the virial radius of the system for NF (left panel) and E-0.7 (right panel). We also show with dots the specific angular momentum for the baryons which end up as cold gas or stars in the central $20\ h^{-1}$ kpc at $z = 0$. The arrows show the specific angular momentum of disk and spheroid stars.

simulated galaxy has a mass of $3.3 \times 10^{10}\ h^{-1}\ M_\odot$, a half-mass radius of $5.7\ h^{-1}$ kpc, a half-mass height of $0.5\ h^{-1}$ kpc, and a half-mass formation time of 6.3 Gyr. The spheroid mass and half-mass formation time are $4.1 \times 10^{10}\ h^{-1}\ M_\odot$ and 2.5 Gyr, respectively. It is clear that the characteristic half-mass times are very different in the two cases, the disk component being formed by younger stars.

In Fig. 3 we show the evolution of the specific angular momentum of the dark matter (within the virial radius) and of the cold gas plus stars (within twice the optical radius) for the no-feedback case (left-hand panel) and for the feedback case E-0.7 (right-hand panel). The evolution of the specific angular momentum of the dark matter component is similar in the two cases, growing as a result of tidal torques at early epochs and being conserved from turnaround ($z \approx 1.5$) until $z = 0$. On the contrary, the cold baryonic components in the two cases differ significantly, in particular at late times. In the no-feedback case (NF), much angular momentum is lost through dynamical friction, particularly through a satellite which is accreted onto the main halo at $z \sim 1$. In E-0.7, on the other hand, the cold gas and stars lose rather little specific angular momentum between $z = 1$ and $z = 0$. Two main factors contribute to this difference. Firstly, in E-0.7 a significant number of young stars form between $z = 1$ and $z = 0$ with high specific angular momentum (these stars form from high specific angular momentum gas which becomes cold at late times); and secondly, dynamical friction affects the system much less than in NF, since satellites are less massive. At $z = 0$, disk stars have a specific angular momentum comparable to that of the dark matter, while spheroid stars have a much lower specific angular momentum.

In Fig 4 we show the oxygen profiles for the no-feedback (NF) and feedback (E-0.7) runs. ¿From this figure we can see that SN feedback strongly affects the chemical distributions. If no feedback is included, the gas is enriched only in the very central regions. Including SN feedback triggers a redistribution of mass and metals through galactic winds and fountains, giving the gas component a much higher level of enrichment out to large radii. A linear fit to this metallicity profile gives a slope of -0.048 dex kpc^{-1} and a zero-point of 8.77 dex, consistent with the observed values in real disk galaxies (e.g. Zaritsky et al. 1994).

Finally, we investigate the effects of SN feedback on different mass systems. For that purpose we have scaled down our initial conditions to generate galaxies of $10^{10}\ h^{-1}\ M_\odot$ and $10^9\ h^{-1}\ M_\odot$ halo mass, and simulate their evolution including the SN feedback

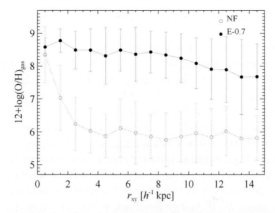

Figure 4. Oxygen abundance for the gas component as a function of radius projected onto the disk plane for our no-feedback simulation (NF) and for the feedback case E-0.7. The error bars correspond to the standard deviation around the mean.

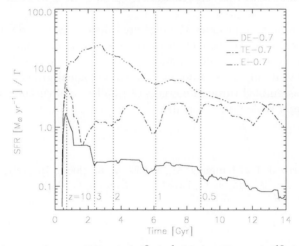

Figure 5. SFRs for simulations DE-0.7 (10^9 h^{-1} M$_\odot$), TE-0.7 (10^{10} h^{-1} M$_\odot$) and E-0.7 (10^{12} h^{-1} M$_\odot$) run with energy feedback. To facilitate comparison, the SFRs are normalized to the scale factor Γ.

model (with the same parameters than E-0.7). These simulations are TE-0.7 and DE-0.7, respectively. In Fig. 5 we show the SFRs for these simulations, as well as for E-0.7, normalized to the scale factor ($\Gamma = 1$ for E-0.7, $\Gamma = 10^{-2}$ for TE-0.7 and $\Gamma = 10^{-3}$ for DE-0.7). From this figure it is clear that SN feedback has a dramatic effect on small galaxies. This is because more violent winds develop and baryons are unable to condensate and form stars. In the smallest galaxy, the SFR is very low at all times because most of the gas has been lost after the first starburst episode. This proves that our model is able to reproduce the expected dependence of SN feedback on virial mass, without changing the relevant physical parameters.

4. Conclusions

We have run simulations of a Milky Way-type galaxy in its cosmological setting in order to investigate the effects of SN feedback on the formation of galaxy disks. We compare two simulations with the only difference being the inclusion of the SN energy feedback model of Scannapieco et al. (2005, 2006). Our main results can be summarized as follows:

- SN feedback helps to settle a self-regulated cycle for star formation in galaxies, through the heating and disruption of cold gas and the generation of galactic winds. The regulation of star formation allows gas to be mantained in a hot halo which can condensate at late times, becoming a reservoir for recent star formation. This contributes significantly to the formation of disk components.
- When SN feedback is included, the specific angular momentum of the baryons is conserved and disks with the correct scale-lengths are obtained. This results from the late collapse of gas with high angular momentum, which becomes available to form stars at later times, when the system does not suffer from strong interactions.
- The injection of SN energy into the interstellar medium generates a redistribution of chemical elements in galaxies. If energy feedback is not considered, only the very central regions were stars are formed are contaminated. On the contrary, the inclusion of feedback triggers a redistribution of metals since gas is heated and expands, contaminating the outer regions of galaxies. In this case, metallicity profiles in agreement with observations are produced.
- Our model is able to reproduce the expected dependence of SN feedback on virial mass: as we go to less massive systems, SN feedback has stronger effects: the star formation rates (normalized to mass) are lower, and more violent winds develop. This proves that our model is well suited for studying the cosmological growth of structure where large systems are assembled through mergers of smaller substructures and systems form simultaneously over a wide range of scales.

References

Abadi, M. G., Navarro, J. F., Steinmetz, M., & Eke, V. R., 2003, *ApJ*, 591, 499
Governato, F., Willman, B., Mayer, L., Brooks, A., Stinson, G., Valenzuela, O., Wadsley, J., & Quinn, T., 2007, *MNRAS*, 374, 1479
Martin, C. L., 2004, *A&AS*, 205, 8901
Navarro, J. F. & Benz, W., 1991, *ApJ*, 380, 320
Navarro, J. F. & White, S. D. M., 1993, *MNRAS*, 265, 271
Robertson, B., Yoshida, N., Springel, V., & Hernquist, L., 2004, *ApJ*, 606, 32
Scannapieco, C., Tissera, P. B., White, S. D. M., & Springel, V., 2005, *MNRAS*, 364, 552
Scannapieco, C., Tissera, P. B., White, S. D. M., & Springel, V., 2006, *MNRAS*, 371, 1125
Scannapieco, C., Tissera, P. B., White, S. D. M., & Springel, V., 2008, *MNRAS*, in press (astro-ph/0804.3795)
Springel, V. & Hernquist, L., 2002, *MNRAS*, 333, 649
Springel, V. & Hernquist, L., 2003, *MNRAS*, 339, 289
Springel, V. 2005, *MNRAS*, 364, 1105
White, S. D. M. & Frenk, C. S., 1991, *ApJ*, 379, 52
Zaritsky, D., Kennicutt, R. C. Jr., & Huchra, J. P., 1994, *ApJ*, 420, 87

Chemodynamical simulations of the Milky Way Galaxy

Chiaki Kobayashi[1]

[1]The Australian National University
Mt. Stromlo Observatory, Cotter Rd., Weston ACT 2611
email: chiaki@mso.anu.edu.au

Abstract. We simulate the chemodynamical evolution of the Milky Way Galaxy, including the nucleosynthesis yields of hypernovae and a new progenitor model for Type Ia Supernovae (SNe Ia). In our nucleosynthesis yields of core-collapse supernovae, we use light curve and spectral fitting to individual supernovae to estimate the mass of the progenitor, the explosion energy, and the iron mass produced. A large contribution of hypernovae is required from the observed abundance of Zn ([Zn/Fe] ~ 0). In our progenitor model of SNe Ia, based on the single degenerate scenario, the SN Ia lifetime distribution spans a range of $0.1 - 20$ Gyr with peaks at both ~ 0.1 and 1 Gyr. A metallicity effect from white dwarf winds is required from the observed trends of elemental abundance ratios (i.e., [(α,Mn,Zn)/Fe]-[Fe/H] relations). In our simulated Milky Way-type galaxy, the kinematical and chemical properties of the bulge, disk, and halo are broadly consistent with observations. 80% of the thick disk stars are older than ~ 8 Gyr and tend to have larger [α/Fe] than in the thin disk.

Keywords. galaxies: abundances, galaxies: evolution, methods: n-body simulations

1. Introduction

While the evolution of the dark matter is reasonably well understood, the evolution of the baryonic component is much less certain because of the complexity of the relevant physical processes, such as star formation and feedback. With the commonly employed schematic star formation criteria alone, the predicted star formation rates are higher than what is compatible with the observed luminosity density. Thus feedback mechanisms are generally invoked to reheat the gas and suppress star formation. Supernovae inject not only thermal energy, but also heavy elements into the interstellar medium, which can enhance star formation. Chemical enrichment must be solved along with energy feedback. "Feedback" is also important for solving the angular momentum problem and the missing satellite problem, and for explaining the existence of heavy elements in the intracluster and intergalactic medium, and possibly the mass-metallicity relation of galaxies (Kobayashi et al. 2007).

In the next decade, high-resolution multi-object spectroscopy (HERMES/AAT and WFMOS/Subaru) and space astrometry (GAIA) will provide kinematics and chemical abundances of a million stars in the Local Group. Since different heavy elements are produced from different supernovae with different timescales, elemental abundance ratios can provide independent information on "age". Therefore, stars in a galaxy are fossils to tell the history of the galaxy (i.e., galactic archeology). To understand how the Milky Way Galaxy form and evolve from these observations, we simulate the formation and chemodynamical evolution of Milky Way-type galaxies, and show the frequency distribution of elemental abundance ratios in each component of the Galaxy; thin and thick disk and bulge.

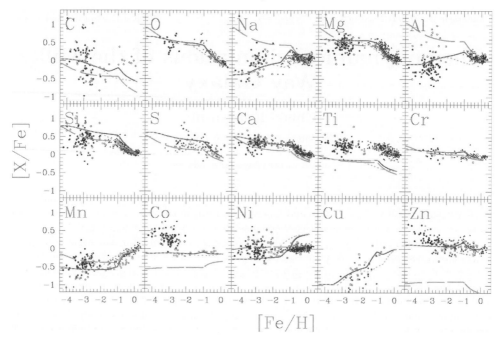

Figure 1. Evolution of heavy-element abundance ratios [X/Fe] vs. [Fe/H] for one-zone models with our new yields (solid line), with only SNe II (dashed line), and with the double-degenerate scenario for SNe Ia (dotted line). The dots are observational data (see K06 for the references).

2. Hypernovae

Although the explosion mechanism of core-collapse supernovae remains under debate, the ejected explosion energy and ^{56}Ni mass (which decays to ^{56}Fe) can be estimated from the observations, i.e., light curve and spectra fitting to individual supernovae. It is then found that hypernovae (HNe), which have more than ten times larger explosion energy ($E_{51} \gtrsim 10$) than normal SNe II, produce a certain amount of iron.

We calculate the nucleosynthesis yields for wide ranges in metallicity ($Z = 0 - Z_\odot$) and explosion energy (normal SNe II and HNe). Assuming that a half of the supernovae with $\leqslant 20 M_\odot$ are HNe, the evolution of the elemental abundance ratios from oxygen to zinc (Fig.1) are in good agreement with observations in the solar neighborhood, bulge, halo, and thick disk (Kobayashi et al. 2006, hereafter K06). Among the α-elements, O, Mg, Si, S, and Ca show a plateau at [Fe/H] $\lesssim -1$, while Ti is underabundant overall. The observed decrease in the odd-Z elements (Na, Al, and Cu) toward low [Fe/H] is reproduced by the metallicity effect on nucleosynthesis. The iron-peak elements (Cr, Mn, Co, and Ni) are consistent with the observed mean values at $-2.5 \lesssim$ [Fe/H] $\lesssim -1$, and the observed trend at the lower metallicity can be explained by the energy effect. Especially, the observed abundance of Zn ([Zn/Fe] ~ 0) can be explained only by such a large contribution of HNe. Since the observed [Zn/Fe] shows an increase toward lower metallicity (Primas et al. 2000; Nisesn et al. 2007), the HN fraction may be larger in the earlier stage of galaxy formation.

HNe play an essential role in suppressing cosmic star formation to match the observations. The HN feedback drives galactic outflows efficiently in low mass galaxies, which results in the observed mass-metallicity relations of galaxies (Kobayashi et al. 2007).

3. Type Ia Supernovae

The progenitors of the majority of Type Ia Supernovae (SNe Ia) are most likely the Chandrasekhar (Ch) mass white dwarfs (WDs). For the evolution of accreting C+O WDs toward the Ch mass, two scenarios have been proposed: One is the double-degenerate (DD) scenario, i.e., merging of double C+O WDs with a combined mass surpassing the Ch mass limit, although it has been theoretically suggested that it leads to accretion-induced collapse rather than SNe Ia. The other is our single-degenerate (SD) scenario, i.e., the WD mass grows by accretion of hydrogen-rich matter via mass transfer from a binary companion.

We construct a new model of SNe Ia, based on the SD scenario, taking account of the metallicity effect of the WD wind (Kobayashi et al. 1998) and the mass-stripping effect in binary systems (Hachisu et al. 2008). Our model naturally predicts that the SN Ia lifetime distribution spans a range of 0.1 − 20 Gyr with double peaks at ∼ 0.1 and 1 Gyr, reflecting the two types of companion stars; main-sequence (MS)+WD and red-giant (RG)+WD systems, respectively. Because of the metallicity effect, i.e., because of the lack of winds from WDs in binary systems, the SN Ia rate in the systems with [Fe/H] $\lesssim$ − 1, e.g., high-z spiral galaxies, is supposed to be very small.

Figure 1 shows the evolution of several heavy-element abundance ratios [X/Fe] vs. [Fe/H] for one-zone models with our new yields (solid line), with only SNe II (dashed line, Nomoto et al. (1997)'s yields adopted), and with the other SN Ia model (dotted line; the lifetime distribution function from the DD scenario is adopted). As time passes, the iron abundance increases, and the abundance ratio for many elements stay constant with a plateau at [Fe/H] $\lesssim$ − 1, which is determined only by SNe II and HNe.

From [Fe/H] ∼ −1, SNe Ia start to produce more Fe than α-elements, and thus [α/Fe] decreases toward the solar abundance. The decreasing [Fe/H] depends on the SN Ia progenitor model. Our SN Ia model can give better reproduction of the [(α, Mn, Zn)/Fe]-[Fe/H] relations than other models such as the DD scenario. With the DD scenario, the typical lifetimes of SNe Ia are ∼ 0.1 Gyr, which results in a too early decrease of [α/Fe] at [Fe/H] ∼ −2. Even with our SD model, if we do not include the metallicity effect, [α/Fe] decreases too early because of the short lifetime of the MS+WD systems, ∼ 0.1 Gyr. In other words, the presence of a young population of SNe Ia strongly favours the presence of a metallicity effect to be consistent with the chemical evolution of the solar neighbourhood. Note that the star formation history and total number of SNe Ia are determined to meet the observed metallicity distribution function (Kobayashi & Nomoto 2008).

While the present SN Ia rate in elliptical galaxies can be reproduced with the old population of the RG+WD systems, the large SN Ia rate in radio galaxies could be explained by a young population of MS+WD systems. We also succeed in reproducing the galactic supernova rates and their dependence on morphological type of the galaxies (Mannucci et al. 2005), and the cosmic SN Ia rate history with a peak at $z \sim 1$. At $z \gtrsim 1$, the predicted SN Ia rate decreases toward higher redshift, and SNe Ia will be observed only in systems that have experience chemical enrichment on a short timescale. This suggests that the evolution effect in supernova cosmology can be small.

4. Chemodynamical Model

We simulate the chemodynamical evolution of the Milky Way Galaxy with our GRAPE-SPH code (Kobayashi 2004), the details of which are summarized as follows. i) The Smoothed Particle Hydrodynamics (SPH) method is adopted, and the gravity is

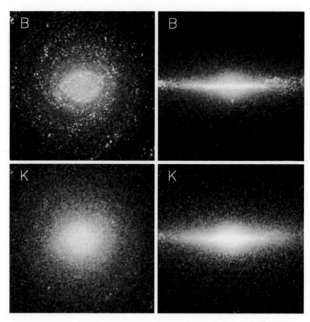

Figure 2. Luminosity Map of our simulated Milky Way-type galaxy at present for the rest B (upper panels) and K (lower panels) bands.

calculated by direct summation using the special purpose computer GRAPE (GRAvity PipE). ii) **Radiative cooling** is computed using a metallicity-dependent cooling function (Sutherland & Dopita 1993). iii) Our **star formation** criteria are the same as in Katz (1992); (1) converging flow; $(\nabla \cdot \boldsymbol{v})_i < 0$, (2) rapid cooling; $t_{\rm cool} < t_{\rm dyn}$, and (3) Jeans unstable gas; $t_{\rm dyn} < t_{\rm sound}$. The star formation timescale is proportional to the dynamical timescale ($t_{\rm sf} \equiv \frac{1}{c} t_{\rm dyn}$), where the star formation timescale parameter $c = 0.1$ is adopted (Kobayashi 2005). If a gas particle satisfies the above star formation criteria, a fractional part of the mass of the gas particle turns into a star particle. Since an individual star particle has a mass of $10^{5-7} M_\odot$, it does not represent a single star, but an association of many stars. The mass of the stars associated with each star particle is distributed according to an initial mass function (IMF). We adopt a Salpter IMF with a slope $x = 1.35$. iv) For the **feedback** of energy and heavy elements, we do not adopt the instantaneous recycling approximation. Via stellar winds, SNe II, and SNe Ia, thermal energy and heavy elements are ejected from an evolved star particle as functions of time and metallicity, and are distributed to all surrounding gas particles out to a constant radius of 1 kpc. v) The **photometric evolution** of a star particle is identical to the evolution of a simple stellar population (SSP). SSP Spectra are taken from Kodama & Arimoto (1997) as a function of age and metallicity.

The **initial condition** is similar to those in Kobayashi (2004, 2005), but with an initial angular momentum of $\lambda \sim 0.1$, total mass of $\sim 10^{12} M_\odot$, and $\sim 120,000$ particles. We choose an initial condition where the galaxy does not undergo major mergers; otherwise no disk galaxy can form. The cosmological parameters are set to be $H_0 = 70$ km s^{-1} Mpc^{-1}, $\Omega_m = 0.3$, and $\Omega_\Lambda = 0.7$.

5. Chemodynamical Evolution of the Milky Way Galaxy

In the CDM scenario, galaxies form through the successive merging of subgalaxies with various masses. In our simulation, the merging of subgalaxies induces an initial starburst

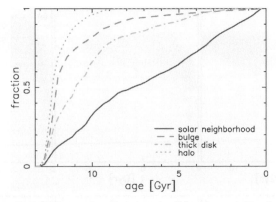

Figure 3. Cumulative distributions of stellar ages in the solar neighborhood (solid line), thick disk (dot-dashed line), bulge (dashed line), and halo (dotted line).

and the bulge forms by $z \gtrsim 3$. 80% of bulge stars are older than ~ 10 Gyr. According to the late gas accretion, the disk structure is seen at $z \lesssim 2$. At present, the bulge and disk structures are well seen in the K and B bands, respectively (Fig. 2). Figure 3 shows the cumulative distributions of stellar ages in the solar neighbourhood ($r = 7.5 - 8.5$ kpc, $|z| \leqslant 0.5$ kpc), bulge ($r \leqslant 1$ kpc), and halo ($r = 5 - 10$ kpc). In the solar neighbourhood, 50% of disk stars are younger than ~ 8 Gyr, and old stars tend to have small rotation velocity v and large velocity dispersion σ. When we define thick disk as $v/\sigma < 1.5$, 80% of thick disk stars are older than ~ 8 Gyr.

The age-metallicity relations are shown in the upper panels of Figure 4. (a) In the solar neighbourhood, [Fe/H] increases to ~ 0 at $t \sim 2$ Gyr, which is broadly consistent with the observations (Nordström et al. 2004). (c) In the thick disk, the relation is similar as in the solar neighborhood, but most stars populated the region with old age and low [Fe/H]. (e) In the bulge, [Fe/H] increases more quickly than in the disks. Metal-rich stars with [Fe/H] ~ 1 appear at $t \sim 2$ Gyr.

The [O/Fe]-[Fe/H] relations are shown in the lower panels of Figure 4, and we obtain similar results for other α elements. (b) In the solar neighborhood, we can reproduce the observational trend (Edvardsson et al. 1993, open circles; Bensby et al. 2004, small filled circles; Gratton et al. 2003, triangles; Cayrel et al. 2004, large filled circles). [α/Fe] decreases because of the delayed iron enrichment by SNe Ia. If we do not include the metallicity effect on SNe Ia, or if we do not include HNe, we cannot reproduce the plateau at [Fe/H] $\lesssim -1$, and the scatter of [α/Fe] at [Fe/H] $\lesssim -1$ is too large. At [Fe/H] $\gtrsim -1$, the scatter is large. This may be because the mixing of heavy elements among gas particles has not been included in our chemodynamical model.

(d) In the thick disk, the chemical enrichment timescale is so short that [α/Fe] tends to be larger than in the thin disk, which is consistent with the observations (Bensby et al. 2004). (f) In the bulge, the chemical enrichment timescale is shorter than in the disks, the [α/Fe] plateau continues to [Fe/H]~ 0, which is consistent with some observations (Zoccali et al. 2008). The star formation has not been terminated in the simulation, and some new stars are forming also in the bulge. Such young stars tend to have large [Fe/H] and low [α/Fe] in our simulation, and the observed stars in Cunha et al. (2007) may be affected by inhomogeneity, or some missing physics. Particularly, if the relations for O and Mg are different, we may have to include non-supernova physics such as strong stellar winds.

We also trace the orbits of the star particles and study the origin of the thick disk. The fraction of stars that have formed in the disk ($z < 1$ kpc) are $\sim 40\%$. The rest,

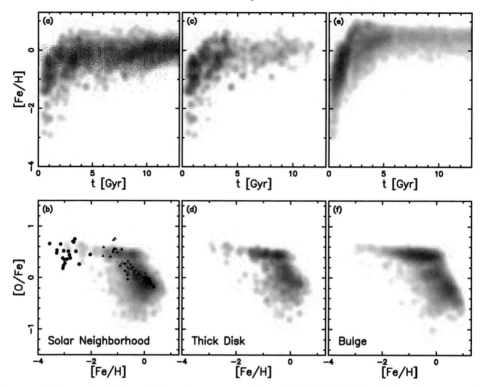

Figure 4. Age-metallicity relations (upper panels) and [O/Fe]-[Fe/H] relations (lower panels) in the solar neighborhood (a and b), thick disk (c and d), and bulge (e and f). The contours show the mass density for the simulation. The dots show the observations of stars in the solar neighborhood (see the text for data sources).

more than half of the thick disk stars, have formed in merging subgalaxies before they accrete onto the disk. In this sense, the CDM picture seems not to conflict with the Milky Way Galaxy. However, it is very hard to find initial condition to form disk galaxies (with this resolution) because major mergers break the disk structure, and late star formation caused by slow gas accretion is not enough to re-generate (contrary to Steinmetz & Navarro 2002). For the frequency of disk galaxies, the CDM picture seems to have a problem, although it should be tested with cosmological simulations.

We succeed in reproducing the kinematical and chemical properties of the Milky Way Galaxy by adopting a Salpeter IMF for all components. From a statistical comparison for the frequency distribution along the relations, chemodynamical models should be tested and improved. Different components of the Milky Way Galaxy have different stellar populations, namely, different elemental abundance patterns. Using future observations of abundance ratios, we can learn when and where the stars formed, and Galactic archaeology can be a powerful tool in chemodynamical simulations.

References

Bensby, T., Feltzing, S., & Lundström, I. 2004, A&A, 415, 155
Kobayashi, C., 2004, MNRAS, 347, 740
Kobayashi, C. & Nomoto, K. 2008, ApJ, submitted, astro-ph/0801.0215
Kobayashi, C., Springel, V, & White, S. D. M. 2007, MNRAS, 376, 1465
Kobayashi, C., Umeda, H., Nomoto, K., Tominaga, N., & Ohkubo, T. 2006, ApJ, 653, 1145
Nordström, B., et al. 2004, A&A, 418, 989
Zoccali, M. et al. 2008, Talk at MPA/ESO/MPE/USM 2008 Conference

On the chemical evolution of the Milky Way

Nikos Prantzos

IAP, 98bis Bd Arago, 75013 Paris; email: prantzos@iap.fr

Abstract. I discuss three different topics concerning the chemical evolution of the Milky Way (MW). 1) The metallicity distribution of the MW halo; it is shown that this distribution can be analytically derived in the framework of the hierarchical merging scenario for galaxy formation, assuming that the component sub-haloes had chemical properties similar to those of the progenitors of satellite galaxies of the MW. 2) The age-metallicity relationship (AMR) in the solar neighborhood; I argue for caution in deriving from data with important uncertainties (such as the age uncertainties in the Geneva-Copenhagen Survey) a relationship between average metallicity and age: derived relationships are shown to be systematically flatter than the true ones and should not be directly compared to models. 3) The radial mixing of stars in the disk, which may have important effects on various observables (scatter in AMR, extension of the tails of the metallicity distribution, flatenning of disk abundance profiles). Recent SPH + N-body simulations find considerable radial mixing, but only comparison to observations will ultimately determine the extent of that mixing.

Keywords. Galaxy: evolution – Galaxy: disk – Galaxy: halo – Galaxy: abundances

1. The MW halo in cosmological context

The regular shape of the metallicity distribution of the Milky Way (MW) halo can readily be explained by the simple model of galactic chemical evolution (GCE) with outflow, as suggested by Hartwick (1976). However, that explanation lies within the framework of the monolithic collapse scenario for the formation of the MW (Eggen, Lynden-Bell and Sandage, 1962). Several attempts to account for the metallicity distribution of the MW halo in the modern framework (hierarchical merging of smaller components, hereafter sub-haloes) were undertaken in recent years, (Bekki and Chiba 2001; Scanapieco and Broadhurst 2001, Font *et al.* 2006, Tumlinson 2006, Salvadori, Schneider and Ferrara 2007). Independently of their success or failure in reproducing the observations, those recent models provide little or no physical insight into the physical processes that shaped the metallicity distribution of the MW halo. Indeed, if the halo was built from a large number of sucessive mergers of sub-haloes, why is its metallicity distribution so well described by the simple model with outflow (which refers to a single system)? And what determines the peak of the metallicity distribution at [Fe/H]=∼–1.6, which is (successfully) interpreted in the simple model by a single parameter (the outflow rate) ?

In this work we present an attempt to built the halo metallicity distribution analytically, in the framework of the hierarchical merging paradigm. Preliminary results have already been presented in Prantzos (2007b) and a detailed account is given in Prantzos (2008). It is assumed that the building blocks were galaxies with properties similar to (but not identical with) those of the Local group dwarf galaxies that we observe today. In that way, the physics of the whole process becomes (hopefully) more clear.

1.1. The halo metallicity distribution and the simple model

The halo metallicity distribution is nicely described by the simple model of GCE, in which the metallicity Z is given as a function of the gas fraction μ as $Z = p\ ln(1/\mu) + Z_0$, where Z_0 is the initial metallicity of the system and p is the *yield* (metallicities and yield are expressed in units of the solar metallicity $Z_\odot$). If the system evolves at a constant mass (closed box), the yield is called the *true yield*, otherwise (i.e. in case of mass loss or gain) it is called the *effective yield*. The *differential metallicity distribution* (DMD) is:

$$\frac{d(n/n_1)}{d(\log Z)} = \frac{\ln 10}{1 - \exp\left(-\frac{Z_1 - Z_0}{p}\right)} \frac{Z - Z_0}{p} \exp\left(-\frac{Z - Z_0}{p}\right) \qquad (1.1)$$

where Z_1 is the final metallicity of the system and n_1 the total number of stars (having metallicities $\leqslant Z_1$). This function has a maximum for $Z - Z_0 = p$, allowing one to evaluate easily the effective yield p if the DMD is determined observationally.

The DMD of field halo stars peaks at [Fe/H]$\sim$−1.6, suggesting an effective halo yield $p_{Halo} \sim 1/40 \, Z_\odot$ for Fe. This has to be compared to the true yield obtained in the solar neighborhood, which is $\sim Z_\odot$ (from the peak of the local MD), but $\sim 2/3$ of that originate in SNIa, so that the true Fe yield of SNII during the local disk evolution was $p_{Disk} \sim 0.32 \, Z_\odot$; consequently, the *effective Fe yield* of SNII during the halo phase (where they dominated Fe production) was $p_{Halo} \sim 0.08 \, p_{Disk}$.

Hartwick (1976) suggested that *outflow* at a rate $F = k \, \Psi$ (where Ψ is the Star Formation Rate or SFR) occured during the halo formation. In the framework of the simple model of GCE such outflow reduces the true yield p_{True} to its effective value $p_{Halo} = p_{True}(1 - R)/(1 + k - R)$ (e.g. Pagel 1997) where R is the Returned Mass Fraction ($R \sim 0.32$-0.36 for most of the modern IMFs, e.g. Kroupa 2002). The halo data suggest then that the ratio of the outflow to the star formation rate was

$$k = (1 - R)\left(\frac{p_{True}}{p_{Halo}} - 1\right) \qquad (1.2)$$

and, by replacing p_{True} with p_{Disk} one finds that $k \sim 7$-8.

Although the siple model of GCE with outflow fits extremely well the bulk of the halo DMD, the situation is less satisfactory for the low metallicity tail of that DMD. Prantzos (2003), noting that the situation is reminiscent of the local G-dwarf problem, suggested that a similar solution should apply, namely an early phase of rapid infall (in a time scale of less than 0.1 Gyr) forming the Milky Way's halo, as illustrated in Fig. 1.

1.2. *The halo DMD and hierarchical merging*

In this work we assume that the MW halo was formed by the merging of smaller units ("sub-haloes"), as implied by the hierarchical merging scenario for galaxy formation. In order to calculate semi-analytically the resulting DMD, it is further assumed that each of the sub-haloes had a DMD described well by the simple model, i.e. by Eq. (1.1). It remains then to evaluate the effective yield $p(M)$ of each sub-halo as a function of its mass, as well as the mass function of the sub-haloes dN/dM.

It is assumed here that each one of the merging sub-haloes has a DMD described by the simple model with an appropriate effective yield. This assumption is based on recent observations of the dwarf spheroidal (dSph) satellites of the Milky Way, as will be discussed below. It is true that the dSphs that *we see today* cannot be the components of the MW halo, because of their observed abundance patterns (e.g. Shetrone, Côté and Sargent 2001; Venn et al. 2004): their α/Fe ratios are typically smaller than the [a/Fe]$\sim$0.4$\sim$const. ratio of halo stars. This implies that they evolved on longer timescales than the Galactic halo, allowing SNIa to enrich their ISM with Fe-peak nuclei and thus to lower the α/Fe ratio by a factor of $\sim$2-3 (as seen from the [O/Fe]$\sim$0 ratio in their highest-metallicity stars). However, the shape of the DMD of the simple model does not

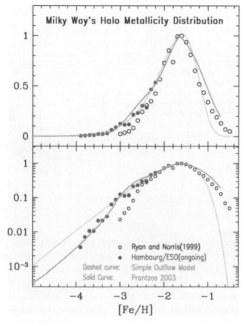

Figure 1. Differential metallicity distribution of field halo stars in linear (*top*) and logarithmic (*bottom*) scales. Data are from Ryan and Norris (1991, *open symbols*), and the ongoing Hamburg/ESO project (*filled symbols*). The two data sets are normalised at [Fe/H]-2.2; above that value, the Hamburg/ESO data are incomplete, while below that metallicity the Ryan and Norris (1991) data set is incomplete. The *dotted curve* is a simple model with instantaneous recycling (IRA) and outflow rate equal to 8 times the star formation rate. The solid curve is obtained as in Prantzos (2003), from a model without IRA, an early phase of rapid infall and a constant outflow rate equal to 7 times the SFR. All curves and data are normalised to max=1.

depend on the star formation history or the evolutionary timescale, only on gas flows into and out of the system, as well as on the initial metallicity (e.g. Prantzos 2007a).

The DMDs of four neraby dSphs (Helmi *et al.* 2006) are displayed as histograms in Fig. 2, where they are compared to the simple model with appropriate effective yields (*solid curves*). The effective yield in each case was simply assumed to equal the peak metallicity (Eq. 1.1). It can be seen that the overall shape of the DMDs is quite well fitted by the simple models. This is important, since i) it strongly suggests that *all* DMDs of small galaxian systems can be described by the simple model and ii) it allows to determine *effective yields* by simply taking the peak metallicity of each DMD (see below).

Before proceeding to the determination of effective yields, we note that the fit of the simple model to the data of dSphs fails in the low metallicity tails. Helmi *et al.* (2006) attribute this to a pre-enrichment of the gas out of which the dSphs were formed. However, early infall is another, equally plausible, possibility, as argued in Prantzos (2008). The recent simulations of Salvadori *et al.* (2008) find both pre-enrichment and early infall for local dSphs.

If the DMDs of all components of the Galactic halo are described by the simple model, then their shape is essentially described by the corresponding effective yield p (and, to a lesser degree, by the corresponding initial metallicity Z_0). Observations suggest that the effective yield is a monotonically increasing function of the galaxy's stellar mass M_* (Fig. 3). In the case of the progenitor systems of the MW halo, however, the effective

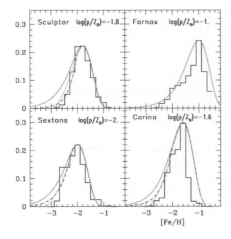 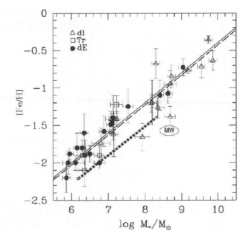

Figure 2. left: Metallicity distributions of dwarf satellites of the Milky Way. Data are in *histograms* (from Helmi *et al.* 2006). *Solid curves* indicate the results of simple GCE models with outflow proportional to the star formation rate; the corresponding effective yields (in $Z_\odot$) appear on top right of each panel. *Dashed curves* are fits obtained with an early infall phase, while *dotted curves* are models with an initial metallicity $\log(Z_0) \sim -3$; both modifications to the simple model (i.e. infall and initial metallicity) improve the fits to the data. **Right:** Stellar metallicity vs stellar mass for nearby galaxies; data and model (*upper curves*) are from Dekel and Woo (2003), with *dI* standing for dwarf irregulars and *dE* for dwarf ellipticals. The *thick dotted* line represents the effective yield of the sub-haloes that formed the MW halo according to this work (i.e. with no contribution from SNIa, see Sec. 3.2). The MW halo, with average metallicity [Fe/H]=−1.6 and estimated mass $\sim 4\times 10^8$ $M_\odot$ falls below both curves.

yield must have been lower by a factor of 2-3, since SNIa had no time to contribute (as seen from the high α/Fe$\sim$0.4 ratios of halo stars). We assume then that the effective yield of the MW halo components (accreted satellites or sub-haloes) is given (in $Z_\odot$) by

$$p(M_*) = 0.005 \left(\frac{M_*}{10^6 M_\odot}\right)^{0.4} \qquad (1.3)$$

i.e. the thick dotted curve in Fig. 3.

Obviously, the stellar mass M_* of each of the sub-haloes should be $M_* < M_H$ where M_H is the stellar mass of the MW halo ($M_H = 4\pm 0.8\ 10^8$ $M_\odot$, e.g. Bell *et al.* 2007).

Hierarchical galaxy formation scenarios predict the mass function of the dark matter sub-haloes which compose a dark matter halo at a given redshift. Several recent simulations find $dN/dM_D \propto M_D^{-2}$ (Diemand *et al.* 2007, Salvadori *et al.* 2007, Giocoli *et al.* 2008). In our case, we are interested in the mass function of the *stellar sub-haloes*, and not of the dark ones. Considering the effects of outflows on the baryonic mass function, Prantzos (20008) finds that

$$\frac{dN}{dM_*} \propto M_*^{-1.2} \qquad (1.4)$$

i.e. *the distribution function of the stellar sub-haloes is flatter than the distribution function of the dark matter sub-haloes*. The normalisation of the stellar sub-halo distribution function is made through

$$\int_{M_1}^{M_2} \frac{dN}{dM_*} M_*\ dM_* = M_H \qquad (1.5)$$

The lower mass limit M_1 is adopted here to be $M_1=1$-$2\ 10^6$ $M_\odot$, in agreement with

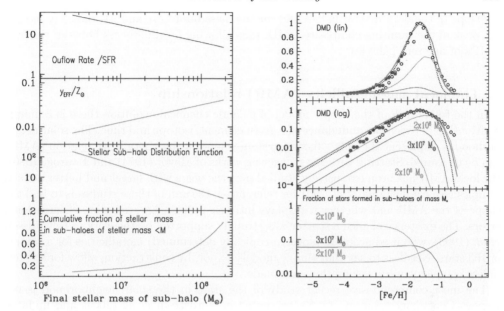

Figure 3. Left:Properties of the sub-haloes as a function of their stellar mass, empirically derived as discussed in Sec. 3. From top to bottom: Outflow rate, in units of the corresponding star formation rate; Effective yield, in solar units; Distribution function; Cumulative fraction of stellar mass contributed by the sub-haloes. The total mass of the MW halo is 4 10^8 M$_\odot$. **Right**:*Top* and *middle* panels: Differential metallicity distribution (in lin and log scales, respectively) of the MW halo, assumed to be composed of a population of smaller units (sub-haloes). The individual DMDs of a few sub-haloes, from 10^6 M$_\odot$ to 4 10^7 M$_\odot$, are indicated in the middle panel, as well as the sum over all haloes (*solid upper curves* in both panels, compared to observations). *Dotted curves* in top and middle panels indicate the results of the simple model with outflow (same as in Fig. 1). Because of their large number, small sub-haloes with low effective yields contribute the largest fraction of the lowest metallicity stars, while large haloes contribute most of the high metallicity stars (*bottom panel*).

the lower mass bound of observed dSphs in the Local group. Such galaxies have internal velocities $V > 10$ km/s. Dekel and Woo (2003) argue that the gas in haloes with $V < 10$ km/s cannot cool to form stars at any early epoch and that dwarf ellipticals form in haloes with $10 < V < 30$ km/s, the upper limit corresponding to a stellar mass $M_* \sim 2$ 10^8 M$_\odot$. This is the upper mass limit M_2 that we adopt here. The main properties of the sub-halo set constructed in this section appear in Fig. 3 (left) as a function of the stellar sub-halo mass M_*. The resulting total DMD is obtained as a sum over all sub-haloes:

$$\frac{d(n/n_1)}{d(logZ)} = \int_{M_1}^{M_2} \frac{d[n(M_*)/n_1(M_*)]}{d(logZ)} \frac{dN}{dM_*} M_* \, dM_* \qquad (1.6)$$

The result appears in Fig. 4 (right, with top panel in linear and middle panel in logarithmic scales, respectively). It can be seen that it fits the observed DMDs at least as well as the simple model à la Hartwick. In summary, under the assumptions made here, the bulk of the DMD of the MW halo results naturally as the sum of the DMDs of the component sub-haloes. It should be noted that all the ingredients of the analytical model are taken from observations of local satellite galaxies, except for the adopted mass function of the sub-haloes (which results from analytical theory of structure formation plus a small modification to account for the role of outflows). Obviously, by assuming different values for

2. The local age-metallicity (AMR) relationship

In the framework of the simple model of galactic chemical evolution there is a unique relationship between the abundance of a given element/isotope and time. The solar neighborhood is at present the only galaxian system where the age-metallicity relation (AMR) can be measured. Starting with the pioneering works of Mayor (1974) and Twarog (1980), the local AMR has been extensively studied over the years with larger and better defined samples (see e.g. Feltzing et al. 2001 for references); the aim of those studies is to find the shape of the AMR and whether it displays intrinsic scatter, i.e. exceeding observational errors. The existence of such a scatter was strongly supported by the work of Edvardsson et al. (1993), who used accurate (spectroscopically determined) metallicities for a sample of 163 stars. Needless to say that simple models do not, by construction, allow for scatter, unless complementary assumptions are made.

The most comprenhesive recent study of the stars in the solar neighborhood is the Geneva-Copenhagen Survey (GCS), presented in Nordström et al. (2004) and updated in Holmberg et al. (2007). It concerns a magnitude limited sample of more than 13000 F and G dwarfs with accurately determined distances and kinematics; metallicities are photometrically determined and have estimated errors $\sim$0.1 dex, while stellar ages suffer from considerable uncertainties, often excceding 50%. The AMR determined by using the large sample, or various sub-samples (to account for different types of biases) displays an upturn for stars of young ages and solar metallicities (which dominate largely the sample), but is essentially flat otherwise (see upper left panel of Fig. 4). Similar results (i.e. an essentially flat AMR, when *average metallicity* is calculated as a function of age) are obtained in the independent study of Soubiran et al. (2007) with a sample of $\sim$700 giant stars (data points with errors bars in the same figure).

If the AMR is indeed as flat as suggested by those studies, this would have important implications for our understanding of the evolution of the solar neighborhood: taken at face value, it would imply that the local Galaxy evolved essentially at constant average metallicity for most of its lifetime ($\sim$0.5 $Z_\odot$), with a small increase in the past few Gyr. None of the present-day models, which satisfy an important number of other observational constraints (e.g. Boissier and Prantzos 1999, Goswami and Prantzos 2000), predict such an evolution; in general, metallicity increases early on and saturates at late times.

Before getting to conclusions based on the AMR one should seriously consider the various biases affecting it. That issue is properly emphasized in Nordström et al. (2004), Haywood (2006), Holmberg et al. (2007). Here we wish to draw attention to an aspect of the AMR which is rarely considered, despite the fact that the quantitative determination of a correlation between two properties of a sample of objects is one of the long standing problems in observational astronomy (and in many other fields of science).

Given the two properties (metallicity and age) of the sample stars, one may ask either what is the *average metallicity at a given age* or *what is the average age at a given metallicity*. If there were a unique AMR (with no scatter, as predicted by the simple model) the answer would be obviously the same in both cases. However, in the presence of scatter in the data, *the answer is not the same*. This can be seen in the bottom left panel of Fig. 4, where average age is calculated for fixed metallicity bins of 0.1 dex (*filled squares*): there is substantial variation of the average age with metallicity in this case, at variance with the average metallicity vs age relationship derived before.

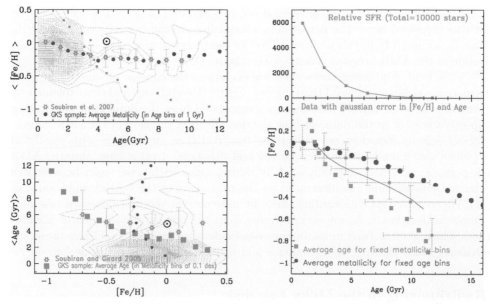

Figure 4. Left: Local Age-Metallicity relationship, in 1 Gyr age bins (*top*) and in 0.1 dex metallicity bins (*bottom*); *isocontours* correspond to the data of GCS (Nordström *et al.* 2004), *large symbols* show the corresponding relationships of age vs mean metallicity (*top*) and Metallicity vs mean age (*bottom*), *small symbols* the complementary relationship, and data with error bars designate results of independant studies with different samples (from Soubiran *et al.* 2007 and Soubiran and Girard 2005). Notice that in both panels the Sun appears to be slightly metal-rich for its age or slightly older than stars of the same metallicity. **Right:** By using simulated data as input for the AMR (points with 1-σ Gaussian error bars in lower panel, as indicated by *dotted lines*) and a star formation rate exponentially increasing with time (top panel), one may derive the corresponding age vs average metallicity (*circles*) and metallicity vs average age (*squares*) relations; the former is flatter than and the latter is steeper than the input relationship (as with the case of the real data, see left pannels). Their average (*solid curve*) is a better approximation to the input data.

Notice that Soubiran and Girard (2005), using a different sample, find a flatter <age> vs metallicity relationship (data points with error bars in lower left panel of Fig. 4), but again orthogonal to the <metallicity> vs age relationship.

In order to better understand that situation, we performed calculations with artificial input data and one of the results is displayed on the right panel of Fig. 4. We adopted a SFR increasing with time (upper right panel in Fig. 4), to simulate the large number of young stars in the GCS sample and we normalised the total number to 10000 stars (comparable to the size of the GCS sample). We adopted an average AMR steadily increasing with time (small points in bottom right panel) with Gaussian scatter of 0.1 dex in metallicity and (age/Gyr) in age, as indicated by the *dotted error bars*. The resulting <age> vs metallicity and <metallicity> vs age relationships are also displayed in the same figure (*thick squares* and *dots*, respectively). They differ considerably from each other and from the input data, converging at young ages, where the number of stars is large. Overall, the figure shows considerable overall similarity with real data analysis on the left. We checked that by reducing the error bars the two derived relationships converge everywhere to the input AMR, as they should, and we tested different SFR with the same input AMR (not displayed in Fig. 4). It turns out that in the case of a strong early SFR the two curves converge at old ages and diverge at young ages, contrary to the case displayed here. The situation is intermediate in the case of a constant SFR.

What do we learn from this analysis? That in the presence of scatter in the data, the derivation of the AMR becomes a delicate enterprise. Of course, if the scatter is intrinsic (due to some kind of inhomogeneous chemical evolution) the concept of <metallicity> vs age is more physicaly meaningful than the one <age> vs metallicity and this explains why in most cases only the former relationship is calculated from the data. However, in case of important scatter in the data, such as the one affecting the ages of the GCS sample, the former relationship does not reflect the true AMR, as shown here with simulated data (which also reproduce convincingly the real situation). In that case, the true AMR could, perhaps, be better approximated by taking some arithmetic mean between the two derived relationships (as illustrated by the *thin curve* in the bottom right panel of Fig. 4). We advance this with caution, since we have not yet thoroughly checked this idea; we simply notice that it is not very different from the ordinary least squares bisector method, which is often used in astronomy, especially after Isobe *et al.* (1990) presented a convincing study showing its advantages over other data analysis methods.

3. Radial mixing in the Milky way disk

The idea that stars in a galactic disk may diffuse to large distances along the radial direction (i.e to distances larger than allowed by their epicyclic motions) was proposed by Wielen *et al.* (1996). They suggested that some of the peculiar chemical properties of the Sun may be explained by the assumption that it was born in the inner Galaxy (i.e. in a high metallicity region, in view of the galactic metallicity gradient) and subsequently migrated outwards. They treated the hypothetical radial migration phenomenologically, acknowledging that the basic mechanism for the gravitational perturbations of stellar orbits is not understood.

Sellwood and Binney (2002, herefter SB02) convincingly argued that stars can migrate over large radial distances, due to resonant interaction with spiral density waves at corotation. Such a migration alters the specific angular momentum of individual stars, but affects very little the overall distribution of angular momentum and thus does not induce important radial heating of the disk. Because high-metallity stars from the inner (more metallic and older) and the outer (less metallic and younger) disc are brought in the solar neighborhood, SB02 showed with a simple toy model that considerable scatter may result in the local age-metallicity relation, not unlike the one observed by Edvardsson *et al.* (1993).

Another obvious implication of the radial migration model of SB02 concerns the flattening of the stellar metallicity gradient in the galactic disk. That issue was quantitatively explored in Lepine *et al.* (2003), who considered, however, the corotation at a fixed radius (contrary to SB02). As a result, the gravitational interaction bassically removes stars from the local disk, "kicking" them inwards and outwards. The abundance profile (assumed to be initially exponential) is little affected in the inner Galaxy, but some flattening is obtained in the 8-10 kpc region. The authors claim that such a flattening is indeed observed (using data of planetary nebulae by Maciel and Quireza 1996) but modern surveys do not find it (see e.g. Cunha *et al.* this meeting).

Finally, Haywood (2008), on the basis of kinematics and abundance observations of a large sample of local stars argues that most of the metal rich stars in the solar neighborhood originate from the inner disk and most of the metal poor ones from the outer disk, and suggests that the local disk started its evolution with a considerably high metallicity of [Fe/H]$\sim$-0.2. Such a large pre-enrichment of the thin disk, however, is difficult to

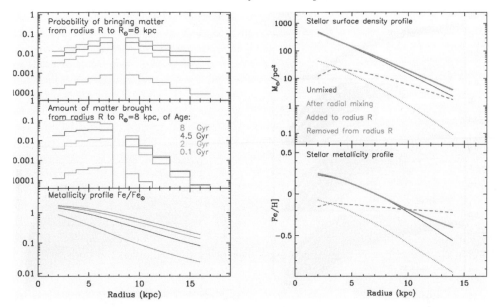

Figure 5. Left: A toy model of radial mixing, applied to a "realistic" disk (as obtained in Boissier and Prantzos 1999). The various histogrammes/curves correspond to material of different ages (8, 4.5, 2 and 0.1 Gyr, respectively, from top to bottom in the *top* and *middle* panels and from bottom to top in the *bottom* panel). **Right:** impact of radial mixing on stellar and metallicity profiles of the disk. *Solid curves* indicate the full profiles before (*thin*) and after (*thick*) mixing: the external regions find their surface density and metallicity enhanced after mixing. *Dashed* and *dotted* curves in both panels indicate material added and removed, respectively from a given zone.

accept. Haywood (2008) attributes it to the thick disk, but despite the relatively large contribution (∼20%) of that component to the local surface density, it cannot have pre-enriched to such a high degree the much more massive thin disk, assuming that the latter evolved as a closed box (as argued in Haywood 2006). And if infall is invoked for the thin disk (despite the fact that is superfluous for the G-dwarf problem if pre-enrichment is assumed), then a large dilution of the initially assumed [Fe/H]∼-0.2 would result. Independantly, however, of his far-reaching conclusions, Haywood (2008) presents convincing arguments that the local stellar population shows evidence for radial migration.

We study some of the implications of radial migration with a toy model similar to the one of SB02, but with boundary conditions based on a "realistic" simple model for the Milky Way disk, which reproduces most of the available observations (Boissier and Prantzos 1999). The key quantity is the probability that a star migrates radially and we adopt here an function decreasing with distance (remote stars have smaller probability of being found in the solar neighborhood that nearby ones) and time (older stars have motre time and thus higher probability of being found here than younger ones). Although the probabilities are intrinsically symmetric in the radial direction for a given radius (Figure 5, top left), the larger surface density of the inner disk results in larger numbers of stars transferred from the inner disk at 8 kpc than from the outer one (Fig. 5 middle left).

We adopt (somewhat arbitrarily) a normalisation of the probability function of radial migration, as to not affect the stellar surface density by more than 10% locally (at $R=8$ kpc). This is, perhaps, an underestimate (as discussed below) but the toy model is presented here only for illustration purposes. In Fig. 5 (top right) it is seen that with

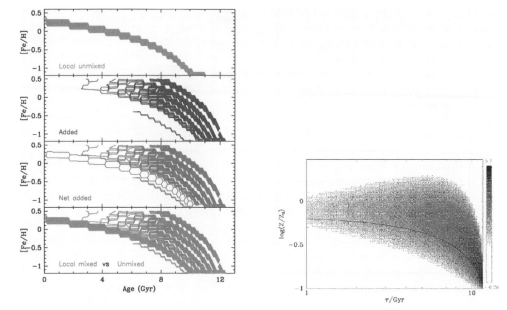

Figure 6. Left: Impact of radial mixing on the local (R=8 kpc) age-metallicity relation accrding to the toy model of the previous figure; from top to bottom: original age-metallicity relation, added material, net added material (after accounting from removed matter) and total age-metallicity relation, with considerable scatter at old stellar ages. The latter should be compared to the figure on the **right**, from the toy model of Sellwood and Binney (2002).

this normalisation, regions at $R > 4$ kpc receive more material (dashed curve) than it is removed from them (dotted curve). The exponential stellar profile of the original model (thin curve) flattens somewhat outwards, with the stellar density almost being doubled (thick curve) at $R=14$ kpc.

The mean metallicity of all the stars received at a given region is always larger than the average metallicity of the stars ever removed from that region (dashed and dotted curves, respectively, in Fig. 5 bottom right) *but not from the original final average (over all stars) metallicity of that same region, except for the zones at $R > 10$ kpc* (thin solid curve in Fig. 5 right bottom). The average metallicity of these zones is affected by the migrated stars and it increases by a factor of 2 (0.3 dex) at $R=14$ kpc.

The implications of radial mixing for the solar neighborhood appear in Fig. 6 (left panel), where the age-metallicity relation appear for the locally born stars (top), for those brought from other regions (middle) and for the total final sampl, i.e. original+added-removed (bottom). A considerable scatter of the AMR is obtained, especially for old stars, confirming the finding of SB02 with a simpler original model. It should be noticed, however, that in both models the final result ressembles little to the stellar sample of the GCS, which displays i) a large number of young stars with considerable metallicity scatter and ii) a flat early AMR with a small upturn at late times. The toy model explored here shows a different trend, both concerning the scatter (important at early times) and the upturn (only at early times). This may imply two things: either i) the GCS sample is seriously biased, favouring excessively young stars, but also underestimating age errors for old stars, in which case it can hardly be used for comparison to chemical evolution models (with our without mixing), or ii) the true AMR, both locally and galaxy-wide,

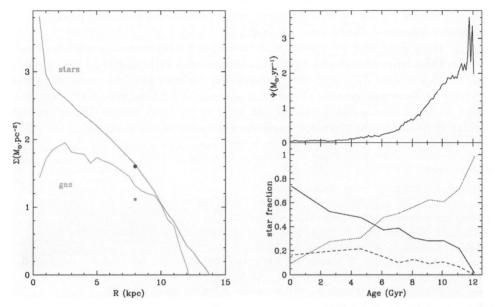

Figure 7. **Left:** Stellar and gaseous surface density profiles (log scale) of the N-body + SPH simulation of Heller *et al.* (2007). Points at 8 kpc represent corresponding quantities in the solar neighborhood. However, the simulation produces an early-type galaxy, as can be seen from the star formation rate vs. age (right top) and its results cannot be directly compared to the Milky way data. Nevertheless,considerable mixing of stars of all ages is found in all radial zones, as can be seen from the situation at 8 kpc (right bottom), where the fraction of locally born stars (*solid)*) is plotted against age and compared to stars that migrated there from the inner disk (*dotted*) and the outer disk (*dashed*).

is very different from the one resulting from standard models and adopted in the radial mixing models presented here.

A convincing case for radial mixing of stars in disks was recently presented in Roskar *et al.* (2008), with high-resolution SPH + N-body simulations of an isolated galaxy growing from dissipational collapse of gas embedded in a spherical dark matter halo. They find a break in star formation in the outer disk (due to the unaivalability of sufficiently cooled gas in those regions) which moves gradually outwards. Despite that break, they find a large number of stars outside the final break radius, which are scattered there from the inner disk on nearly circular orbits by collisions with spiral arms.

We explored the issue of radial mixing by analysing the results of another SPH + N-body simulation, performed by Heller *et al.* (2007). It concerns a galaxy growing in an isolated dark matter halo and developing final stellar and gaseous profils not unlike the corrswponding ones of the Milky Way (Fig. 7 left panel). However, it is an early type galaxy, since most of its star formation occurs early on (top right panel in Fig. 7), and the results cannot be compared in detail to observations of the Milky Way. Considerable mixing of stars is found at $R=8$ kpc, mostly originating from the inner disk (Fig. 7 right bottom); in fact, most of the local stars of age >7 Gyr come from the inner disk in that similation, while the fraction of stars coming in from the outer disk never exceeds 20%. The complete analysis of these results and their impact on the metallicity distribution are presented in Bellil *et al.* (2008, in preparation). We expect the effect of radial mixing on local properties to be smaller for a MW-type galaxy, because of its slower star formation, which leaves less time for stellar transport.

A last, but important point, concerning radial mixing: independently of theoretically appealing arguments, observations will ultimately determine its importance. In that respect, a strong constraint to consider is the resulting [O/Fe] ratio as a function of metallicity: in the solar neighborhood, that ratio displays very little scatter (if any at all). If a large number of local stars originated in inner disk regions (with large early metallicities *and* large O/Fe ratios, because of the early absence of Fe producing SNIa), one would expect to find a substantial scatter in that ratio in the local sample. We are currently working to quantify that effect and constrain the extent of radial mixing in the MW.

References

Bekki, K. & Chiba, M., 2001, ApJ 558, 666
Bell, E., Zuker, D., & Belokurov, V., 2008, ApJ 680, 295
Boissier, S. & Prantzos, N., 1999, MNRAS 307, 857
Edvardsson, B., Andersen, J., Gustaffson B., *et al.*, 1993, A&A 275, 101
Dekel, A. & Woo, J., 2003, MNRAS 344, 1131
Diemand, J., Kuhlen, M., & Madau, P., 2007, ApJ 667, 859
Feltzing, S., Holmberg, J., & Hurley, J. R., A&A 377, 911
Font, A., Johnston, K., Bullock, J., & Robertson, B., 2006, ApJ, 638, 585
Giocoli, C., Pieri, L., & Tormen, G. 2008, MNRAS 387, 689
Goswami, A. & Prantzos, N., 2000, A&A 359, 151
Hartwick, F., 1976, ApJ 209, 418
Haywood, M., 2006, MNRAS 371, 176
Haywood, M., 2008, MNRAS 388, 1175
Heller, C., Shlosman, I., & Athanassoula, E., 2007, ApJ 671, 226
Helmi, A., Irwin, M., Tolstoy, E., *et al.*, 2006, ApJ 651, L121
Holmberg, J., Norström, B., & Andersen, J., 2007, A&A 475, 519
Kroupa, P., 2002, Science 295, 82
Lepine, J. R. D., Acharova, I. A., & Mishurov, Y. N., 2003, ApJ 589, 210
Mayor, M., 1974, A&A 32, 321
Nordström, B., Mayor, M., & Andersen, J., *et al.*, AA418, 989
Pagel, B., 1997, "Nucleosynthesis and galactic chemical evolution" (Cambidge University Press)
Prantzos, N. 2003, A&A 404, 211
Prantzos, N. 2007a, in "Stellar Nucleosynthesis: 50 years after B2FH", C. Charbonnel and J.P. Zahn (Eds.), EAS publications Series (arXiv:0709.0833)
Prantzos, N. 2007b, in "CRAL-2006: Chemodynamics, from first stars to local galaxies", E. Emsellem *et al.* (Eds.), EAS publications Series Vol. 24, p. 3 (arXiv:astro-ph/0611476)
Prantzos, N. 2008, A&A in press (arXiv:0807.1502)
Roskar, R., Debattista, V., Stinson, G., *et al.*, 2008, ApJ 675, L65
Ryan S. & Norris J., 1991, AJ 101, 1865
Salvadori, S, Schneider, R., & Ferrara, A., 2007, MNRAS 381, 647
Salvadori, S, Ferrara, A., & Schneider, R., 2008, MNRAS 361, 348
Scanapieco, E. & Broadhurst, T., 2001, ApJ 550, L39
Sellwood, J. & Binney, J., 2002, MNRAS 336, 785
Shetrone, M., Coté, P., & Sargent, W., 2001, ApJ 548, 592
Soubiran, C. & Girard, P., 2005, A&A 438, 139
Soubiran, C., Bienaymé, O., Mishenina, T. V., & Kovtyukh, V. V. 2008, A&A 480, 91
Tumlinson, J., 2006, ApJ 641, 1
Twarog, B. A., 1980, ApJ 242, 242
Venn, K., Irwin, M., & Shetrone, M., *et al.*, 2004, ApJ 128, 1177
Wielen, R., Fuchs, B., & Dettbarn, C., 1996, A&A 314, 438

Chemical evolution of the Galaxy disk in connection with large-scale winds

Takuji Tsujimoto[1], Joss Bland-Hawthorn[2], and Kenneth C. Freeman[3]

[1]National Astronomical Observatory, Mitaka-shi, Tokyo 181-8588, Japan
email: taku.tsujimoto@nao.ac.jp

[2]Institute of Astronomy, School of Physics, University of Sydney, NSW 2006, Australia

[3]Research School of Astronomy and Astrophysics (RSAA), Australian National University, Cotter Road, Weston Creek, ACT 2611, Australia

Abstract. Comparison of elemental abundance features between old and young thin disk stars may reveal the action of ravaging winds from the Galactic bulge, which once enriched the whole disk, and set up the steep abundance gradient in the inner disk ($R_{GC} \lesssim 10\text{--}12$ kpc) and simultaneously the metallicity floor ([Fe/H]$\sim$ -0.5) in the outer disk. After the end of a crucial influence by winds, chemical enrichment through accretion of a metal-poor material from the halo onto the disk gradually reduced the metallicity of the inner region, whereas an increase in the metallicity proceeded beyond a solar circle. This results in a flattening of abundance gradient in the inner disk, and our chemical evolution models confirm this mechanism for a flattening, which is in good agreement with the observations. Our scenario also naturally explains an observed break in the metallicity floor of the outer disk by young stars since the limit of self-enrichment in the outer disk is supposed to be [Fe/H]$\lesssim$ -1 and inevitably incurs a direct influence of the dilution by a low-metal infall whose metallicity is [Fe/H]$\sim$ -1. Accordingly, we propose that the enrichment by large-scale winds is a crucial factor for chemical evolution of the disk, and claim to reconsider the models thus far for the disk including the solar neighborhood, in which the metallicity is predicted to monotonously increase with time. Furthermore, we anticipate that a flattening of abundance gradient together with a metal-rich floor in the outer disk are the hallmark of disk galaxies with significant central bulges.

Keywords. Galaxy: bulge – Galaxy: disk – Galaxy: evolution – ISM: jets and outflows

1. Introduction

How do disk galaxies form and how do they evolve through cosmic time? Clear answers to these questions continue to elude us, and it may be many years before we converge on a successful physical model. Disks lie at the forefront of galaxy formation and evolution, not least because most stars are in disks today (Benson *et al.* 2007, Driver *et al.* 2007). It is widely recognized that N-body simulations of galaxy formation within cold dark matter (CDM) cosmology fail to produce realistic galactic disks (Navarro & White 1994, Navarro *et al.* 1995, Steinmetz & Müller 1995). But a useful aspect of these incomplete models has been to highlight the possible role of feedback in shaping galaxies. Vigorous feedback in the early phase of galaxy formation is a critical solution to the angular momentum problem by preventing baryons from loosing their specific angular momentum too much through the interaction with dark matter (Fall 2002). Several recent developments have provided the impetus for the present work. In particular, disk galaxies are now being traced over 10 or more optical scale lengths to reveal an [Fe/H] abundance gradient that declines before it flattens off in the outer disk (Yong *et al.* 2006 and also see contribution by Vlajic, this volume). We believe that this abundance transition is a fossil record of long-term feedback and accretion processes, as we discuss.

Here we consider the effects of large-scale winds in the chemical evolution of disks. Several new observations suggest outflows are important in the lifecycle of galaxies. First, Tremonti et al. (2004) find evidence for chemical enrichment trends throughout star-forming galaxies over three orders of magnitude in stellar mass. Indeed, there is evidence that galaxies with the mass of the Milky Way or higher manage to retain a large fraction of their metals, in contrast to lower mass galaxies where metal loss appears to be anticorrelated with baryonic mass. Recent theoretical work has proved that a low-metal content in dwarf galaxies is attributable to efficient metal-enriched outflows (Dalcanton 2007). Secondly, there is now strong evidence for large-scale outflows in the Galaxy across the electromagnetic spectrum (Bland-Hawthorn & Cohen 2003, Fox et al. 2005, Keeney et al. 2006, Everett et al. 2008). Finally, the intergalactic medium at all redshifts shows signs of significant metal enrichment consistent with the action of winds (Cen & Ostriker 1999, Madau et al. 2001, Ryan-Webber et al. 2006, Davé et al. 2008).

Therefore it is important to consider the influence of large-scale winds within Galactic chemical evolution (GCE) models. The key observational constraint that we consider is the Galactic abundance gradient determined from different astrophysical sources (e.g., K giants, HII regions, etc.). However, it confronts the fact that the simple GCE models in which the disk is divided into independently evolving rings successfully reproduce the present abundance gradient along the disk (e.g., Boissier & Prantzos 1999, Hou et al. 2000, Chiappini et al. 2001) and the necessity to invoke winds for the chemical evolution of the Galaxy disk has not been claimed. Thus, in fact, the connection of abundance gradient with winds has been disregarded so far. Recent finding on the time evolution of abundance gradient has imposed the new task on GCE models. The observed gradient has flattened out by half in the last 5 Gyr or so (e.g., Daflon & Cunha 2004, Maciel et al. 2006). Such a large change in abundance gradient seems beyond the predictions by any GCE models so far, and a search for its potential mechanism will invoke some ingredient to be considered in the GCE model. Then, we first propose that the observed flattening phenomenon is attributable to the once enrichment by large-scale winds.

2. Chemical signatures of large-scale winds

Here we consider possible evidence for large-scale winds buried within stellar abundance data. In summary, the compelling evidence for large-scale winds we insist is that we commonly see the remarkable enrichment for some period in each local region of the disk, subsequently followed by a decrease in abundances.

2.1. Solar neighborhood

Detailed elemental abundances of stars and gas in the solar neighborhood indicate that the presence of metal-rich stars with [Fe/H] $\gtrsim 0$ is the end result of additional enrichment by large-scale winds several Gyr ago, in terms of the following three aspects.
1. stellar abundance distribution function

The abundance distribution function (ADF) of solar neighborhood disk stars has been studied by many authors thus far, and the location of its peak at $[Fe/H]_{peak} \sim -0.2$---0.1 has been firmly established. On the other hand, the metal-rich end of ADF extends at least to [Fe/H]$\sim$ +0.2, and the fraction of stars with [Fe/H]>0 is roughly 20%. Spectroscopic observations of elemental abundances for metal-rich disk stars (Feltzing & Gustafsson 1998, Bensby et al. 2005) have confirmed that chemical enrichment in the solar neighborhood has continuously proceeded until [Fe/H]$\sim$ +0.4. Tsujimoto (2007) has claimed that the simultaneous reproduction of both the presence of stars with [Fe/H] $\gtrsim$ +0.2 and $[Fe/H]_{peak}$ <0 is hard to realize through the conventional scheme of local enrichment

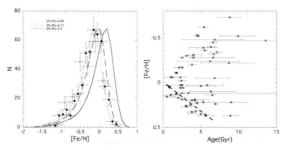

Figure 1. *left panel*: Abundance distribution function of disk stars against the iron abundance. Three curves are the predictions of the model with different input values of the present [Fe/H] abundance. The calculated distributions are convolved using the Gaussian with a dispersion of 0.15 dex in [Fe/H]. Open circles and crosses represent data taken from Edvardsson et al. (1993) and Wyse & Gilmore (1995), respectively. *right panel*: Age-metallicity relation of disk stars taken from Bensby et al. (2005). The trend for metal-richs stars is indicated by the dashed arrow.

under a continuous low-metal infall from the halo (see left panel of Fig. 1), and that the presence of metal-richer stars than the solar is a crucial evidence for the enrichment by outflow from the bulge.

2. *offset with the present abundance*

There is little evidence that the present local gas metallicity is as high as [Fe/H] $>+0.2$. Samples of nearby young stars such as Cepheids (Lemasle et al. 2007) or OB stars (Daflon & Cunha 2004) have a mean of around solar, including the stars in the Orion association (Cunha et al. 1998), with little or no offset with the HII region (e.g., Simon-Díaz 2006). Therefore, there exists a clear discrepancy between the presence of local metal-rich F/G stars with [Fe/H] up to $\sim +0.2$-0.4 and the mean metallicity of gas in the local disk as long as we take a view of the standard chemical evolution in which the metallicity increases with time. This offset should be explained by the scheme that considers a one-time, early phase rapid enrichment and a subsequent dilution.

3. *stellar age*

In the observed age-metalliicty relation of local disk stars (Bensby et al. 2005, Reid et al. 2007), there seems a lack of young ages for metal-rich stars, implying that these stars are not formed in the recent star formation but a few Gyr ago (see right panel of Fig. 1). This observed fact is compatible with the offset discussed above.

2.2. Flattening

The radial metallicity gradient roughly from R_{GC}=4 to R_{GC}=14 kpc has flattened out in the last several Gyr with the change in a slope of $\sim$ -0.1 dex kpc^{-1} to $\sim$ -0.04 dex kpc^{-1} (e.g., Daflon & Cunha 2004, Maciel et al. 2004). Chemical evolution models predict contradictory time evolution of metallicity gradient, i.e., a steepening (Chiappini et al. 2001) or a flattening (Hou et al. 2000). Putting aside their contradiction, these two models in common predict a monotonous increase in abundances for each region, and the predicted absolute change in gradient with time is too small as compared with the observation (see Fig. 10 of Maciel et al. 2006). Thus, it is likely that we have missed some mechanism for a flattening, and we propose that a new light on this issue is brought by a comparison of the metallicity distribution for open clusters with that for Cepheids, which tells us a hint on how the metallicity gradient flattened with time, as indicated by Figure 2; The once steep gradient became shallower owing to a decrease in metalllicity at inner regions, while an increase at outer regions.

This poses the question of whether there exists a theoretical scheme to explain such a puzzling change in abundance gradient. Our proposed idea is that large-scale winds from the bulge once enriched the disk. While wind entrainment is not well understood, it is now well established that winds can carry local ISM gas into the flow (Veilleux et al. 2005). It seems plausible that much of the metal-enriched gas does not travel far and falls close to the bulge. Alternatively, some of the wind explosions may not develop fully and collapse before they progress very far. In either case, metals from the enriched core can be spread over the inner disk with its abundance gradient steeper. Since the steep gradient is an outcome of the action of winds, its temporal gradient got back to a shallow one after the event of ravaging winds while each region experienced chemical enrichment under a continuous low-metal accretion from the halo.

2.3. Outer disk

In the outer disk, we see the abundances in giants (Yong et al. 2005, Carney et al. 2005) flatten with a metallicity of [Fe/H]$\sim$ -0.5 in a spectacular fashion (see Fig. 16 of Yong et al. 2006). On the other hand, the abundances in Cepheids (Yong et al. 2006) does not flatten off as much, and in fact they are offset downwards in [Fe/H] with a scatter of -1 $\lesssim$ [Fe/H] $\lesssim$ -0.5. According to our view, this reversed age-metallicity relation between old and young stars in the outer disk is attributable to the dilution by a low-metal infall during the last several Gyr after a setup of the metallicity floor by large-scale winds. Without the enrichment by winds, the metallicity would not reach even [Fe/H]$\sim$ -1 in the outer disk. Thus, this too much enrichment by winds for the outer disk would result in little reflection of self-enrichment in situ on stellar abundances until the present. This consideration gives a reasonable answer to the puzzling abundance feature of the enhanced [α-element/Fe] ratio exhibited by all outer disk stars (Yong et al. 2006), since both large-scale winds and a low-metal infall from outside the Galaxy should exhibit a SN-II like elemental feature. Yong et al. (2006) claimed that Cepheids with the metallicities lower than red giants should be the comer cannibalized in a recent merger event. However, in fact, at some region such as a so-called Cloud 2 in the outer disk ($R_{GC} \sim$ 20 kpc), the star formation is now ongoing (Yasui et al. 2008) in the molecular cloud with its metallicity of [O/H]$\sim$ -0.7 (Ruffle et al. 2007), implying [Fe/H]$\sim$ -1.

3. Modeling of a flattening

We try to reproduce a flattening of abundance gradient along the disk, by incorporating the enrichment by large-scale winds into the model. The basic picture is that the disk was formed through a continuous low-metal infall of material from outside the disk region based on the inside-out formation scenario (Matteucci & François 1989), that is, the disk is formed by an infall of gas occurring at a faster rate in the inner region than in the outer ones, and for a specific period heavy elements carried by winds from the bulge dropped and enriched the disk. Here we calculate chemical evolutions at three regions with their Galactocentric distances R=4, 8 (solar vicinity), and 12 kpc. Details of our model are described in Tsujimoto et al. (2008, in preparation).

The left panel of Figure 2 shows the predicted age-metallicity relations at three different R, and the corresponding [Fe/H] ratios at T=4 Gyr and the present are shown as a function of R on the right panel. You see that our proposed mechanism for a flattening is a decrease and an increase in abundances at inner and outer regions, respectively, after the construction of a temporally steep gradient by wind enrichment. Predicted somewhat higher [Fe/H] abundances at a solar circle than the observed data are the results calculated with the model which is devised to be in tune with the observed chemical quantities of the solar neighborhood stars. Figure 3 shows the resultant ADF and correlation of [Mg/Fe] with [Fe/H], which are in good agreement with the observations. It

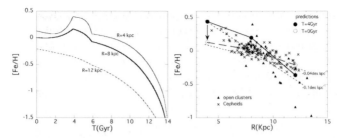

Figure 2. *left panel*: Predicted age-[Fe/H] relations at the distances of $R_{GC}=4$, 8, and 12 kpc. The age of galaxy is assumed to be 14 Gyr. *right panel*: Observed [Fe/H] distributions as a function of the distance from the center for Cepheids (crosses; Andrievsky *et al.* 2004) and open clusters (filled triangles; Friel 2006), compared with the predicted values at $T = 4$ Gyr (filled circles) and the present (open circles), which are extracted from the results of the left panel. The [Fe/H] ratios for Cepheids are shifted by -0.16 dex as performed by Yong *et al.* (2006). The directions of change in [Fe/H] from $T = 4$ Gyr to $T = 0$ are denoted by arrows for each region. The observationally implied [Fe/H] gradients are assigned by dotted lines for the present (-0.04 dex kpc^{-1}) and several Gyr ago (-0.1 dex kpc^{-1}).

should be stressed that the upturning feature of [Mg/Fe] for [Fe/H] $\gtrsim 0$ strongly implies the enrichment by large-scale winds, and the subsequent decrease in abundances well explains an offset of abundances between the present gas and metal-rich stars. Furthermore, the origin of metal-rich stars in the solar neighborhood cannot be attributable to contaminants coming from the inner disk or the Galactic bulge as a result of radial mixing (Sellwood & Binney 2002), since their elemental features represented by an upturning feature are at odds with those exhibited by the stars orbiting near the Galactic center.

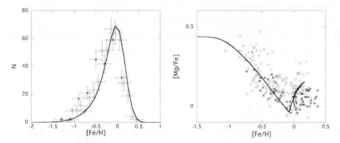

Figure 3. Features of chemical evolution in the solar neighborhood. *left panel*: Predicted abundance distribution function of thin disk stars against the iron abundance. *right panel*: Correlation of [Mg/Fe] with [Fe/H] for thin disk stars. The open circles and crosses are taken from Edvardsson *et al.* (1993) and Bensby *et al.* (2005), respectively.

4. Conclusions

We claim that large-scale winds are the crucial factor for the chemical evolution of disks. Our proposed scenario is (i) winds once set up a steep abundance gradient and a metal-rich floor in the outer disk owing to heavy elements entrained in the winds, and (ii) later evolution leads to a flattening of abundance gradient through chemical enrichment with a low-metal infall from the halo. Accordingly, we predict that a flattening and metal-rich floor are the hallmark of disk galaxies with significant central bulges, and thereby the variation in wind signature among galaxies imprinted on the disk should be correlated with the size of their bulges. Our scenario poses the challenging issues which will be verified by future work. One is about the veiled star formation history of the Galactic bulge. We believe that the bulge has a complex history involving at least one major starburst episode around 5 Gyr ago in addition to the initial burst, and hopefully it will be unveiled with an infrared eye by the Japanese future astrometric satellite JASMINE (Gouda *et al.* 2008). Another issue is on the question of whether and how the chemical evolution of the Galactic disk inside a solar circle during the last Gyr has deviated from the standard picture predicted by GCE models.

References

Andrievsky, S. M., Luck, R. E., Martin, P., & Lépine, J.R.D. 2004, *A&A*, 413, 159
Bensby, T., Feltzing, S., Lundström, I., & Ilyiin, I. 2005, *A&A*, 433, 185
Benson, A. J., Džanović, D., Frenk, C. S., & Sharples, R. 2007, *MNRAS*, 379, 841
Sellwood, J. A. & Binney, J. J. 2002, *MNRAS*, 336, 785
Bland-Hawthorn, J. & Cohen, M. 2003, *ApJ*, 582, 246
Boissier, S. & Prantzos, N. 1999, *MNRAS*, 307, 857
Carney, B. W., Yong, D., de Almeida, L., & Seitzer, P. 2005, *AJ*, 130, 1111
Cen, R. & Ostriker, J. P. 1999, *ApJ*, 514, 1
Chiappini, C., Matteucci, F., & Romano, D. 2001, *ApJ*, 554, 1044
Cunha, K., Smith, V. V., & Lambert, D. L. 1998, *ApJ*, 493, 195
Daflon, S. & Cunha, K. 2004, *ApJ*, 617, 1115
Dalcanton, J. J. 2007, *ApJ*, 658, 941
Davé, R., Oppenheimer, B. D., & Sivanandam, S. 2008, submitted to MNRAS
Driver, S. P., Allen, P. D., Liske, J., & Graham, A. W. 2007, *ApJ*, 657, L85
Edvardsson, B. *et al.* 1993, *A&A*, 275, 101
Everett, J. E. *et al.* 2008, *ApJ*, 674, 258
Fall, S. M. 2006, in: E. Athanassoula, A. Bosma, & R. Mujica (eds.) *Disks of Galaxies: Kinematics, Dynamics and Perturbations* (San Francisco: ASP), p. 389
Feltzing, S. & Gustafsson, B. 1998, *A&AS*, 129, 237
Friel, E. D. 2006, in: L. Pasquini & S. Randich (eds.) *Chemical Abundances and Mixing in Stars in the Milky Way and its Satellites* (Berlin: Springer), p. 3
Fox, A. J. *et al.* 2005, *ApJ*, 630, 332
Gouda, N. *et al.* 2008, in: W.-J. Jin, I. Platais, & M. Perryman (eds.), *A Giant Step: from Milli- to Micro-arcsecond Astrometry*, Proc. IAU Symp. No. 248 (San Francisco: ASP), p. 248
Hou, J. L., Prantzos, N., & Boissier, S. 2000, *A&A*, 362, 921
Keeney, B. A. *et al.* 2006, *ApJ*, 646, 951
Lemasle, B. *et al.* 2007, *A&A*, 467, 283
Maciel, W. L., Lago, L. G., & Costa, R. D. D. 2006, *A&A*, 453, 587
Madau, P., Ferrara, A., & Rees, M.J. 2001, *ApJ*, 555, 92
Matteucci, F. & François, P. 1989, *MNRAS*, 239, 885
Navarro, J. F. & White, S. D. M. 1994, *MNRAS*, 267, 401
Navarro, J. F., Frenk, C. S., & White, S. D. M. 1995, *MNRAS*, 275, 56
Reid, I. N., Turner, E. L., Turnbull, M. C., Mountain, M., & Valenti, J. A. 2007, *ApJ*, 665, 767
Ruffle, P. M. E. *et al.* 2007, *ApJ*, 671, 1766
Ryan-Weber, E. V., Pettini, M., & Madau, P. 2006, *MNRAS*, 371, L78
Simón-Díaz, S. 2006, astro-ph/0611513
Steinmetz, M. & Müller, E. 1995, *MNRAS*, 276, 549
Tremonti, C. A. *et al.* 2004, *ApJ*, 613, 898
Tsujimoto, T. 2007, *ApJ*, 665, L115
Veilleux, S., Cecil, G., & Bland-Hawthorn, J. 2005, *ARA&A*, 43, 769
Wyse, R. F. G. & Gilmore, G. 1995, *AJ*, 110, 2771
Yasui, C., Kobayashi, N., Tokunaga, A. T., Terada, H., & Saito, M. 2008, *ApJ*, 675, 443
Yong, D., Carney, B. W., & de Almeida, L. 2005, *AJ*, 130, 597
Yong, D., Carney, B. W., Luísa, M., de Almeida, L., & Pohl, B. L. 2006, *AJ*, 131, 2256

Session 5: Disk Galaxy Meets ΛCDM Cosmology

Session chair: Laura Portinari

Joss Bland-Hawthorn emphasising a point during his talk.

Ken Freeman and James Binney in good company during the reception at Carlsberg.

Formation and evolution of disk galaxies

Joseph Silk

Dept. of Physics, Oxford Univ., Denys Wilkinson Bldg., 1 Keble Rd., Oxford OX1 3RH UK
email: silk@astro.ox.ac.uk

Abstract. Global star formation is the key to understanding galaxy disk formation. This in turn depends on gravitational instability of disks and continuing gas accretion as well as minor merging. A key component is feedback from supernovae. Primary observational constraints on disk galaxy formation and evolution include the Schmidt-Kennicutt law, the Tully-Fisher relation and the galaxy luminosity function. I will review how theory confronts phenomenology, and discuss future prospects for refining our understanding of disk formation.

Keywords. stars: formation, galaxies: evolution, ISM, starburst

1. Introduction

There are three interrelated empirical scaling functions that relate all disk galaxies and are fundamental to understanding their formation. These are the Tully-Fisher law, the Schmidt-Kennicutt law, and the galaxy luminosity function. If we can ever understand these, we will have come a long way towards understanding disk galaxy formation. I will review some of the issues that arise when theory confronts phenomenology. Star formation and mass assembly are the primary issues to be understood. Although intimately connected with star formation, I will mostly avoid chemical evolution constraints in this discussion, in part because of their strong sensitivity to adoption of an IMF and because they are covered elsewhere in this conference.

2. Tully-Fisher relation

The Tully-Fisher relation is a long known correlation between disk luminosity and maximum rotational velocity. The principal current issues in understanding the Tully-Fisher (TF) relation involve angular momentum loss and adiabatic contraction. The I-band Tully-Fisher relation, preferred to the B-band data because of the lower scatter, provides a challenge for understanding current epoch disks. The contribution of turbulence and rotation to line widths reduces TF scatter and is measured to $z \sim 1$. No significant evolution in the TF relation is seen (Kassin et al. 2007). Even galaxies as far back as $z \sim 2$ (Bouché et al. 2007) remain on the TF law as measured locally.

Loss of angular momentum of the baryons to the dark halo was discovered in numerical simulations (Steinmetz & Navarro 1999). Strong SN feedback can reduce angular momentum loss (Governato et al. 2007). The observed normalisation can be obtained in the inner disk. This success comes at a price, however. One concern is that of the implications for the galaxy luminosity function at low luminosities. Another is disk thickness. The simulated disks are thicker than observed disks although the numerical resolution is inadequate for a real confrontation with data.

The radial structure is a more pressing issue. Lack of evolution is observed in disk sizes at $z < 1$, in contrast to the remarkable compactness found at $z \sim 2-3$ (Buitrago et al. 2008). Adiabatic contraction of the dark halo due to the self-gravity of the collapsing baryons helps account for this by modifying the effective size evolution at early epochs

(Somerville *et al.* 2008). However the response of the dark halo to baryon dissipation by adiabatic contraction means that once the rotation curves are properly corrected, the asymptotic rotation velocity fails to match the observed TF zero point (Dutton *et al.* 08). An equivalent symptom of the TF issue is that the predicted mass-to-light ratio is too high.

3. The Schmidt-Kennicutt law

The disk star formation rate is coupled to the gas surface density by the empirical relation

$$SFR = 0.02(GAS\ SURFACE\ DENSITY)/t_{dyn}.$$

Remarkably, this relation fits quiescent as well as star-bursting galaxies. The individual star-forming complexes in M51 also lie along this line (Kennicutt *et al.* 2008). This is more of a mystery and suggests that local processes control the efficiency of star formation.

We may regard the Schidt-Kennicutt (SK) law as specifying the disk mode of star formation. It is motivated globally by the gravitational instability of cold disks. Feedback from supernovae is crucial to avoid too high a star formation rate. Momentum balance yields

$$SFE = \sigma_{gas} v_{cool} m^*_{SN}/E^{SN}_{initial} \approx 0.02.$$

The global injection of momentum is balanced by local dissipation. Here m^*_{SN} is the mass in forming stars required to produce a supernova of Type II, approximately 150 M$_\odot$ for a Chabrier IMF, and $E^{SN}_{initial} \approx 10^{51}$ ergs is the initial kinetic energy in the supernova explosion.

In this way, a near steady state of cloud motions is obtained at the disk interstellar cloud velocity dispersion of $\sigma \approx 10$ km/s. This corresponds to the peak in the diffuse ISM cooling curve. The local control of momentum input and dissipation is to be found in exploring cloud disruption once massive stars form, develop HII regions and explode. The explanation is derived from the local physics of cloud efficiency which controls the efficiency of star formation, that is, the fraction of gas converted into stars within a specified time-scale. The relevant local time-scale is molecular cloud lifetime. The combination [(SFR × dynamical time) /gas density is approximately constant as a function of gas density, and amounts to about 0.02 (Krumholz & Tan 2007).

There are several unexpected developments for the Schmidt-Kennicutt (SK) law. Star-forming complexes in M51 individually lie on the SK law, albeit with increased dispersion and a slight steepening of the slope (Kennicutt *et al.* 2007). Extreme starbursts fit the SK law both in the nearby universe (Sajina *et al.* 2006) and to $z \sim 2$ for SMGs (Bouché *et al.* 2007). The SK law is dominated by the molecular content of the galaxies. This is not too surprising as stars form in molecular clouds. The total molecular mass measured via CO amounts to ~ 10 % of the HI in LMC, M33. In the MWG, the atomic and molecular gas masses are comparable, and many ultraluminous infrared-emitting galaxies are molecule-dominated (Gao & Solomon 2004). Use of molecular indicators that trace the densest gas such as HCN reveal a linear law whereas use of CO, which traces the least dense molecular gas, prefers the SK law (Komugi *et al.* 2005). The change of slope is most simply understood in terms of the excitation thresholds (Krumholz & Thompson 2007), with the dominant mass in molecular gas, traced by CO, representing the physics of the self-gravitating cold disk instabilities and cloud collisions that drive the SK law.

A more quantitative model of ISM turbulence comprises cloud collisions, with stirring by gravitational instabilities and by SNe momentum injection. Let a typical cloud have pressure p_{cl} and surface density Σ_{cl}. We expect star-forming clouds to be marginally self-gravitating, so that $p_{cl} = \pi G \Sigma_{cl}^2$. They are also marginally confined by ambient gas pressure, $p_g = \rho_g \sigma_g^2$. Clouds certainly form this way, and this state may be maintained if the cloud covering factor is of order unity. This guarantees that cloud collisions occur on a local dynamical time-scale. We have

$$p_g = \pi G \Sigma_g \Sigma_{tot} = p_{cl} = \pi G \Sigma_{cl}^2, \quad (3.1)$$

from which $\Sigma_{cl} = (\Sigma_{tot} \Sigma_g)^{1/2}$. Then the covering factor Ω_{cl} of clouds in the disk is directly inferred to be $\Omega_{cl} = \Sigma_g/\Sigma_{cl} = (\Sigma_g/\Sigma_{tot})^{1/2}$. The cloud collision time-scale is $t_{coll} = (\Sigma_{cl} H)/(\Sigma_g \sigma_g)$, where the scale height $H^{-1} = (\pi G \Sigma_{tot})/\sigma_g^2$, and σ_g is the cloud velocity dispersion. The collision time can also be expressed as $t_{coll} = \Omega_{cl}^{-1} t_{cross}$ with $t_{cross} = H/\sigma_g$, which becomes $t_{coll} = (\Sigma_{tot}/\Sigma_g)^{1/2} (H/\sigma_g)$.

Assume the disk star formation rate is self-regulated by supernova feedback which drives the cloud velocity dispersion. Momentum balance gives

$$\dot{\Sigma}_*(E_{SN}/(m_{SN}^* v_c)) = f_c \Sigma_g \sigma_g / t_{coll}. \quad (3.2)$$

Here f_c is the cloud volume filling factor, which can be expressed in terms of porosity Q as $f_c = e^{-Q}$. Also v_c is the velocity at the onset of strong cooling. Canonical numbers used throughout are $m_{SN}^* = 150 M_\odot$ and $v_c = 400 \, \mathrm{km s^{-1}}$.

We can rewrite this expression as

$$\dot{\rho}_* = \epsilon_{SN} f_c \sqrt{G \rho_g} \rho_g \quad (3.3)$$

with $\epsilon_{SN} = (m_{SN}^* v_c \sigma_g) E_{SN}^{-1} (p_g/p_{cl})^{1/2}$. For disks, we obtain

$$\dot{\Sigma}_* = f_c[(\pi G)(\Sigma_{tot})^{1/2}((m_{SN} v_c)/E_{SN})][p_g/p_{cl}]^{\frac{1}{2}} \Sigma_g^{3/2}$$
$$= \epsilon_{SN} f_c \sqrt{f_g} (R/H)^{\frac{1}{2}} \Sigma_{gas} \Omega,$$

where $\Sigma_{tot} = \Sigma_g + \Sigma_*$. Here the disk gas fraction is $f_g \sim 0.1$, and we use the disk radius-to-scale-height relation $H/R = (\sigma_g/v_r)^2$ for a disk rotating at v_r with $\Omega^2 = G\Sigma_{tot}/R$. Remarkably, although the preceding formula ignores the multi-phase nature of the interstellar medium and the possibility of gas outflows (see below), one nevertheless manages to fit the Schmidt-Kennicutt relation (Silk & Norman 2008). Mergers, which certainly drive starbursts, will increase the cloud velocity dispersion. However there is self-regulation between star formation efficiency $\epsilon_{SN} \propto \sigma$ and and disk scale-height $H \propto \sigma^2$. This accounts for the robustness of the SK law fit.

There is one important addition to the above discussion. The disk heats up dynamically unless a supply of cold gas is provided. Cold gas accretion is needed in order to maintain global disk instability and continuing star formation. Major mergers cannot provide the gas source, as they would be too destructive for thin disks. Minor mergers provide a motivated source. Observations provide strong justification for an extensive reservoir of neutral gas, as seen in the M33 group (Grossi et al. 2008). Many nearby spirals are found to be embedded within extensive envelopes of HI. In some cases, such as that of NGC 6946, there is direct evidence of angular momentum loss as the HI accretes into the star-forming inner galaxy as well as of extensive fountain activity (Boomsma et al. 2008). In all cases, the low star formation efficiency is plausibly due to SNe feedback, although this may not be the only mechanism at work. For example, magnetic pressure may also play a role in globally controlling gas flows via the Parker instability.

3.1. The rate of star formation

In reality, we find a multiphase interstellar medium which consists of hot phase ($\sim 10^6$ K gas) as well as atomic and molecular gas. The heating and cooling of this gas, as well as mass transfer between the different phases, controls the rate of star formation. A simple porosity description of supernova feedback in a multiphase ISM (Silk 2001) provides an expression for the star formation rate in which porosity-driven turbulence is the controlling factor: $\dot{\rho}_* = Q m_{SN} (4\pi/3 R_a^3 t_a)^{-1}$. The shell evolution is generally described by (Cioffi, Bertschinger and McKee 1988) $t = t_0 E_{51}^{3/14} n_g^{-4/7} (v_c/v)^{10/7}$ and $R = R_0 E_{51}^{2/7} n_g^{-3/7} (t/t_0)^{3/10}$, where $v_0 = 413$km/s, $R_0 = 14$pc and $t_0 = 1.3 \times 10^4$yr. Here $v_c = 413 E_{51}^{1/8} n_g^{1/4} \lambda^{3/8}$km s^{-1} where the cooling time-scale within a SN-driven shell moving at velocity v_c is $t_c = v_c/\lambda \rho$, $\lambda^{-1} = 3 m_p^{3/2} k^{1/2} T^{1/2}/\Lambda_{ff}$ and $\Lambda_{eff}(T)$ is the effective cooling rate ($\propto t^{-1/2}$ over the relevant temperature range).

The SNR expansion is limited by the ambient turbulent pressure to be $\rho_g \sigma_g^2$, and we identify the ambient turbulent velocity dispersion v_a with σ_g. We obtain

$$\dot{\rho}_* = Q c_0^{-1} m_{SN}^* n_g^{13/7} E_{51}^{-15/14} (\sigma_g/v_0)^{19/7}$$

where $c_0 = \frac{4\pi}{3} R_0^3 t_0$. Rearranging, we have

$$\dot{\rho}_* = Q \sqrt{G \rho_g} \rho_g (\sigma_g/\sigma_{fid})^{19/7}, \quad (3.4)$$

where

$$\sigma_{fid} = (c_0 G^{1/2} m_p^{3/2} v_0^{19/7} E_{51}^{62/49} {m_{SN}^*}^{-1})^{7/19} n_g^{-1/14}$$
$$\approx 20 n_g^{-1/14} m_{SN,100}^{-0.37} E_{51}^{0.47} \text{km/s},$$

where m_{SN}^* is normalized to 100 M$_\odot$ and E_{SN} to 10^{51} ergs. We rewrite the star formation rate as $\dot{M}_* = \epsilon_Q M_g/t_d$, where $\epsilon_Q = Q(\sigma_g/\sigma_{fid})^{2.7} f_g^{1/2}$. Now feedback physics compels to identify ϵ_Q with ϵ_{SN}. It follows that $Q \propto \sigma^{-1.7}$. Hence porosity decreases with increasing turbulence.

Simulations of star formation with SN feedback in a multiphase ISM confirm that porosity provides a good description of feedback. The analytical formula fits the numerical simulations performed at high enough resolution to follow the motions of OB stars prior to explosion (Slyz et al. 2006). The star formation rate prescription has been tested in numerical simulations of a merger. In the case of the Mice (NGC 4676 a,b), a pure density law fails to account for the extended nature of star formation. A turbulence prescription for the SK law gives a better fit to the observed distribution of young stars (Barnes 2004).

3.2. Application to starbursts

Starbursts are characterised by high turbulent velocities and strong concentrations of molecular gas. Locally, the gas density where stars form is characterised by molecular clouds. It is presumably the cloud concentration that is enhanced by the dynamics of tidal interactions, which includes both stirring by bars and mergers. Now the porosity ansatz, if porosity self-regulates, leads to an expression for the rate of star formation:

$$SFR = POROSITY \times (TURBULENT\ PRESSURE)^{1.36}.$$

We have already noted that if turbulent velocities are high, the porosity must be low. This conspiracy, together with that between gas disk scale-height and star formation efficiency parameter, helps explain why even extreme merger-induced starbursts stay on the Schmidt-Kennicutt law.

It turns out that the molecular gas fraction is regulated by the turbulent pressure (Blitz and Rosolowsky 2006).The molecular fraction is found to be approximately proportional to the pressure, so that the star formation rate empirically is

$$\Sigma_{SFR} = 0.1\epsilon\Sigma_g \left(\frac{p_{mol}}{p_0}\right)^{0.92} \text{M}_\odot\text{pc}^{-2}\text{Gyr}^{-1}.$$

Here ϵ is an empirical SK law fit parameter. The molecular fraction is $(p_{mol}/p_0)^{0.92\pm0.10}$, where $p_0/k = 4.3 \pm 0.6 \times 10^4 \text{cm}^3\text{K}$. Since $\Sigma_g \propto p_{mol}^{0.5}$, the inferred star formation rate dependence on pressure is similar to that obtained at constant porosity. Disk simulations that include a prescription for star formation in molecular clouds (Robertson and Kravtsov 2008) can reproduce the slope and dispersion of the SK law.

It is plausible that porosity should self-regulate. Low Q is associated with enhanced turbulent pressure. The supernova remnant expansion is blocked, further enhancing the pressure and squeezing the clouds. This triggers more star formation and supernovae that result in blow-out of fountains from the disk. Blow-out drives high Q. If Q is high, we obtain lower pressure and infall is allowed. The infall in turn reduces Q, and drives further accretion that eventually results in star formation and supernova remnants. This in turn enhances Q.

Another reason that Q self-regulates is that it is insensitive to the strength of the turbulence. The volume filling fraction of cold gas depends only logarithmically on the Mach number (Wada and Norman 2007).

The resulting star formation is not expected to be monotonic in a porosity-driven prescription. The time-scale for variation is of order the life-time of a molecular cloud complex to disruption by SN feedback, of order 10^7–10^8 yr. A local burst-like behaviour is indeed seen in the solar vicinity (Rocha-Pinto et al. 2000). The inner star-forming regions of nearby galaxies also reveal non-monotonic behaviour. For example, M100 has undergone a sequence of bursts (Allard et al. 2006).

Supernova-driven outflows lead to entrainment and loading of the cold ISM into the fountain and/or wind. The degree of loading depends on the ability to resolve Kelvin-Helmholtz instabilities. This is completely inadequate even in state-of-the-art simulations. Empirically, the load factor is a few. as measured in NGC 1569 (Martin et al. 2002).

The fate of the outflowing gas is uncertain. The outflow rate is proportional to Q times the star formation rate times the load factor. Since Q is suppressed as $\sigma^{-1.7}$, it follows that outflows are suppressed in massive galaxies. One still expect fountains to recirculate gas from disk to halo and back to disk, as the halo gas cools.

Star formation is observed to continue below the Toomre threshold for disk gravitational instabilities. This is naturally explained in the cloud collision model where even the outer regions are dynamically populated with orbiting disk clouds. The star formation rate is most likely modulated by the molecular gas fraction which itself is controlled by the local UV radiation field (Schaye 2004).

3.3. Some current issues in understanding the Schmidt-Kennicutt law

We still need to attain a deeper understanding of the universality of the SK law. It prevails both locally as well as globally, and at high as well as at low redshift. It applies in regions of extremely high star formation, such as extreme starbursts, and in regions of very low average star formation rate, albeit somewhat suppressed, such as the outer regions of disks and DLAs.

4. Galaxy luminosity function

The final diagnostic, and relic, of disk formation that I will discuss is the galaxy luminosity function. The case for outflows has been made in the context of the cold dark model, as a means of suppressing the numbers of dwarfs (Dekel and Silk 1986) and for driving early chemical enrichment of the intergalactic medium as inferred from the LBG "missing metals" problem (Pettini et al. 1999) and wind-driven enrichment simulations (Cen & Ostriker 1999, Oppenheimer & Dave 2006). Early numerical simulations of individual winds (Mac Low & Ferrara 1999) argued for incomplete mixing of supernova ejecta, and this problem has not yet been resolved via multiphase simulations. Suppression of winds in massive galaxy potential wells was demonstrated by Springel and Hernquist (2003). Ejection of baryons may be accomplished either in the assembly stage via dwarf winds, as is relevant here, or from massive spheroids via AGN-triggered outflows. The latter is a separate topic that itself merits an entire review.

An alternative, and complementary, means of dwarf suppression appeals to reionisation inducing mass loss from low mass dwarfs, with escape velocity $\lesssim$ 20 km/s. This has been implemented in simulations but fails to account for the shape of the observed Milky Way luminosity function of dwarf satellites within 280 kpc (Koposov et al. 2007). The observed effective power-law index is $\alpha = 0.25$ from $M_V = -2.5$ to -18. Either there is a poor fit to the observed luminosity function (Somerville 2002), or the predicted survivors have far too high a central surface brightness (Benson et al. 2002). Tidal stripping cannot account for the discrepancy (Penarrubia, Navarro & McConnachie 2008).

There is an equally serious problem for giant galaxies. The problem here is that accretion of gas continues over a Hubble time, and the galaxies continue to grow. The net result is too many and too blue massive galaxies. Supernova feedback cannot resolve these problems, but AGN/quasar feedback is considered to provide the needed panacea by heating the infalling gas and quenching star formation.The relevant feedback is that associated with supermassive black hole growth, and is destined to yield an explanation of the Magorrian relation between spheroid velocity dispersion and black hole mass.

This is not the only problem for massive galaxies and their halos: there are others. For example, cooling flows occur in galaxy clusters at a rate that exceeds the observed rate of cool gas deposition. Additional entropy input, most likely in the form of preheating, is needed to reduce their role, in order to avoid an excessive rate of star formation in the brightest cluster galaxies. AGN are again the most likely culprit. Late feeding of AGN and jet-driven outflows provide a non-localised heat source that complements the early role of quasars in spheroid formation.

CDM-motivated theory does account for at least one observational result, namely the characteristic galactic mass of the Schechter function fit, $L_* \sim 3 \times 10^{10} L_\odot$. This is based on the requirement that cooling within a dynamical time is a necessary condition for efficient star formation. The inferred mass-to-light ratio is reasonable, although one is left with the need to hide a significant fraction of the baryons.These are probably in the form of intracluster gas (Gonzalez, Zaritsky & Zabludoff 2007), within the observational errors, albeit some may be hidden at large radii.

Modelling has succeeded in giving an approximate fit to the galaxy luminosity function. A relatively significant amount of dust is needed for the massive galaxies in addition to both quasar and AGN feedback, as well as reionisation and SNe feedback for the low mass galaxies (Bower et al. 2006; Croton et al. 2006; De Lucia & Blaizot 2007). The models are tuned to fit optical data. The fitting process is challenged as new, improved data becomes available. For example, the recent NIR data (Smith et al. 2008) reveals how fragile semi-analytical modelling has become. The UKIDSS SEDs yield improved stellar mass estimates that are not particularly well fit by published models, for either low mass or massive galaxies.

5. SEEKING A UNIFIED THEORY

Feedback is essential for slowing down or even quenching star formation. One can make the case that feedback, in some circumstances, accelerates star formation. One of the most worrying issues is that of the IMF. Comparison of the UV rest frame-measured star formation history and the NIR-measured stellar mass assembly histories agrees at low redshift but diverges towards high redshift (Wilkins et al. 2008). The explanation is unlikely to be due to uncertainties in the dust extinction law at high redshift, where dust corrections are relatively small, A variation in the IMF is one possible explanation, the IMF becoming progressively top-heavy towards high redshift.

There is a long history of top-heavy IMF discussions, largely centered around the G dwarf problem. This explanation is now largely discredited, in favour of a metal-free gas accretion history. The chemical abundance ratio signatures require that the Milky Way IMF was 'nearly invariant with time, place and metallicity (Gilmore & Wyse 2004). However more recently, similar explanations invoking a top-heavy IMF have been advanced to explain the excess in the submillimetre galaxy counts (Lacey et al. 2008), the enrichment of the intracluster gas (Nagashima et al. 2005), the failure of semi-analytical modelling to reproduce the star formation efficiency enhancement towards high redshift (Davé 2008), the extragalactic light background (Fardal et al. 2007), and the luminosity and colour evolution of massive cluster galaxies (van Dokkum 2008). All of these interpretations, some of which have been given esoteric names by their proponents such as 'bottom-light' or 'paunchy', involve a partial suppression of the mass-carrying stars, below a solar mass, relative to those responsible for the bulk of the light and/or chemical yields.

The best direct evidence for a truncated or top-heavy IMF was in the Arches star cluster (Stolte et al. 2005) but this disappeared with deeper data (Kim et al. 2007). Only within the central 0.1pc of our galaxy, within the gravitational influence of the central supermassive black hole, is there a confirmed deficit of low mass stars relative to the observed massive stars (Figer 2008). This does perhaps support arguments for a top-heavy IMF in extreme situations, such as in the broad emission line regions of quasars where supersolar metallicities are observed and inferred to have been generated at $z \gtrsim 8$ (Mathias and Hamann 2008). However the gas masses involved, $\lesssim 10^4 M_\odot$ (Baldwin et al. 2003), seem to fall well short of what is needed if gas enriched by a top-heavy IMF were to have a global impact on galaxy evolution.

Rather than add new parameters to the semi-analytical box of tricks, the future is more likely to bring greatly refined numerical treatments. It is clear that we are woefully resolution-limited when it comes to the 3-dimensional mixing of multiphase media that involves energy and momentum injection in one or more of the gas phases. Our understanding of feedback is likely to change dramatically once our simulations match our aspirations. Consider the track-record: our insights into star formation, galaxy harassment and ram pressure stripping all underwent near 180 degree reversals once adequate adaptive grid power was brought to bear. I suspect that we are barely scratching the surface when it comes to understanding angular momentum transfer and AGN or even SNe feedback in a multiphase interstellar medium embedded within a protogalaxy.

Acknowledgements

I thank my collaborator Colin Norman for many discussions which have inspired much of the work described here.

References

Allard, E. *et al.* 2006, MNRAS, 371, 1087
Baldwin, J. *et al.* 2003, ApJ, 582, 590
Barnes, J. 2004, MNRAS, 350, 798
Benson, A. J., Frenk, C. S., Lacey, C. G., Baugh, C. M., Cole, S. 2002, MNRAS, 333, 177
Blitz, L. & Rosolowsky, E. 2006, ApJ, 650, 933
Boomsma, R. *et al.* 2008, A&A in press, preprint arXiv:0807.3339
Bouché, N. *et al.* 2007, ApJ, 671, 303
Bower, R. *et al.* 2006, MNRAS, 370, 645
Buitrago, F *et al.* 2008, preprint arXiv:0807.4141
Cen, R. & Ostriker, J. P. 1999, ApJ, 519, L109
Cioffi, D., McKee, C., & Bertschinger, E. 1988, ApJ, 334, 252
Croton, D. *et al.* 2006, MNRAS, 365, 11
Davé, R. 2008, MNRAS, 385, 147
Dekel, A. & Silk, J. 1986, ApJ, 303, 39
De Lucia, G. & Blaizot, J. 2007, MNRAS, 375, 2
Dutton, A., van den Bosch, F., & Courteau, S. 2008, preprint arXiv0801.1505
Fardal, M. *et al.* 2007, MNRAS, 379, 985
Figer, D., 2008, preprint arXiv0803.1619
Gao, Y. & Solomon, P. 2004 ApJ, 606, 258
Gilmore, G. & Wyse, R. 2004, arXiv:astro-ph/0411714, proceedings of the ESO/Arcetri-workshop on "Chemical Abundances and Mixing in Stars", Sep. 2004, Castiglione della Pescaia, Italy, L. Pasquini, S. Randich (eds.)
Gonzalez, A., Zaritsky, D., & Zabludoff, A. 2007, ApJ, 666, 147N U
Governato, F. *et al.* 2007, MNRAS, 374, 1479
Grossi, M. *et al.* 2008, preprint arXiv0806.0412
Kassin, S, 2007, ApJ, 660, L35
Kennicutt, R. *et al.* 2007, ApJ, 671, 333
Kim, S. *et al.* 2006, ApJ, 653, L113
Komugi, S. *et al.* 2005 PASJ, 57, 733
Koposov, S. *et al.* 2007 2007arXiv0706.2687
Krumholz. M. & Tan, J. 2007, ApJ, 654, 304
Krumholz, M. & Thompson, T.. 2006, ApJ, 669, 289
Lacey, C. *et al.* 2008, MNRAS, 385, 1155
Mac Low, M.-M. & Ferrara, A. 1999, ApJ, 513, 142
Martin, C., Kobulnicky, H., & Heckman, T. 2002, ApJ, 574, 663
Matthias, D. & Hamann, F. 2008, RevMexAA (Serie de Conferencias), 32, 65
Oppenheimer, B. D. & Dave, R. 2006, MNRAS, 373, 1265
Nagashima, M. *et al.* 2005, MNRAS, 358, 1247
Penarrubia, J., Navarro, J., & McConnachie, A. 2008, ApJ, 673, 226
Pettini, M. 1999, in proc. ESO Workshop: Chemical Evolution from Zero to High Redshift, eds. J. Walsh, & M. Rosa, astro-ph/9902173
Robertson, B. & Kravtsov, A. 2008, ApJ, 680, 1083
Rocha-Pinto, H. *et al.* 2000, A&A, 358, 869
Sajina, A. *et al.* 2006, MNRAS, 369, 939
Schaye, J. 2004, ApJ, 609, 667
Silk, J. 2001, MNRAS, 323, 313
Silk, J. & Norman, C. 2008, ApJ, submitted
Slyz, A. *et al.* 2006, MNRAS, 356, 737
Smith, A. *et al.* 2008, preprint arXiv0806.0343
Somerville, R. S. 2002, ApJ, 572, L23
Somerville, R. S. *et al.* 2008, ApJ, 672, 776
Springel, V. & Hernquist, L. 2003, MNRAS, 339, 312
Steinmetz, M. & Navarro, J. 1999, ApJ, 513, 555

Stolte, AQ. 2005ApJ...628L.113
van Dokkum, P. 2008, ApJ, 674, 29
Wada, K. & Norman, C. 2007, ApJ, 660, 276
Wilkins, S., Trentham, N., & Hopkins, A. 2008, MNRAS, 385, 687

James Binney giving his review paper.

Birgitta Nordström welcoming participants on opening day.

Helge Kragh, Bengt Gustafsson and Poul Erik Nissen exchanging anecdotes about Bengt Strömgren at Carlsberg.

Disk Sizes in a ΛCDM Universe

Qi Guo[1] and Simon White[2]

Max Planck Institut for Astrophysics,
Postbus 85741, Garching, Germany
[1] email: guoqi@mpa-garching.mpg.de
[2] email: swhite@mpa-garching.mpg.de

Abstract. We introduce a model which uses semi-analytic techniques to trace formation and evolution of galaxy disks in their cosmological context. For the first time we model the growth of gas and stellar disks separately. In contrast to previous work we follow in detail the angular momentum accumulation history through the gas cooling, merging and star formation processes. Our model successfully reproduces the stellar mass–radius distribution and gas-to-stellar disk size ratio distribution observed locally. We also investigate the dependence of clustering on galaxy size and find qualitative agreement with observation. There is still some discrepancy at small scale for less massive galaxies, indicating that our treatment of satellite galaxies needs to be improved.

Keywords. Cosmology, Gas Disk, Stellar Disk, Correlation Function

1. Introduction

Understanding the origin of galaxy disks is an important aspect of the study of galaxy formation and evolution. Almost all observed star formation occurs either in galaxy disks or in material which came from galaxy disks (e.g. in starbursts). The size of disks is closely related to their gas surface density which in turn is critical in setting the star formation rate. Galaxy sizes also correlate with many other physical properties, stellar mass, luminosity, circular velocity, etc. These relations, as well as their evolution, pose strong constraints on galaxy formation models.

A long-standing problem for cosmological simulations has been to reproduce galaxies with the proper size distribution. The high efficiency of gas cooling at early times, causes the later assembly of disks to proceed by coalescence of cold gas clumps. As these clumps merge onto the main galaxy, they lose much of their initial angular momentum through dynamical friction. The resulting disks are then substantially smaller than observed and contain relatively little stellar masss. Much effort has been directed to solving this problem, most of it invoking some form of feedback to delay collapse (Sommer-Larsen et al. 1999, Thacker and Couchman 2001), or adopting an alternate initial fluctuation spectrum with reduced small-scale power (Sommer-Larsen et al. 2001). Some authors (Governato et al. 2004) claim the problem can be significantly reduced by improving the numerical resolution, but there is no consensus yet on the true solution.

In this work we implement a new treatment of disk formation within the Munich galaxy formation model used in De Lucia & Blaizot (2007), hereafter DLB07. Our model distinguishes between gas and stellar disks and treats their growth separately. We model the accumulation of mass and angular momentum in a gas disk by conserving the specific angular momentum of the cooling gas and the cold gas component of accreted satellite galaxies. When star formation occurs in the disk, we further assume that the cold gas which is converted into stars has the typical specific angular momentum of the current gas disk. We apply our model to the Millennium Simulation and we compare the stellar mass–size relation of disks and the distribution of the ratio of stellar and gas disk sizes to observation. The dependence of galaxy clustering on galaxy size is also investigated.

2. Galaxy Formation Models

2.1. *Simulation*

The *Millennium Simulation* is one of the largest simulations of cosmic structure evolution so far carried out. It follows $N = 2160^3$ particles within a comoving box of side-length 685 Mpc from redshift $z = 127$ to $z = 0$. Each particle has a mass of $8.6 * 10^8 M_\odot$. The simulation assumes the concordance ΛCDM cosmology with parameters consistent with a combined analysis of the 2dFGRS (Colless *et al.* 2001) and the first-year WMAP data (Spergel *et al.* 2003): $\Omega_m = 0.25$, $\Omega_b = 0.045$, $h = 0.73$, $\Omega_\Lambda = 0.75$, $n = 1$, and $\sigma_8 = 0.9$, where the Hubble constant is parameterized as $H_0 = 100 h\,\mathrm{kms}^{-1}\mathrm{Mpc}^{-1}$. Galaxy formation models are then implemented on halo merger trees constructed from the stored output of this dark matter simulation. A detailed description can be found in Springel *et al.* (2005).

As the basis for our work, we use the galaxy formation model of DLB07, changing only the treatment of galaxy sizes. We refer the reader to the original papers (Croton *et al.* 2006 and DLB07) for a detailed description of this model.

2.2. *Disk size*

Here we introduce a new disk model by tracking the transfer of angular momentum between the hot gas, the cold gas and the stellar component.

We assume that the hot gas cools onto the centre with specific angular momentum which matches the current value for the dark matter halo. The change in angular momentum of the gas disk can then be expressed as

$$\delta J_g = \dot{M}_{cool} \frac{J_{DM}}{M_{DM}} \delta t \qquad (2.1)$$

where J_g is the total angular momentum of the gas disk, $\dot{M}_{cool}$ is the cooling rate, δt is the corresponding time interval, and J_{DM} and M_{DM} are the total angular momentum and total mass of the dark matter halo, respectively.

In a minor merger, when the mass ratio of the two merging galaxies is larger than 3, we assume that any cold gas in the satellite galaxy is added to the disk of the central galaxy carrying a specific angular momentum equal to the current value for the dark matter halo. The stars are added to the bulge. The angular momentum change is thus

$$\delta J_g = M_{sat,gas} \frac{J_{DM}}{M_{DM}} \qquad (2.2)$$

where $M_{sat,gas}$ is the cold gas mass of the satellite galaxy. The final angular momentum of the gas disk is a vector sum of its original angular momentum, the changes due to gas cooling and accretion from minor mergers, and the angular momentum lost to the stellar disk through star formation (as described below).

For the stellar disk, we assume the star formation rate to be proportional to gas density excess above some threshold. When cold gas is converted into stars we assume it carries the average specific angular momentum.

$$\delta J_* = \dot{M}_* \frac{J_g}{M_g} \delta t = -\delta J_g \qquad (2.3)$$

where J_* is the total angular momentum of the stellar disk, M_g is the total mass of the gas disk, $\dot{M}_*$ is the star formation rate and δt is the corresponding time interval. As for the gas disk, the angular momentum of the stellar disk is the vector sum of the original angular momentum and the change due to star formation.

We assume the cold gas and stellar disks to be thin, rotationally supported systems with exponential surface density profiles. For gas disk we have

$$\Sigma(R_g) = \Sigma_{g0} exp(-R_g/R_{gd}) \quad (2.4)$$

and for stellar disk we have

$$\Sigma(R_*) = \Sigma_{*0} exp(-R_*/R_{*d}) \quad (2.5)$$

where R_{gd} and R_{*d} are the scale-lengths and Σ_{g0} and Σ_{*0} are the central surface densities for the gas and stellar disks, respectively. Assuming the circular velocity to be constant, the galaxy to reside in an isothermal dark matter halo, and the gravity of the galaxy to be negligible, the scale-lengths can be calculated as

$$R_{g(*)d} = \frac{J_{g(*)}/M_{g(*)}}{2V_{cir}} \quad (2.6)$$

where $M_{g(*)}$ is the total mass of gas (stellar) disk and V_{cir} is the circular velocity. Here we adopt the maximum velocity of the dark matter halo as V_{cir} (Croton *et al.* 2006).

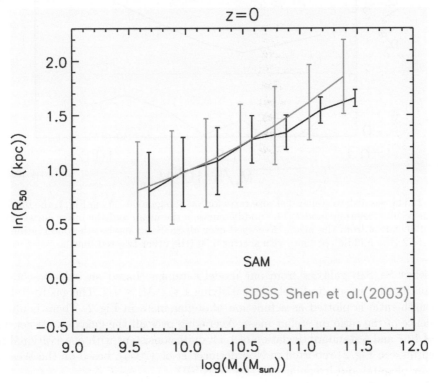

Figure 1. Stellar mass vs. half stellar mass radius relation for spiral galaxies at the local universe. The black curve and black error bars are from our model catalogue and the red curve and red error bars are from SDSS data by Shen *et al.* 2003.

3. Results

In this section we show our predictions for the relation between stellar mass and size, for the distribution of gas-to-stellar disk size ratios, and for the dependence of galaxy clustering on galaxy size, and we compare them to observation.

3.1. *Galaxy Stellar Mass and Size*

Fig. 1 shows the relation between stellar mass and the half stellar mass radius for spiral galaxies which are selected from our model catalogue using the criteria $1.5 \leqslant \triangle M_I \leqslant 2.6$ ($\triangle M_I = M_{Ibulge} - M_{Itotal}$). Our model reproduces quite well the power-law dependence of galaxy radius on stellar mass. Both the median value and the scatter match well for galaxies less massive than $\sim 10^{10.5} M_{sun}$. For more massive galaxies, the median value is still close to the observed value, but our scatter is much smaller than observed.

3.2. *Gas-to-Stellar Size Ratio*

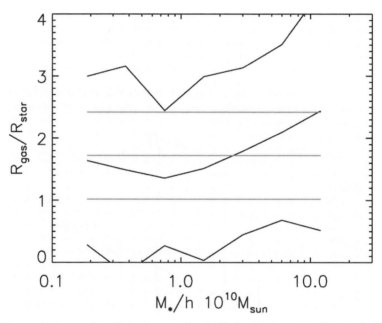

Figure 2. The gas disk to stellar disk size ratio for Sa/Sab galaxies. As in Fig. 1, the black curves represent results from our model. The middle curve is the mean and the outer curves show one standard deviation from the mean. Averaged over all Sa/Sab galaxies, the observational mean value is 1.72 (the middle red line) with scatter 0.70 (the other two red lines).

We select Sa/Sab galaxies from our model catalogue based on bulge-to-total ratio. More specifically, we select all galaxies satisfying $1 \leqslant \triangle M_I \leqslant 1.4$. The gas-to-stellar disk scale-length ratio is plotted as a function of stellar mass in Fig. 2. There is almost no dependence of the ratio on stellar mass. Averaging over all the galaxies, we get a mean value of 1.67 and a standard deviation of 1.44, quite consistent with observational results (the red lines in Fig. 2) reported in Noordermeer *et al.* (2005), based on the Westerbork HI survey of spiral and irregular galaxies (WHISP).

3.3. *Correlation Functions*

The dependence of clustering on galaxy size is shown in Fig. 3. We divide the galaxies into different mass bins and in each bin we further divide the galaxies into two populations

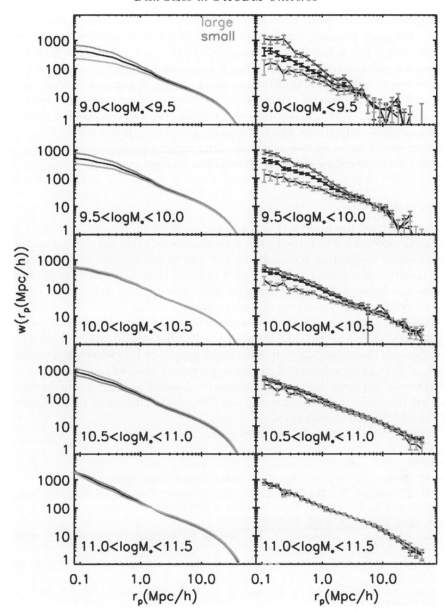

Figure 3. Projected two point correlation function as a function of stellar mass and galaxy size. The left column is for our model and the right column is for SDSS data. The galaxy stellar mass bin is indicated at the bottom left corner of each panel. The projected correlation functions are represented by black curves. In each mass bin the galaxies are further divided into large/low surface density (green) and small/high surface density (red) populations according to their half-mass radius/stellar surface mass density.

according to their sizes. On scales larger than a few Mpc, there is no dependence of clustering on size for any stellar mass. On smaller scales and for low-mass galaxies, the clustering of small galaxies is much stronger than of large ones. The size dependence becomes weaker with increasing stellar mass. For galaxies more massive than $\sim 10^{10} M_{sun}$, there is almost no dependence on galaxy size.

These differences reflect size differences between central and satellite galaxies of similar mass. For low-mass galaxies, centrals tend to be larger than satellites of similar mass. This difference is much less for more massive systems and probably reflects the facts that low-mass galaxies tend to survive much longer as satellites than massive systems. The latter merge much more quickly because of shorter dynamical friction times. The size dependence of our model galaxy is qualitatively consistent with the SDSS results taken from Li *et al.* (2006). Quantitatively, however, the dependence is stronger in the observational data and extends to more massive galaxies. A possible way to remove this discrepancy would be to include the tidal stripping of satellite galaxies in our model. This would reduce the sizes of long-lived satellites further.

4. Conclusion

We have modeled the growth of gas and stellar disks in galaxies separately using a semi-analytic approach which tracks the accumulation of their angular momentum. The differing angular momentum accumulation histories of galaxies and of dark matter halos make it possible for a galaxy to have higher or lower specific angular momentum than, and to be misaligned with, its surrounding halo. The gas and stellar disks can also be misaligned, allowing the modelling of galaxy warps.

We show that our model can reproduce many observed size-related relations including the stellar mass vs. size relation and the gas-to-stellar disk size ratio distribution. Both the mean values and the scatter are well reproduced, especially for galaxies less massive than $10^{10.5} M_{sun}$. For the dependence of clustering on galaxy size, we qualitatively reproduce observed trends, but some discrepancy remains on the small scale for less massive galaxies. Further improvement of our treatment of satellite galaxies, including tidal stripping, may help to clarify the reasons behind this.

References

Colless, M., Dalton, G., Maddox, S., & etc. 2001, *MNRAS*, 328, 1039
Croton, D. J., Springel, V., White, S. D. M., De Lucia, G., Frenk, C. S, Gao, L., Jenkins, A., Kauffmann, G., Navarro, J. F., & Yoshida, N. 2006, *MNRAS*, 365, 11
De Lucia, G. & Blaizot, J. 2007, *MNRAS*, 375, 2
Governato, F., Mayer, L., Wadsley, J., Gardner, J. P., Willman, B., Hayashi, E., Quinn, T., Stadel, J., & Lake, G *ApJ*, 607, 688
Li, C., Kauffmann, G., Jing, Y. P., White, S. D. M., Börner, G., & Cheng, F. Z. 2006, *MNRAS*, 368, 21
Noordermeer, E., van der Hulst, J. M., Sancisi, R., Swaters, R. A., & van Albada T. S. 2005, *A&A*, 442, 137
Shen, S., Mo, H. J., White, S. D. M., Blanton, M. R., Kauffmann, G., Voges, W., Brinkmann, J., & Csabai, I. 2003, *MNRAS*, 343, 978
Sommer-Larsen, J., Gelato, S., & Vedel, H. 1999, *ApJ*, 519, 501
Sommer-Larsen, J. & Dolgov, A. 2001, *ApJ*, 551, 608
Spergel, D. N., Verde, L., Peiris, H. V., & etc. 2003, *ApJS*, 148, 175
Springel, V., White, S. D. M., Jenkins, A., Frenk, C. S., & etc. 2005, *Nature*, 435, 629
Thacker, R. J. & Couchman, H. M. P. 2001, *ApJ*, 555, 17

Cold Dark Matter Substructure and Galactic Disks

Stelios Kazantzidis[1], Andrew R. Zentner[2], and James S. Bullock[3]

[1] Center for Cosmology and Astro-Particle Physics, The Ohio State University,
191 West Woodruff Avenue, Columbus, OH 43210, USA
email: stelios@mps.ohio-state.edu

[2] Department of Physics & Astronomy, University of Pittsburgh,
100 Allen Hall, 3941 O'Hara Street, Pittsburgh, PA 15260, USA
email: zentner@pitt.edu

[3] Center for Cosmology, Department of Physics & Astronomy,
The University of California at Irvine, 4168 Reines Hall, Irvine, CA 92697, USA
email: bullock@uci.edu

Abstract. We perform a set of high-resolution, fully self-consistent dissipationless N-body simulations to investigate the influence of cold dark matter (CDM) substructure on the dynamical evolution of thin galactic disks. Our method combines cosmological simulations of galaxy-sized CDM halos to derive the properties of substructure populations and controlled numerical experiments of consecutive subhalo impacts onto initially-thin, fully-formed disk galaxies. We demonstrate that close encounters between massive subhalos and galactic disks since $z \sim 1$ should be common occurrences in ΛCDM models. In contrast, extremely few satellites in present-day CDM halos are likely to have a significant impact on the disk structure. One typical host halo merger history is used to seed controlled N-body experiments of subhalo-disk encounters. As a result of these accretion events, the disk thickens considerably at all radii with the disk scale height increasing in excess of a factor of 2 in the solar neighborhood. We show that interactions with the subhalo population produce a wealth of distinctive morphological signatures in the disk stars, many of which resemble those being discovered in the Milky Way (MW), M31, and in other disk galaxies, including: conspicuous flares; bars; low-lived, ring-like features in the outskirts; and low-density, filamentary structures above the disk plane. These findings highlight the significant role of CDM substructure in setting the structure of disk galaxies and driving galaxy evolution.

Keywords. cosmology: theory, dark matter, galaxies: formation, galaxies: dynamics, galaxies: structure, methods: numerical

1. Introduction

The currently favored cold dark matter (CDM) paradigm of hierarchical structure formation (e.g., Blumenthal *et al.* 1984), predicts significant dark matter halo substructure in the form of small, dense, self-bound *subhalos* orbiting within the virialized regions of larger host halos (e.g., Klypin *et al.* 1999). Observational probes of substructure abundance thus constitute fundamental tests of the CDM model. Due to the fact that most subhalos associated with galaxy-sized host halos lack of a significant luminous component, a constraint on the amount of substructure in these systems may be obtained via their gravitational influence on galactic disks. If there is a considerable subhalo population, it may produce strong tidal effects and induce distinctive gravitational signatures which might be imprinted on the structure and kinematics of the host galactic disk. Thus, establishing the role of substructure in shaping the fine structure of galactic disks may prove fundamental in informing our ideas about global properties of galaxy formation and evolution.

Significant theoretical effort has been devoted to quantifying the resilience of galactic disks to infalling satellites (e.g., Quinn & Goodman 1986; Velazquez & White 1999; Font et al. 2001; Gauthier et al. 2006; Read et al. 2008; Villalobos & Helmi 2008). Despite their usefulness, most earlier investigations suffered basic shortcomings that limited their applicability. For example, some considered encounters of single satellites with galactic disks, a set-up which is at odds with CDM predictions of multiple, nearly contemporaneous accretion events. Other studies made ad hoc assumptions about the orbital parameters and internal structures of the infalling systems. Consequently, it remains uncertain whether these earlier investigations faithfully captured the responses of galactic disks to halo substructure in a cosmological context. Here we address this issue using a hybrid approach that combines cosmological simulations to derive the merger histories of galaxy-sized CDM halos with controlled numerical experiments of consecutive subhalo impacts onto N-body realizations of fully-formed disk galaxies.

2. Methods

A thorough description of our methods is presented in Kazantzidis et al. (2007) and we summarize them here. First, we analyze cosmological simulations of the formation of four galaxy-sized halos in the ΛCDM cosmology. The simulations were performed with the Adaptive Refinement Tree (ART) N-body code (Kravtsov 1999). All of these halos accrete only a small fraction of their final masses and experience no major mergers at $z \lesssim 1$ (a look-back time of ≈ 8 Gyr), and therefore may reasonably host a disk galaxy. Second, we identify subhalos in these hosts and select the massive substructures that pass near the center of the host halo where they may interact appreciably with a galactic disk for further consideration. Finally, we use a representative subset of these accretion events from one of the host halos to seed controlled N-body simulations of satellite impacts onto an initially-thin disk galaxy.

While the present work is informed by many past numerical investigations of satellite-disk interactions, our methodology is characterized by at least three major improvements. First, we consider satellite populations whose properties are extracted directly from the cosmological simulations of galaxy-sized CDM halos. This eliminates many assumptions regarding the internal properties and impact parameters of infalling systems inherent in many previous studies. Second, we employ primary disk galaxy models that are both self-consistent and flexible enough to permit detailed modeling of actual galaxies such as the MW and M31 by fitting to a wide range of observational data sets (Widrow & Dubinski 2005). In this work, we employ galaxy model MWb of Widrow & Dubinski (2005) which reproduces many of the observed characteristics of the MW galaxy. This galaxy model comprises an exponential stellar disk with a sech^2 scale height of $z_d = 400$ pc, a Hernquist model bulge, and an NFW dark matter halo.

Lastly, and most importantly, we incorporate for the first time a model in which the infalling subhalo populations are representative of those that impinge upon halo centers since $z \sim 1$, instead of the $z = 0$ *surviving* substructure present in a CDM halo. Previous studies utilized the *present-day* properties of a large ensemble of dark matter subhalos in order to investigate the dynamical effects of substructure on galactic disks (Font et al. 2001; Gauthier et al. 2006). Successes notwithstanding, this methodology has the drawback of eliminating from consideration those massive satellites that, prior to $z = 0$, pass very close to the central regions of their hosts, where the galactic disk resides. These systems can potentially produce strong tidal effects on the disk, but are

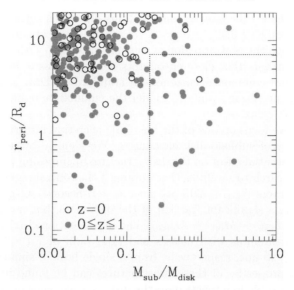

Figure 1. A scatter plot of mass versus pericentric distance for satellites identified in four galaxy-sized halos formed in the ΛCDM cosmology. Subhalo masses and orbital radii are presented in units of the mass, $M_{\rm disk} = 3.53 \times 10^{10} M_\odot$, and radial scale length, $R_d = 2.82$ kpc, respectively, of the galactic disk in the controlled simulations. *Filled* symbols show results for subhalos that pass closer than an infall radius of $r_{\rm inf} = 50$ kpc of their host halo center since a redshift $z = 1$. *Open* symbols refer to the $z = 0$ population of surviving substructures. Accretions of massive subhalos onto the central regions of their hosts, where the galactic disk resides, since $z \sim 1$ should be common occurrences in standard ΛCDM.

unlikely to constitute effective perturbers at $z = 0$ as they suffer substantial mass loss during their orbital evolution precisely because of their forays into the central halo (e.g., Zentner & Bullock 2003).

Figure 1 illustrates the importance of accounting for subhalo infall over time. This figure is a scatter plot of mass versus pericentric distance for two different satellite populations within all four galaxy-sized host CDM halos. The masses and distances in Figure 1 have been scaled to the mass, $M_{\rm disk} = 3.53 \times 10^{10} M_\odot$, and radial scale length, $R_d = 2.82$ kpc, of the stellar disk in the primary galaxy model used in the satellite-disk encounter simulations. The dotted line encloses an area in the $M_{\rm sub} - r_{\rm peri}$ plane corresponding to subhalos more massive than $0.2 M_{\rm disk}$ with pericenters of $r_{\rm peri} \lesssim 20$ kpc ($r_{\rm peri} \lesssim 7 R_d$). We refer to this area as the "danger zone". Satellites within this area are expected to constitute effective perturbers and may cause considerable damage to the disk, but we intend this as a rough criterion to aid in illustrating our point.

The first satellite population in Figure 1 consists of the $z = 0$ surviving substructures. The second subhalo population consists of systems that approach the central regions of their hosts since a redshift $z = 1$. These subhalos cross within a (scaled) infall radius of $r_{\rm inf} = 50$ kpc from the host halo center. This selection is fixed empirically to identify orbiting substructure that are likely to have a significant dynamical impact on the structure of the disk (Kazantzidis et al. 2007). The masses associated with this group of satellites are defined at the simulation output time nearest the inward crossing of $r_{\rm inf}$. Pericenters are computed from the orbit of a test particle in a static NFW potential whose properties match those of the host CDM halo at the time of $r_{\rm inf}$.

Figure 1 demonstrates that the $z = 0$ subhalo populations contain very few massive systems on potentially damaging orbits. In fact, statistics of all four galaxy-sized host halos indicate that only *one* satellite can be identified inside the danger zone in this case. On the other hand, the danger zone contains numerous substructures that passed through or near the galactic disk since $z = 1$. On average, ~ 5 satellites more massive than $0.2 M_{\rm disk}$ cross through the central region of a galaxy-sized halo with $r_{\rm peri} \lesssim 20$ kpc during this period. This suggests that close encounters between massive subhalos and galactic disks since $z = 1$ are common occurrences in standard ΛCDM. Thus, it is important to account for such accretion events to model the cumulative dynamical effects of halo substructure on disk galaxies.

In what follows, we focus on one of the host halo accretion histories to seed controlled N-body experiments of subhalo-disk encounters. We identify target satellites that are likely to have a substantial effect on the disk structure by imposing two selection criteria. First, we limit our search to satellites that approach the central region of their host with small orbital pericenters ($r_{\rm peri} \lesssim 20$ kpc) since $z = 1$. Second, we restrict re-simulation to subhalos that are a significant fraction of the disk mass, but not more massive than the disk itself ($0.2 M_{\rm disk} \lesssim M_{\rm sub} \lesssim M_{\rm disk}$). The aforementioned criteria resulted in six accretion events of satellites with masses and tidal radii of $7.4 \times 10^9 \lesssim M_{\rm sub}/M_\odot \lesssim 2 \times 10^{10}$, and $r_{\rm tid} \gtrsim 20$ kpc, respectively, from a single host to simulate over a ~ 8 Gyr period. Additional properties of these substructures can be found in Kazantzidis et al. (2007). We modeled subhalo impacts onto the disk as a sequence of encounters. Starting with the first satellite, we included subsequent systems at the epoch when they were recorded in the cosmological simulation.

We extracted the density structures of these cosmological subhalos and followed the procedure outlined in Kazantzidis et al. (2004) to construct self-consistent, N-body realizations of satellites models. Each system was represented with $N_{\rm sat} = 10^6$ particles and a gravitational softening length of $\epsilon_{\rm sat} = 150$ pc. For the primary disk galaxy, we used $N_d = 10^6$ particles to represent the disk, $N_b = 5 \times 10^5$ in the bulge, and $N_h = 2 \times 10^6$ in the dark matter halo, and softenings of $\epsilon_d = 50$ pc, $\epsilon_b = 50$ pc, and $\epsilon_h = 100$ pc, respectively. All satellite-disk encounter simulations were carried out using PKDGRAV (Stadel 2001). The "final" disk discussed in the next sections has experienced all six subhalo impacts and was evolved in isolation for ~ 4 Gyr after the last interaction. Finally, we compute all disk properties and show all visualizations of the disk morphology after centering the disk to its center of mass and rotating it to a new coordinate frame defined by the three principal axes of the total disk inertia tensor.

3. Global Disk Morphology

Figure 2 depicts the transformation of the global structure of a thin galactic disk that experiences a merging history of the kind expected in the ΛCDM paradigm of structure formation. This figure shows face-on and edge-on views of the initial and final distribution of disk stars.

Figure 2 demonstrates that encounters with CDM substructure are responsible for generating several distinctive morphological signatures in the disk. The final disk is considerably thicker (or "flared") compared to the initial distribution of disk stars and a wealth of low-density features have developed both in and above the disk plane as a consequence of these disturbances. Particularly intriguing is the fact that a high-density, in-plane structure survives after the satellite bombardment. A standard "thin-thick" disk decomposition analysis for the final disk indicates that this feature would be recognized as a thin disk component (Kazantzidis et al. 2007). The edge-on view of the final disk also reveals additional filamentary structures and other complex configurations above the disk plane. These structures bear some resemblance to tidal streams, but are in fact disk stars that have been excited by the subhalo impacts. Interestingly, the same image shows

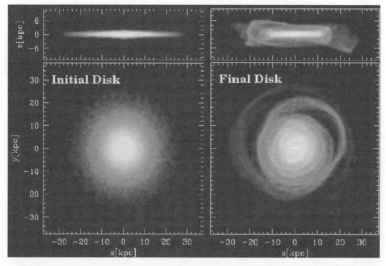

Figure 2. Density maps of disk stars illustrating the global morphological transformation of a galactic disk subject to a ΛCDM-motivated satellite accretion history. The *left* panel shows the initial disk assuming that the sequence of satellite-disk interactions initiates at $z = 1$, while the *right* panel depicts the disk after the last satellite passage, evolved in isolation for additional ~ 4 Gyr, so that the evolution of disk stars is followed from $z = 1$ to $z = 0$. The edge-on (*upper panels*) and face-on (*bottom panels*) views of the disk are displayed in each frame. Satellite-disk interactions of the kind expected in ΛCDM models produce several distinctive signatures in galactic disks including: long-lived, low-surface brightness, ring-like features in the outskirts; conspicuous flares; bars; and faint filamentary structures above the disk plane that (spuriously) resemble tidal streams in configuration space. These morphological features are similar to those being discovered in the Milky Way, M31, and in other disk galaxies.

a characteristic "X" shape in the bright central disk, a finding also reported by Gauthier *et al.* (2006). This feature is often linked to secular evolution of galaxies driven by the presence of a bar when it buckles as a result of becoming unstable to bending modes.

The face-on image of the final disk illustrates the formation of a moderately strong bar and extended ring-like features in the outskirts of the disk. The existence of these features indicate that the axisymmetry of the disk has been destroyed by the encounters with the infalling subhalos. We emphasize that the aforementioned structures are persistent, surviving for a considerable time after the satellite passages (~ 4 Gyr), and that the bar is induced in response to the accretion events, not by amplified noise.

4. Disk Thickening

Among the most striking signatures induced by the subhalo accretion events in our simulations is the pronounced increase in disk thickness. A quantitative analysis of disk thickening is presented in Figure 3. This figure shows that the initial disk thickens considerably at all radii as a result of the substructure impacts. Remarkably, the scale height of the disk near the solar radius increases in excess of a factor of 2. The outer disk is much more susceptible to damage by the infalling satellites: at $R = R_d$ the scale height grows by $\sim 50\%$ compared to approximately a factor of 3 increase at $R = 4R_d$. The larger binding energy of the inner, exponential disk and the presence of a massive bulge ($M_b \sim 0.3 M_{\rm disk}$) that acts as a sink of satellite orbital energy are responsible for the robustness of the inner disk.

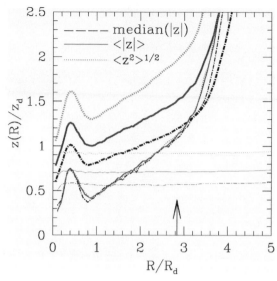

Figure 3. Disk thickening. Thickness profiles, $z(R)$, for the disk initially (*thin lines*) and after the satellite passages (*thick lines*). Lines of *intermediate* thickness show the fractional increase in disk thickness caused by the subhalo bombardment. Profiles are normalized to the initial disk scale height, z_d, and are plotted as a function of projected radius in units of the disk radial scale length, R_d. Different lines correspond to different measures of disk thickness. The arrow indicates the location of the solar radius, $R_\odot$. The galactic disk thickens considerably at all radii as a result of the encounters with CDM substructure.

Font et al. (2001) and Gauthier et al. (2006) performed similar numerical studies of the dynamical response of disks to CDM subhalos. Both investigations reported negligible tidal effects on the global structure of the disk. In contrast, Figure 3 indicates substantial disk thickening due to substructure bombardment. The primary reason for this discrepancy is that we followed the formation *history* of a host halo since $z \sim 1$, whereas Font et al. (2001) and Gauthier et al. (2006) considered the $z = 0$ population of surviving substructure present in a CDM halo.

References

Blumenthal, G. R., Faber, S. M., Primack, J. R., & Rees, M. J. 1984, *Nature*, 311, 517
Font, A. S., Navarro, J. F., Stadel, J., & Quinn, T. 2001, *Ap. Lett.*, 563, L1
Gauthier, J.-R., Dubinski, J., & Widrow, L. M. 2006, *ApJ*, 653, 1180
Kazantzidis, S., Bullock, J. S., Zentner, A. R., Kravtsov, A. V., & Moustakas, L. A. 2007, *ApJ* accepted (astro-ph/0708.1949)
Kazantzidis, S., Magorrian, J. & Moore, B. 2004, *ApJ*, 601, 37
Klypin, A., Kravtsov, A. V., Valenzuela, O., & Prada, F. 1999, *ApJ*, 522, 82
Kravtsov, A. V. 1999, PhD thesis, New Mexico State University
Quinn, P. J. & Goodman, J. 1986, *ApJ*, 309, 472
Read, J. I., Lake, G., Agertz, O., & Debattista, V. P. 2008, *MNRAS* accepted (astro-ph/0803.2714)
Stadel, J. G. 2001, Ph.D. Thesis, Univ. of Washington
Velazquez, H. & White, S. D. M. 1999, *MNRAS*, 304, 254
Villalobos, Á. & Helmi, A. 2008, *MNRAS* submitted (astro-ph/0803.2323)
Widrow, L. M. & Dubinski, J. 2005, *ApJ*, 631, 838
Zentner, A. R. & Bullock, J. S. 2003, *ApJ*, 598, 49

# The Galaxy and its stellar halo - insights from a hybrid cosmological approach

Gabriella De Lucia[1] and Amina Helmi[2]

[1] Max–Planck–Institut für Astrophysik,
Karl–Schwarzschild–Str. 1, D-85748 Garching, Germany
email: gdelucia@mpa-garching.mpg.de

[2] Kapteyn Astronomical Institute, University of Groningen,
P.O. Box 800, 9700 AV Groningen, Netherlands
email: ahelmi@astro.rug.nl

Abstract. We use a series of high-resolution N-body simulations of a 'Milky-Way' halo, coupled to semi-analytic techniques, to study the formation of our own Galaxy and of its stellar halo. Our model Milky Way galaxy is a relatively young system whose physical properties are in quite good agreement with observational determinations. In our model, the stellar halo is mainly formed from a few massive satellites accreted early on during the galaxy's lifetime. The stars in the halo do not exhibit any metallicity gradient, but higher metallicity stars are more centrally concentrated than stars with lower abundances. This is due to the fact that the most massive satellites contributing to the stellar halo are also more metal rich, and dynamical friction drags them closer to the inner regions of the host halo.

Keywords. Methods: N-body simulations, Galaxy: evolution, Galaxy: formation, Galaxy: halo

1. Introduction

Our own galaxy - the Milky Way - is a fairly large spiral galaxy consisting of four main stellar components: (1) the thin disk, that contains most of the stars with a wide range of ages and on high angular momentum orbits; (2) the thick disk, that contains about 10-20 per cent of the mass in the thin disk and whose stars are on average older and have lower metallicity than those in the thin disk; (3) the bulge, which contains old and metal rich stars on low angular momentum orbits; and (4) the stellar halo which contains only a few per cent of the total stellar mass and whose stars are old and metal poor and reside on low angular momentum orbits.

While the Milky Way is only one Galaxy, it is the one that we can study in unique detail. Over the past years, accurate measurements of ages, metallicities and kinematics have been collected for a large number of individual stars, and much larger datasets will become available in the next future thanks to a number of ongoing and planned astrometric, photometric and spectroscopic surveys. This wealth of detailed and high-quality observational data provides an important benchmark for current theories of galaxy formation and evolution.

In the following, we outline the main results of a recent study of the formation of the Milky Way and of its stellar halo in the context of a hybrid cosmological approach which combines high-resolution simulations of a 'Milky Way' halo with semi-analytic methods. We refer to De Lucia & Helmi (2008) for a more detailed description of our method and of our results.

2. The simulations and the galaxy formation model

We use the re-simulations of a 'Milky-Way' halo (the GA series) described in Stoehr et al. (2002) and Stoehr et al. (2003), with an underlying flat Λ-dominated CDM cosmological model. The candidate halo for re-simulations was selected from an intermediate-resolution simulation (particle mass $\sim 10^8$ $M_\odot$) as a relatively isolated halo which suffered its last major merger at $z > 2$. The same halo was then re-simulated at four progressively higher resolution simulations with particle mass $\sim 1.7 \times 10^8$ (GA0), $\sim 1.8 \times 10^7$ (GA1), $\sim 1.9 \times 10^6$ (GA2), and $\sim 2.1 \times 10^5$ $M_\odot$ (GA3). Simulation data were stored in 108 outputs from $z = 37.6$ to $z = 0$, and for each simulation output we constructed group catalogues (using a standard friends-of-friends algorithm) and substructure catalogues (using the SUBFIND algorithm developed by Springel et al. 2001). Substructure catalogues were then used to construct merger history trees for all self-bound haloes as described in Springel et al. (2005) and De Lucia & Blaizot (2007). Finally, these merger trees were used as input for our semi-analytic model of galaxy formation.

3. Physical properties and metallicity distributions

Fig. 1 shows the evolution of different mass components for the model Milky Way galaxies in the four simulations used in our study (lines of different colours). The histories shown in the different panels have been obtained by linking the galaxy at each time-step to the progenitor with the largest stellar mass. Fig. 1 shows that approximately half of the final mass in the dark matter halo is already in place (in the main progenitor) at $z \sim 1.2$ (panel a) while about half of the final total stellar mass is only in place at $z \sim 0.8$ (panel

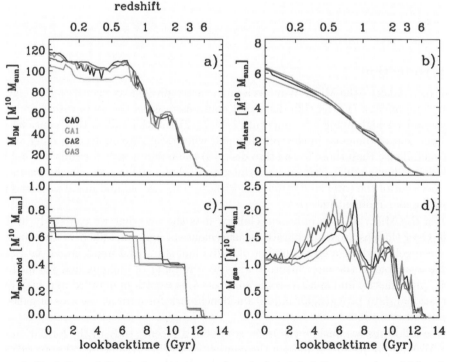

Figure 1. Evolution of the dark matter mass (panel a), total stellar mass (panel b), spheroid mass (panel c), and cold gas mass (panel d) for the model Milky Way in the four simulations used in our study (different colours).

b). About 20 per cent of this stellar mass is already in a spheroidal component (panel c). The mass of the spheroidal component grows in discrete steps as a consequence of our assumption that it grows during mergers and disk instability episodes, and approximately half of its final mass is already in a spheroidal component at $z \sim 2.5$. In contrast, the cold gas mass varies much more gradually.

Interestingly, the model produces consistent evolutions for all four simulations used in our study, despite the large increase in numerical resolution. Some panels (e.g. panel b) do not show perfect convergence, due to the lack of complete convergence in the N-body simulations (see panel a). Fig. 1 also shows that the total stellar mass of our model Galaxy (6×10^{10} M$_\odot$) is very good agreement with the estimated value $\sim 5-8 \times 10^{10}$ M$_\odot$. The mass of the spheroidal component is instead slightly lower than the observed value (assumed to be about 25 per cent of the disk stellar mass), while our fiducial model gives a gas mass which is about twice the estimated value.

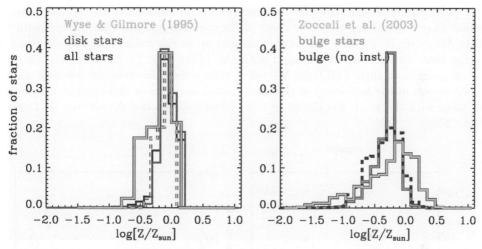

Figure 2. Metallicity distribution for stars in the disk (blue histogram in the left panel) and spheroid (red histogram in the right panel) of the model Milky Way from the highest resolution simulation in our study. The solid black histogram in the left panel shows the metallicity distribution for all stars in the model galaxy, while the dashed black histogram in the right panel shows the metallicity distribution of stars in the spheroidal component for our fiducial model if spheroid growth through disk instability is suppressed. The solid orange histograms show observational measurements by Wyse & Gilmore (1995, left panel) and Zoccali et al. (2003, right panel). The dashed orange histogram in the left panel has been obtained converting the [Fe/H] scale of the original distribution by Wyse & Gilmore into an [O/H] scale by using the observed [O/H]-[Fe/H] relation for thin disk stars by Bensby, Feltzing & Lundström (2004).

Fig. 2 shows the metallicity distributions of the stars in the disk and spheroid of our model Milky Way from the highest resolution simulation used in our study. The left panel shows the metallicity distribution of all stars (black) and of the stars in the disk (blue) compared to the observational measurements by Wyse & Gilmore (1995). The right panel shows the metallicity distribution of the spheroid stars in our fiducial model (red) and in a model where the disk instability channel is switched off (dashed black). Model results are compared to observational measurements by Zoccali et al. (2003). The metallicity distribution of disk stars in our model peaks at approximately the same value as observed, but it exhibits a deficiency of low metallicity stars. When comparing model results and observational measurements, however, two factors should be considered: (1) the observational measurements have some uncertainties (~ 0.2 dex) which tend to broaden the true underlying distributions; (2) the observational measurements provide *iron* distributions, and iron is not well described by our model that adopts an instantaneous recycling approximation. In order to show the importance of this second caveat we have converted the measured [Fe/H] into [O/H] using a linear relation, obtained by fitting data for thin disk stars from Bensby, Feltzing & Lundström (2004). The result of this conversion is shown by the dashed orange histogram in the left panel of Fig. 2. The observed [O/H] metallicity distribution is now much closer to the modelled $\log[Z/Z_\odot]$ distribution. The same caveats applies to the comparison shown in the right panel, which indicates that our model spheroid is significantly less metal rich than the observed Galactic bulge.

4. The stellar halo

In order to study the structure and metallicity distribution of the stellar halo, we assume that it builds up from the cores of the satellite galaxies that merged with the Milky Way over its lifetime. The stars that end up in the stellar halo are identified by tracing back all galaxies that merged with the Milky Way progenitor, until they are central galaxies of their own halo. We select then a fixed fraction (10 per cent for the results shown in the following) of the most bound particles of their parent haloes, and tag them with the mean metallicity of the central galaxies (for details, see De Lucia & Helmi 2008).

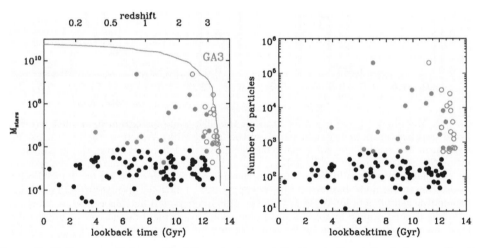

Figure 3. Stellar mass (left panel) of the galaxies contributing to the stellar halo, as a function of the lookback time of galaxy's merger. The right panel shows the number of particles associated to the dark matter haloes at the time of accretion. Red symbols correspond to objects with more than 500 particles. Open symbols correspond to red filled circles but are plotted as a function of the time of accretion.

Fig. 4 shows the stellar mass of the galaxies contributing to the stellar halo as a function of the lookback time of the galaxy's merger (left panel), and the number of particles associated to the dark matter haloes at the time of accretion (right panel). Most of the accreted galaxies lie in quite small haloes and only a handful of them are attached to relatively more massive systems, which are accreted early on during the galaxy's lifetime. These are the galaxies that contribute most to the build-up of the stellar halo. The results illustrated in Fig. 4 are in good agreement with those by Font *et al.* (2006) who combined

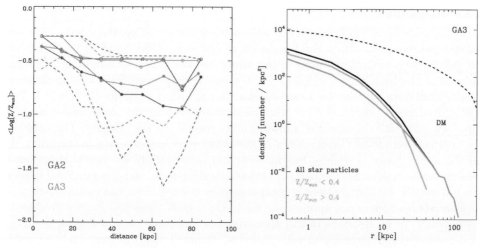

Figure 4. Left panel: Mean (filled circles) and median (empty circles) metallicity of star particles as a function of the distance from the most bound particle in the Milky Way halo, for the simulations GA2 (blue) and GA3 (red). Dashed lines correspond to the 15th and 85th percentiles of the distribution. Right panel: Projected density profile of the stellar halo (solid black line) and of the dark matter halo (dashed black line) for the simulation GA3. The solid green and orange lines show the projected density profiles for star particles with metallicity smaller and larger than $0.4\,Z_\odot$ respectively.

mass accretion histories of galaxy-size haloes with a chemical evolution model for each accreted satellites to study the formation of the stellar halo.

Fig. 4 shows the metallicity of star particles as a function of the distance from the most bound particle in the Milky Way halo for the simulations GA2 (blue) and GA3 (red). For both simulations, the mean metallicity decreases from $\mathrm{Log}[Z/Z_\odot] \sim -0.4$ at the centre to ~ -0.8 at a distance of $\sim 40\,\mathrm{kpc}$. The median and upper 85th percentile of both distributions are approximately flat around ~ -0.5. Note that the metallicity of our stellar halo is higher than what is known for the Galactic halo near the Sun. We note also that both distributions are dominated in number by star particles associated to one or a few accreted galaxies with relatively high metallicity (hence the flat behaviour of the median and upper percentile of the distribution). The lower percentile declines with increasing distance from the centre, suggesting that the inner region is largely dominated by high-metallicity stars while the contribution from lower metallicity stars becomes more important moving to the outer regions. This is shown more explicitly in the right panel of Fig. 4 which shows the projected density profile of the stellar halo (black) for the simulation GA3. The solid orange and green lines in this panel show the projected profiles of star particles with metallicity larger and smaller than $0.4\,Z_\odot$ respectively. High metallicity stars are more centrally concentrated than stars with lower abundances, suggesting that the probability of observing low-metallicity stars increases at larger distances from the Galactic centre ($\gtrsim 10-20\,\mathrm{kpc}$), where the contribution from the inner more metal-rich stars is less dominant. Interestingly, this result appears to be in qualitative agreement with recent measurements by Carollo et al. (2007).

In our model, the 'dual' nature of the stellar halo originates from a correlation between the stellar metallicity and the stellar mass of accreted galaxies. Since the most massive galaxies decay through dynamical friction to the inner regions of the halo, this is where higher metallicity stars will be found preferentially.

5. Conclusions

We have combined high-resolution resimulations of a 'Milky Way' halo with semi-analytic techniques to study the formation of our own Galaxy and of its stellar halo. The galaxy formation model used in our study has been used in a number of previous studies and has been shown to provide a reasonable agreement with a large number of observational data both in the local Universe and at higher redshifts (De Lucia & Helmi 2008 and references therein). Our study demonstrates that the same model is able to reproduce quite well the observed physical properties of our own Galaxy. The agreement is not perfect: our model Galaxy contains about twice the gas observed in the Milky Way, and the model bulge is slightly less massive and substantially less metal rich than the Galactic bulge. A detailed comparison between model results and observational measurements of metallicity distributions is complicated by the use of an instantaneous recycling approximation which is not appropriate for the iron-peak elements, mainly produced by supernovae Type Ia. Relaxing of this approximation in future work, will allow us to carry out a more detailed comparison with observed chemical compositions, and to establish similarities and differences between present-day satellites and the building blocks of the stellar halo.

Our model stellar halo is made up of very old stars (older than ~ 11 Gyr) with low metallicity, although higher than what is known for the stellar halo of our Galaxy. Most of the stars in the halo are contributed by a few relatively massive satellites accreted early on during the galaxy's lifetime. The building blocks of the stellar halo lie on a well defined mass-metallicity relation. Since the most massive galaxies are dragged closer to the inner regions of the halo by dynamical friction, this produces a stronger concentration of more metal rich stars, in qualitative agreement with recent observational measurements. The numerical resolution of our simulations is too low for the study of spatially and kinematically coherent structures in model stellar halo. Higher resolution simulations are needed for this kind of study.

References

Bensby, T., Feltzing, S., & Lundström, I. 2004, *A&A*, 415, 155
Carollo, D., Beers, T. C., Lee, Y. S., Chiba, M., Norris, J. E., Wilhelm, R., Silvarani, T., Marsteller, B., Munn, J. A., Bailer-Jones, C. A. L., Fiorentin, P. R., & York, D. G. 2007, *Nature*, 450, 1020
De Lucia, G. & Blaizot, J. 2007, *MNRAS*, 375, 2
De Lucia, G. & Helmi, A. 2008, *MNRAS* submitted, arXiv:0804.2465
Font, A. S., Johnston K. V., Bullock J. S., & Robertson, B. E. 2006, *ApJ*, 638, 585
Springel, V., White, S. D. M., Tormen, G., & Kauffmann, G. 2001, *MNRAS*, 328, 726
Springel, V., White, S. D. M., Jenkins, A., Frenk, C. S., Yoshida, N., Gao, L., Navarro, J., Thacker, R., Croton, D., Helly, J., Peacock, J. A., Cole, S., Thomas, P., Couchman, H., Evrard, A., Colberg, J., & Pearce, F. 2005, *Nature*, 435, 629
Stoehr, F., White, S. D. M., Tormen, G., & Springel, V. 2002, *MNRAS*, 335, L84
Stoehr, F., White, S. D. M., Springel, V., Tormen, G., & Yoshida, N. 2003, *MNRAS*, 345, 1313
Wyse, R. F. G. & Gilmore, G. 1995, *AJ*, 110, 2771
Zoccali, M., Renzini, A., Ortolani, S., Greggio, L., Saviane, I., Cassisi, S., Rejkuba, M., Barbuy, B., Rich, R. M., & Bica, E. 2003, *A&A*, 399, 931

Numerical simulations of galaxy evolution in cosmological context

Marie Martig[1,2], Frédéric Bournaud[1,2] and Romain Teyssier[1,2]

[1]CEA, IRFU, SAp. F-91191 Gif-sur-Yvette, France
[2]Laboratoire AIM, CNRS, CEA/DSM, Université Paris Diderot. F-91191 Gif-sur-Yvette, France

Abstract. Large volume cosmological simulations succeed in reproducing the large-scale structure of the Universe. However, they lack resolution and may not take into account all relevant physical processes to test if the detail properties of galaxies can be explained by the CDM paradigm. On the other hand, galaxy-scale simulations could resolve this in a robust way but do not usually include a realistic cosmological context.

To study galaxy evolution in cosmological context, we use a new method that consists in coupling cosmological simulations and galactic scale simulations. For this, we record merger and gas accretion histories from cosmological simulations and re-simulate at very high resolution the evolution of baryons and dark matter within the virial radius of a target galaxy. This allows us for example to better take into account gas evolution and associated star formation, to finely study the internal evolution of galaxies and their disks in a realistic cosmological context.

We aim at obtaining a statistical view on galaxy evolution from $z \simeq 2$ to 0, and we present here the first results of the study: we mainly stress the importance of taking into account gas accretion along filaments to understand galaxy evolution.

Keywords. galaxies: evolution, galaxies: interactions

1. Introduction

The morphology of galaxies in the Local Universe is well constrained by observations, but is still largely unexplained. Indeed, large volume cosmological simulations fail to reproduce realistic galaxies. For instance, the disks formed are often too concentrated : it is the "angular momentum problem", well known since the early work of Navarro & Benz (1991). It is still unclear whether this is an intrinsic problem of the ΛCDM paradigm or if something (i.e. resolution, physical processes...) is missing in these simulations.

Another puzzle is the question of disk survival till $z=0$ (Koda *et al.*, 2007). For instance, Kautsch *et al.* (2006) study a large sample of edge-on spiral galaxies in the SDSS and find that a significant fraction of them (i.e. roughly one third) are bulgeless or "superthin". This is still unexplained by cosmological models. Indeed, ΛCDM predicts that galaxy interactions are frequent (see e.g. the recent work by Stewart *et al.*, 2007). More exactly, major mergers, that are well known to destroy disks to form ellipticals (Barnes & Hernquist, 1991) are rather rare, but minor mergers are much more common. These minor mergers can thicken disks, and if frequent enough could even form elliptical galaxies (Bournaud *et al.*, 2007). The problem is then to find whether ΛCDM predicts too many mergers, or if the satellites have properties and orbital parameters such that they have little influence on the galactic disks. Also, gas accretion along filaments could fuel a thin disk and counteract the effect of mergers (Dekel & Birnboim, 2005, Keres *et al.*, 2005, Ocvirk *et al.*, 2008).

To study the properties of galaxies at low and high redshift, it thus seems necessary to take the full cosmological context into account. Large scale cosmological simulations could of course achieve this goal and give a statistical view on galaxies at each redshift, but for now they mainly lack resolution at the galactic scale. On the contrary, small volume cosmological simulations like the one performed by Naab et al. (2007) can resolve galactic scales in detail but are so time-consuming that obtaining a statistical sample is for now a challenge.

A first method to solve these problem is to use semi-analytical models, i.e. extracting merger trees from cosmological simulations and using different recipes to infer physical properties of galaxies (Somerville et al., 2001, Hatton et al., 2003, Khochfar & Burkert, 2005). The drawback is that approximations are necessary.

Another possibility has been explored by Kazantzidis et al. (2007), Read et al. (2007) and Villalobos & Helmi (2008) : they extract merger histories from cosmological simulations and re-simulate these histories at higher resolution. Nevertheless, they perform collisionless simulations with no gas component, neither in the main galaxy, nor in satellites, nor in filaments.

We here present a new approach where we re-simulate at high resolution a history given by a cosmological simulation, using self consistent realistic galaxies (the main galaxy and the satellites have a gas disk, a stellar disk and a dark matter halo), and we also take into account gas accretion from cosmic filaments. Our goal is to obtain a statistical sample of merger and accretion histories in a ΛCDM context to simulate the resulting galaxies and to compare our results to observations at various redshifts.

After a description of the technique used, we will present our first results and emphasize the importance of gas accretion along filaments to understand galaxy evolution.

2. Method

2.1. Analysis of the cosmological simulation

Merger histories and accretion data are extracted from a dark matter only cosmological simulation performed with the AMR code RAMSES (Teyssier, 2002). This simulation has an effective resolution of 512^3 and a comoving box length of 20 h^{-1} Mpc. The mass resolution is 6.9×10^6 $M_\odot$, so that a Milky Way type halo is made of a few 10^5 particles. The cosmology is set to ΛCDM with $\Omega_m = 0.3$, $\Omega_\Lambda = 0.7$, $H_0 = 70 km.s^{-1}.Mpc^{-1}$ and $\sigma_8 = 0.9$.

In this simulation, halos are detected with the HOP algorithm (Eisenstein & Hut, 1998), with $\delta_{peak} = 240$, $\delta_{saddle} = 200$ and $\delta_{outer} = 80$ (the minimal number of particles per halo is fixed to 10). In the following, we also take into account particles that do not belong to a halo, and we consider them as diffuse accretion.

The halo of which we want to build the merger and accretion history is then chosen in the final snapshot of the simulation (at $z = 0$) and is traced back to higher redshift (typically $z \simeq 2$) : we will call it the main halo. From $z \simeq 2$ to $z = 0$, each halo or particle (in the case of diffuse accretion) entering a sphere around the main halo (the radius of this sphere is the virial radius of the main halo at $z = 0$) is recorded, with its mass, position, velocity and spin (spin is of course omitted for diffuse accretion).

2.2. High resolution re-simulation

2.2.1. The PM code

The history that has been extracted from the cosmological simulation is re-simulated with a particle-mesh code (Bournaud & Combes, 2002). Gas dynamics is modeled with

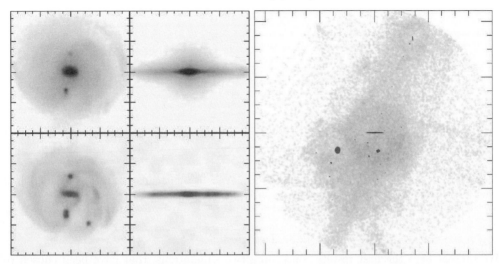

Figure 1. Left : Initial distribution of stars (top panel) and gas (bottom panel) for the main galaxy, seen face-on and edge-on (each panel is 40 kpc × 40 kpc in size). Right : large scale view of the gas distribution in a simulation box (the panel is 440 kpc × 440 kpc in size).

a sticky-particle scheme with $\beta_r = 0.8$ and $\beta_t = 0.7$, and star formation is computed according to a Kennicutt law with an exponent 1.5.

The maximum spatial resolution is 130 pc. For the two simulations shown hereafter, the mass resolution varies from 1.2×10^4 M$_\odot$ to 2.1×10^4 M$_\odot$ for gas particles, from 6×10^4 M$_\odot$ to 1.4×10^5 M$_\odot$ for stellar particles and from 1.2×10^5 M$_\odot$ to 4.4×10^5 M$_\odot$ for dark matter particles. This allows to have a total number of particles of the order of 15×10^6 at the end of both simulations.

2.2.2. Model galaxies

Each halo of the cosmological simulation (i.e. the main halo as well as all the interacting satellites) is replaced with a realistic galaxy, having a disk, a bulge and of course a dark matter halo. The total mass of the galaxy is divided in 20% of baryons and 80% of dark matter (the mass of dark matter being given by the cosmological simulation). The dark matter halo follows a Burkert profile extended to its virial radius, with a core radius chosen to follow the scaling relations given in Salucci & Burkert (2000). The disk radius of each galaxy is proportional to the square root of its mass so that the surface density is constant from one galaxy to another.

The gas fraction in the disk is 30% for galaxies that have a halo mass lower than 10^{11} M$_\odot$. For galaxies that have a greater halo mass, the gas fraction is set to 30% at high redshift (z>0.8) and 15% at low redshift.

Figure 1 (left side) shows for example the initial distribution of gas and stars in the main galaxy.

2.2.3. Diffuse accretion

Each dark matter particle that is considered as diffuse accretion in the cosmological simulation is replaced with a small blob of particles, containing in mass 20% of gas and 80% of dark matter.

The right side of figure 1 shows an example of simulation where the main galaxy (edge-on) is surrounded by accreted gas (clearly in a filament) and a few satellite galaxies.

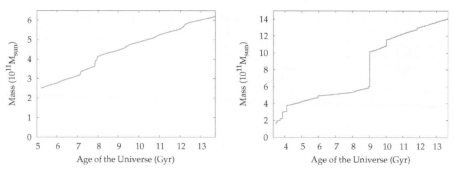

Figure 2. Evolution of the total mass of dark matter in the simulation box as a function of time for the two simulations studied here : in the left case, the mass growth is dominated by accretion, and in the right one by mergers.

2.3. Two examples

We present here the first results concerning two simulations, that have been chosen to have a mass at $z=0$ of the order of magnitude of the mass of the Milky Way. They have very different histories.

In the the first one, the mass growth of the galaxy is dominated by diffuse accretion (at a mean rate of ~ 5 $M_\odot$/yr). Only some very minor mergers take place, the most important of these mergers having a mass ratio of 12:1 (see on the left panel of figure 2 the mass evolution as a function of time). We will call this simulation *"the calm case"*.

The other simulation also contains diffuse accretion, but is mainly dominated by mergers. There is a first period of repeated minor and major mergers (mass ratios 8:1, 10:1, 3:1 and 4:1) at the very beginning of the simulation, then a calm phase and finally a major merger (mass ratio 1.5:1) at low redshift (see right panel of figure 2). We will call it *"the violent case"*.

3. Results

3.1. The calm case

The evolution of the distribution of gas and stars is shown in figure 3. Gas is smoothly accreted around the galaxy and falls onto the disk. Minor mergers are not strong and frequent enough to destroy the stellar disk. They only slightly heat it, and a thin stellar disk is rebuilt thanks to gas from diffuse accretion along the filaments. The thin disk is mainly formed from stars younger than 4 Gyr, and has a well-defined structure with two spiral arms.

3.2. The violent case

In this case, the evolution of the morphology of the galaxy is totally different (see figure 4). The disk is destroyed early by the first series of mergers. In fact, after the first of these mergers (which has a mass ratio of 8:1) the disk is already very perturbed, and the following mergers contribute to the transformation of the galaxy into an elliptical.

Nevertheless, thanks to gas accretion that takes place along a filament, a gas disk is gradually re-built into the elliptical galaxy (this would not happen if only mergers were taken into account in the simulation). New stars form in this disk, forming a young stellar disk inside the old spheroid (see figure 5), this disk being in a perpendicular plane with respect to the initial disk. Finally, the last major merger (with a mass ratio of 1.5:1) destroys this disk and the galaxy becomes elliptical again.

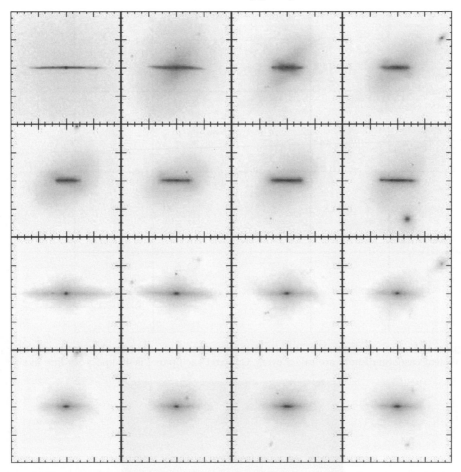

Figure 3. Evolution of the distribution of gas (top panels) and stars (bottom panels) for the calm case. Snapshots are taken every Gyr and each panel is 40 kpc × 40 kpc in size.

4. Conclusion

In order to study galaxy evolution in cosmological context, we have successfully developed a technique that allows us to perform high resolution simulations taking into account realistic merger and gas accretion histories.

The first two simulations shown here do not allow us to draw any general conclusion on galaxy evolution in a ΛCDM context. Nevertheless, we can already confirm that even low mass satellites can thicken disks and that ellipticals form both through repeated minor mergers and major mergers. We also emphasize that gas accretion from filaments can allow to rebuild a thin disk in a galaxy, which proves the absolute necessity to take this accretion into account to understand galaxy evolution.

References

Barnes, J. E. & Hernquist, L. E. 1991, ApJL, 370, L65
Bournaud, F. & Combes, F. 2002, A&A, 392, 83
Bournaud, F., Jog, C. J., & Combes, F. 2007, A&A, 476, 1179
Dekel, A. & Birnboim, Y. 2005, MNRAS, 368, 2
Eisenstein, D. J. & Hut, P. 1998, ApJ, 498, 137

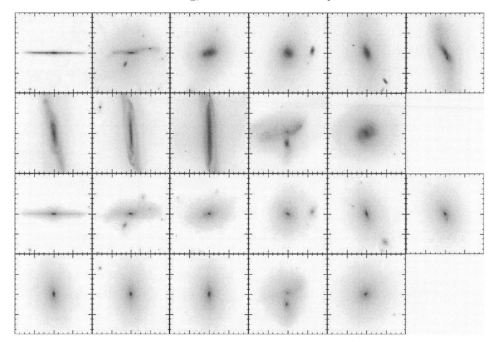

Figure 4. Evolution of the distribution of gas (top panels) and stars (bottom panels) for the violent case. Snapshots are taken every Gyr and each panel is 40 kpc × 40 kpc in size.

Figure 5. Projected stellar mass density at z = 0.2 for the violent case.

Hatton, S., Devriendt, J. E. G., Ninin, S., Bouchet, F. R., Guiderdoni, B., & Vibert, D. 2005, MNRAS, 343, 75

Kautsch, S. J., Grebel, E. K., Barazza, F. D., & Gallagher, J. S., III. 2006, A&A, 445, 765

Kazantzidis, S., Bullock, J. S., Zentner, A. R., Kravtsov, A. V., & Moustakas, L. A. 2007, preprint (arXiv:0708.1949)

Keres, D., Katz, N., Weinberg, D. H., & Dave, R. 2005, MNRAS, 363, 2

Khochfar, S. & Burkert, A. 2005, MNRAS, 359, 1379

Koda, J., Milosavljevic, M., & Shapiro, P. R. 2007, preprint (arXiv:0711.3014)

Naab, T., Johansson, P. H., Ostriker, J. P., & Efstathiou, G. 2007, ApJ, 658, 710

Navarro, J. F. & Benz, W. 1991, ApJ, 380, 320

Ocvirk, P., Pichon, C., & Teyssier, R.. 2008, preprint (arXiv:0803.4506)

Read, J. I, Lake, G., Agertz, O., & Debattista, V. P. 2008, preprint (arXiv:0803.2714)

Salucci, P. & Burkert, A. 2000, ApJL, 537, L9

Somerville, R. S., Primack, J. R., & Faber, S. M. 2001, MNRAS, 320, 504

Stewart, K. R., Bullock, J. S., Wechsler, R. H., Maller, A. H., & Zentner, A. R. 2007, preprint (arXiv:0711.5027)

Teyssier, R. 2002, A&A, 385, 337

Villalobos, A. & Helmi, A. 2008, preprint (arXiv:0803.2323)

Session 6: Surveys, Challenges and Prospects for the Future

Session chairs: James Binney, Eva Grebel and Joss Bland-Hawthorn

The Saturday group excursion on the ferry on the way to Hven. In front: Daniel Christlein, Michel Grenon, and Marie Martig; in the background, David Latham and Birgitta Nordström.

Visiting the old church of Skt. Ibb on Hven, Tycho Brahe's parish church.

The Challenge of Modelling Galactic Disks

Andreas Burkert[1]

[1] University Observatory, University of Munich,
Scheinerstr. 1, D-81679, Munich, Germany
email: burkert@usm.lmu.de

Abstract. Detailed models of galactic disk formation and evolution require knowledge about the initial conditions under which disk galaxies form, the boundary conditions that affect their secular evolution and the microphysical processes that drive the multi-phase interstellar medium and regulate star formation. Unfortunately, up to now, most of these ingredients are still poorly understood. The challenge therefore is to still construct realistic models of galactic disks with predictive power.

Keywords. Galaxy: disk, galaxies: formation, galaxies: evolution, ISM: general, ISM: clouds, hydrodynamics, turbulence

1. Initial and Boundary Conditions: The Cosmological Angular Momentum Problem

The radial surface density profiles of galactic disks are determined by the gravitational potential of the galaxy that is dominated in the outer parts by dark matter and the specific angular momentum distribution of the infalling gas that dissipates its potential and kinetic energy while settling into centrifugal equilibrium in the inner regions of a dark matter halo. In addition one has to consider the secular evolution of galactic disks. Viscous angular momentum redistribution and selective gas loss in galactic winds strongly affects the evolution of disks making it difficult to infer the initial conditions from their presently observed structure.

Angular momentum is an important ingredient in order for galactic disks to form. It is generally assumed that, before collapse, gas and dark matter are well mixed and therefore acquire a similar specific angular momentum distribution (Peebles 1969; Fall & Efstathiou 1980; White 1984). If angular momentum would be conserved during gas infall, the resulting disk size should be directly related to the specific angular momentum λ' of the surrounding dark halo where (Bullock *et al.* 2001)

$$\lambda' = \frac{J}{\sqrt{2} M_{vir} V_{vir} R_{vir}} \quad (1.1)$$

with R_{vir} and $V_{vir}^2 = G M_{vir}/R_{vir}$ the virial radius and virial velocity of the halo, respectively, and M_{vir} its virial mass. Adopting a flat rotation curve, the disk scale length is (Mo *et al.* 1998; Burkert & D'Onghia 04)

$$R_{disk} \approx 8 \left(\frac{\lambda'}{0.035} \right) \left(\frac{v_{max}}{200 km/s} \right) kpc \quad (1.2)$$

where v_{max} is the maximum rotational velocity in the disk.

Figure 1 shows the correlation between the disk scale length R_{disk} and the maximum rotational velocity v_{max} for massive spiral galaxies (Courteau 1997) which is consistent with a mean value of $\lambda' \approx 0.025$. This value is somewhat smaller than the theoretically

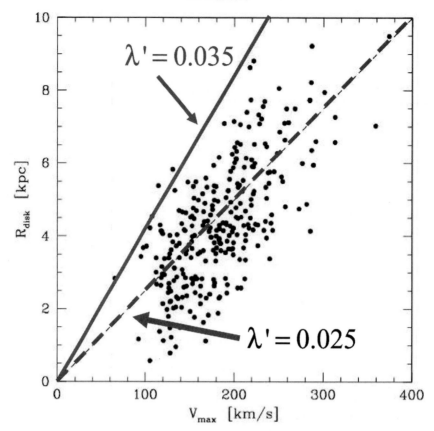

Figure 1. The observed scale lengths versus the maximum rotational velocities of galactic disks are shown for the Courteau (1997) sample. The sold line shows the theoretically predicted correlation for $\lambda' = 0.035$. The dashed curve corresponds to $\lambda' = 0.025$.

predicted value of $\lambda' = 0.035$, indicating that the gas could on average have lost some amount of angular momentum during the phase of decoupling from the dark component and settling into the equatorial plane.

This result is promising. The situation is however far less satisfactory when we consider more detailed numerical models of gas infall and accumulation in galactic disks. Many simulations of galactic disk formation suffer from catastrophic angular momentum loss which leads to disks with unreasonably small scale lengths and surface densities that are too large. The origin of this problem has been attributed to a strong clumping of the infalling gas which looses angular momentum by dynamical friction within the surrounding dark matter halo (Navarro & Benz 1991, Navarro & Steinmetz 2000). Other possibilites are low numerical resolution (Governato et al. 2004, 2007), the effect of substantial and major mergers (d'Onghia et al. 2006) or artificial secular angular momentum transfer from the cold disk to its hot surrounding (Okamoto et al. 2003). Over the years many groups have tried to solve this problem by including star formation and energetic feedback (e.g. Sommer-Larsen et al. 2003, Abadi et al. 2003, Springel & Hernquist 2003, Robertson et al. 2004, Oppenheimer & Dave 2006, Dubois & Teyssier 2008). The results are however confusing. First of all the origin of the angular momentum problem is not clearly understood. Secondly, no reasonable, universally applicable feedback prescription has been found that would lead to the formation of large-sized, late-type disks, not only for one special case, but in general.

Progress in our understanding of the cosmological angular momentum problem has recently been achieved by Zavala et al. (2008) who confirmed that the specific angular momentum distribution of the disk forming material follows closely the angular momentum evolution of the dark matter halo. The dark matter angular momentum grows at early times as a result of large-scale tidal torques, consistent with the prediction of linear theory and remains constant after the epoch of maximum expansion. During this late phase however angular momentum is redistributed within the dark halo with the inner dark halo regions loosing up to 90% of their specific angular momentum to the outer parts. The process leading to this angular momentum redistribution is not discussed in details. It is however likely that substantial mergers with mass ratios less than 10:1 that are expected to occur frequently even at late phases during galaxy formation perturb the halo and generate global asymmetries in the mass distribution that are known to be an efficient mechanism for angular momentum transfer. Small satellite infall is probably of minor importance. It would be interesting to study the role of major and minor mergers in this process in greater details.

It is then likely that any gas residing in the inner regions during such an angular momentum redistribution phase will also loose most of its angular momentum, independent of whether the gas resides already in a protodisk, is still confined to dark matter substructures or is in an extended, diffuse distribution. Zavala et al. (2008) (see also Okamoto et al., 2005 and Scannapieco et al., 2008) show that efficient heating of the gas component can prevent angular momentum loss, probably because most of the gaseous component resides in the outer parts of the dark halo during its angular momentum redistribution phase. The gas would then actually gain angular momentum rather than loose it and could lateron settle smoothly into an extended galactic disk in an ELS-like (Eggen, Lynden-Bell & Sandage 1962) accretion phase.

Unfortunately little is known about the energetic processes that could lead to such an evolution. Obviously, star formation must be delayed during the protogalactic collapse phase in order for the gas to have enough time to settle into the plane before condensing into stars. However star formation is also required in order to heat the gas and preventing it from collapsing prior to the angular momentum redistribution phase. Scannapieco et al. (2008) show that their supernova feedback prescription is able to regulate star formation while at the same time pressurizing the gas. Their models are however still not efficient enough in order to produce disk-dominated, late-type galaxies. Large galactic disks are formed. The systems are however dominated by a central, massive, low-angular momentum stellar bulge component. This is in contradiction with observations which indicate a large fraction of massive disk galaxies with bulge-to disk ratios smaller than 50% (Weinzirl et al. 2008) that cannot be produced currently by numerical simulations of cosmological disk formation.

2. Energetic Feedback and Star Formation

As discussed in the previous section, star formation and energetic feedback plays a dominant role in understanding the origin and evolution of galactic disks and in determining the morphological type of disk galaxies. Scannapieco et al. (2008) for example demonstrate that the same initial conditions could produce either an elliptical or a disk galaxy, depending on the adopted efficiency of gas heating during the protogalactic collapse phase. We do not yet have a consistent model of the structure and evolution of the multi-phase, turbulent interstellar medium and its condensation into stars. This situation is now improving rapidly due to more sophisticated numerical methods and fast computational platforms that allow us to run high-resolution models, incorporating a large number of possibly relevant physical processes (Wada & Norman 2002, Krumholz & McKee 2005, Tasker & Bryan 2008, Robertson & Kravtsov 2008). Most cosmological simulations however have up to now adopted simplified observationally motivated descriptions of star formation that are based on the empirical Kennicutt relations (Kennicutt 1998, 2007) that come in two different versions. The first relation (K1) represents a correlation between the star formation rate per surface area Σ_{SFR} and the gas surface density Σ_g, averaged over the whole galaxy

$$\Sigma_{SFR}^{(K1)} = 2.5 \times 10^{-4} \left(\frac{\Sigma_g}{M_\odot/pc^2}\right)^{1.4} \frac{M_\odot}{kpc^2 \; yr} \qquad (2.1)$$

The second relation (K2) includes a dependence on the typical orbital period τ_{orb} of the disk

$$\Sigma_{SFR}^{(K2)} = 0.017 \left(\frac{\Sigma_g}{M_\odot/pc^2}\right)\left(\frac{10^8 yrs}{\tau_{orb}}\right) \frac{M_\odot}{kpc^2 \; yr}. \qquad (2.2)$$

These relationships have been derived from observations as an average over the whole disk. They are however often also used as theoretical prescriptions for the local star formation rate which appears observationally justified if the total gas surface densitiy Σ_g is replaced by the local surface density of molecular gas. The origin of both relationships is not well understood yet. For example, Li et al. (2005, 2006) ran SPH simulations of a gravitationally unstable gaseous disks, confined by the gravitational potential of a surrounding dark matter halo. Gravitationally bound gas clumps form in their disks and are replaced by accreting sink particles. The authors assume that 30% of the mass of these particles is in stars with the rest remaining gaseous. However, no stellar feedback or a destruction mechanism of the partly gaseous sink particles is adopted. The star formation surface density is investigated for different galactic disk models with different rotational velocities and initial gas surface densities. The authors find a good agreement with the first Kennicutt relation (K1) if they correlate Σ_{SFR} with Σ_g at a time when the star formation rate has decreased by a factor of 2.7 with respect to the initial value which in their model typically corresponds to an evolutionary time of a few 10^7 yrs. The significance of this result is however not clear. Obviously, the galaxies studied by Kennicutt are much older and in a phase of self-regulated star formation that cannot be considered in models without energetic feedback. In addition, the authors cannot reproduce the second relation (K2), indicating that K2 is not directly related to K1 but instead represents a second constraint for theoretical models.

We can combine K1 and K2 and derive a relationship between the average gas density in galactic disks and their orbital period

$$\Sigma_g \sim \tau_{orb}^{-2.5} \sim \left(\frac{v_{rot}}{R_{disk}}\right)^{2.5} \qquad (2.3)$$

where v_{rot} and R_{disk} are the rotational velocity and the size of the galactic disk, respectively. This result is puzzling as it is not clear why the kinematical properties of galactic disks should correlate with their gas surface densities especially in galaxies of Milky Way type or earlier where the gas fraction is small compared to the mass in stars. Recent detailed hydrodynamical simulations of disk galaxies by Robertson & Kravtsov (2008), including low-temperature gas cooling and molecular hydrogen physics can indeed reproduce both Kennicutt relations. The authors however note themselves that the physical reason for the origin of the K2-relation in their simulations is unclear. They argue that in disk galaxies with exponential density profiles the disk surface density should scale with

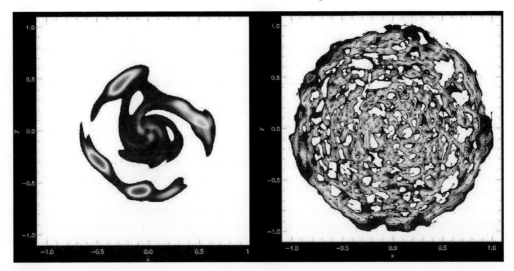

Figure 2. Gas surface density of an initially homogeneous, gravitationally unstable gas-rich galactic disk, embedded in a dark matter halo. The left panel shows the gas density distribution if star formation is suppressed. The disk forms a few massive gaseous clumps that spiral into the center by dynamical friction. The situation is however very different if star formation and stellar energy feedback is included. In this case, supernova explosions efficiently disrupt dense clumps before they can merge into giant cloud complexes while at the same time generating a highly turbulent and filamentary multi-phase interstellar medium (Burkert et al. 2009).

the orbital period as $\Sigma_d \sim \tau_{orb}^{-2}$. In this case, K2 requires that $\Sigma_g \sim \Sigma_d^{1.2} \sim (\Sigma_* + \Sigma_g)^{1.2}$ with Σ_* the stellar surface density. It is not clear why this relation should hold, especially for disks with $\Sigma_* > \Sigma_g$.

3. Secular Evolution and Turbulence in Galactic Disks

Bullock et al. (2001) demonstrated that dark halos have a universal angular momentum distribution that should also be characteristic for the infalling gas component. Van den Bosch et al. (2001) lateron showed that this angular momentum distribution is not consistent with the observed distribution of exponential galactic disks indicating that viscous angular momentum redistribution in galactic disks must have played an important role. The viscosity is likely driven by interstellar turbulence which is a result of stellar energetic feedback processes (see Fig. 2) or global disk instabilities (magneto-rotational instability or gravitational instability). Note, that viscous effects will make the angular momentum problem worse as viscosity in general removes angular momentum from the dominant mass component in the disk.

The viscous formation of exponential stellar disks from gas disks with various different surface density distributions has been studied e.g. by Slyz et al. (2002). Their numerical simulations show that exponential disks form if the star formation timescale is of order the viscous timescale. Genzel et al. (2008) derived a timescale for turbulent viscosity in galactic disks of

$$\tau_{visc} = \frac{1}{\alpha} \left(\frac{v_{rot}}{\sigma}\right)^2 \tau_{orb} \qquad (3.1)$$

where α is of order unity. $\tau_{visc} \approx 10^{10}$ yrs for disks like the Milky Way with $\sigma \approx 10$-20 km/s and self-regulated low star formation rates. Hα integral field spectroscopy has however detected $z \sim 2$ star forming disk galaxies with large random gas motions of order 40 km/s to 60 km/s and viscous timescales of less than 10^9 yrs (Genzel et al., 2006, 2008, Förster-Schreiber et al. 2006). Interestingly, for these objects, the star formation timescales are again similar to the viscous timescales, leading to star formation rates of 100 M$_\odot$/yr and confirming that galactic disk gas turbulence, star formation and secular evolution are intimately coupled. The origin of the clumpiness and high turbulence in redshift 2 disks is not well understood yet. It seems likely that it is a result of substantial filamentary gas inflow (Dekel et al. 2008), combined with gravitational instabilites in the disk (Bournaud et al. 2007).

Turbulence seems to regulate star formation not only on large galactic scales but also on local cloud scales. Most of the molecular gas in the Milky Way is found in giant molecular clouds with masses of order 10^4–$10^6 M_\odot$, temperatures of order 10 K and average densities of order 100 cm^{-3}. As their Jeans mass is of order 20 M$_\odot$ which is much smaller than their total mass one would expect that molecular clouds should collapse and condense into stars on a local free-fall time which is of order 5×10^6 yrs. Adopting a total molecular mass of $M_{H_2} \approx 3 \times 10^9 M_\odot$ and assuming that a fraction $\eta_{SF} \approx 0.1$ of the molecular cloud's mass forms stars, the inferred mean star formation rate in the Milky Way is

$$SFR = \eta_{SF} \frac{M_{H_2}}{\tau_{ff}} \approx 60 M_\odot/yr \qquad (3.2)$$

which is an order of magnitude larger than observed. A possible solution of this problem is turbulence. Molecular clouds are observed to be driven and shaped by supersonic turbulence that might strongly affect their stability and star formation rate. The origin of this turbulent motion and its impact on the cloud's lifetime and star formation process is not well understood yet. It is however likely that large-scale disk turbulence is the seed for turbulence in molecular clouds which again affects the star formation rate that in turn drives again large scale disk turbulence and by this also the viscous secular evolution of galactic disks.

4. Summary

We are currently living in a very exciting time where the various complex processes that can affect galactic disk formation and evolution are being uncovered and studied observationally and theoretically. Combined with the now well established cold dark matter structure formation scenario the time seems ripe for self-consistent models of galaxy formation with predictive power. Given the high capacities of present-day supercomputers it is understandable that one tries to including as many processes as possible, most of which being however not well understood. These models not only suffer from a large number of free parameters. They also do not necessarily lead to insight as they are so complex and depend on so many different implemented physical aspects that it is impossible to clearly understand what in the end the origin of a certain result will be.

I wonder whether one needs a high complexity in order to understand important questions of galactic disk evolution, like the two KS laws, the origin of turbulence in the diffuse interstellar medium or in molecular clouds or the origin of the strong correlation between the viscous timescale and the star formation timescale.

Let us try to solve simple questions first before we focus on the complex puzzles that involve many processes that are not well understood yet.

References

Abadi, M. G., Navarro, J. F., Steinmetz, M., & Eke, V. R. 2003, *ApJ*, 591, 499
Bournaud, F., Elmegreen, B. G., & Elmegreen, D. M. 2007, *ApJ*, 670, 237
Bullock, J. S., Dekel, A., Kolatt, T. S., Kravtsov, A. V., Klypin, A. A., Porciani, C., & Primack, J.R. 2001, *ApJ*, 555, 240
Burkert, A. & d'Onghia, E. 2004, in *Penetrating Bars through Masks of Cosmic Dust: The Hubble Tuning Fork Strikes a New Note*, eds. D.L. Block, I. Puerari, K.C. Freeman, R. Groess and E.K. Block, ASSL 319, 341
Courteau, S. 1997, *AJ*, 114, 2402
Dekel, A. et al. 2008, *Nature, in press*, arXiv:0808.0553
Dubois, Y. & Teyssier, R. 2008, *AA*, 477, 79
D'Onghia, E., Burkert, A., Murante, G., & Khochfar, S. 2006, *MNRAS*, 372, 1525
Eggen, O. J., Lynden-Bell, D., & Sandage, A. R. 1962, *ApJ*, 136, 748
Fall, S. M. & Efsthatiou, G. 1980, *MNRAS*, 193, 189
Förster-Schreiber, N. M. et al. 2006, *ApJ*, 645, 1062
Genzel, R. et al. 2006, *Nature*, 442, 786
Genzel, R. et al. 2008, *Apj, submitted*, arXiv:0807.1184
Governato, F. et al. 2004, *ApJ*, 607, 688
Governato, F. et al. 2007, *MNRAS*, 374, 1479
Kennicutt, R. C., Jr. 1998, *ApJ*, 498, 541
Kennicutt, R. C., Jr. et al. 2007, *ApJ*, 671, 333
Krumholz, M. R. & McKee, C. F. 2005, *ApJ*, 630, 250
Li, Y., Mac Low, M.-M., & Klessen, R. S. 2005, *ApJ*, 620, L19
Li, Y., Mac Low, M.-M., & Klessen, R. S. 2005, *ApJ*, 639, 879
Mo, H. J., Mao, S., & White, S. D. M. 1998, *MNRAS*, 295, 319
Navarro, J. & Benz, W. 1991, *ApJ*, 380, 320
Navarro, J. & Steinmetz, M. 2000, *ApJ*, 538, 477
Okamoto, T., Jenkins, A., Eke, V. R., Quilis, V., & Frenk, C. S. 2003, *MNRAS*, 345, 429
Okamoto, T., Eke, V. R., Frenk, C. S., & Jenkins, A. 2005, *MNRAS*, 363, 1299
Oppenheimer, B. D. & Dave, R. 2006, *MNRAS*, 373, 1265
Peebles, P. J. E. 1969, *ApJ*, 155, 393
Robertson, B., Yoshida, N., Springel, V., & Hernquist, L. 2004, *ApJ*, 606, 32
Robertson, B. & Kravtsov, A. V. 2008, *ApJ*, 680, 1083
Scannapieco, C., Tissera, P. B., White, S. D. M., & Springel, V. 2008, *MNRAS*, 389, 1137
Slyz, A. D., Devriendt, J. E. G., Silk, J., & Burkert,A. 2002, *MNRAS*, 333, 894
Sommer-Larsen, J., Götz, M., & Portinari, L. 2003, *ApJ*, 596, 47
Springel, V. & Hernquist, L. 2003, *MNRAS*, 339, 289
Tasker, E. J. & Bryan, G. L. 2008, *ApJ*, 673, 810
Van den Bosch, F. C., Burkert, A., & Swaters, R. A. 2001, *MNRAS*, 326, 1205
Wada, K. & Norman, C. A. 2007, *ApJ*, 660, 276
Weinzirl, T., Jogee, S., Khochfar, S., Burkert, A., & Kormendy, J. 2008, *ApJ*, submitted (arXiv:0807.0040)
White, S. D. M. 1984, *MNRAS*, 286, 38
Zavala, J., Okamoto, T., & Frenk, C. S. 2008, *MNRAS*, 387, 839

Starting the meeting: The welcome reception at Copenhagen University. Frank Grundahl, Ken Freeman and Sofia Feltzing in the foreground.

– and ending it: The Saturday excursion on the way to Uranienborg in one of the characteristic tractor-powered buses on Hven.

Hydrodynamical Adaptive Mesh Refinement Simulations of Disk Galaxies

Brad K. Gibson[1]†, Stéphanie Courty[1], Patricia Sánchez-Blázquez[1], Romain Teyssier[2], Elisa L. House[1], Chris B. Brook[1], and Daisuke Kawata[3]

[1] Centre for Astrophysics, University of Central Lancashire, Preston, PR1 2HE, UK
email: bkgibson@uclan.ac.uk

[2] Service d'Astrophysique, CEA Saclay, Batiment 709, 91191 Gif sur Yvette, France

[3] Carnegie Observatories, 813 Santa Barbara St., Pasadena, CA, 91101, USA

Abstract. To date, fully cosmological hydrodynamic disk simulations to redshift zero have only been undertaken with particle-based codes, such as GADGET, Gasoline, or GCD+. In light of the (supposed) limitations of traditional implementations of smoothed particle hydrodynamics (SPH), or at the very least, their respective idiosyncrasies, it is important to explore complementary approaches to the SPH paradigm to galaxy formation. We present the first high-resolution cosmological disk simulations to redshift zero using an adaptive mesh refinement (AMR)-based hydrodynamical code, in this case, RAMSES. We analyse the temporal and spatial evolution of the simulated stellar disks' vertical heating, velocity ellipsoids, stellar populations, vertical and radial abundance gradients (gas and stars), assembly/infall histories, warps/lopsideness, disk edges/truncations (gas and stars), ISM physics implementations, and compare and contrast these properties with our sample of cosmological SPH disks, generated with GCD+. These preliminary results are the first in our long-term Galactic Archaeology Simulation program.

Keywords. galaxies: formation, galaxies: evolution, methods: n-body simulations

1. Introduction

The ability to form and evolve (correctly!) a disk galaxy with the aid of massively parallel computers and optimised algorithms remains an elusive challenge for astrophysicists. Ameliorating the non-physical effects associated with overcooling, overmerging, angular momentum loss, and the capture of accurate phenomenological prescriptions for the sub-grid physics governing galaxy evolution (star formation, feedback, etc.) has been achieved through rapid advancements in both hardware and software algorithms, but their complete elimination has yet to be realised.

Fully self-consistent cosmological hydrodynamic simulations of Milky Way-like disk galaxies, with sufficient resolution ($\lesssim$500 pc) to decompose and analyse various galactic sub-components (eg. halo, bulge, and thin + thick disks) have only really appeared over the past $\sim$5 years (Sommer-Larsen *et al.* 2003; Abadi *et al.* 2003; Governato *et al.* 2004, 2007; Robertson *et al.* 2004; Bailin *et al.* 2005; Okamoto *et al.* 2005).

A common thread linking these studies is the use of a particle-based approach to representing and solving the equations of hydrodynamics - usually through the use of a smoothed particle hydrodynamics (SPH) code, such as GADGET, Gasoline, or GCD+. Where there is no disputing the impact that SPH has had on the field, it is important to be aware of both the strengths *and* weaknesses of any specific approach - as O'Shea *et al.* (2005) and Agertz *et al.* (2007) have shown, both subtle *and* overt differences can be

† http://www.uclan.ac.uk/~bkg/

introduced when employing a particle-based, as opposed to a mesh-based (or grid-based) approach (and *vice versa*), when simulating galaxy formation and evolution.

To address these concerns, we have initiated a long-term Galactic Archaeology Simulation programme aimed at complementing the aforementioned particle-based studies (including our own) with a comprehensive suite of simulations generated with a grid-based N-body + hydrodynamical code employing adaptive mesh refinement (AMR) - our software tool of choice has been RAMSES (Teyssier 2002). *These simulations represent (to our knowledge) the first to be generated through to redshift zero, with a grid code, within a fully cosmological and hydrodynamic framework.*[†]

In this contribution, we provide a brief summary of the methodology adopted, and highlight several *preliminary* results associated with our analyses of the simulations' disk kinematics, chemistry, disk edges / truncations, and assembly / infall histories.

2. Methodology

From a parent 20 h^{-1} Mpc ΛCDM collisionless particle simulation, several representative $\sim 5-8 \times 10^{11}$ $M_\odot$ halos were identified for higher-resolution re-simulation with the full baryonic physics capabilities of RAMSES. Unlike previous studies, we placed essentially no *a priori* restrictions during the halo selection process - ie, we did *not* purposefully select isolated, median-spin halos, with particularly quiescent assembly histories, in an attempt to bias the selection towards a "Milky Way-like" halo.

The parent dark matter simulation was re-centred on the halo of interest with now three nested areas of different mass resolution. Only the central 512^3 coarse grid was then refined, up to 7 additional levels, with the full suite of baryonic physics included (eg. star formation, blast-wave supernovae feedback parametrisation, chemical enrichment, UV background, metal-dependent cooling, etc.), resulting in a formal spatial (baryonic mass) resolution of 435 pc (10^6 $M_\odot$) at $z=0$.[‡] The resolution is roughly a factor of two better than that employed in our earlier SPH work (Brook *et al.* 2004; Bailin *et al.* 2005).

3. Basic Characteristics

Our first two $\sim L^*$ disks (Ramses1 and Ramses2, respectively) both ended as fast-rotating massive (7.6×10^{11} $M_\odot$ and 5.5×10^{11} $M_\odot$, respectively) galaxies in low-spin ($\lambda=0.02$) halos. Bandpass-dependent bulge-to-disk (B/D) decompositions show, not surprisingly, B/D ratios in the range of $\sim 0.4-0.8$, signatures of the same overcooling / overcentralisation "problems" which plague traditional SPH disk simulations. The stellar bulge has a $V/\sigma \sim 0.5$, reflecting its ~ 70 km/s rotation.[¶] That said, the simulated I-band images (edge-on and face-on) for Ramses1 (middle and right panels of Fig 1) are more than encouraging. The left-most panel of Fig 1 shows the gas density distribution (different projection); the tell-tale warp can be traced to a neighbouring satellite.

The overall star formation histories for the two RAMSES disks are not dissimilar to those associated with our two GCD+ SPH disks - GCD1 (Bailin *et al.* 2005: fully-cosmological, using the Abadi *et al.* (2003) initial conditions; GCD2 (Brook *et al.* 2004) semi-cosmological (Fig 2). Each shows the tell-tale peak in star formation between $z \approx 2-4$, with an associated exponential decline over the past ~ 10 Gyrs to a present-day rate of $\sim 1-2$ $M_\odot$/yr.

[†] The beautiful simulations of Ceverino & Klypin (2008), generated with the ART grid code were not (again, to our knowledge) run to redshift zero.
[‡] At the time of writing, an additional level of refinement has been completed, taking the resolution to ~ 200 pc.
[¶] Similar to that of the Milky Way, although we should emphasise again that Ramses1 is not supposed to be a one-to-one "clone" of the Milky Way.

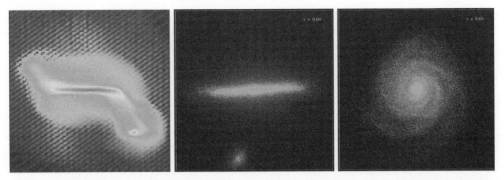

Figure 1. Gas (left) and stellar light (I-band: middle and right) distributions (60×60 kpc) for the Ramses1 disk galaxy.

In detail, the star formation histories of both RAMSES disks compare very favourably to that inferred from semi-numerical Galactic Chemical Evolution models, for the Milky Way as a whole; indeed, Ramses2 and the Milky Way model of Fenner *et al.* (2005), are extremely similar in their global star formation histories.

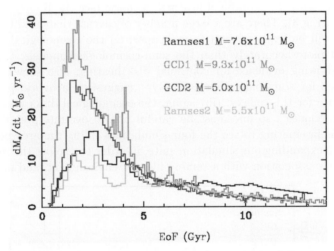

Figure 2. Time evolution of the star formation rate (within 30 kpc) for the two RAMSES disks described here, alongside those for our two earlier GCD+ SPH disk galaxies.

4. Chemistry

Our current implementation of chemistry within RAMSES is restricted to the global metal content (Z), under the assumption of the instantaneous recycling approximation. We are in the midst of porting to RAMSES the more sophisticated chemical evolution modules from our GEtool (95 isotopes; 34 elements: Fenner *et al.* 2005) and GCD+ (9 isotopes; 9 elements: Kawata & Gibson 2003) codes. Having said that, there are useful and important chemical characteristics which can be extracted and examined, including the degree of azimuthal variation in the global metallicity (both stellar and gas-phase) and the overall metallicity gradients in the thin and thick stellar disk components.

For example, within the Milky Way, the local peak-to-peak scatter in the gas-phase abundances is $\sim$0.4 dex (Cescutti *et al.* 2007; Fig 3). Examining the Ramses1 simulation, we find a comparable azimuthal variation in the gas metallicities ($\sim$0.2$\rightarrow$$-$0.5 $Z_\odot$) at radii of 8–10 kpc. We find a mid-plane (thin disk) abundance gradient of $dZ/dR$$\sim$$-$0.03 dex/kpc, comparable to that observed locally (Cescutti *et al.*; Tbl 5) and consistent with an "inside-out" disk growth scenario (Fenner *et al.* 2005); the gradient in the thick disk is a factor of two shallower over the same galactocentric distance.

5. Kinematics

One of the pillars of Galactic Archaeology has been the suggestion that within the Milky Way, vertical disk heating saturates at $\sigma_Z \sim 20$ km/s for stars of ages $\sim 2-10$ Gyrs (Quillen & Garnett 2001); for older stars, a discrete jump to $\sigma_Z \sim 45$ km/s is apparent which could be a signature of the thick disk. These conclusions have been questioned by Holmberg *et al.* (2007), who claim that the evidence instead supports a picture in which the disk has undergone continual heating throughout this period. These opposing scenarios are represented in schematic form by the yellow lines in Fig 3.

Our semi-cosmological models (Brook *et al.* 2004) appear to be more consistent with the "disk saturation" scenario (see the blue GCD2 curve of Fig 3), with the older, hotter, stars being associated with the *in situ* formation of the thick disk during the intense gas-rich merger phase at high-redshift. The cosmological disks (both the new RAMSES pair, and our Bailin *et al.* 2005 SPH galaxy) though appear to be more consistent (or at least not inconsistent) with the "continual heating" scenario (see the Ramses1, Ramses2, and GCD1 curves of Fig 3). There are a large number of caveats that need to be noted here, each of which will be explored in a future paper: (i) the cosmological models capture late-time infall more accurately than the semi-cosmological models; (ii) the Ramses2 model, while showing evidence for continual disk heating, also shows evidence of an "impulsive" step for stars older than ~ 10 Gyrs, suggesting a hybrid picture might be more appropriate for this galaxy; (iii) again, the cosmological disks were not chosen to be Milky Way "clones", so one must be careful not to overinterpret the simulations. Regardless, it is fascinating to see the four simulations filling the area between the two extrema; we are expanding our simulation suite, in order to explore the range of heating scenarios, and its association with assembly history, environment, and mass.

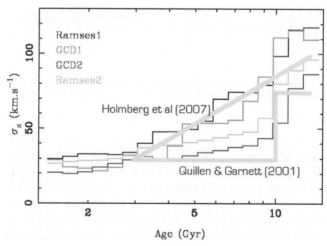

Figure 3. Age-velocity dispersion relation in the vertical direction for "local" disk stars in our suite of RAMSES and GCD+ simulations. The suggested behaviour in the solar neighbourhood under the "disk saturation" (Quillen & Garnett 2001) and "continual heating" (Holmberg *et al.* 2007) scenarios are shown in schematic form.

We can go one step further and examine the spatial variation of σ_Z (over $\sim 1-4$ disk scale lengths) for intermediate-age stars (bottom right panel of Fig 5). The exponential decline observed is consistent with that observed by Herrmann & Ciardullo (2008) in six nearby face-on galaxies. This is the expected behaviour for disks with constant M/L; the relatively high dispersions seen in the simulations (and in the observational data) beyond ∼2−3 scale lengths is likely due to the combination of disk heating and flaring.

Finally, we note in passing that the thick disks in our simulations lag those of their respective thin disks, much like the spirals in Yoachim & Dalcanton (2008) (by ∼70−100 km/s at galactocentric radii of 8−10 kpc and 3−6 kpc above the mid-plane) This is encouraging, but we next need to explore how this scales with mass (cf. the mass-dependent lag claimed by Yoachim & Dalcanton 2008), whether the thick disk scale lengths are consistently greater than those of the thin disk (Yoachim & Dalcanton 2006), and whether these are independent of environment, as claimed by Santiago & Vale (2008). Our full suite of simulations will be brought to bear on these problems.

6. Disk Edges / Truncation

6.1. Gas Disks

The Ramses1 disk has been simulated with a range of ISM physics treatments, represented by models with and without a polytropic equation of state. Taking our simulation without the polytrope, we examined the distribution of both the neutral and ionised gas of the disk (see also Fig 1). Ramses1 possesses a lopsided HI disk with a truncation/break near 19 kpc, where the HI column density is $\sim 2\times 10^{19}$ cm^{-2} (red curve of Fig 4), consistent with that observed by the THINGS team (Portas *et al.* 2008). Admittedly, the break is not as clear as that observed empirically (Portas *et al.* Fig 3), but that reflects the fact that we have azimuthally averaged over 2π radians, as opposed to splitting into twelve $\pi/6$ segments and aligning at the "break"; as such, the lopsidedness "smears" the break in Fig 4 from 19 kpc to a range of radii spanning 19−26 kpc. The ionised disk extends ∼30−50% beyond the neutral disk, before being "lost" in the background corona, similar to that observed (Bland-Hawthorn *et al.* 1997).

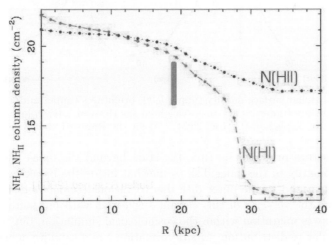

Figure 4. Neutral and ionised gas density distributions for the "no polytrope" simulation of Ramses1. The edge of the HI disk occurs near 19 kpc (vertical line), but is smeared out to ∼26 kpc by having averaged azimuthally over 2π radians.

6.2. Stellar Disks

The origin of the apparent truncations to the exponential disks seen in the surface brightness distributions of spirals both locally and at high-z is one of the most exciting areas of disk galaxy "astrophysics" today (eg. Bakos et al. 2008 (B08); Roskar et al. 2008 (R08); and many references therein). The relative roles of star formation thresholds and radial migration / re-distribution of stars due to secular effects remains hotly debated.

In Fig 5 (ignoring the bottom right panel, which refers only to § 5), we show a series of panels, based on our analysis of the Ramses1 simulation, which should be examined beside the idealised simulation of R08 (Fig 1) and the observational data of B08 (Fig 1). There are a number of similarities, and tantalising differences, between the various datasets.

First, the two upper left panels show that while a break is seen at ∼10 kpc in both the B- and K-band, there is little (if any) evidence for a break in the stellar surface density. In addition, the bottom middle panel shows the disk colour becomes blue with increasing radius prior to the break, but becomes redder beyond it. Each of these are in agreement with the inferences derived empirically by B08; this is also more-or-less in agreement with the conclusions derived by R08 from a non-cosmological (but higher resolution) simulation, although R08 find a break in the stellar surface density that we (and B08) do not. Much like the colour becoming redder beyond the break, we find an associated increase in the stellar age of these stars (bottom left panel), in agreement with R08.

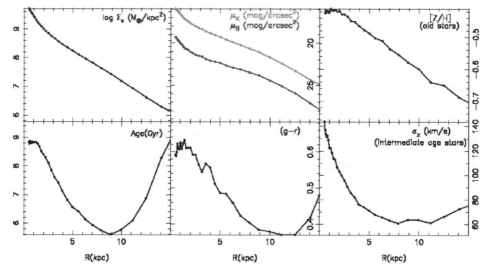

Figure 5. Radial stellar surface density (upper left), brightness (upper middle), mass-weighted metallicity for old stars (upper right), mass-weighted age (bottom left), colour (middle bottom), and vertical velocity dispersion (bottom right: § 5) for the Ramses1 stellar disk.

Within the picture proposed by R08, the stars beyond the break today are primarily old, born primarily in the inner disk (somewhat inside the break radius at the time of birth), and migrated to the outer disk due to various secular re-distribution effects. We are exploring the veracity of this elegant suggestion with our suite of cosmological simulations; there is migration within the cosmological simulation, but the relative contribution of secular re-distribution, *in situ* formation,† and satellite debris,‡ beyond the break, needs careful examination.

† Which does occur, as there is a not insignificant number of gas cells in the disk with densities in excess of the 0.1 cm^{-3} star formation threshold.
‡ ∼50% of stars beyond the break formed since $z=2.3$, the highest redshift for which we could accurately identify and align the disk; of this fraction, ∼30% formed at galactocentric radii in excess of the present-day break; indeed, much of this disk "debris" formed in satellites distributed fairly uniformly in radius within the virial radius of the host halo.

One issue which needs to be addressed, which has perhaps not been appreciated previously, is that shown in the upper right panel to Fig 5. Here, we show the metallicity gradient of the old (>7 Gyrs) stars in our Ramses1 simulation. From the bottom left panel, we see that the outer disk stars are old; from the upper right panel, we see that they are also relatively metal-poor; in and of itself, this seems consistent with R08. The potential problem which arises is that from the upper right panel, we also see that stars of the same age in the inner disk are a factor of two more metal-rich. *One must ask why it is that old metal-poor stars from the inner disk get re-distributed to the outer disk, but old metal-rich stars of the same age from the inner disk do not get re-distributed.* We need to examine the metallicity distribution functions as a function of space and time within the simulations (both our's and those of R08) to better understand the situation.

7. Gas Accretion / Infall

We defer a detailed discussion of the gas accretion history to the future, but felt it worth noting here that we have measured the spatial and temporal infall of cold, warm, and hot gas both into the halo as a whole and onto the disk itself. For Ramses1, the flux of gas across a sphere of radius 30 kpc since redshift $z\sim2$ is $\sim1.5-0.5$ $M_\odot$/yr; this can be contrasted with the inferred infall within semi-numerical Galactic Chemical Evolution models: for example, for our Milky Way model described in Fenner *et al.* (2005), averaging over the entire disk, for the same redshift range, results in a predicted infall rate of $\sim5-1$ $M_\odot$/yr. We can also determine the metallicity distribution of this infalling gas; for this particular 30 kpc surface at $z=0$, $<Z_g>\sim0.02$ $Z_\odot$, with essentially no component in excess of ~0.05 $Z_\odot$, again consistent with the Fenner *et al.* semi-numerical models (which assume the infalling gas is $\lesssim0.1$ $Z_\odot$). We have also laid virtual slabs ±6 kpc above and below the mid-plane of the simulated disk and measured the vertical gas flux through these surfaces, finding a rate of ~1 $M_\odot$/yr. We have conducted the same experiment with virtual slabs at ±1 kpc above/below the plane, and found fluxes $\sim10-50$ × greater, reflecting the much greater "ISM circulation flux" near the plane dominating over the flux from cosmological infall, consistent with observations.

8. Summary and Future Directions

We have realised (to the best of our knowledge) the first fully self-consistent cosmological hydrodynamic disk simulations to $z=0$ with a mesh code; the resolution attained is 435 pc. Several preliminary results include:
- the saturated vertical disk heating seen in semi-cosmological SPH simulations has not yet been clearly replicated in our cosmological simulations;
- the neutral gas disks show "edges" at comparable column densities to those observed; ionised gas disks extend beyond the neutral gas, again in agreement with those observed;
- the stellar surface brightness profiles show "breaks" in the exponential profiles, with associated increases (reddening) in the age (colour) of the stellar populations beyond the break, in agreement with observation; little evidence is seen for an associated break in the stellar surface density profile, also as inferred from observations; stars of the same age beyond and interior to the break do not appear to have the same metallicity, which may prove problematic for radial migration scenarios;
- gas accretion is not smooth, but does appear to be more-of-less "inside-out";
- the disk-halo "circulation flux" is $\sim10-50\times$ that of the "infalling flux" (again, consistent with the broad numbers associated with the Milky Way).

Beyond the analysis of the extant simulations, we have a number of planned enhancements, including full chemical evolution / tagging, a ten-fold increase in the number of simulations (to examine scaling relations, environmental dependencies, and assembly history variations), a range of ISM physics implementations (various polytropic equations of state, blast wave parametrisations), quantifying warp and lopsidedness statistics (Mapelli et al. 2008), 2d IFU Lick index-style maps, dusty radiative transfer, high-velocity clouds, radial gas flows, and detailed SPH vs AMR comparisons with identical initial conditions.

Acknowledgements

The support of the UK's Science & Technology Facilities Council (ST/F002432/1) and the Commonwealth Cosmology Initiative are gratefully acknowledged. We also wish to thank Ignacio Trujillo and Judit Bakos for their guidance. Simulations and analyses were carried out on COSMOS (the UK's National Cosmology Supercomputer) and the University of Central Lancashire's High Performance Computing Facility. The parent N-body simulation was performed within the framework of the Horizon collaboration (http://www.projet-horizon.fr).

References

Abadi, M. G., Navarro, J. F., Steinmetz, M., & Eke, V. R. 2003, *ApJ*, 591, 499
Agertz, O., Moore, B., & Stadel, J., et al. 2007, *MNRAS*, 380, 963
Bailin, J., Kawata, D., & Gibson, B. K., et al. 2005, *ApJ*, 627, L17
Bakos, J., Trujillo, I., & Pohlen, M. 2008, *ApJ*, 683, L103
Bland-Hawthorn, J., Freeman, K. C., & Quinn, P. J. 1997, *ApJ*, 490, 143
Brook, C. B., Kawata, D., Gibson, B. K., & Freeman, K. C. 2004, *ApJ*, 612, 894
Cescutti, G., Matteucci, F., Francois, P., & Chiappini, C. 2007, *A&A*, 462, 943
Ceverino, D. & Klypin, A. 2008, *ApJ*, submitted
Fenner, Y., Murphy, M. T., & Gibson, B. K. 2005, *MNRAS*, 358, 468
Governato, F., Mayer, L., & Wadsley, J. et al. 2004, *ApJ*, 607, 688
Governato, F., Willman, B., & Mayer, L., et al. 2007, *MNRAS*, 374, 1479
Herrmann, K. A. & Ciardullo, R. 2008, in: J. G. Funes & E. M. Corsini (eds.), *Formation and Evolution of Galaxy Disks* (ASP Conf Ser), in press
Holmberg, J., Nordström, B., & Anderson, J. 2007, *A&A*, 475, 519
Kawata, D. & Gibson, B. K. 2003, *MNRAS*, 346, 135
Mapelli, M., Moore, B., & Bland-Hawthorn, J. 2008, in: J. Anderson, J. Bland-Hawthorn & B. Nordström (eds.), *The Galaxy Disk in Cosmological Context* (CUP), in press
Okamoto, T., Eke, V. R., Frenk, C. S., & Jenkins, A. 2005, *MNRAS*, 363, 1299
O'Shea, B. W., Nagamine, K., Springel, V., Hernquist, L., & Norman, M. L. 2005, *ApJS*, 160, 1
Portas, A., Brinks, E., & Usero, A., et al. 2008, in: J. Anderson, J. Bland-Hawthorn & B. Nordström (eds.), *The Galaxy Disk in Cosmological Context* (Cambridge University Press), in press
Quillen, A. C. & Garnett, D. R. 2001, in: J. G. Funes & E. M. Corsini (eds.), *Galaxy Diks and Disk Galaxies* (ASP Conf Ser), p. 87
Robertson, B., Yoshida, N., Springel, V., & Hernquist, L. 2004, *ApJ*, 606, 32
Roskar, R., Debattista, V. P., & Stinson, G. S., et al. 2008, *ApJ*, 675, L65
Santiago, B. X. & Vale, T. B. 2008, *A&A*, 485, 21
Sommer-Larsen, J., Götz, M., & Portinari, L. 2003, *ApJ*, 596, 47
Teyssier, R. 2002, *A&A*, 385, 337
Yoachim, P. & Dalcanton, J. 2006, *AJ*, 131, 226
Yoachim, P. & Dalcanton, J. 2008, *ApJ*, 682, 1004

Present state and promises to unravel the structure and kinematics of the Milky Way with the RAVE survey

M. Steinmetz[1], A. Siebert[2], T. Zwitter[3] and the RAVE collaboration

[1] Astrophysikalisches Institut Potsdam, Potsdam, Germany
[2] Observatoire de Strasbourg, Strasbourg, France
[3] University of Ljubljana, Faculty of Mathematics and Physics, Ljubljana, Slovenia

Abstract. The RAdial Velocity Experiment (RAVE) is an ambitious survey to measure the radial velocities, temperatures, surface gravities, metallicities and abundance ratios for up to a million stars using the 1.2-m UK Schmidt Telescope of the Anglo-Australian Observatory (AAO), over the period 2003–2011. The survey represents a big advance in our understanding of our own Milky Way galaxy. The main data product will be a southern hemisphere survey of about a million stars. Their selection is based exclusively on their I–band colour, so avoiding any colour-induced bias. RAVE is expected to be the largest spectroscopic survey of the Solar neighbourhood in the coming decade, but with a significant fraction of giant stars reaching out to 10 kpc from the Sun. RAVE offers the first truly representative inventory of stellar radial velocities for all major components of the Galaxy. Here we present the first scientific results of this survey as well as its second data release which doubles the number of previously released radial velocities. For the first time, the release also provides atmospheric parameters for a large fraction of the second year data, making it an unprecedented tool to study the formation of the Milky Way. Plans for further data releases are outlined.

Keywords. catalogs, stars: fundamental parameters, surveys, Galaxy: stellar content, Galaxy: kinematics and dynamics

1. Introduction

It is now widely accepted that the Milky Way galaxy is a suitable laboratory to study the formation and evolution of galaxies. Despite the fact that the Galaxy is one unique system, understanding its formation holds important keys to study the broader context of disc galaxy formation. Thanks to the past and ongoing large surveys such as Hipparcos, SDSS, 2MASS or DENIS, we have access to data which allow us to refine our knowledge of Galaxy formation. However, with the exception of the SDSS survey, which mainly samples the halo of the Galaxy, the full description of the 6D phase space, i.e. the combination of the position and velocity spaces, is not available due to the missing radial velocity and/or distance.

With the advent of multi-fiber spectroscopy, combined to the large field of view of Schmidt telescopes, it is now possible to acquire in a reasonable amount of time spectra for a large sample of stars that is representative for the different populations of the Galaxy. Spectroscopy enables us to measure the generally missing radial velocity, which in turn allows us to study the details of Galactic dynamics. Spectroscopy also permits to measure the abundance of chemical elements in a stellar atmosphere which holds important clues on the initial chemical composition and its subsequent metal enrichment. The measurement of the radial velocity and of the chemical abundances as well as the derivation of stellar temperature and gravity in order to complement existing catalogues is the main purpose of the RAVE project.

RAVE is using the 6dF multi-fiber spectroscopic facility at the UK Schmidt telescope of the Anglo-Australian Observatory in Siding Spring, Australia. The 6dF enables us to collect up to 150 spectra in one single pointing, with an average resolution of 7500 in the Calcium triplet region around 8500 Å. This medium resolution allows the measurement of accurate radial velocities (~ 2 km s^{-1}) as well as atmospheric parameters (T_{eff}, $\log g$, [M/H]) and chemical abundances. In Section 2 we present the first scientific outcomes obtained using the RAVE data, while Section 3 presents the second data release of the RAVE project (DR2) and discusses the current status and prospects of the project.

2. First Results from the Survey

In this section we outline some of the recently published results based on RAVE data.

2.1. The Escape Velocity of the Galaxy

In Smith et al. (2007) we revisited the local escape speed of our Galaxy by combining a sample of high velocity stars detected by the RAVE survey with previously known ones. Using a maximum likelihood technique, we find $498 < v_{esc} < 608$ km s^{-1} at the 90% confidence level, with a median likelihood of 544 km s^{-1}. This result demonstrates the presence of a dark halo in the Milky Way, but simultaneously argues for a halo of relatively low circular velocity ($v_c \approx 140$ km s^{-1}).

2.2. Streams (or lack thereof) in the Solar Neighborhood

In Seabroke et al. (2008) we searched the CORAVEL and RAVE survey for signatures of vertically infalling stellar streams in the Solar vicinity. Using a Kuiper test, we demonstrated that the Solar neighborhood is empty of any vertical streams containing more than a few hundreds of stars. Therefore, we confirm recent simulations that are favoring a model in which the Sagittarius stream is not entering the Solar neighborhood. We also argue against the Virgo overdensity crossing the disc near the Sun.

2.3. The Vertical Structure of the Galactic Disc

Combining RAVE data with photometric and astrometric catalogues, in Veltz et al. (2008) we were using G– and K–type stars towards the Galactic poles in order to identify whether there is a kinematic discontinuity between the thin and thick discs. We conclude that such a discontinuity indeed exists, which is a strong constraint on the formation scenario of the thick disc: it is arguing against continuous processes such as scattering by spiral stellar or molecular arms, but favoring violent processes such as the accretion of or violent heating by a satellite.

2.4. The Tilt of the Velocity Ellipsoid

Using RAVE red clump giants towards the South Galactic pole, in Siebert et al. (2008) we have measured the inclination of the velocity ellipsoid at 1 kpc below the Galactic plane. The value of the tilt, $7.3 \pm 1.8°$, is consistent with either a short scale length for the disc ($R_d \sim 2$ kpc) if the halo is oblate, or a long scale length ($R_d \sim 3$ kpc) if the halo is prolate. Combined to independent measurements of the minor-to-major axis ratio of the halo, which prefers an almost spherical halo. A scale length of the disc in the range [2.5–2.7] kpc is preferred.

2.5. Diffuse Interstellar Bands

In Munari *et al.* (2008) we used spectra of hot stars from the RAVE survey to investigate the properties of 5 diffuse interstellar bands (DIB) in the Ca triplet region. Our findings indicate that the DIB at 8620.4 Å is strongly correlated to reddening and follows the relation $E(B-V) = (2.72 \pm 0.03) \times \text{EW}$, where EW is the equivalent width of the DIB in Å. This DIB is thus a suitable tracer of general Galactic reddening in stellar spectra. On the other hand the existence of the DIB at 8648 Å is confirmed, but its intensity or equivalent width does not appear to correlate with reddening.

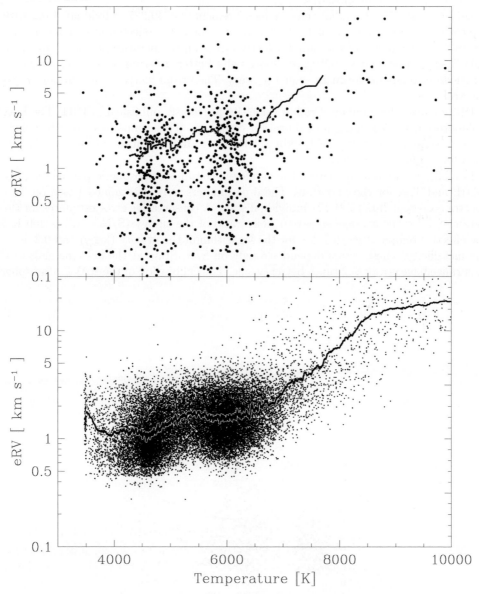

Figure 1. Radial velocity errors for stars in the second RAVE data release as a function of stellar temperature. The error is the internal velocity error (bottom) and the standard deviation of velocities determined from repeated observations of the same object (top). Lines trace a smoothed average dependence, with a boxcar width of 100 points.

3. Second data release: Current Status and Ongoing efforts

3.1. Second Data Release

In July 2008, the RAVE collaboration has released the 2nd catalogue of reduced RAVE data (DR2) (Zwitter et al. 2008). DR2 covers an area of $\sim 7200\,\mathrm{deg}^2$ in the southern hemisphere at galactic latitudes larger than $25°$. As for the first data release (DR1) (Steinmetz et al. 2006), the magnitude range is $9 \leqslant I_{IC} \leqslant 12$, where I_{IC} is the I magnitude of input catalogue. It equals the SuperCosmos photographic I magnitude for faint stars ($11 < I_{IC} < 12$), while for bright stars ($I_{IC} < 11$) it was derived from the Tycho-2 B_T and V_T magnitudes.

DR2 contains 51,829 radial velocity measurements for 49,327 individual stars, with a typical radial velocity error of 1.3 to 1.7 km s^{-1} for data collected after DR1. Figure 1 shows the dependence of the radial velocity errors on temperature, as determined from RAVE spectra. Radial velocities are marginally better determined for cool giants than for medium temperature main sequence stars. The errors increase for stars of an early spectral type.

DR2 doubles the number of radial velocities previously published in DR1. The RAVE collaboration is continuing its calibration campaign. The comparison to external data shows a standard deviation of 1.3 km s^{-1} for the radial velocities, which is twice better than for the first data release.

In addition to radial velocities, the catalogue contains atmospheric parameters, $\log g$, [M/H] and T_{eff}, for the first time. These parameters were determined based on 22,407 spectra corresponding to 21,121 individual stars. A conservative error estimates for these parameters for the average signal–to–noise ratio of the survey (S/N ~ 40) is 400 K for the effective temperature, 0.5 dex for the logarithm of the gravity ($\log g$) and 0.2 dex for the metallicity. These errors depend strongly on S/N. Stars at the extreme ends of the S/N range have errors ~ 2 times better/worse. The calibration of the RAVE atmospheric

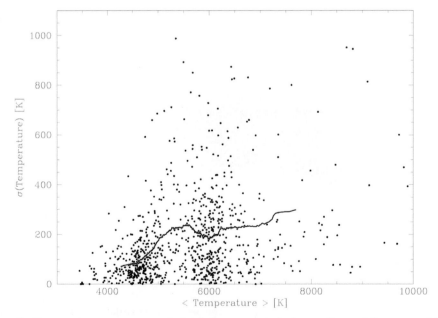

Figure 2. Temperature accuracy as judged from repeated observations of the same object in the second data release. Standard deviation of temperature is given as a function of its average value. The line is a smoothed average using the adjacent 50 cooler and 50 hotter objects.

parameters is obtained using a devoted validation campaign where high-resolution spectra of standard stars have been taken with various instruments and compared to spectra obtained by RAVE.

Figures 2 to 4 show standard deviations of temperature, gravity, and metallicity as obtained from repeated observations where a certain star was observed more than once by RAVE. The results show that typical errors are only $\lesssim$ 200 K in effective temperature, $\sim$ 0.25 dex in gravity and $\lesssim$ 0.15 dex in metallicity.

The catalogue is available from the RAVE website www.rave-survey.org or at the CDS using the VizieR database.

3.2. Pilot Survey

The RAVE pilot survey (data collected between April 2003 and February 2006) is now completed. A data release is scheduled for mid-2009. It contains about 85,000 radial velocity measurements as well as measurements of stellar atmospheric parameters for a large fraction of the sample. The release of the pilot survey will mark a major step forward in the RAVE project, and subsequent data releases will be based on a new input catalogue drawn from the DENIS survey.

3.3. Current Status

RAVE currently operates at its full potential, observing is scheduled for 25 nights per lunation. The project entered its main survey phase in March 2006. The main survey relies on a new input catalogue based on DENIS I-band magnitudes. RAVE has so far collected approximately 215,000 spectra in its main phase, which adds to a total of 301,000 spectra for 270,000 stars when combining the two phases of the project. Figure 5 plots the density of spectra per square degree observed before 6th of July 2008. The survey now covers almost the entire Southern hemisphere with the exception of the Galactic plane where only a few test observations have been obtained.

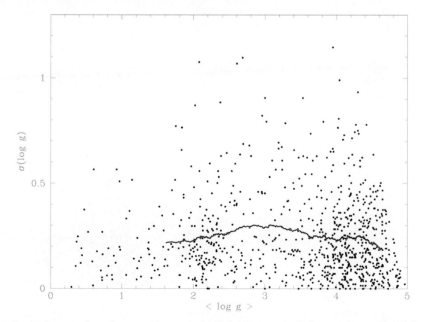

Figure 3. Accuracy of surface gravity as judged from repeated observations. Plot follows the style of Figure 2.

3.4. Ongoing Efforts

RAVE's primary goal is to measure accurate radial velocities for stars in the Southern hemisphere. The quality of the RAVE spectra also enables us to measure atmospheric parameters ($\log g$, [M/H] and T_{eff}) which can be used to select subsample of the catalogue with specific properties. For example, the combination of colours and $\log g$ measurements from RAVE permits an accurate selection of red clump giants (Veltz et al. 2008, Siebert et al. 2008). For this population, the narrow luminosity function enables us to obtain distances to 20% from the apparent magnitude alone.

However, the measurement of the atmospheric parameters is non-trivial at RAVE resolution and the transformation from the measured parameters to the true parameters relies on calibration data. The RAVE collaboration constantly acquires data for standard and pseudo-standards stars, using both the 6dF instrument and high-resolution spectra from other instruments, with the aim to refine our calibration.

Despite the medium resolution of the RAVE spectra, chemical abundances for about 12 elements can be measured with a good accuracy ($\sim$ 0.2 dex) in the high signal to noise spectra ($S/N > 80$). The RAVE collaboration puts a particular effort on measuring and validating these abundances. For this purpose, RAVE also acquires spectra for nearby stars for which accurate abundances have been measured using high-resolution spectroscopy. So far, abundances have been derived for more than 20,000 RAVE targets, and the resulting measurements will be published in a seperate catalogue. Furthermore, we plan to publish a list of fast rotating (and generally hot) stars where the rotational broadening can be detected at the resolving power of RAVE spectra.

The knowledge of the 6D phase space requires that we can transform the proper motions into spatial velocities. This step relies on good distance estimates for individual stars, which can be obtained from comparison of apparent and absolute magnitudes, the latter inferred from values of atmospheric parameters. The availability of distances in the near future will allow to exploit the full potential of the RAVE catalogue, permitting to

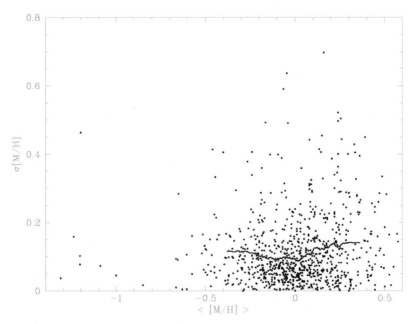

Figure 4. Accuracy of metallicity as judged from repeated observations. Plot follows the style of Figure 2.

study the detailed shape of the phase space and reveal new details of the formation of the Solar neighbourhood.

4. Conclusions

The RAVE collaboration has released a second catalogue in July 2008, reporting radial velocities for 51,829 spectra and 49,327 different stars, randomly selected in the magnitude range of $9 < I < 12$ and located more than $25°$ away from the Galactic plane. This release doubles the size of the previously published catalogue. The typical error of the published radial velocities is between 1.3 and 1.7 km s^{-1}.

In addition to radial velocities, the catalogue contains, for the first time, atmospheric parameter measurements for more than 20,000 spectra. Uncertainties for a typical RAVE star are of the order of 400 K in temperature, of 0.5 dex in gravity, and of 0.2 dex in metallicity, but the error depends on the S/N and can vary by a factor of ~ 2 for stars at the extreme ends of the S/N range. Comparison of parameter measurements for repeated observations of the same targets indicates that these estimates are conservative and that the true errors may be smaller.

The survey continues to collect spectra. Currently more than 300,000 spectra have been collected and are currently being processed. Acquisition of new calibration data is also underway, enabling us to refine the atmospheric parameter measurements as well as radial velocities, and represents an import ongoing effort. Other current activities include the measurement of chemical abundances and distances which will be released as companion catalogues in the future.

The RAVE catalogue can be used to study the formation of the Milky Way and, for example, the collaboration succeeded in refining significantly the measurement of the

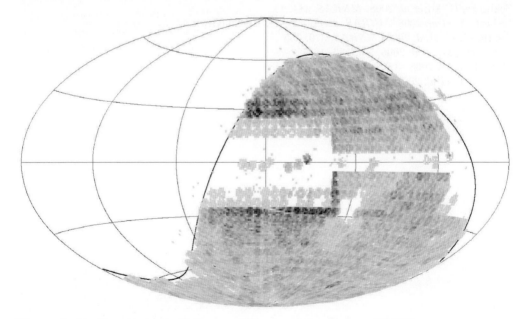

Figure 5. Aitoff projection in Galactic coordinates of the number of RAVE spectra per square degree. Only the spectra observed before 6th of July 2008 and which successfully passed the quality controls are included. They cover 16,730 square degrees of the Sourhern sky, with a mode of 8 spectra per square degree. Density is linearly coded in shades of grey from zero in white to the maximum of 93 spectra per square degree in black. Spectra on the Galactic plane were observed for calibration and test purposes. The S-shaped curve marks the celestial equator.

local escape velocity using the high velocity stars found in the RAVE catalogue. Further studies carried out by the collaboration include the search for vertical stream of matter in the Solar neighborhood; a new determination of the vertical structure of the Galactic disc; the study of diffuse interstellar bands in the Galactic plane and their correlation to the interstellar extinction; the measurement of the inclination of the velocity ellipsoid towards the Galactic plane at 1 kpc below the plane.

As the survey progresses and new data become available, the RAVE catalogue will enable us to refine our view on the structure and formation of the Milky Way, paving the way for new insights on the formation of galaxies.

Acknowledgements

Funding for RAVE has been provided by the Anglo-Australian Observatory, the Astrophysical Institute Potsdam, the Australian Research Council, the German Research foundation, the National Institute for Astrophysics at Padova, The Johns-Hopkins University, the Netherlands Research School for Astronomy, the Natural Sciences and Engineering Research Council of Canada, the Slovenian Research Agency, the Swiss National Science Foundation, the National Science Foundation of the USA (AST-0508996), the Netherlands Organisation for Scientific Research, the Particle Physics and Astronomy Research Council of the UK, Opticon, Strasbourg Observatory, and the Universities of Basel, Cambridge, and Groningen. The RAVE Web site is at www.rave-survey.org.

References

Munari, U. *et al.* 2008, *A&A*, 488, 969
Seabroke, G. M. *et al.* 2008, *MNRAS*, 384, 11
Siebert, A. *et al.* 2008, *MNRAS*, accepted (arXiv:0809.0615)
Smith, M. C. *et al.* 2007, *MNRAS*, 379, 755
Steinmetz, M. *et al.* 2006, *AJ*, 132, 1645
Veltz, L. *et al.* 2008, *A&A*, 480, 753
Zwitter, T. *et al.* 2008, *AJ*, 136, 421

SEGUE, and the future of large scale surveys of the Galaxy

Timothy C. Beers[1], Young Sun Lee[1], and Daniela Carollo[2]

[1]Department of Physics & Astronomy, CSCE: Center for the Study of Cosmic Evolution, and JINA: Joint Institute for Nuclear Astrophysics, Michigan State University,
East Lansing, MI 48824, USA
email: beers@pa.msu.edu, lee@pa.msu.edu

[2]Research School of Astronomy and Astrophysics, Australian National University, Mount Stromlo Observatory, Cotter Road, Weston, ACT 2611, Australia, and INAF-Osservatorio Astronomico di Torino, 10025 Pino Torinese, Italy
email: carollo@mso.anu.edu.au

Abstract. The Sloan Extension for Galactic Exploration and Understanding (SEGUE) has now been completed. This is one of three surveys that were executed as part of the first extension of the Sloan Digital Sky Survey (SDSS-II), which consist of LEGACY, SUPERNOVA SURVEY, and SEGUE. The SEGUE program has obtained over 3600 square degrees of *ugriz* imaging of the sky outside the original SDSS-I footprint. The regions of sky targeted for SEGUE imaging were primarily at lower Galactic latitudes ($|b| < 35°$), in order to better sample the disk/halo interface of the Milky Way. SEGUE also obtained medium-resolution ($R = 2000$) spectroscopy, over the wavelength range 3800-9200 Å, for over 200,000 stars in 200 selected areas over the sky available from Apache Point, New Mexico. We discuss the determination of stellar atmospheric parameters ($T_{\rm eff}$, log g, and [Fe/H]) for these stars, and highlight several of the scientific results obtained to date. The proposed second extension of SDSS, known as SDSS-III, will include SEGUE-2, a program to roughly double the numbers of stars with available spectroscopy, as well as APOGEE, a program to obtain high-resolution ($R = 20000$) near-IR spectroscopy for over 100,000 stars in the disk, bulge and halo populations of the Galaxy. Other massive spectroscopic surveys of interest to Galactic science are also briefly discussed.

Keywords. Astronomical data bases: surveys; techniques: spectroscopic; methods: data analysis; stars: fundamental parameters; Galaxy: disk, halo

1. Introduction

We are now firmly entrenched in the era of massive photometric and spectroscopic surveys of the stellar populations of the Galaxy, as highlighted by the Sloan Digital Sky Survey (SDSS-I), and its first extension, SDSS-II, as well as by the contemporaneous survey RAVE. These surveys, and others that are planned for the near future, promise to completely revolutionize our understanding of the formation and evolution of the Milky Way, and provide new insight into the formation of the chemical elements.

Here we summarize the status of the SEGUE (Sloan Extension for Galactic Exploration and Understanding) program, executed as part of SDSS-II, and foreshadow the goals of SEGUE-2, to be carried out during the proposed next extension of SDSS, known as SDSS-III. We also briefly mention several additional ongoing high-resolution spectroscopic surveys, as well as a number of other massive surveys planned for the near (APOGEE, LAMOST) and the more distant future (WFMOS).

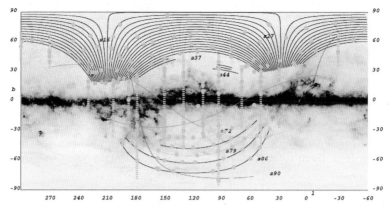

Figure 1. Map of the imaging and spectroscopic coverage obtained during the SDSS/SEGUE program. The black lines correspond to regions imaged during SDSS-I, while the colored lines indicate regions of SEGUE imaging. The green filled dots indicate lines of sight where SEGUE obtained over 1200 individual spectra (in two spectroscopic plug-plates, a bright plate and a faint plate). The image of the Milky Way is derived from the dust maps of Schlegel *et al.* (1998).

2. SEGUE

The SEGUE survey took its last data in June, 2008. Over 3600 square degrees of new imaging was obtained, and a total of over 400 spectroscopic plug-plates were obtained, covering some 200 lines of sight probing directions of interest for exploration of the thick disk and halo populations of the Milky Way, from 0.5 kpc to almost 100 kpc from the Sun (see Fig. 1). The medium-resolution ($R = 2000$) spectroscopy obtained for roughly 240,000 program stars (in the magnitude range $14.0 < g < 20.5$), in combination with the *ugriz* photometry from the imaging, form the inputs for determinations of radial velocities (accurate to between 5 and 25 km/s, depending on the color of the star and the S/N of the spectrum) and, most crucially, the fundamental stellar atmospheric parameters ($T_{\rm eff}$, log g, and [Fe/H]) for the majority of the program objects. The full data set (and derived parameters) is planned to be released to the public in October, 2008.

The atmospheric parameters are derived from a set of techniques, based on a number of different calibrations, in an effort to provide robust determinations that remain valid over the large range of parameter space and S/N explored by SEGUE. These approaches, which collectively are applied by the SEGUE Stellar Parameter Pipeline (SSPP), include techniques for finding the minimum distance (parameterized in various ways) between observed spectra and grids of synthetic spectra (e.g., Allende Prieto *et al.* 2006), non-linear regression models (e.g., Re Fiorentin *et al.* 2007, and references therein), correlations between broadband colors and the strength of prominent metallic lines, such as the CaII K line (Beers *et al.* 1999), auto-correlation analysis of a stellar spectrum (Beers *et al.* 1999, and references therein), obtaining fits of spectral lines (or summed line indices) as a function of broadband colors (Wilhelm *et al.* 1999), or the behavior of the CaII triplet lines as a function of broadband color (Cenarro *et al.* 2001).

Each of the methods employed by the SSPP exhibit optimal behavior over restricted temperature and metallicity ranges; outside of these regions they are often un-calibrated, suffer from saturation of the metallic lines used in their estimates at high metallicity or low temperatures, or lose efficacy due to the weakening of metallic species at low metallicity or high temperatures. The techniques that make use of specific spectral features are susceptible to other problems, e.g., the presence of emission in the core of the CaII K line for chromospherically active stars, or poor telluric line subtraction in the region of the CaII triplet. Because SDSS stellar spectra cover most of the entire optical wavelength range, one can apply several approaches, using different wavelength regions, in order to glean optimal information on stellar parameters. The combination of multiple techniques results in estimates of stellar parameters that are more robust over a much wider range of $T_{\rm eff}$, log g, and [Fe/H] than those that might be produced by individual methods.

Details of these procedures and tests of the validity of the resulting parameter estimates are presented in a series of three papers, Lee et al. (2008a), Lee et al. (2008b), and Allende Prieto et al. (2008), to which the interested reader is referred.

The precision of the parameter estimates varies with the S/N of the spectra. At the median S/N of the SEGUE spectra (roughly 20/1), the estimates have typical errors of $\delta T_{\rm eff}$ = 150 K, δlog g = 0.30 dex, and δ[Fe/H] = 0.25 dex, respectively. Tests on the accuracy of the atmospheric parameter estimates indicate that there exist negligible zero-point offsets over the majority of the parameter space.

3. Summary of early results from SEGUE

Although the final data set has only just been obtained, numerous scientific results from SEGUE have already emerged. A few of these include:

• The identification of ultra low luminosity dwarf spheroidals, and halo debris streams (Belokurov et al. 2006, Zucker et al. 2006). It has been argued, based on new high-resolution spectroscopic estimates of [Fe/H] (see Kirby et al. 2008), that these newly discovered dwarfs indeed include substantial numbers of stars with [Fe/H]< −3.0, strengthening the possible association of the formation of the (outer) halo with the accretion and destruction of similar objects in the past.

• Confirmation of the inner/outer halo structure of the Milky Way (Carollo et al. 2007). Although it had been long speculated that the halo of the Milky Way might be structurally complex, and comprise more than a single stellar population, the calibration stars from SDSS/SEGUE have been used to demonstrate convincingly that the halo is indeed clearly divisible into two broadly overlapping structural components, an inner and an outer halo, which exhibit different spatial density profiles, stellar orbits, and stellar metallicities. While the inner halo has a modest net prograde rotation, the outer halo exhibits a net retrograde rotation and a peak metallicity three times lower ([Fe/H] = −2.2) than that of the inner halo ([Fe/H] = −1.6). These properties indicate that the individual halo components likely formed in fundamentally different ways, through successive dissipational (inner) and dissipationless (outer) mergers and tidal disruption of proto-Galactic clumps. Work is now in progress (Carollo et al., in prep.) to derive the velocity ellipsoids and metallicity distribution functions of the individual populations. See Fig. 2.

• Measurement of a new (lower) mass for the Galaxy (Xue et al. 2008). These authors isolated a nearly pure sample of some 2400 blue horizontal-branch stars with distances up to 60 kpc from the Galactic center, and compared their observed radial velocities with cosmologically motivated galaxy formation scenarios to obtain estimates of the Milky Way's circular velocity curve, implying $M(< 60 \text{ kpc}) = 4.0 \pm 0.7 \times 10^{11} M_\odot$. The associated virial mass of the dark halo of the Milky Way is estimated to be $M_{\rm vir} = 0.93 \pm 0.25 \times 10^{12} M_\odot$, which is lower than many previous estimates. This estimate implies that nearly 40% of the baryons within the virial radius of the Milky Way's dark matter halo reside in the stellar components of our Galaxy.

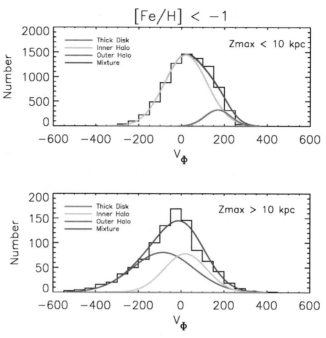

Figure 2. An example maximum-likelihood decomposition of the contribution of thick disk, inner-halo, and outer-halo populations to the overall observed distribution of V_Φ, the rotation with respect to the Galactic center, based on an analysis of the SDSS/SEGUE calibration stars from DR-6 (Adelman-McCarthy et al. 2008). All stars in the figure have [Fe/H] < −1.0. The upper panel applies to stars with $Z_{max} < 10$ kpc; the lower panel applies to stars with $Z_{max} > 10$ kpc, where Z_{max} refers to the maximum distance from the Galactic plane reached by an individual star during the course of its orbit. Note that the inner-halo population completely dominates the observed distribution in the upper panel, while a higher dispersion, net retrograde outer-halo population is required to account for the extended low velocity tail in the lower panel.

- Construction of a metallicity map for millions of stars in the disk and halo populations of the Milky Way (Ivezic et al. 2008). Based on the spectroscopically calibrated stars from SDSS/SEGUE, these authors have derived photometric estimates of [Fe/H] for some 2.5 million F-type stars in the color range $0.2 < g − r < 0.4$, with accuracies only slightly less than the spectroscopic determinations, and exploring distances up to 9 kpc from the Sun. Such a technique will be used in the future, e.g., with data from LSST and SkyMapper, to obtain metallicity maps of the halo of the Galaxy out to 100,000 kpc, with substantially improved accuracy. See Fig. 3.
- Measurement of [α/Fe] ratios for tens of thousands of stars (Lee et al., in prep.). As part of his PhD thesis, Lee has calibrated techniques to obtain estimates of the [α/Fe] ratio for high-S/N SDSS/SEGUE spectra (S/N > 25/1). The method is precise to on the order of δ[α/Fe] = 0.1 dex, which is sufficient to provide information concerning the nature of the sub-Galactic fragments (such as their mass distributions) involved with the assembly of the thick disk, and the inner- and outer-halo populations. See Fig. 4.
- The identification of over 100,000 F-G-K stars with [Fe/H] < −1.0, over 15,000 with [Fe/H] < −2.0, and several hundred with [Fe/H] < −3.0 (Beers et al., in prep.)
- The identification of many thousands of Carbon-Enhanced Metal-Poor (CEMP) Stars (Sivarani et al., in prep.)

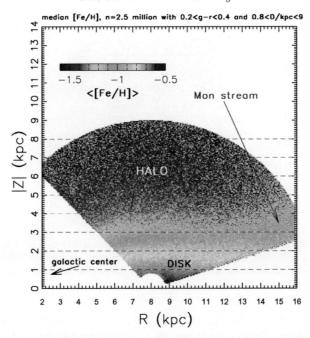

Figure 3. Photometric metallicities for 2.5 million blue F-type stars, based on the calibration described by Ivezic et al. 2008. The median occupancy of the 40,000 pixels shown in this map is 33 stars, with a minimum of 5 stars per pixel. The color-coded metallicities clearly indicate the presence of the disk, thick-disk, and inner-halo populations of the Galaxy. The presence of the Monoceros stream in the region close to the disk plane is indicated.

4. Future surveys

SEGUE-2 is one of four surveys planned as part of the proposed next extension to the SDSS, known as SDSS-III. SEGUE-2 will use the existing SDSS spectrograph to obtain medium-resolution spectra for roughly an additional 200,000 stars, during a period of one year starting in August, 2008. Building on what has been learned from SEGUE, SEGUE-2 will specifically target stars (in a variety of categories) that explore the transition from the inner to the outer halo, concentrating on stars at the main-sequence turnoff, distant giants and horizontal-branch stars, and on stars selected to be likely low-metallicity objects.

There exist a number of large high-resolution spectroscopic surveys that are just now getting underway, which should greatly enlarge the numbers of very metal-poor (VMP, [Fe/H] < −2.0) stars with available elemental abundance information. The Chemical Abundances of Halo Stars (CASH) survey is making use of the Hobby-Eberly Telescope to obtain moderately high-resolution ($R = 15000$) spectroscopy for up to 1,000 VMP stars identified during the course of the HK-I, HK-II, and HES efforts, with a large number of additional stars from SDSS/SEGUE. Aoki and collaborators have recently been awarded Key Project status on the Subaru Telescope in order to obtain $R = 50000$ spectroscopic observations for up to 200 VMP stars, the majority of which will be drawn from SDSS/SEGUE targets. Both of these surveys have as one of their primary aims to test for the existence (or not) of chemical signatures that might be associated with the inner/outer halo dichotomy reported by Carollo et al. (2007).

The Apache Point Observatory Galactic Evolution Experiment (APOGEE), another survey to be conducted as part of SDSS-III, will use high-resolution ($R = 20000$), high

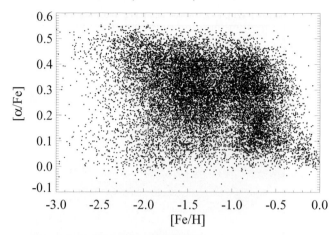

Figure 4. Distribution of $[\alpha/\text{Fe}]$ for over 18,000 stars from SDSS/SEGUE, as a function of [Fe/H]. There is an apparent bifurcation of this distribution at low metallicity, into predominantly high alpha abundances and low alpha abundances. The full implications of these observations will require detailed modeling to better understand.

signal-to-noise ratio (S/N = 100/1), H-band (1.6 μm) spectroscopy to penetrate the dust that obscures the inner Galaxy from our view, observing 100,000 red giant stars across the full range of the Galactic bulge, bar, disk, and halos. The high spectral resolution of APOGEE will allow element-by-element measurements of chemical abundances, which can be used to reconstruct the history of star formation that produced these elements. Together, SEGUE-2 and APOGEE will provide a picture of the Milky Way that is unprecedented in scope, richness, and detail. The combined data set will play a central role in "near-field cosmology" tests of galaxy formation physics and the small scale distribution of dark matter.

The Large Area Multi-Object Spectroscopy Telescope (LAMOST) project, funded by the Chinese National Academy of Science, will employ a massive 4000 fiber system mounted on a wide-field 4m meridian Schmidt telescope. It is expected that this ambitious project will obtain up to several million medium-resolution ($R = 2000$) spectroscopic observations of stars in the Milky Way and Local Group galaxies. With the planned addition of $R = 5000/10000$ gratings, follow-up observations at moderately high spectral resolution will also become possible.

Of course, we all look forward to the possible execution of the WFMOS (Wide Field Multi-Object Spectrograph) survey of up to one million stars at resolving power $R = 50000$. Currently, this Gemini instrument is expected to be mounted on the prime focus of the Subaru telescope, in order to take advantage of its wide field of view. The hope, and expectation, is that this survey will finally reveal the rich set of elemental abundances for stars that probe the entire history of chemical evolution throughout the Galaxy.

Acknowledgements

Funding for the SDSS and SDSS-II has been provided by the Alfred P. Sloan Foundation, the Participating Institutions, the National Science Foundation, the U.S. Department of Energy, the National Aeronautics and Space Administration, the Japanese Monbukagakusho, the Max Planck Society, and the Higher Education Funding Council for England. The SDSS Web Site is http://www.sdss.org/.

This work received partial support from grants AST 07-07776 and PHY 02-15783; Physics Frontier Center / Joint Institute for Nuclear Astrophysics (JINA), awarded by the US National Science Foundation.

References

Adelman-McCarthy, J. K., Agüeros, M. A., Allah, S. S, Allende Prieto, C., & Anderson, K. S. J., et al. 2008, *ApJS*, 175, 297

Allende Prieto, C., Beers, T. C., Wilhelm, R., Newberg, H. J., & Rockosi, C. M., et al. 2006, *ApJ*, 636, 804

Allende Prieto, C., Sivarani, T., Beers, T. C., Lee, Y. S., & Koesterke, L. et al. 2008, *AJ*, in press (arXiv:0710.5780)

Beers, T. C., Rossi, S., Norris, J. E., Ryan, S. G., & Shefler, T. 1999, *AJ*, 117, 981

Belokurov, V., Zucker, D. B., Evans, N. W., Gilmore, G., & Vidrih, S., et al. 2006, *ApJ*, 642, L137

Carollo, D., Beers, T. C., Lee, Y. S., Chiba, M., & Norris, J. E., et al. 2007, *Nature*, 450, 1020

Cenarro, A. J., Cardiel, N, Gorgas, J., Peletier, R. F., Vazdekis, A., & Prada, F. 2001, *MNRAS*, 326, 959

Ivezic, Z., Sesar, B., Juric, M., Bond, N., & Dalcanton, J., et al. 2008, *ApJ*, in press (arXiv:0804.3850)

Kirby, E. N., Simon, J. D., Geha, M., Guhathakurta, P., & Frebel, A. 2008, *ApJ*, submitted (arXiv:0807.1925)

Lee, Y. S., Beers, T. C., Sivarani, T., Allende Prieto, C., & Koesterke, L., et al. 2008, *AJ*, in press (arXiv:0710.5645)

Lee, Y. S., Beers, T. C., Sivarani, T., Johnson, J. A., & An, D., et al. 2008, *AJ*, in press (arXiv:0710.5778)

Re Fiorentin, P., Bailer-Jones, C. A. L., Lee, Y. S., Beers, T. C., & Sivarani, T., et al. 2007, *A&A*, 467, 1373

Schlegel, D. J., Finkbeiner, D. P., & Davis, M. 1998, *ApJ*, 500, 525

Wilhelm, R., Beers, T. C., & Gray, R. O. 1999, *AJ*, 117, 2308

Xue, X.-X., Rix, H.-W., Zhao, G., Re Fiorentin, P., & Naab, T., et al. 2008, *ApJ*, in press (arXiv:0801.1232)

Zucker, D. B., Belokurov, V., Evans, N. W., Wilkinson, M. I., Irwin, M. J., et al. 2006, *ApJ*, 643, L103

Ole Strömgren giving his speech at the closing dinner. From the left around the table: Line Kragh, Flemming Besenbacher, John Renner Hansen, Gerry Gilmore, Ole Strömgren, Birgitta Nordström, Joss Bland-Hawthorn, and Johannes Andersen (from back). Photo: Lars Buchhave.

Enthusiastic participants dancing to the live Big-Band after dinner.

Galaxy And Mass Assembly (GAMA)

Simon P. Driver[1] and the GAMA team[†]

[1](SUPA) School of Physics and Astronomy, University of St Andrews
St Andrews, Fife, KY16 8RS, Scotland
email: spd3@st-and.ac.uk

Abstract. The GAMA survey aims to deliver 250,000 optical spectra (3–7Å resolution) over 250 sq. degrees to spectroscopic limits of $r_{AB} < 19.8$ and $K_{AB} < 17.0$ mag. Complementary imaging will be provided by GALEX, VST, UKIRT, VISTA, HERSCHEL and ASKAP to comparable flux levels leading to a definitive multi-wavelength galaxy database. The data will be used to study all aspects of cosmic structures on 1kpc to 1Mpc scales spanning all environments and out to a redshift limit of $z \approx 0.4$. Key science drivers include the measurement of: the halo mass function via group velocity dispersions; the stellar, HI, and baryonic mass functions; galaxy component mass-size relations; the recent merger and star-formation rates by mass, types and environment. Detailed modeling of the spectra, broad SEDs, and spatial distributions should provide individual star formation histories, ages, bulge-disc decompositions and stellar bulge, stellar disc, dust disc, neutral HI gas and total dynamical masses for a significant subset of the sample ($\sim$ 100k) spanning both the giant and dwarf galaxy populations. The survey commenced March 2008 with 50k spectra obtained in 21 clear nights using the Anglo Australian Observatory's new multi-fibre-fed bench-mounted dual-beam spectroscopic system (AAΩ).

Keywords. galaxies: general, galaxies: structure, galaxies: formation, galaxies: evolution

1. GAMA Motivation

Galaxy And Mass Assembly (GAMA) is a major expansion of the Millennium Galaxy Catalogue (MGC) survey (Liske *et al.* 2003; Allen *et al.* 2006, Driver *et al.* (2005) and a natural extension of the extremely productive nearby "Legacy" surveys (e.g., SDSS, 2MASS, HIPASS etc). In comparison to the superb SDSS survey GAMA will only sample 250 sq degrees of sky but will extend to significantly fainter spectroscopic limits (12× the redshift density of SDSS main, 5× stripe 82), to higher spatial (0.6″ FWHM) and spectral (3–7Å) resolutions, as well as moving to a far broader wavelength coverage (UV to Radio). GAMA has come about by parallel technological developments leading to a suite of new facilities whose survey sensitivities, resolutions, and capabilities are reasonably well matched. Until now the study of galaxies has generally been restricted to either large samples of limited wavelength data or multi-wavelength studies of small (and often biased) samples. However galaxy systems are extremely complex and diverse, exhibiting strong environmental and mass dependencies and containing distinct but interlinked

---

† I.K. Baldry (LJMU), S. Bamford (Nott), J. Bland-Hawthorn (USyd), T. Bridges (AAO), E. Cameron (StA), C. Conselice (Nott), W.J. Couch (Swinburne), S. Croom (USyd), N.J.G. Cross (Edin), S.P. Driver (StA), L. Dunne (Nott.), S. Eales (Cardiff), E. Edmondson (Ports), S.C. Ellis (USyd), C.S. Frenk (Durham), A.W. Graham (Swinburne), H. Jones (AAO), D. Hill (StA), A. Hopkins (USyd), E. van Kampen (Inns), K. Kuijken (Leiden), O. Lahav (UCL), J. Liske (ESO), J. Loveday (Sussex), B. Nichol (Ports.), P. Norberg (Edin), S. Oliver (Sussex), H. Parkinson (Edin), J.A. Peacock (Edin), S. Phillipps (Bristol), C.C. Popescu (UCLan), M. Prescott (LJMU), R. Proctor (Swinburne), R. Sharp (AAO), L. Staveley-Smith (UWA), W. Sutherland (QMW), R.J. Tuffs (MPIK), S. Warren (Imperial).

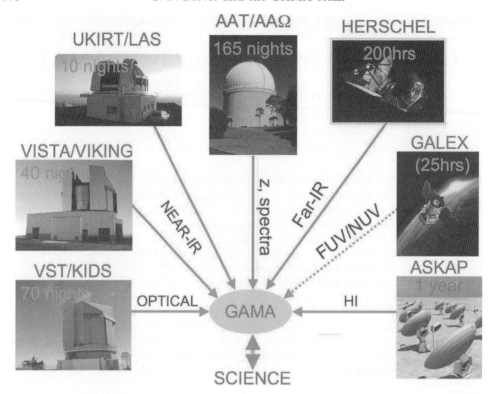

Figure 1. Facilities contributing to the final GAMA database.

components (AGN, nucleus, bulge, pseudo-bulge, bar, disc etc) and constituents (SMBH, plasma, stars, gas, dust etc). It then follows that a clear understanding of galaxy formation and evolution may only come about via the construction of a comprehensive survey which simultaneously samples all of these facets. The GAMA team aims to provide this data. In addition to the provision of a generic galaxy database, the GAMA project also includes a number of more focussed science goals, in particular:

1. Measurement of the Halo Mass Function via virialised group velocity dispersions to directly test the *numerical* prediction from CDM (and WDM) simulations.

2. Measurement of the dynamic, baryonic, HI and stellar mass functions to LMC masses versus redshift, environment, type, and component (as well as higher order relations, e.g., mass-spin $[M - \lambda]$).

3. Measurement of the recent merger rates and star formation rates versus type, mass and environment over a 3–4 Gyr baseline.

2. Facilities Contributing to GAMA

Fig. 1 shows the facilities currently contributing to the GAMA project along with the approximate time allocations within the GAMA sky regions (see Fig. 2 and Tables 1 & 2). The expected source resolution and detection sensitivities (5σ point source) are shown in Fig. 3 (upper and lower) in arcseconds and milliJanskys. Overlaid on the lower panel is the modelled NGC891 spectra (Popescu *et al.* 2000) with a weak AGN added and transposed to $z \approx 0.1$. The UKIRT data is provided courtesy of the UKIDSS LAS Public

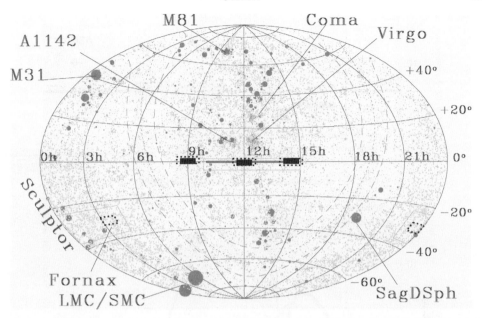

Figure 2. (black rectangles) Survey regions of first year data obtained in 21 clear nights as compared to the 2dFGRS (green outline) and MGC (blue band) surveys. Also shown are galaxies within 10Mpc (red dots), abell clusters (yellow), bright stars (cyan) and the NGC catalogue (magenta). The dashed regions show the final GAMA survey extent.

Survey while the VST and VISTA data are provided via the KIDS and VIKING ESO Public Surveys (whose teams include GAMA members). The Herschel data is provided as part of the broader Herschel-ATLAS survey and a proposal is currently pending to complete GALEX medium depth observations of the GAMA regions (~50% already covered with MIS). A major advancement over previous surveys will be the inclusion of radio data via ASKAP (Australian Square Kilometer Array Pathfinder, see Johnston et al. 2007) which should allow HI mass, dynamical mass and continuum measurements for all GAMA galaxies with high or normal neutral gas content. The deep ASKAP pointing (a single $6^o \times 6^o$ field of 1 year integration) is predicted to have an n(z) distribution comparable to that derived for GAMA (see Fig. 4). While the exact location of the ASKAP deep pointing has not been finalised it is highly likely, given the similarity in the n(z) distributions that one of the GAMA fields will be adopted (nominally the 12hr field).

Table 1. The current extent of the GAMA survey showing the year 1 regions and two possible expansion options currently under consideration.

GAMA Field ID	Year 1 Regions		Extension 1		Extension 2	
	RA(deg)	δ(deg)	RA(deg)	δ(deg)	RA(deg)	δ(deg)
G09	**129.0 − 141.0**	**-1 − +2**	129.0 − 141.0	-1 − +3	129.0 − 141.0	-3 − +3
G12	**174.0 − 186.0**	**-2 − +1**	174.0 − 186.0	-2 − +2	129.0 − 186.0	-3 − +3
G14	**211.5 − 223.5**	**-1 − +2**	211.5 − 223.5	-2 − +2	211.5 − 223.5	-3 − +3
G03	−	−	45.0 − 57.0	-28 − -31	−	−
G22	−	−	348.0 − 360.0	-28 − -31	−	−

472   S. P. Driver and the GAMA team

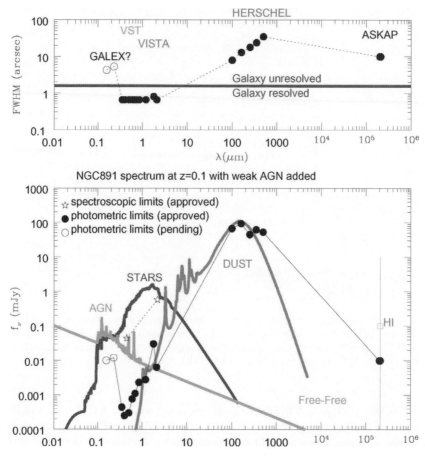

Figure 3. (lower) Spectroscopic, photometric and (upper) resolution sensitivities in mJsy and arcseconds respectively. Overlaid are the typical AGN (green), stellar (blue) and dust emissions (red; based on NGC891 at $z \approx 0.1$ with weak AGN added).

3. First Light

The survey commenced March 2008 with 50k spectra obtained in 21 clear nights using the Anglo Australian Observatory's new multi-fibre-fed bench-mounted dual-beam spectroscopic system (AAΩ). Fig. 5 shows the areas of sky surveyed (upper) and the resulting cone plot (lower). This includes the existing 25k redshifts within these regions from the MGC, SDSS and 2dFGRS. AAΩ represents an upgrade of the pre-existing 2dF system using the same fibre positioner/tumbler but replacing the two telescope mounted spectrographs with a single bench-mounted, double-beam spectrograph (see Sharp et al. 2006 or the AAO website). The facility can be used for both multi-fibre and integral field spectroscopy and in multi-fibre mode is capable of obtaining 350–400 spectra in a single 2° diameter field. During an 8hr observation period the system is capable of obtaining ~3000 spectra. Data are reduced in real-time and redshifts also obtained in real-time via cross-correlation with a template library. All data are later re-reduced and processed with GANDALF (see Schawinski et al. 2007) to obtain line indices and velocity dispersion measurements. The GAMA survey at the AAT uses the 580V and 385R gratings yielding a resolution of 1300 or 3–7ÅFWHM.

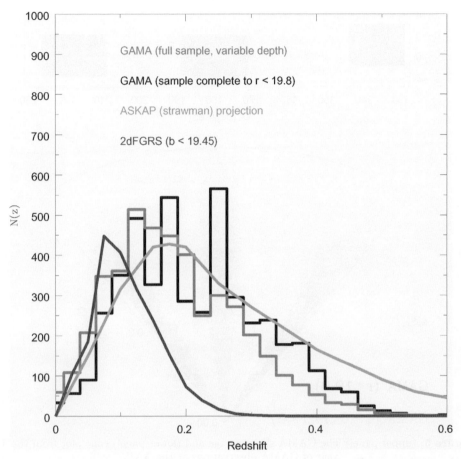

Figure 4. The GAMA n(z) distribution compared to that expected by the ASKAP 1 year deep stare, aribtrarily normalised to give a comparable height peak.

Table 2. Time allocations, resolutions and sensitivities of the facilities contributing to the GAMA survey. Limits are for 5σ detections in AB mag or mJsky ($1Jsky = 3631 \times 10^{-0.4m_{AB}}$).

Window	Facility	Collab.(Time)	Detection limits					Resol.	GAMA Fields
UV	GALEX	MIS + 25hrs pend.	FUV 23.0		NUV 23.0			FUV/NUV 4–5″	G09,G12,G15
Opt	SDSS	DR6	u 22.0	g 22.0	r 22.2	i 21.3	z 20.5	u-z 1.0″-2.5″	G09,G12,G15
	VST	KIDS (70n)	24.8	25.4	25.2	24.2	–	0.6″ – 1.0″	All
Near-IR	UKIRT	LAS (10n)	Z –	Y 20.9	J 20.6	H 20.3	K 20.1	Z-K 0.6″-1.0″	G09,G12,G15
	VISTA	VIKING (40n)	23.1	22.3	22.1	21.5	21.1	0.6″-0.8″	All
Far-IR	Herschel	ATLAS (200hrs)	110μm 67mJ	170μm 94mJ	250μm 45mJ	350μm 62mJ	500μm 53mJ	110–500μm 8–35″	All
Radio	ASKAP	DEEP (1 year)	0.7-1.8 GHz 10μJ					0.7-1.8GHz ∼10″	G12?

4. Summary

The GAMA project has commenced with data flows imminent from a number of international facilities. The survey will allow for a comprehensive study of structure on 1kpc to 1Mpc scales as well as the subdivision of the galaxy population into its distinct

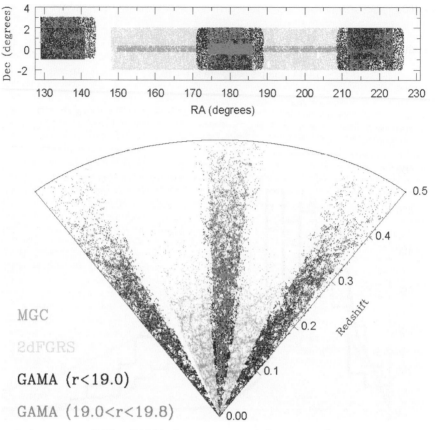

Figure 5. (upper panel) The GAMA sky coverage and (lower panel) cone plot from the first year of GAMA observations at the AAT.

components and constituents. Progress and data releases (1st data release forecast for Dec 2009) can be monitored via the GAMA website: http://www.eso.org/~jliske/gama/ and anyone interested in further details should contact Simon Driver at spd3@st-and.ac.uk.

The GAMA team acknowledges all staff of the Anglo Australian Observatory for provision of the superb AAΩ facility and the continued smooth running of the Anglo Australian Telescope. SPD thanks the IAU254 organisers for a most enjoyable meeting.

References

Allen, P. D., Driver, S. P., Graham, A. W., *et al.*, 2006, MNRAS, 371, 2
Driver, S. P., Liske, J., Cross, N. J. G., *et al.*, 2005, MNRAS, 360, 81
Driver, S. P., Popescu, C. C., Tuffs, R. J., *et al.*, 2008, ApJL, 678, 101
Johnston, S., *et al.*, 2007, PASA, 24, 174
Liske, J., Lemon, D. J., Driver, S. P., *et al.*, 2003, MNRAS, 344, 307
Popescu, C. C., Misiriotis, A., Kylafis, N. D., *et al.*, 2000, A&A, 362, 138
Sharp, R., *et al.*, 2006, SPIE, 6269, 14
Schawinski, K., Thomas, D., Sarzi, M., *et al.*, 2007, MNRAS, 382, 1415

What will Gaia tell us about the Galactic disk?

Coryn A. L. Bailer-Jones

Max-Planck-Institut für Astronomie, Königstuhl 17, 69117 Heidelberg, Germany
email: calj@mpia.de

Abstract. Gaia will provide parallaxes and proper motions with accuracy ranging from 10 to 1000 microarcsecond on up to one billion stars. Most of these will be disk stars: for an unreddened K giant at 6 kpc, it will measure the distance accurate to 2% and the transverse velocity to an accuracy of about 1 km/s. Gaia will observe tracers of Galactic structure, kinematics, star formation and chemical evolution across the whole HR diagram, including Cepheids, RR Lyrae, white dwarfs, F dwarfs and HB stars. Onboard low resolution spectrophotometry will permit – in addition to an effective temperature estimate – dwarf/giant discrimination, metallicity measurement and extinction determination. For the first time, then, Gaia will provide us with a three-dimensional spatial/properties map and at least a two-dimensional velocity map of these tracers. (3D velocities will be obatined for the brighter sources from the onboard RV spectrograph). This will be a goldmine of information from which to learn about the origin and evolution of the Galactic disk. I briefly review the Gaia mission, and then show how the expected astrometric accuracies translate into distance and velocity accuracies and statistics. I then briefly examine the impact Gaia should have on a few scientific areas relevant to the Galactic disk, specifically disk structure and formation, the age–metallicity–velocity relation, the mass–luminosity relation, stellar clusters and spiral structure. Concerning spiral arms, I note how a better determination of their locations and pattern speed from their OB star population, plus a better reconstruction of the Sun's orbit over the past billion years (from integration through the Gaia-measured gravitational potential) will allow us to assess the possible role of spiral arm crossings in ice ages and mass extinctions on the Earth.

Keywords. space vehicles; telescopes; astrometry; stars: general; Galaxy: structure, stellar content, spiral arms; ice ages and mass extinctions

1. Gaia overview

Gaia is a high-accuracy astrometric satellite to be launched by ESA at the end of 2011. By measuring the positions of stars tens of times over a five year baseline, it can derive the mean position, the parallax and two-dimensional proper motion of each star. These are five components of the six dimensional phase space, the sixth component – the radial velocity – being provided for the brighter stars by an onboard spectrograph. Gaia delivers absolute parallaxes tied to an inertial (quasar–based) reference frame. Gaia, which is currently under construction, represents a major step beyond the enormously successful Hipparcos mission, and is currently the only large-scale astrometric mission beyond the planning stage.

Gaia will perform a survey of the entire sky complete to magnitude G=20 (V=20–22). This covers some 10^9 stars, a million quasars and a few million galaxies. The sky coverage is complete to this magnitude, bar about 1% of the sky area where Gaia is confusion limited. Gaia will achieve an astrometric accuracy of 12–25 μas at G=15 (providing a distance accuracy of 1–2% at 1 kpc) and 100–300 μas at G=20. These numbers are also the approximate parallax accuracy in μas and the proper motion accuracy in μas/year. The accuracy range reflects a colour dependency: better accuracy is achieved for redder sources (as more photons are collected). Astrometry and photometry are done in a broad ("white light") band (G). Gaia will also measure radial velocities to a precision of 1–15 km/s for stars with V=17 via R=11 500 resolution spectroscopy around the CaII triplet (the "Radial Velocity Spectrograph", RVS). To characterize all sources (which are detected in real time), each is observed via low dispersion prism spectrophotometry over 330–1000 nm with a dispersion between 3 and 30 nm/pixel (the "BP/RP" instrument). From this we will estimate the "usual" astrophysical parameters, T_{eff}, $\log g$ and [Fe/H], but also the line-of-sight extinction to stars individually, plus perhaps also [α/Fe] in some cases. Spectra from the radial velocity spectrograph will help the parameter determination for some stars brighter than about $V = 14$, and will also allow the detection of emission lines and abundance anomalies.

Gaia has a nominal mission duration of five years (plus a possible one year extension), and 2–3 years following the end of operations are planned to complete the data processing. (The astrometry is self-calibrating, so the data must be reduced globally/simultaneously to get the final solutions and best astrometric accuracy.) The final catalogue will be available in about 2020, proceeded by earlier data releases. For more information on the satellite, science and data processing see http://www.rssd.esa.int/Gaia and the proceedings volume by Turon et al. (2005) (also available from the website).

2. Distance and velocity accuracy and statistics

Distances are vital in every area of astronomy. We need them to convert 2D angular positions to 3D spatial coordinates, allowing us to reveal the internal structure of stellar clusters or map the location of the spiral arms, for example. Knowing the distance we can convert 2D (angular) proper motions to physical velocities, and the apparent luminosity to the intrinsic luminosity, a fundamental quantity in stellar structure and evolution studies. Parallaxes are the only method of direct distance determination which does not make assumptions about the target source. We can measure parallaxes to virtually anything (not just, say, eclipsing binaries) and virtually all other rungs in the distance ladder are ultimately calibrated by them.

2.1. How accuracy varies with distance

Apart from very bright sources, the parallax error from Gaia is dominated by photon statistics. Thus the parallax uncertainty, $\sigma(\varpi)$, varies with the measured flux, f, as $\sigma(\varpi) \propto 1/\sqrt{f}$. Therefore, for a source of given intrinsic luminosity, the parallax error is linearly proportional to distance, $\sigma(\varpi) \propto d$. The parallax itself of course also decreases with increasing distance, $\varpi = 1/d$, so the fractional parallax error, $\sigma(\varpi)/\varpi$, increases as distance squared, $\sigma(\varpi)/\varpi \propto d^2$. Converting a parallax to a distance is actually nontrivial, due to the need to account for Lutz-Kelker biases (see Binney & Merrifield 1998 for a discussion). However, we can roughly equate this fractional parallax error with the fractional distance error, hereafter abbreviated as **fde**.

The proper motion uncertainty varies with distance in the same way as the parallax, and the measured proper motion itself, for given space velocity, decreases linearly with increasing distance. Therefore, the transverse velocity uncertainty (not fractional), also increases with the square of the distance, $\sigma(v) \propto d^2$ (for a source with fixed M_V). A useful relation to remember is that a proper motion of 1 mas/yr for a source at 1 kpc corresponds to a transverse velocity of about 5 km/s. (Note that this is by definition because the parallax definition is related to the mean velocity of the Earth around the Sun.)

2.2. Accuracies for certain types of stars

The above relations are useful, because they enable us to see to what distance certain Galactic structure tracers can still be observed with useful accuracy. Here are some examples. A K1 giant, with an intrinsic luminosity of $M_V = 1.0$, will have an apparent magnitude of about $G = 15$ when at a distance of 6 kpc (assuming zero interstellar extinction). (For simplicity I ignore the colour dependence of both the V to G relationship and the astrometric accuracy, adopting the lower accuracy blue limit for the latter.) The fde is 1.5% and $\sigma(v) = 0.75$ km/s. This clearly allows us to trace this tracer with high accuracy in phase space.

The astrometric accuracy, e.g. $\sigma(\varpi)$, is a factor of $\sqrt{10}$ worse for a source which is 2.5 mag fainter; similarly the astrometric accuracy is a factor of two worse for a source which is 1.5 mag fainter (equivalent to doubling the distance for a source of constant M_V).

A G3 dwarf with $M_V = 5.0$ has $G = 16.5$ at a distance of 2 kpc, and so a parallax accuracy of about 40 μas. The parallax itself is 500 μas, so the fde is 8%. The transverse velocity accuracy is just 0.4 km/s Moving it to 4 kpc reduces the accuracy and parallax to 80 μas and 250 μas respectively, giving an fde of 30%.

The velocity accuracies in particular are very good when one considers typical velocities within the Galaxy (e.g. the rotation of the LSR about the Galactic centre is about 220 km/s). So as the distance increases, it is parallaxes rather than the proper motions which first become less useful, although at large distances the transverse velocity uncertainty becomes dominated by the parallax error. The distance at which the parallax error is 50% is 5 kpc for the G3 dwarf and 35 kpc for the K1 giant, where the latter has an apparent magnitude of $G = 18.8$. Thus we see that beyond a distance of about 40 kpc (depending on intrinsic magnitude), the Gaia parallaxes on *individual* objects are no longer much good, so we will have to rely on spectroscopic parallaxes (derived from the onboard spectrophotometry), in particular to convert proper motions to velocities.

2.3. Accuracy statistics

While it's easy to throw out these numbers, we must not forget the enormous improvement they represent in accuracy above what is currently possible. The improvement with respect to Hipparcos comes primarily through the larger field of view: Gaia simulaneously observes tens of thousands of stars on a large CCD focal plane, whereas Hipparcos could only observe one at a time (with a photomultiplier tube, PMT) (Lindegren 2005). In addition, the larger QE detector (ca. 75% for CCDs compared to around 1% for a PMT), the larger collecting area ($0.7\,m^2$ compared to $0.03\,m^2$) and the broader effective bandpass together provide a much larger detectable photoelectron flux per star.

The value of Gaia is not just accuracies of individual objects, but also the fact that it is an all sky survey with a relatively deep limiting magnitude (G=20), so these accuracies are achieved on large numbers of objects. (Note that the bright – saturation – limit of Gaia is $G = 6$, so it can just reach stars visible to the naked eye.) Roughly how many can be found via a Galaxy model. We find that Gaia will provide distances with an accuracy of 1% or better for 11 million stars. Taking 12 μas as the lower limit to the Gaia parallax accuracy, this means that all of these stars must lie within 800 pc (i.e. have a parallax less than 1250 μas). Some 150 million stars will have a distance accuracy better than 10% (all within 8 kpc) and some 100 000 stars better than 0.1% (all within 80 pc). For comparison, currently fewer than 200 stars have had their parallaxes measured to better than 1%. All are within 10 pc and all were measured with Hipparcos.

3. Astrophysical parameter estimation

Gaia does not use an input catalogue, so the properties of the sources are not known a priori. Gaia only observes point sources and the onboard processing generally bins the CCD data around a source in one-dimension (so that only the higher spatial resolution along-scan direction remains). Consequently there is little morphological information (although some can be reconstructed from the multiple scans of a given source which occur at a range of positions angles). There are, therefore, two main classification tasks with the Gaia data. The first is a discrete source classification, to discriminate between stars, galaxies and quasars (and perhaps also unresolved binaries and other types). Identification of a clean quasar sample (Bailer-Jones & Smith 2008) is necessary in particular for the global astrometric solution.

The second step is to estimate the stellar astrophysical parameters (APs). This is done primarily using the low resolution prism spectrophotometry (BP/RP), as well as the parallax when available. The classification software is still under development, but we have preliminary estimates of the performance of a stellar parametrizer based on support vector machine classifiers applied to synthetic spectra convolved with a Gaia instrument and noise model. At $G = 15$, the RMS uncertainties in the derived APs are as follows: 1–5% in T_{eff} for a wide range of effective temperatures (OB to K stars); line-of-sight extinction, A_V, to 0.05-0.1 mag for hot stars; < 0.2 dex in [Fe/H] for spectral types later than F down to -2.0 dex metallicity; $\log g$ to between 0.1 and 0.4 dex for all luminosity classes, but to better than 0.1 dex for hot star (OBA stars). It should be noted that the performance varies a lot with the physical parameters themselves, as well as with G magnitude (i.e. signal-to-noise ratio). Work in this area is ongoing.

The inferred stellar parameters are tied, of course, to the stellar models used to simulate the synthetic spectra on which the classification models are trained. (If the models are instead trained directly on real spectra, then these must somehow be parametrized using models, so our results are always tied to physical stellar models). In addition to developing classification algorithms, the Gaia DPAC (Data Processing and Analysis Consortium) is also working on improving the stellar input physics and model atmospheres. A ground-based observing programme will provide some data to calibrate the fluxes of the models.

4. Gaia science

I now outline a few areas of disk astrophysics where Gaia should make significant contributions.

4.1. Disk structure

Many aspects of the thin and thick disk structure are not well known (see Vallenari et al. 2005 for a summary). For example, thin disk scale height estimates vary between 200 and 330 pc. The inferred scale length varies between 2–3 kpc (as measured by star counts and integrated light) and 2–4 kpc (as measured by kinematics). The vertex deviation and the vertial tilt are not well known. All of these measurements of the velocity elipsoids are based on relatively small samples. Gaia will provide much larger samples with 3D spatial structure and 2D or 3D velocity measures. Transverse velocities accurate to about 1 km/s will be achieved for some 50 million stars out to a few kpc.

4.2. Substructure and mergers

One of the most important questions Gaia will address is that of how and when the Galaxy formed. ΛCDM models of galaxy formation predict that galaxies are built up by the hierarchical merger of smaller components and some suggest that the halo is composed primarily of merger remnants (for a review see Freeman & Bland-Hawthorn 2002). ¿From extragalactic observations there is good evidence for both the accretion of small components and for the merging of similar-sized galaxies. Within our own Galaxy, recent surveys – 2MASS and SDSS in particular – have found the fossils of past and ongoing mergers in the halo and possibly also the disk of our Galaxy. They have been identified primarily as spatial overdensities in two-dimensional (angular) photometric maps of large areas of the sky (sometimes accompanied with radial velocities). In a few cases, distance measures have been included by taking magnitude as a distance proxy (Belokurov et al. 2006) or by examining the 2D density of a limited range of spectral types (i.e. using a spectroscopic parallax of some tracers) (Yanny et al. 2000). But because merging satellites are disrupted by the Galactic potential and the material spread out after several orbits, density maps are a limited means of finding substructure. Without an accurate distance the interpretation of 2D maps is plagued by projection effects. Even with perfect 3D maps, the contrast against the background (including other streams) is often low (Brown et al. 2005). To improve this, we need 3D kinematics, i.e. radial velocities and proper motions (combined with distances). In an axisymmetric potential the component of angular momentum parallel to this axis (L_z) of a merging satellite is an integral of motion. In a static potential, the energy (E) is also an integral of motion (Binney & Tremaine 1987). Thus while a merging satellite could be well-mixed spatially, it would remain (approximately) unperturbed in (L_z, E) space. Of course, the Galactic potential is neither time-independent nor perfectly axisymmetric and Gaia has measurement errors, but simulations have demonstrated that Gaia will be able to detect numerous streams 10 Gyr or more (i.e. many orbits) after the start of their disruption (Helmi & de Zeeuw 2000).

Gaia will perform a 5D phase space survey over the whole sky, with the sixth component – radial velocity – being available for stars brighter than V=17. At this magnitude, spectral types A5III, A0V and K1III (which all have $M_V \simeq 1.0$) are seen at a distance of 16 kpc (for zero extinction). The corresponding proper motion accuracy is about 50 μas/yr, or 4 km/s. In addition to the 5D or 6D phase space information, Gaia provides abundances and ages for individual stars. Search for patterns in this even higher dimensional space permits an even more sensitive (or reliable) search for substructure.

4.3. Dark matter

Two distinct aspects of the Gaia mission permit us to study the mass and distribution of dark matter in our Galaxy. First, from the 3D kinematics of selected tracer stars, Gaia will map the total gravitational potential (dark and bright) of our Galaxy, in particular the disk. Second, from its parallaxes and photometry, Gaia will make a detailed and accurate measurement of the stellar luminosity function. This may be converted to a (present-day) stellar mass function via a Mass–Luminosity relation (see section 4.5). From this we can infer a stellar mass distribution. Subtracting this from the total mass distribution obtained from the kinematics yields the dark matter distribution.

4.4. Stellar structure and stellar clusters

Stellar luminosity is a fundamental property, so its measurement across a range of masses, ages and abundances is a critical ingredient for testing and improving stellar models. In open and globular clusters an accurate determination of luminosities and effective temperatures (which Gaia also provides) gives us the HR diagram for different stellar populations. (To derive an accurate luminosity we also need an accurate estimate of the line-of-sight extinction. This will be obtained star-by-star from the Gaia spectrophotometry.) We may then address fundamental questions of stellar structure, such as the bulk Helium abundance (which is not observable in the spectrum), convective overshooting and diffusion. One of the main uncertainties in the age estimation in clusters is accurately locating the main sequence (for open clusters) or main sequence turn off (for globular clusters). Gaia's accurate parallaxes and unbiased (magnitude-limited) survey will greatly improve this.

In addition to using clusters as samples for refining stellar structure and evolution, we can also study them as populations. Gaia will observe many hundreds of clusters, allowing us to determine the (initial) mass function into the brown dwarf regime and examine its dependence on parameters such as metallicity, stellar density and environment. There are perhaps 70 open clusters and star formation regions with 500pc. For these, Gaia will provide individual distances to stars brighter than G=15 to better than 1%. This will permit us, for the first time, to map the 3D spatial structure of many clusters, with a depth accuracy as good as 0.5–1 pc for clusters at 200 pc. From the 3D kinematics we can likewise study the internal dynamics of a cluster. A G=15 star will have its proper motion measured with an accuracy of 20 μas/yr, corresponding to a speed uncertainty of 0.1 km/s at this distance (half this for a red star). The speed uncertainty varies linearly with the distance for a fixed magnitude, so at 200 pc the speed uncertainty is just 20 m/s. With this accuracy we can measure the internal kinematics of the cluster and so investigate the phenomena of mass segregation, low mass star evaporation and the dispersion of clusters into the Galactic field. (I have assumed that the uncertainty in the transverse velocity is dominated by the proper motion error and not the distance error. For the sake of the proper motion to velocity conversion we can assume all stars to be at the same distance, and therefore average over many stars to get a more accurate distance.)

At larger distances Gaia will be able to say less about the internal structure of an open cluster. However, using the proper motions to help define membership, the average distance to the cluster, plus its age and metallicity, can be measured accurately. In this way, we can use a few thousand open clusters over distances of tens of kpc as tracers of the disk abundance gradient.

Just as Gaia should be able to identify the fossils of past mergers from their phase space substructure (section 4.2), so the 6D phase space data plus astrophysical parameters for tens of millions of stars will allow Gaia to detect new stellar clusters, associations or moving groups based on their clustering in a suitable multi-dimensional parameter space. It can likewise confirm or refute the existence of controversial clusters.

4.5. *Stellar mass–luminosity relation*

Gaia will detect many binary systems. These are found in a number of ways, as summarized by Arenou & Söderhjelm (2005). Most are found via the astrometry as astrometric solutions which do not fit the standard 5-parameter model. These include both Keplerian fits to the data (for periods up to 10 years) and nonlinear motions (accelerated proper motions, periods up to a few hundred years). Very wide binaries can be found as common proper motion pairs. In addition, bright binaries can be found with the radial velocity spectrograph and as eclipsing binaries. Unresolved binaries of different spectral types with brightness ratios not too far from unity can be found as part of the classification work, from the identification of two spectral energy distributions.

For those systems with orbital periods of about ten years or less, Gaia can solve for the orbital elements and for the total mass of the system. If the components of the system are spatially resolved then we may determine their individual masses. Gaia furthermore measures accurate intrinsic luminosities. Together these allow us to determine the stellar Mass–Luminosity relation, and to do it with more stars and over a wider mass range that has yet been performed. (As Johannes Andersen pointed out after my talk, it may not be useful to talk about achieving an accurate Mass–Luminosity relation because such a thing does not exist, in the sense that the luminosity of a star is not uniquely determined by its mass, but also by abundance, rotation etc. Gaia could nonetheless shed some light on these extra dependencies.)

5. Spiral arms, ice ages and mass extinctions

Several papers have reported on a correlation and possible causal connection between spiral arm passages and ice ages and/or mass extinctions on the Earth. One possible mechanism is the exposure to massive stars in star forming regions: the increased cosmic ray flux from type Ib/II supernovae within 10 pc during such passges could increase terrestrial cloud cover (through water drop nucleation) and thus lower global temperatures for millions of years (e.g. Ellis & Schramm 1995), and/or the OB star UV flux could destroy the ozone layer and cause widespread extinction. A second mechanism is that the passage through regions of larger stellar and gas density found in spiral arms could perturb the Oort cloud and send minor bodies into the inner solar system, where they could impact on the Earth.

Through accurate measurements of the 3D kinematics of sars, Gaia will map the gravitational potential of the Galaxy and accurately determine the velocity of the Sun. Via numerical integration we can then reconstruct the path of the Sun through the Galaxy over the last few hundred million years and examine whether past ice ages and mass exinctions coincide with spiral arms passages. This has been done by Gies & Helsel (2005) using the potential and solar motion derived from Hipparcos by Dehnen & Binney (1998a,b). While they found some correlation between arm passages and ice ages, the results depended heavily on the poorly known position and velocity (pattern speed) of the spiral arms.

With Gaia we can dramatically improve this analysis: the astrometric accuracies are better than Hipparcos by a factor of 500 (12-25 μas compared to 1000 μas), it includes many more stars (1 billion compared to 120 000) and extends to fainter magnitudes (20 rather than 12.4). With these data we can determine the gravitational potential at higher spatial resolution and therefore reconstruct the solar motion more accurately. Moreover, we can measure the position and velocities of the spiral arms themselves from the Gaia observations of their OB star population without assuming a rotation curve or needing to know the extinction. For an OB star at a distance of 4 kpc from the Sun observed through 4 magnitudes of extinction, Gaia will determine its distance to an accuracy of 13%, its space velocity to about 1 km/s. Gaia can do this for some 50 000 OB stars within a few kpc, allowing to build a more accurate model for the spiral arm kinematics and thus make more conclusive statements about the correlation between arm passages and Earth cataclysms.

Finally, Gaia will make a survey of Near-Earth Objects (NEOs). Gaia does real-time onboard object detection, so is sensitive to transient phenomena and fast moving objects. Gaia is predicted to detect some 16 000 NEOs (Mignard 2002). While ground-based surveys will discover many more in the coming years, Gaia can derive accurate orbits and is sensitive to parts of the orbital parmeter space which cannot easily be reached from the ground. It should make a significant contribution to the census of potential Earth impactors.

References

Arenou, F. & Söderhjelm, S. 2005, in: Turon, C., O'Flaherty, K. S., Perryman, M. A. C. (eds.), *The Three-Dimensional Universe with Gaia*, ESA, SP-576, p. 557

Bailer-Jones, C. A. L., Smith, K. S., Gaia Technical Note, GAIA-C8-TN-MPIA-CBJ-036, available from `http://www.mpia.de/GAIA`

Belokurov, V., Zucker, D. B., & Evans, N. W., *et al.* 2006, *ApJ* 642, L137

Binney, J. & Merrifield, M. 1998, Galactic Astronomy, Princeton University Press

Binney, J. & Tremaine, S. 1987, Galactic Dynamics, Princeton University Press

Brown, A. G. A, Velázquez, H. M. & Aguilar, L. A. 2005, *MNRAS* 359, 1287

Dehnen, W. & Binney J. 1998a, *MNRAS* 294, 429

Dehnen, W. & Binney J. 1998b, *MNRAS* 298, 387

Ellis, J. & Schramm, N. 1995, Proc. Natl. Acad. Sci. 92, 235

Freeman, K. & Bland-Hawthorn, J. 2002, *ARAA* 40, 487

Gies, D. R. & Helsel, J. W. 2005, *ApJ* 626, 844

Helmi, A. & de Zeeuw, P. T. 2000, *MNRAS* 319, 657

Migarnd, F. 2002, *A&A* 393, 727

Lindegren, L. 2005, in: Turon, C., O'Flaherty, K. S., Perryman, M. A. C. (eds.), *The Three-Dimensional Universe with Gaia*, ESA, SP-576, p. 29

Turon, C., O'Flaherty, K. S., & Perryman, M. A. C. (eds.), 2005, *The Three-Dimensional Universe with Gaia*, ESA, SP-576

Vallenari, A., Nasi, E., Bertelli, G., & Chiosi, C., S. 2005, in: Turon, C., O'Flaherty, K. S., Perryman, M. A. C. (eds.), *The Three-Dimensional Universe with Gaia*, ESA, SP-576, p. 113

Yanny, B., Newberg, H., & Kent, S., *et al.* 2000, *ApJ* 540, 825

A Roadmap for Delivering the Promise of Gaia

T. Prusti[1], C. Aerts[2], E. K. Grebel[3], C. Jordi[4], S. A. Klioner[5], L. Lindegren[6], F. Mignard[7], S. Randich[8] and N. A. Walton[9]

[1]ESA, ESTEC, The Netherlands
email: tprusti@rssd.esa.int

[2]Katholieke Universiteit Leuven, Faculteit Wetenschappen, Belgium

[3]Universität Heidelberg, Zentrum für Astronomie, Germany

[4]Universitat de Barcelona, Facultat de Fisica, Spain

[5]Technische Universität Dresden, Germany

[6]Lund Observatory, Sweden

[7]Observatoire de la Côte d'Azur, France

[8]INAF, Osservatorio Astrofisico di Arcetri, Italy

[9]University of Cambridge, Institute of Astronomy, United Kingdom

Abstract. Gaia development is in full speed aiming to the launch in December 2011. While the entities formally responsible to make Gaia happen are very focused to be ready in time, it is necessary to explore the readiness of the wider scientific community to exploit Gaia data in the future. The GST sees a role for itself in building and enabling the community to use the Gaia catalogues when they become available. This paper gives the background for the current activities the GST is taking to promote the community to start preparations for the exploitation of Gaia data.

Keywords. space vehicles, surveys, astrometry

1. Introduction

Gaia is an ESA space astrometry mission with main goals related to the origin, structure and evolutionary history of our Galaxy. The scientific questions are addressed by Gaia with an all sky survey covering in addition to astrometry also photometry and spectroscopy. The astrometric and photometric survey will be down to the 20th magnitude leading to measurements of about one billion stars. The spectroscopic instrument is less sensitive providing an all sky survey down to the 17th magnitude. Due to the all sky aspect the scientific results from Gaia will reach beyond those of the primary goals. Gaia will contribute in many areas of astronomy, stellar astrophysics, solar system studies and general relativity just to mention a few. The details of the scientific performance and applications to the topic of this meeting are covered by the accompanying Gaia presentation by Bailer-Jones (2008) in these proceedings.

This contribution will address, as the title suggests, the way the scientific results will be optimally achieved. The roadmap to deliver Gaia has two elements. The functional part of building the spacecraft and data processing system to provide the results are essential pre-conditions to make Gaia a success. The status and organisational structure for this are briefly summarised in this contribution. The main emphasis of this paper is on the less direct organisational aspect: how to involve the "astronomical world" outside Gaia to maximise the scientific output from the mission.

2. Overview of the Gaia project

At the top level the Gaia project can be divided into ESA, industrial and scientific community elements. In the case of Gaia the whole spacecraft including the payload is built by industry with EADS Astrium as the prime industrial contractor. Another industrial component in the mission is the launcher which is the Soyuz/Fregat managed by Starsem. The launch will take place in Kourou. The ESA role is the traditional management of the industrial contracts and participation in the operations. The community has a crucial role in Gaia by providing the Data Processing and Analysis Consortium (DPAC) which is responsible of producing the immediate scientific output of the mission: the Gaia catalogue. In order to ensure the optimal scientific operations, the ESA Science Operations Centre has been integrated into DPAC to form a single entity responsible of the scientific operations and data processing.

All the above mentioned mission elements have been selected and are in full speed to complete their part of the project. EADS Astrium passed successfully the Preliminary Design Review (PDR) 2007 and is currently in Phase C/D completing the details of the design and building the flight hardware. The Data Processing and Analysis Consortium was formally approved 2007 and the same year passed successfully the System Requirements Review (SRR) together with the ESA Mission Operations Centre (MOC). Currently Gaia is on schedule for a launch in December 2011.

In addition to the entities responsible of implementing Gaia, the project, like every ESA project, has also advisory bodies providing guidance and recommendations to all aspects of the missions. For Gaia the closest advisory body is the Gaia Science Team (GST), who are authoring this roadmap paper, with the responsibility to provide the scientific advise for all aspects of the project.

3. Gaia project and the astronomical community

Although the Gaia project elements already in place form a complete entity that will make Gaia happen, the scientific part of Gaia cannot be separated from the overall scientific environment. The astronomers in DPAC are part of the overall scientific community, but in addition Gaia needs the scientific world outside the strict boundaries of the project. By definition this interface cannot be formalised, but neither can it be left on its own with the hope that the matter sorts out by itself as a natural process between scientists.

The classical example of (difficult) project interaction with the rest of the astronomical world is the need of calibration observations. Calibration observations for another project is not exactly the favourite proposal for any time allocation committee. In Gaia the ground based observations needs are coordinated by the DPAC entity GBOG (Ground Based Observations for Gaia). GBOG is facing the typical problem of calibration proposal writing where the proposal needs to be scientifically motivated rather than by Gaia calibration needs. While many calibration observations can be obtained in observing projects with a direct scientific goal next to the Gaia calibration needs, there are always calibration observations needed which simply cannot be embedded in a regular observing proposal. Yet these observations are needed for the optimal science from Gaia. This problem is generic and by no means unique to Gaia. In this topic the attempt to coordinate ground based facilities in Europe under the ASTRONET consortium is an interesting development and it is desirable that any concept coming out from that process will also consider the need for calibration observations.

When moving away from the interface between Gaia and ground based calibration observations, the needs become less clear. In principle the baseline is very clear. Gaia is based on data release policy without any proprietary rights. This means that Gaia needs to document the catalogue and its contents when it is published, and the work after that is up to the scientific community. Undoubtedly this way the scientific harvest will also follow, but it is equally sure that with good preparations the exploitation is not only quicker, but also with more depth and breadth. This is an area where the GST has the role to bridge the gap between Gaia and the outside world for maximum scientific return.

In addition to the final Gaia catalogue, intermediate catalogues and science alerts will be published in the course of the mission. It is obvious that any scientific follow up based on the Gaia results is left to the community at large. The issue is that some follow up observations may be time consuming to obtain. In this kind of cases it is better not to wait till the publication of the Gaia catalogue, but rather to inform the community to start "follow up" preparations already earlier. Another example of early preparation need is e.g. a requirement to follow up with facilities which are available now but won't be available in the Gaia era. A pre-requisite of this activity is to provide the community with information of Gaia performances, including biases or shortcomings, and schedule also in the context of other contemporary survey projects. Rather than assuming the step of informing the community to be a task to be completed as soon as possible, we should be looking into an iterative and interactive process. A positive side effect of conducting this process interactively is the community building aspect. We must not only worry about having all the follow up preparations done in time, but also ensure a wide enough community to exploit the data. The GST is going to be actively involved in this process.

As an example of a concrete action taken at the time when these proceedings are written, the GST is currently (September 2008) soliciting expressions of interest for the GREAT programme (Gaia Research for European Astronomy Training; see details in http://www.ast.cam.ac.uk/GREAT/). The aim is to build a research network for the promotion of topical workshops, conferences, training events, exchange visits, publications and outreach activities addressing the major scientific issues that the Gaia satellite will impact upon.

4. Conclusions

Gaia is a mission scheduled for launch in December 2011 with anticipated first release of an intermediate catalogue some years after. Who are going to use these data? What other data should be obtained already now? These are issues which can be addressed together with the wider scientific community. GST sees its role in this community building and enabling process and is at the moment of this proceedings being written taking concrete steps toward the community by probing their interest to join the effort under the GREAT concept.

References

Bailer-Jones, C. 2008, *these proceedings*, p. 475

From Kapteyn to Gaia: Adriaan Blaauw reviewing 70 years of Galactic research, including half a century with Bengt Strömgren.

The Science of Galaxy Formation

Gerard Gilmore

Institute of Astronomy, Madingley Road, Cambridge CB3 0HA, UK

Abstract. Our knowledge of the Universe remains discovery-led: in the absence of adequate physics-based theory, interpretation of new results requires a scientific methodology. Commonly, scientific progress in astrophysics is motivated by the empirical success of the "Copernican Principle", that the simplest and most objective analysis of observation leads to progress. A complementary approach tests the prediction of models against observation. In practise, astrophysics has few real theories, and has little control over what we can observe. Compromise is unavoidable. Advances in understanding complex non-linear situations, such as galaxy formation, require that models attempt to isolate key physical properties, rather than trying to reproduce complexity. A specific example is discussed, where substantial progress in fundamental physics could be made with an ambitious approach to modelling: simulating the spectrum of perturbations on small scales.

Keywords. Galaxy: formation, Galaxy: disk, sociology of astronomy, elementary particles

1. The Scientific Method

Astrophysics challenges the limits of our scientific methodologies. We have no control over what Nature allows us to 'observe', and much of what we can observe involves complex non-linear physics. At the same time, astrophysics challenges the limits of our concepts of "reality", so that our adopted methodolgy is important. Significant astrophysical queries include the form(s) of the dominant types of matter in the Universe, the nature of zero-point energy, and, what may be related, the interpretation of the observed acceleration of the expansion of the Universe, among other Big Questions. The appropriate scientific methodology with which to address such questions is itself problematic: how does one apply what many consider the "traditional scientific method", involving objective analysis of independent repeated experiments as a test of theory, when the Universe does not allow us to experiment, in the traditional laboratory physics sense; when we have no useful predictive theory for much of astrophysics; and when the nature of the Universe may restrict our observation to only a very small part of an unobservable larger whole? More specifically, is the observational test of prediction how science actually operates? Is that how astrophysics operates?

The scientific method as popularly conceived is essentially the application of reason to experience, independent of authority. This concept has a long and complex evolutionary history, with many notable figures in its history, from classical Greece, through Ibn Tufayl (see e.g. Cerda-Olmedo 2008), William of Occam's "Entia non sunt multiplicanda praeter necessitatem", Francis Bacon's discourse in his "Novum Organum", Copernicus, Galileo and many more great scientists and philosophers. In his paper to the Royal Society in November 1801, "On the theory of light and colours", Thomas Young updates Newton's "Hypotheses non fingo" in his introduction by "Although the invention of plausible hypotheses, independent of any connection with experimental observations, can be of very little use in promotion of natural knowledge...", before introducing what we now know as one of the great successes and great challenges of the scientific method, that light behaves as both a wave and a particle. Niels Bohr, when becoming a Knight of the Elephant in 1947, adopted the motto "Contraria sunt Complementa" (opposites are complementary), recognising the more general importance of wave-particle duality in quantum mechanical descriptions of Nature.

This raises two of the more unexpected consequences of application of the scientific method - is there such a concept as a single "answer", and do the resulting theories describe how the world "really is"? How can they, if apparently inconsistent descriptions are both valid? Is there such a thing as "truth" in science or Nature? Again to quote Bohr "It is wrong to think that the task of physics is to find out how Nature is. Physics concerns what we say about Nature". Or, among many hundreds of similar discussions of the meaning of probability and the role of the observer in quantum mechanics, von Neumann notes the prime requirement of a model is that "it is expected to work". It may well be that abandoning the classical notion of "realism" is the latest step we must take in our Copernican path to remove observer-specific influence and authority from our application of reason to some generalised concept of experience (*cf* the discussion in Leggatt 2008).

Astrophysicists are traditionally proud of their special role in what is often called the "Copernican Principle", the scientific methodology which applies scepticism to any model of a phenomenon in which there is a special role for the observer and/or interpreter. This methodology in astrophysics, and the name, is derived from empirical "success". Removing the special place for Mankind as the focus of all creation led to a sequence of models, ranging from Newtonian gravity, through general relativity, to modern precision cosmology. Along the way the Earth lost its central place in the Universe, followed by the Sun, then the Milky Way Galaxy. The concept of absolute time vanished, baryonic matter was dethroned by dark matter, mass-energy became secondary compared to dark energy. This last step is a significant extension of the Copernican Principle. If current speculations on long-term futures in a Universe dominated by dark energy, Multiverses, and so on, are relevant to "reality", the Universe may well be a concept in which what we see, and what we are, is a temporary fluctuation on what, for most of space-time, may be very, very different. Cosmic variance becomes not a consideration but the dominant factor limiting understanding. We, as observers, may be seeing - or may only be able to see - an extremely unusual, temporary, microstate, and have no direct knowledge of a much, much, larger macroscopic "reality".

For practising scientists, it is a matter of scientific habit that a "theory" which predicts a previously-unobserved phenomenon is considered supported by experiment. This overstates the case. While a positive outcome is certainly not neutral, in that the opposite outcome would lead to quite different reactions, no set of experiments can ever establish the "truth" of any theory. Even if theory **T** predicts outcome **O**, and **O** is observed, **T** *is not* proven. If **O** were outlandish, but observed, it is commonly assumed that **T** is more likely to be correct. While a successful test justifies continued use, and future testing, of that theory, **T** remains unproven. Supporting the correctness of **T** given the observation of **O** is the fallacy of "affirmation of the consequent" (cf. Leggatt (2008) for further discussion).

There is no fundamental theory supporting the validity of application of the "Copernican Principle". It is an assumption, whose future validity, and whose valid range of applications, is unknowable. It may well be limited. There is certainly no objective justification for its application in fields beyond those few where it has proven utility. As an illustration, public reaction to evolutionary biology, and the scientific realisation that modern, Cro-Magnon man has been painting caves and doing science for some 10^{-5} of the age of the Earth, remains of considerable complexity, and illustrates well the difficulty many people have in acting as dispassionate "Copernican" observers. There are indeed fields of intellectual enquiry where objective analysis, independent of the concept of authority, is inappropriate. A particularly interesting example is the debate in legal and political circles of the role of the US Supreme Court in interpreting the US Constitution. Many distinguished legal theorists insist that a positivist interpretation of what is written, free from the preferences of specific judges, is most appropriate. Others disagree. This debate intriguingly combines the concepts of an authoritative document, and an objective observer and interpreter. The creativity, sophistication, and continuation, of this debate illustrates the complexity of the issues. In micro-physics the meaning and role of the "observer" in Young's "experimental observations" and the concept of uniqueness, and/or completeness, of possible observations, have become more complex with developments in quantum mechanics. The continuing public interest in debating the validity of string theory as a science (eg Cartwright & Frigg 2007) is yet another illustration of both the importance of the questions, and the incompleteness, or at least complexity, of current interpretations of the terms "science", "scientific method", "theory" and "truth".

2. The Scientific Method in Galaxy Formation

With that context, it is perhaps unsurprising that astrophysics is implemented in a practical approximation to the philosophic ideal. Many great names in the development of twentieth century science declared, in essence, "don't worry too much about the philosophy, just find, and use, equations which calculate observables". Preferably previously un-predicted observables. In that context, what do we do, and what should we do, in astrophysics.

In practise, we adopt a paradigm, or set thereof, develop it/them in so far as is possible, testing against, and - hopefully - predicting, new observables. In that context significant advances have been made. In astro-particle physics, the interplay between solar structure models and neutrino astronomy is an exceptional example, as is the limitation of the numbers and masses of neutrinos from large scale structure studies. Steady State cosmology is another exceptional example – predictions were made, tested against observations, and the model found to be inappropriate as a description of the Universe. Science at its best. Such examples are however rare. Much of astrophysics either has no *ab initio* theory, or involves complex non-linear physics, so that robust and unique prediction is impossible. Given our experimental inability to isolate and test models of individual physical processes, since we are unable to experiment, we cannot "test" the outcome of a theory in astrophysics.

Much of what we do in astrophysics is similar to weather forecasting: weather forecasts use observations as boundary conditions, implement the most sophisticated available physics essentially as an interpolation (in space, in time, ...), exploit heroic achievements in computing, and extrapolate the observables to other places and times. Sometimes this is accurate, sometimes not, in which case the differences between prediction and observation are analysed to allow the forecasting system to be improved. Eventually, given enough data, and enough complexity in the model, weather forecasts will become, asymptotically, as accurate as the predictability of the system allows. They will reach a physical accuracy limit. But no weather forecast can ever be "right" or "wrong", in the sense that a scientific theory can be. A forecast may be accurate, or less accurate. It is unlikely even that any forecasting system could ever be unique, since there may well be many physical processes whose effects are comparable in amplitude to measurement error. A system with considerable complexity, and inevitable approximation, will invariably have many statistically-indistinguishable solution maxima.

Coming specifically to models of galaxy formation, we have a similar situation. There is an interesting distinction between (some) Galaxy (i.e., Milky Way) models and (some) galaxy (i.e. generic) models. Substantial progress is being made in development of specific models of the Milky Way Galaxy, particularly in preparation for Gaia. Gaia will produce information from which we expect to determine the current state of the Milky Way Galaxy in some detail, and hence to deduce something of how the Milky Way in particular, and, *modulo* cosmic variance, disk galaxies in general, formed and evolved. A systematic approach to modelling, analysing and interpreting the anticipated Gaia data is underway. The adopted strategy is to proceed through a sequence of models of increasing complexity, guided at each stage by analysis of mis-matches between the current model and available simulations, real data and on-going surveys, such as RAVE (see e.g. Binney (2002), as one example of the many underway). This process is intended to develop what is essentially a tool-kit for investigation of the Gaia dataset, and hence the Milky Way Galaxy. This modelling approach is, in a real sense, equivalent to a laboratory experiment, rather than being development of a theory.

Formation models for galaxies in general are very different in approach and ambition. They adopt analyses of the properties of the early universe, derived from observations of the cosmic microwave background, and supplementary data, as boundary conditions. These boundary conditions are unconstrained by observations on small scales, and so are extrapolated [usually as a simple power-law spectrum of fluctuations] down to as-yet unobserved physical length scales. This extrapolated set of boundary conditions is then evolved forward in time, requiring considerable sophistication and heroic achievements in computing. Approximations to the behaviour of baryons, and hence the properties of most observables, are then added. Comparison with observations of real galaxies, when made, has so far invariably identified gross discrepancies, indicating perhaps that more complex baryon physics is needed. Or different physics: perhaps the extrapolation of the observational boundary conditions is inappropriate? Unfortunately, analysis and interpretation of the predictions of these inevitably highly idealised models is complex.

In order to calculate "observables" *ad hoc* prescriptions for the key baryonic physics must be added by hand. Star formation, chemical elements, black holes and so on are added using some observationally motivated recipe. After unsuccessful comparison to observation, the complexity is increased, including both plausibly anticipated and some quite *ad hoc* effects – bias, scale-dependant bias, feedback, AGN feedback, ... etc, are included. The complexities of "post-formation" dynamical evolution (or even survival) must all be approximated. And so on. Considerable current effort is involved in adjusting the non-linear aspects of the baryonic physics to try to regain consistency with observation. Consistency with observation is not a natural feature of extant models of galaxy formation.

The development of the currently available sophisticated galaxy models is a powerful and extremely impressive achievement. Is it developing a theory? There is no *ab initio* theory, no first-principles calculation, of many of the physical processes. It is feasible that a model can be identified, with eventual sufficient complexity, which is able to reproduce all extant observables. This will not be a theory. It will never be "right" or "wrong". Until key parameter space is investigated, no model will even be unique within its limited starting points and methodology. That is, there is a fundamental distinction between development of a model/tool-kit which is appropriate to investigate Gaia-like data sets, and modelling galaxy formation from linear perturbations early in the Universe. The latter models can never be compared to data, except after 'processing' through complex non-linear processes, which are themselves neither understood nor quantified.

So is there any point in devoting effort to building complex models of galaxy formation, when they are inherently untestable and not unique? Yes! In fact, such modelling can, or could, already be used for important investigations of some key assumptions in general astrophysics and cosmology. Galaxy formation models, given their present (impressive) sophistication, are valuable tools to investigate hypotheses. As yet, however, the models are incapable of testing hypotheses as complex as the formation of a galaxy. Appropriate hypotheses to test are more fundamental than the highly specific challenge of adding complexity to a recipe to become not-inconsistent with extant observations. Galaxy formation models are, as yet, not very helpful tools to determine the details of the complex mix of non-linear physics which describes the evolution of baryons and dark matter on small scales in a galaxy. Galaxy formation models could however, if applied appropriately, be a very valuable tool-kit to investigate much more fundamental physics.

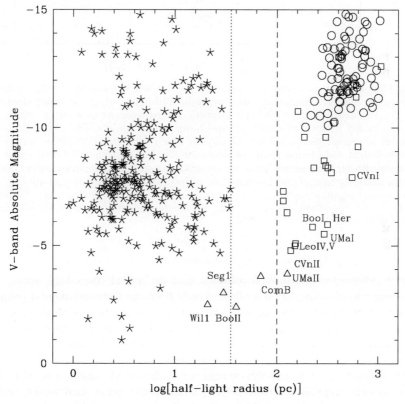

Figure 1. The relation between absolute luminosity and luminous half-light radius for small stellar systems in the local Universe. Globular clusters from several host galaxies, Ultra Compact Dwarfs, and galactic nuclei star clusters, are represented as asterisks. Local Group dSph galaxies, with the most newly discovered identified by name, are shown as open squares. Galaxies from the Local Volume survey of Sharina *et al.* (2008) are shown as open circles. Milky Way satellites of unknown equilibrium status are shown as open triangles (see Fig 2). All equilibrium galaxies have half-light radii larger than the minimum size line at 100pc. All apparently purely stellar systems have half-light radii smaller than about 30pc. Further details are in Gilmore, Wilkinson, Wyse, *et al.* (2007).

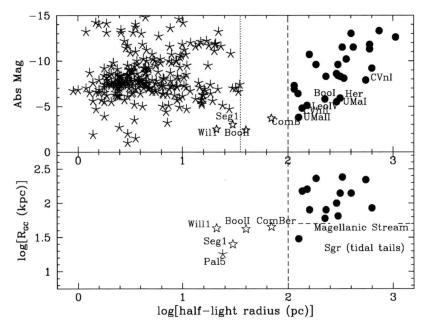

Figure 2. The observed structural properties of the smallest galaxies and stellar systems. Top: relation between absolute luminosity and luminous half-light radius. Globular clusters, Ultra Compact Dwarfs, and galactic nuclei star clusters, are represented as asterisks. dSph galaxies, with the most newly discovered identified, are shown as solid points. Objects of unknown equilibrium status are shown as open stars. All objects which show robust evidence for dark matter halos have half-light radii larger than the minimum size line at 100pc. All stellar systems have half-light radii smaller than about 30pc, and none shows evidence for dark matter. Bottom: the uncertain dynamical state of the intermediate objects is emphasised by considering size as a function of Galacto-centric distance. All uncertain objects are in a region where Galactic tides are expected to be important, and so may have time-dependent structures. Further details are in Gilmore, Wilkinson, Wyse, *et al.* (2007).

2.1. *What galaxy formation models could tell us*

There are fundamental questions in physics, and in ΛCDM cosmology, which *are* best addressed using galaxy formation models. Applications to the fundamental properties of neutrinos are mentioned above. To give just one more important example, the standard model of particle physics is known to be incomplete. Extension to a more general theory requires guidance from observations. Recently, most of these new observations have come from astrophysics - neutrino masses, baryogenesis, matter-anti-matter asymmetry, the dominance of dark matter, the importance of dark energy, are among this list. The minimal super-symmetric extension of the particle physics standard model, which does not even encompass all the complexity required to address all the items on this list, has more than 120 free parameters. Hopefully, in the near future, CERN will advance measurement of aspects of the parameter space. Astrophysics has done so – limits on neutrino masses from large scale structure are a superb example – but can do much more, including investigating aspects of physics at a more fundamental scale than is possible with accelerators.

Perhaps the best and most immediate example is in testing the small-scale extension of the spectrum of perturbations. At present, ΛCDM models adopt the spectrum of perturbations from analysis of CMB and other observations, and extend this to zero scale. The extension is unphysical, in being ultraviolet divergent. Suppression of the divergence is provided essentially by numerical smoothing ("finite resolution") in cosmological simulations. It is unlikely that Nature does it that way. Rather, the small-scale power spectrum may well be where astroparticle physics comes into action on observable scales. Testing this is arguably much more interesting than is applying ingenuity to fine-tune outcomes of the models to make them not-inconsistent with already known observations.

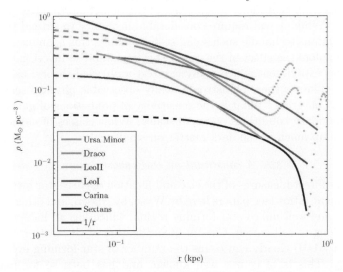

Figure 3. Derived internal mass profiles for the well-studied dSph galaxies. In each case a cored dark matter distribution is preferred by the kinematic data, with a scale radius comparable to that of the luminous scale shown in Figures 1 & 2. The similarity of this scale in all cases studied implies it is an inherent property of dark matter itself. Further details are in Gilmore, Wilkinson, Wyse, *et al.* (2007).

A huge literature is available considering the implications of specific possibilities and elementary particle types in astrophysics. Ostriker & Steinhardt (2003) provide an accessible summary of the interplay between small-scale structure and classes of physically motivated explanations for suppression of the ultraviolet divergence. The key is that physical effects may be expected on scales of the smallest galaxies, the dSph. It is perhaps not coincidence that it is on these scales that the bottom-up galaxy formation models have proven to be most inconsistent with observations. The list of relevant issues is long, and well-known. Examples from a long list include the existence of old red massive galaxies, the Tully-Fisher relation being in place by redshift unity, the frequency of large old cold disks in galaxies, and the satellite problem. Such a long list of observations all inconsistent with apparently fundamental features of galaxy formation models suggests two approaches. In one approach, new complex physics ("feedback") must be added, to "improve" agreement with observation. The appearances are to be saved. In another, common assumptions in the galaxy simulations could be examined further.

The recent observational status of the small-scale problem is described in Gilmore, Wilkinson, Wyse, *et al.* (2007). An updated summary is presented in figures 1, 2, and 3 here. There are two key results: the lowest luminosity galaxies are all very dark-matter dominated, and all have an equilibrium minimum half-light optical radius of $>$ 100pc. The largest star clusters have half-light radius less than 30pc. No equilibrium objects are known in the local Universe with half-light size between 30pc and 100pc (Figures 1,2). Dynamical studies of stellar kinematics in very low-luminosity galaxies all prefer a dark-matter distribution which is cored, with a mass scale length comparable to the luminosity scale length (Figure 3). Standard assumptions adopted in simulations of galaxy formation have no presumed physical scale (ie, the UV divergence), so featureless smooth distributions are a natural feature. A specific length scale is not anticipated, but is seen. This physical scale seems to be special.

It may well be that we are discovering a physics-based solution to the medley of challenges to galaxy formation models: the divergence of the small-scale extrapolation of the perturbation spectrum is at fault. The physics of the mix of dark matter particles may be the explanation. While it will require considerable ingenuity to extend the resolution of numerical simulations to handle such small scales reliably, this is an example where simulations could explore the effect of the power spectrum, and so investigate a new regime. That is a physics experiment which really can test a physical theory: using observations of galaxies as a guide, is there an astrophysically observable physical scale at which the power spectrum converges? What is the sensitivity of predictions of galaxy formation to the assumed boundary conditions on small scales? What classes of elementary particles must then make up much of the dark matter on small scales?

2.2. *A constraint on early substructure*

Among the most direct measures of the size and location of early star formation is scatter in chemical element ratios (see papers here by Wyse, by Nissen, and others). Small scatter requires a large and well-mixed star-forming region, which has an independent existence for sufficiently long to self-enrich. Careful quantification of the scatter in element ratios as a function of [M/H] clearly can count the number of star-forming events in the early Galaxy directly. This is of course well known, and has been so for many years. An interesting extension of this analysis can be applied to the light elements Beryllium (and Lithium) which are made, fully for Be, partly for Li, by cosmic ray spallation, probably with CNO nuclei as cosmic ray primaries spallating onto H-nuclei (cf Pasquini, this meeting, and Gilmore, Gustafsson, Edvardsson & Nissen 1992). This spallation involves very high-energy heavy nuclei, probably accelerated by the same supernovae in which they were created. Such high-energy particles have a very long mean free path. They cannot be retained inside a small or short-lived star forming event, such as a small, transient, dark matter halo. Thus, inevitably, any stars formed in such small halos will have little or no Be. Their Li abundances will also provide a robust determination of the relative contributions of BBN and later spallation to their Li abundances. Determination of the range of Be abundances in field halo stars, and – ideally – in either a low-mass dSph galaxy or a verifiable kinematic stream, will provide extremely interesting constraints on the range of places where early star formation occurred.

3. An historical lesson

This meeting celebrates the centenary of Bengt Strömgren. Openness to the implications of observations, and an ability to move beyond preconceptions, was one of his great attributes. My personal example involved his long-standing research interest in the formation and evolution of the Galactic disk. After decades of work, developing the Strömgren photometric system, and acquiring vast data sets, Bengt Strömgren, in retirement(!!) was near to finalising his major study of the distribution of stellar ages and abundances near the Sun. In 1983 Gilmore & Reid announced their discovery of the Galactic thick disk. The thick disk stellar population is old and relatively metal-rich (Gilmore & Wyse 1985), with a main-sequence turn-off to the red of the F-star range which was at the time being studied in Strömgren's survey. Seeing this result, Bengt Strömgren invited me to visit, rapidly persuaded himself that his extant survey was biased by being based on a too-restrictive assumption on the past age-metallicity relation, and so extended his survey to include redder stars. He did this knowing that he might well not live to see the outcome of his lifetime research project. An impressive example of scientific objectivity, indeed. Fortunately, his colleagues worked hard, the weather was good, so Bengt Strömgren was able to present the first results of his expanded survey in his last scientific paper, Strömgren (1987).

References

Binney, J. 2002, *EAS Publications Series, Volume 2, Proceedings of "GAIA: A European Space Project"*, held 14-18 May, 2001 Les Houches, France. Edited by O. Bienaymé and C. Turon. EDP Sciences pp. 245-256
Cartwright, N. & Frigg, R. 2007, *Physics World*, 20, 14 (September)
Cerda-Olmedo, E. 2008, *European Review*, 16, 159
Gilmore, G., Gustafsson, B., Edvardsson, B., & Nissen, P. E. 1992, *Nature*, 357, 379
Gilmore, G. & Reid, I. N. 1983, *MNRAS*, 202, 1025
Gilmore, G. & Wyse, R. F. G. 1985, *AJ*, 90, 2015
Gilmore, G., Wilkinson, M. I., Wyse, R. F. G., Kleyna, J. T., Koch, A., Evans, N. W., & Grebel, E. K. 2007, *ApJ*, 663, 948
Leggatt, A. J. 2008, *RPP*, 71, 1
Ostriker, J. P. & Steinhardt, P. 2003 *Science*, 300, 1909
Sharina, M. E., Karachentsev, I. D., & Dolphin, A. E., *et al.* 2008 *MNRAS*, 384, 1544
Strömgren, B. 1987 *in 'The Galaxy'*, eds G Gilmore and R Carswell, CUP, p. 229.

Marija Vlajic presenting her contribution.

Gerry Gilmore settling an important issue in Galaxy Formation with Burkhard Fuchs ...

... and another with Preben Grosbøl.

New mechanisms for international coordination of large observing projects

Johannes Andersen

The Niels Bohr Institute, University of Copenhagen
and Nordic Optical Telescope, La Palma, Spain
Juliane Maries Vej 30, DK-2100 Copenhagen, Denmark
E-mail: ja@astro.ku.dk

Abstract. Progress on many fronts in the study of the Milky Way requires major new imaging and/or spectroscopic surveys as well as coordinated ground-based programmes in support of space missions. The present uncoordinated planning and operation of the 2-4m (and some of the larger) telescopes across the world are an impediment to the efficient conduct of such programmes and the overall cost-effectiveness of the investments in astronomy. We briefly report on recent initiatives to improve the situation, notably the part played by the ASTRONET consortium in setting up a comprehensive long-term plan for the development of European astronomy.

Keywords. sociology of astronomy

1. Background

As shown in many contributions to this session, progress on several fronts in our understanding of the origin and evolution of disk galaxies require homogeneous high-quality data for ever larger samples of objects – be they galaxies in the distant Universe or individual stars or gas clouds in the Milky Way and its neighbours. Some such programmes are self-contained ground- or space-based projects, but many involve a combination of observations from space and from the ground.

In practical terms, such projects typically require hundreds of nights on 2-4m-class ground-based telescopes, with optimised instrumentation, data reduction pipelines, and the staff to run it all efficiently. As a rule, coordination is needed, either to ensure spectroscopic coverage of objects selected in imaging surveys; to ensure that the space- and ground-based data are obtained and combined to reach the scientific objectives in a timely manner; or simply to achieve all-sky coverage.

The scientific advantages of such coordinated programmes are obvious. Yet, their practical implementation is impeded by the way our telescopes are traditionally organised, operated and funded: On an institutional, national or at best regional basis (e.g. ESO). Observing time is allocated to individual P.I.s, based on proposals ranked by individual scientific merit rather than by the global scientific returns of the national investments, and with no overall coordination of instrumentation or scheduling.

Coordination of ground-based follow-up with space missions has been especially weak in Europe. As just one example, the Geneva-Copenhagen Survey of the Solar Neighbourhood (Nordström *et al.* 2004) provided photometry and radial velocities for $\sim$15% of the stars observed by Hipparcos, several years after the astrometry was complete and after 20 years of struggle by a small team with private telescopes. Complete, homogeneous ground-based data for all the stars could have been ready in time for $\sim$2% of the cost of the Hipparcos mission itself, but rigid policies prevented this from happening, at great loss in scientific returns from the investment in the satellite.

Conversely, the Sloan Survey owes its success to the coherent planning of the imaging and spectroscopic parts of the survey, with a dedicated telescope and the resources needed to provide the processed data in a timely manner. Yet, the Sloan Survey is limited to the northern hemisphere, just as the RAVE survey is currently limited to the southern hemisphere for lack of a suitable northern telescope with the required funding. The scientific drawbacks of this situation are obvious.

2. A new start - hopefully!

Recent initiatives have been taken to improve this situation. Planning exercises both in the USA and Europe are aiming to review the role and organisation of medium-size ground-based telescopes as part of the overall complement of facilities needed to reach overarching scientific goals.

In the USA, the NOAO has appointed the ReSTAR and ALTAIR committees to review the future role and organisation of the US system of 2-4m and 6-10m telescopes, respectively. In Europe, similar reviews are part of the recent, comprehensive ASTRONET planning initiative. Because ASTRONET is probably not yet familiar to most of the community, a short account is provided here.

ASTRONET is a consortium of national funding agencies for astronomy in Europe, plus the European Southern Observatory (ESO) and the European Space Agency (ESA). It was founded in 2005 and is supported by the European Commission with a four-year, 2.5 MEuro grant under the ERA-NET scheme. Its goal is to develop a long-term plan for the *science-based, rational, and coordinated* development of European astronomy – somewhat akin to the Decadal Surveys in the USA, but with significant differences.

Notably, the science drivers for the plan were first collected, prioritised, and published as a separate exercise. The report, *A Science Vision for European Astronomy* (de Zeeuw & Molster 2007), focused the discussion on four broad questions:

Do we understand the extremes of the Universe?
How do galaxies form and evolve?
How do stars and planets form? and
How do we fit in?

Extensive community input on a first draft of the *Science Vision* was collected at a large open Symposium and via a web-based forum both before and after the Symposium.

With the *Science Vision* as the scientific basis, a comprehensive *Infrastructure Roadmap* was then developed (Bode *et al.* 2008) to describe the resources needed to reach these scientific goals. The *Roadmap* is also structured according to a few major headings:

High energy astrophysics, astroparticles and gravitational waves,
UV, optical, IR and radio/mm facilities, including survey instruments,
Solar telescopes and "in situ" Solar System missions,
Theory, computing facilities, networks, Virtual Observatory, and
Education, recruitment and training, and public outreach.

The *Roadmap* lists *and prioritises* the major new research facilities required – at all wavelengths of the electromagnetic spectrum as well as for particles and gravitational waves – and in space as well as on the ground. It also includes the concomitant investments in laboratory astrophysics and in theory, computing, networks, and archiving, including the Virtual Observatory.

Last, but not least, it also assesses the measures needed to provide the necessary skilled humans – education, recruitment, and outreach – and it includes not only the established ESO and ESA member states, but *all* of Europe as we know it today. As for the *Science Vision*, community input was actively involved in its preparation.

The *Roadmap* was developed in close cooperation with the ERA-NET ASPERA for astroparticle physics, and the infrastructure coordination networks OPTICON, RadioNet, ILIAS, and EuroPlaNet, which support the hands-on implementation of European cooperation on the facilities in their respective domains. Close contact to the funding agencies was also maintained to ensure that realistic costs of construction, operation, and exploitation were included in the budget estimates, and that the overall cost and spending profiles remained in touch with reality.

In parallel, ASTRONET worked to include all astronomical communities in its endeavour. Currently, 25 Member and Associated States of the European Union, with a combined population of over 500 million people, are associated with ASTRONET in some form or other.

The publication of the *Roadmap* immediately marks the start of the implementation phase, and initiatives are already being taken to carry out its recommendations. To prepare the future decisions on the largest infrastructure projects, contacts are being taken to the European Strategy Forum for Research Infrastructures (ESFRI) and the OECD Global Science Forum (GSF). As regards the smaller, existing facilities, action is being taken by ASTRONET itself, as described below.

3. Coordination of existing telescope facilities

In order to review the existing European complement of 2-4m telescopes and propose innovative measures to organise them in a more cost-effective manner, the ASTRONET Board appointed a *European Telescope Strategy Review Committee* in June 2008. Its report is due in the autumn of 2009 and will address a number of key questions:

- Identify those goals of the ASTRONET Science Vision that are most effectively delivered by 2-4m-class optical/infrared telescopes;
- Identify which observational capabilities (site, field of view, instrumentation, modes of operation) are required; and
- Develop a realistic roadmap, including technical developments and upgrades, and organisational/financial arrangements, which would enable a set of European 2-4m-class telescopes to deliver the best scientific output for European astronomy in a cost-effective manner.

The benefits expected to accrue from this exercise include:

Ability to implement large (all-sky) programmes (e.g. RAVE+)
Ability to implement coordinated ground/space projects (e.g. Gaia)
Include new partners, new and more flexible funding sources, and
More effective coordination with world-wide partners.

In short, the goal is to do more and better science through more intelligent international planning and cooperation. The experience gained from this review will subsequently be used in similar reviews of the existing European 8-10m telescopes, and later of the radio telescopes as well.

The impending decision points on some of the large projects (ELTs, SKA, ...) will provide a good opportunity to re-think the organization of the smaller facilities as well. Hopefully, the funding agencies will then follow up this review with far-sighted decisions, which will enable us to undertake the large future programmes described in the *Science Vision*. In a world of finite resources, such a gain will come at the expense of some of the traditional P.I.-driven science, so interaction with the community during the review process will again be crucial.

The ASTRONET review is not conceived in an atmosphere of isolation: Its goals are similar to the ReSTAR and ALTAIR reviews in the USA, and contacts are maintained to identify potential areas of synergy. Having clear overviews of the aims and capabilities of both sides will be useful in the next step, optimising the overall scientific returns on a global basis. Obvious opportunities for cooperation are:

Science cases for individual 2-4m telescopes,
Hands-on training of the next generation of astronomer/engineers,
Complementarity in instrumentation and sky coverage,
Instrument/detector/software development,
Standardised telescope performance criteria,
Time domain astronomy and distributed facilities, and
Optimising common use of observatory sites and coordinate operations.

Recent developments and funding decisions affecting our present complement of medium-size telescopes have not conformed particularly accurately to this ideal. But if astronomers cannot lead the way to effective global scientific cooperation, who can?

References

Bode, M. F., Cruz, M. D., & Molster (Eds.) 2008, *An Infrastructure Roadmap for European Astronomy*, ASTRONET
de Zeeuw, P. T. & Molster, F. J. (Eds.) 2007, *A Science Vision for European Astronomy*, ASTRONET
ALTAIR: *http://www.noao.edu/system/altair/*
ASPERA: *http://www.aspera-eu.org/*
ASTRONET: *http://www.astronet-eu.org*
ESFRI: *http://cordis.europa.eu/esfri/*
EuroPlaNet: *http://www.europlanet-eu.org*
Nordström, B., Mayor, M., & Andersen, J., *et al.* 2004, *A&A* 418, 989
OECD Global Science Forum: *http://www.oecd.org/*
RadioNet: *http://www.radionet-eu.org/*
ReSTAR: *http://www.noao.edu/system/restar/*

Summary, Conclusions and Recommendations

Rosemary F. G. Wyse

Department of Physics & Astronomy, Johns Hopkins University,
Baltimore, MD 21218, USA
email: wyse@pha.jhu.edu

1. Legacies of Bengt Strömgren

Studying the disk of the Milky Way in its cosmological context, as reflected in the title of this conference, was pioneered by Bengt Strömgren, in whose honour we are gathered here. A significant legacy of his is the understanding that advances are achieved by making *connections* among fields that are tenuously related on initial inspection. When the tools to achieve this were not available, Bengt Strömgren developed the necessary techniques, such as Strömgren photometry to quantify the local stellar population.

This points to another legacy of Bengt Strömgren: the appreciation that many approaches, tools and techniques provide complementary information, and one should use them all. Examples, all of which formed lively discussions here, include

• The direct study of systems at high redshift, where one is limited to interpretation of the integrated light, compared to more detailed study in local systems of individual old stars, which formed long ago.
• Numerical simulations compared to analytic theory
• Indeed, theoretical predictions confronted with observations
• Properties of stellar populations compared with those of the gaseous components
• The different observational evidence provided by spectroscopy, photometry and astrometry

2. Connections

The presentations and discussions at this meeting identified several areas where connections are apparent, albeit that we are not yet clear of the nature of the connection, or the physics behind the connection. As in any lively interchange of ideas, as many questions were raised as were answered.

2.1. High redshift ⇔ low redshift

What are the descendants of the turbulent '(thick?) disk' galaxies at redshifts ~ 2? These have very high inferred gas densities and star formation rates; could this be more suggestive of bulge formation that of thick disk formation?

Where are the first stars (Population III), and their immediate descendants, stars that are enriched by Population III and should show distinctive elemental abundance patterns if predictions of a top-heavy stellar initial mass function are correct?

What were/are the first galaxies? Have they all been subsumed into larger systems? The limitation to integrated light in high redshift systems leads to inevitable degeneracies such as between the inferred star formation rate and stellar IMf, and between age and metallicity. The more detailed information available for resolved stellar populations locally allows one to break these degeneracies.

2.2. Local star formation ⇔ global star formation efficiency

How effective is feedback from supernovae? Type Ia and Type II? Is it always negative? Do supernovae actually act coherently to drive winds or do they simply lower the star formation efficiency within bound gas clouds? What sets the constant of proportionality in the scaling relation between gas content and star formation rate (the 'Schmidt-Kennicutt' law)? Why does this relation remain valid over a very large range of physical conditions and scales? The molecular gas fraction is clearly important in determining star formation rates and efficiencies, given that stars are observed forming in molecular clouds. What sets this fraction?

Supernovae – both core collapse, Type II, and thermonuclear, Type Ia – clearly act to disperse metals, which aid cooling and later star formation. But how this mixing occurs within a galaxy's interstellar medium, and what fraction of these metals are mixed into the intergalactic medium, remain much debated.

2.3. Bulge formation ⇔ (thick) disk formation

At least in the Milky Way Galaxy, these two stellar components have similar stellar age distributions, being dominantly old, typically 10–12 Gyr. The more limited data for external galaxies also suggest old ages, certainly significantly older than the mean age of stars in thin disks.

Both components also have relatively high metallicity, which, combined with the old mean age and narrow age range, implies (relatively) high star formation rates and enrichment efficiency. This points to high gas densities and a deep potential well and could be suggestive of the turbulent stellar systems at redshift ~ 2 (look-back time of 10.3 Gyr in the 'concordance' cosmology) referred to above. It does not suggest significant mass build-up in the form of stars accreted from former satellite galaxies. Mergers can however drive gas to the central regions, to fuel a dissipative starburst, and perhaps a merger both created the thick disk and triggered bulge star formation.

The (present) thin disk would be largely established later, plausibly in the form of gas accretion/infall. This provides a connection to another 'connection' below, gas ⇔ stars, as several speakers pointed out the uncertainties in gas reservoir and accretion rates.

In all interpretations of stellar populations, age determinations are critical (and one of the goals of Strömgren photometry).

2.4. First stars ⇔ stars now

Arguments based on cooling inefficiency and slow rates of molecular gas formation at zero metallicity lead to the expectation that the first stars should have a mass function that is biased to massive stars, perhaps even to the exclusion of any stars below the most massive stars seen today. However, the same arguments lead to the expectation that the stellar IMF is invariant above some low threshold metallicity. Observational evidence for an invariant IMF comes from star counts within the Milky Way and its companion galaxies: these are consistent with the same low-mass IMF in the central bulge (high metallicity, high SFR, high density) as in globular clusters (low metallicity) as in the Ursa Minor dwarf spheroidal galaxy (low metallicity, low SFR, low density), as in the local disk now (high metallicity, moderate SFR, moderate density).

The past high-mass IMF is constrained by the elemental abundance patterns in the long-lived, low-mass stars that those high-mass stars enriched: yields of α-elements from Type II supernovae tend to increase with the (main sequence) mass of the progenitor, while the iron yield varies little with progenitor mass. These trends result in higher predicted values of $[\alpha/\mathrm{Fe}]$ in stars enriched by massive stars with an IMF biased to the highest mass, massive stars. Observations of elemental abundances in halo stars down to below one-thousandth of the solar iron abundance have revealed very little scatter in the value of the 'Type II' plateau: this low scatter implies that independent of metallicity, the same massive-star IMF was sampled – and that there was good mixing, and that the entire IMF was sampled, both of these points themselves posing puzzles.

2.5. Substructure ⇔ galactic scale structure

Substructure, whether in the form of CDM pre-galactic systems, dynamical instabilities or even black holes, can make its presence known in several ways. Disk-heating associated with the merging of satellite galaxies is the focus of most discussion, but gravitational interactions with a various classes of substructure could e.g. disrupt wide binaries, disturb thin tidal arms from disrupting globular clusters and satellites, cause microlensing events etc. Keeping substructure dark does not hide it, unless it is kept far away from any stellar component.

Interactions of stars with giant molecular clouds and spiral arms have long been implicated in underlying the age-velocity dispersion relation for thin disk stars (the thick disk is too hot for this mechanism to be responsible for its vertical structure), albeit imperfectly understood. It has recently become clear that stellar disks are rather dynamic, and that resonances within the disk can interact with disk stars to create 'moving groups', consisting of stars with a common motion but often a rather broad age range and metallicity distribution. These need to be distinguishable and distinguished from debris from any putative satellite system accreted into the thin disk plane. Connecting large scales, resonances can induce radial migration of stars within the thin disk, over essentially a disk scale-length; this effect needs to be quantified.

Disk instability can also form a central bar, which in turn can buckle to form a bulge; this phenomenon has been know for years, but the duty cycle, prevelance, signatures and indeed criteria are still not well understood. How bars interact with a background dark halo is also a topic of much research, in particular whether or not the mass profile of the dark halo could be significantly modified. Interactions with substructure, such as satellites, bars, and spiral arms can cause transport of angular momentum on larger scales, and again the lively discussions point to an active research area.

Stellar streams in the (outer) stellar halo of the Milky Way and other galaxies are testament to interactions between substructure and the host system. Dynamical times are longer in the lower density outer parts, allowing streams in coordinate space to persist and be identified. Indeed, leading and trailing debris from more than one periGalactic passages of the Sagittarius dwarf spheroidal can be traced, and this one system accounts for a significant fraction of the detected substructure in the outer halo. A similar situation may occur in M31, where one satellite can in principle provide all the observed structure in the halo.

2.6. Dark matter ⇔ light matter

The Tully-Fisher relation is a fundamental observational relationship that all theories of galaxy formation need to reproduce from first principals. The fewer 'baryonic physics' knobs that are tuned, the better. However, there is some observational evidence for different star-formation modes (e.g. starburst *vs* quiescent; stochastic *vs* triggered) in different mass/redshift haloes, and this needs to be understood.

The motions of light matter (stars and gas) within dominant dark matter haloes can be used to constrain the nature of dark matter by, e.g., identification of the signatures of merging. In this context, large surveys of Galactic stars are very important (as recognized by Bengt Strömgren!).

2.7. Gas ⇔ stars

The 'Schmidt-Kennicutt Law' of star formation clearly lacks understanding. While one can straightforwardly argue that any scaling of star formation rate should involve the amount of fuel (gas) divided by a timescale (dynamical time, going like one over the square root of gas volume density for self-gravitating gas), it is neither observationally nor theoretically clear whether this criterion should invoke the molecular gas, or the atomic gas, or the total gas, and whether it should be volume density or surface density, or one power or two (collisionally induced?) and on what scale it is appropriate to average.

Fuel for star formation, i.e. gas, apparently needs to be added to galaxies as they evolve, and there was much discussion of accretion and gas cycling among different physical phases. The evidence for accretion depends on analysis of high velocity clouds in HI maps of our own and external galaxies, and is complicated by the possibility of large-scale 'fountain' effects and distance uncertainties. It remains unclear what forms the reservoir – is it the 'missing baryons' postulated to be in a diffuse, warm 'cosmic web'? The accretion rate as a function of time and place is poorly constrained, as are the angular momentum content and metallicity of any accreted gas. Whether or not there are always associated dark matter lumps, or one has smooth dissipational settling of baryonic material is unclear.

2.8. Chemical evolution ⇔ dynamical evolution

On large scales, the existence of the well-established mass-metallicity relation for galaxies of a range of morphological types argues strongly that chemical evolution and dynamics are linked – the most straightforward explanation is that the bulk of stars formed in galaxies with a potential well depth very similar to the one in which the stars are today.

Within galaxies, one cannot ignore the effects of dynamical instabilities and resonances in the thin disk when calculating e.g. chemical abundance gradients, and stellar age-velocity dispersion relations. There could be significant radial re-arrangement and mixing of stars, in addition to gas flows.

It is also clear that one cannot model chemical evolution of the galaxy assuming an extension of the Simple Model that maintains homogeneity and a one-to-one correspondence between time and metallicity, even for the solar neighbourhood. Stochastic star formation, possibly reviving metal-enhanced star formation ideas, should be incorporated. However, one needs to maintain the remarkable lack of scatter in elemental abundance patterns in each of the different stellar components, at all metallicities.

Cosmological boundary conditions should also be applied, rather than invoking gas inflow/outflow *ad hoc*, with little or no consideration to angular momentum conservation or cooling times. Progress is being made, and this is a very active field, as envisaged by Bengt Strömgren.

3. Conclusions & Recommendations

Bringing together people from across the larger astrophysical community with the same 'big picture' science goals, as achieved with this meeting, is very fruitful, and certainly fosters the discussions and connections among researchers that is needed to make progress. I recommend that we should repeat this gathering!

Acknowledgements

I would like to thank Birgitta Nordström and Johannes Andersen for their extraordinary commitment and enthusiasm that played a pivotal role in the success of this meeting. All the members of the SOC and LOC are also due a vote of thanks.

Poster papers

Posters are available on-line at: journals.cambridge.org/IAU_S254

O. Agertz, G. Lake, B. Moore, L. Mayer, R. Teyssier & A.B. Romeo
  Turbulence in galactic disks: The role of self-gravity and supernova feedback
D.M. Allen, S.G. Ryan & S.A. Tsangarides
  S/R ratios in carbon-enhanced metal-poor stars
B. Anguiano, K. Freeman, M. Steinmetz, E. Wylle de Boer, A. Siebert & the RAVE collaboration
  RAVE: The Age-Metallicity-Velocity relation in the nearby disk
T. Antoja, F. Figueras, B. Pichardo, D. Fernández, E. Moreno, J. Torra & O. Valenzuela
  Local stellar kinematic constraints on the Galactic potential
A.S. Árnadottir, S. Feltzing & I. Lundström
  Properties of the Milky Way stellar disks in the direction of the Draco dSph galaxy
M. Aumer & J.J. Binney
  Stellar kinematics and the history of the solar neighbourhood – revisited
J. Bakos, I. Trujillo & M. Pohlen
  Probing outer disk stellar populations
D.G. Banhatti
  Newtonian mechanics & gravity fully model disk galaxy rotation curves without dark matter
M. Barker, A. Ferguson, A. Cole, R. Ibata, M. Irwin, G. Lewis, T Smecker-Hane & N. Tanvir
  Probing the fossil record of disk galaxy evolution in M33
M. Barker & A. Sarajedini
  Chemical evolution and inflow history in M33's outskirts
I. Bikmaev, S. Melnikov, N. Sakhibullin & A. Galeev
  RTT150 spectroscopic study of the Galaxy disk stars selected from Geneva-Copenhagen Survey
S. Buehler, A. Ferguson, M. Irwin, N. Arimoto, P. Jablonka & D. Mackey
  The stellar outskirts of the disk galaxy NGC 4244
J. A. Carballo-Bello & D. Martínez-Delgado
  Photometric survey of the Galactic Anticenter: the Canis Major debate
L. Casagrande, C. Flynn & M. Bessell
  Fundamental parameters of M dwarfs
Li Chen, X.H. Gao & J.L. Zhao
  Open clusters and the radial abundance gradient of Galactic disc
E. Colavitti, F. Matteucci & G. Murante
  The evolution of a Milky-Way-like galaxy in a cosmological context
R.D.D. Costa, O. Cavichia & W.J. Maciel
  Chemical abundances of planetary nebulae in the disk-bulge connection
A. Curir, P. Mazzei & G. Murante
  Bars driven by the cosmology in stellar-gaseous disks
R.S. de Jong, D.J. Radburn-Smith & J.N. Sick
  GHOSTS - Bulges, halos, and the resolved stellar outskirts of massive disk galaxies
S. Demers, P. Battinelli & H. Forest
  Outer MW kinematics from Carbon stars

E. Gardner, K.A. Innanen & C. Flynn
  Dynamical effects of the long bar in the Milky Way
I. Goldman
  Hα velocity fluctuations in NGC 5033
P. Grosbøl & H. Dottori
  Birthplaces of very young stellar clusters in nearby disk galaxies
L. Gutiérrez, A. Tamm, J.E. Beckman, L. Abrahamson, P. Erwin & M. Guittet
  Where have all the bulges gone?
U. Heiter & R.E. Luck
  How unique is the local region of the Galaxy disk?
S. Herbert-Fort, D. Zaritsky, J. Moustakas, C. Engelbracht, X. Fan, R.W. Pogge, B. Weiner & A. Zabludoff
  The spatial extent of nearby outer disks
J. Hou, J. Yin, S. Boissier, N. Prantzos, R.X. Chang & L. Chen
  Properties and chemical evolution of the M31 disk: Comparison with the Milky Way
Erik Høg
  Bengt Strömgren and modern astrometry
H.R. Jacobson, E.D. Friel & C.A. Pilachowski
  Exploring the Milky Way abundance transition zone $R_{GC} \sim 10$ kpc with open clusters
A. Just, S. Vidrih & H. Jahreiss
  The local disk model and high latitude SDSS/SEGUE data
C. Juul Hansen, B. Nordström et al.
  Tracing SN yields in metal-poor HB stars
X. Kang
  Modeling the Milky-Way satellite galaxies
G. Kerekes, I. Csabai, L. Dobo & G. Herczeg
  A new approach for photometric parallax estimation
R. Klement, B. Fuchs & H.-W. Rix
  Identifying stellar streams in the 1st RAVE Public Release data
J. Knude
  Depth of the Lupus I - VI complex
A. Kučinskas, V. Dobrovolskas, A. Černiauskas & T. Tanabé
  A new photometric metallicity calibration and metallicities of star clusters in the Magellanic Clouds
A. Kučinskas, H.-G. Ludwig, A. Ivanauskas & E. Caffau
  Observable properties of late-type giants predicted by 3D hydrodynamical and 1D stellar atmosphere models
W.J. Maciel & R.D.D. Costa
  Abundance gradients in the galactic disk: Space and time variations
V. Makaganiuk, A. Korn & B. Edvardsson
  Taking MARCS to the next level: Ca IR triplet lines in NLTE
M. Mapelli, B. Moore & J. Bland-Hawthorn
  The origin of lopsidedness in galaxies
M. Mapelli, B. Moore, E. Ripamonti, L. Giordano, L. Mayer, M. Colpi & S. Callegari
  Are ring galaxies the ancestors of giant low surface brightness galaxies?
C. Mateu, K. Vivas & R. Zinn
  The QUEST RR Lyrae survey of the Canis Major overdensity
L. Mattsson
  N/O trends in late-type galaxies: AGB-stars, IMFs, abundance gradients and the origin of nitrogen

L. Mattsson & N. Bergvall
  Correcting emission line data in the SDSS for underlying stellar absorption
P.L. Nedialkov, A.T. Valcheva, V.D. Ivanov & L. Vanzi
  Dust properties of nearby disks: M31 case
M.S. Oey, T. Bensby & S. Feltzing
  Abundances in the old thin disk of the Milky Way
Sang Hoon Oh, W.-T. Kim, H.M. Lee & J. Kim
  Physical properties of tidal features in interacting disk galaxies
Seungkyung Oh, P. Kroupa, H. Baumgardt & K. Menten
  The dynamics of massive stars in star clusters
S. Pedrosa, P.B. Tossera & C. Scannapieco
  Effects of SN feedback on the dark matter distribution
F. Piontek & M. Steinmetz
  A systematical study of the formation of disk galaxies II: Star formation and feedback
N.A. Popescu
  Environmental effects on disk galaxies in mixed pairs
A. Portas, E. Brinks, A. Usero, F. Walter, W.J.G. de Blok & R.C. Kennicutt, Jr.
  The edges of THINGS
L. Portinari, C. Flynn, J. Holmberg, B. Fuchs & H. Jahreiss
  M/L ratio of the Galactic disk from the optical to the NIR
I. Puerari & L. Aguilar
  Formation of rings by galactic collisions
I. Puerari, H.M. Hernández-Toledo, J.A. García-Barreto, M. Cano-Díaz, O. Valenzuela & E. Moreno
  Optical and near-infrared morphology of the barred galaxy NGC 3367
J. Rasmussen, T.J. Ponman, L. Verdes-Montenegro, M.S. You & S. Borthakur
  The evolution of galaxy disks in dense environments – Lessons from compact groups
A. Rawat, Y. Wadadekar & D. de Mello
  Rest frame UV vs. optical morphologies of galaxies: Important implications to high-z results
A. Recio-Blanco, P. de Laverny, G. Gilmore, R. Wyse, O. Bienaymé, A. Vallenari, A. Bijaoui, M. Zoccali, G. Bono, J. Norris & M. Wilkinson
  How did a typical galaxy form? - Clues from the old stellar populations of the Milky Way thick disc
B.E. Reddy & D.L. Lambert
  Searching for the metal-weak thick disc in the solar neighbourhood
P. Re Fiorentin, C.A. Bailer-Jones, T.C. Beers, T. Zwitter, Y.S. Lee & X. Xue
  Toward constraints on galaxy formation scenarios: Stellar properties from Galactic surveys
S.J. Ribas, E. Figueras & J. Torra
  A kinematical study of the Galactic disk using Red Clump stars
J. Richardson, A. Ferguson, R. Johnson, M. Irwin, N. Tanvir, R. Ibata, K. Johnston, G. Lewis & D. Faria
  The stellar populations of halo substructure in M31
A.B. Romeo, O. Agertz, B. Moore & J. Stadel
  Discreteness effects in ΛCDM simulations
R. Roškar, V.P. Debattista, T.R. Quinn, G.S. Stinson, J. Wadsley & T. Kaufmann
  Clues to radial migration from the properties of outer disks
C. Ruhland, E.F. Bell, B. Häusler, E.N. Taylor, M. Barden & D.H. McIntosh
  Transformation of disk galaxies to ellipticals

K. Saha
  On the dynamical nature of lopsidedness and the Milky Way's dark halo
S.E. Sale, J.E. Drew, R. Greimel, Y.C. Unruh & the IPHAS consortium
  High spatial resolution empirical 3D extinction mapping with IPHAS
L.V. Sales, J.F. Navarro, A. Helmi, J. Schaye, C. Della Vecchia & V. Springel
  Simulated disk galaxies at $z = 2$ in OWLS
W.J. Schuster, L. Parrao & P.E. Nissen
  Scientific successes of the Strömgren photometer at San Pedro Mártir
Z. Shao, Q. Xiao & S. Shen
  Inclination-dependent LF of disc galaxies as a constraint of dust extinction law
J. Simmerer, S. Feltzing, F. Primas & R. Johnson
  Spectroscopy of the thick disk globular cluster NGC 5927
M. Takamiya, C. Willmer, M. Young & M. Chun
  Disk morphologies at $z \sim 0.7$
A. Tamm, E. Tempel & P. Tenjes
  Luminous and dark matter in the Andromeda galaxy
M. Valdez-Gutiérrez & I. Puerari
  Morphological studies and the OAUBSGS
Grazina Tautvaišiene & E. Puzeras
  Red clump stars in the Galactic field
Á. Villalobos & A. Helmi
  Formation of thick disks
T.K. Wyder & the GALEX Science Team
  Star formation in low surface brightness galaxies
X. Xue, G. Zhao, H.-W. Rix, P. Re Fiorentin, T. Naab, M. Steinmetz, F.C. van den Bosch, T.C. Beers, R. Wilhelm, Y.S. Lee, E-F. Bell, C. Rockosi, B. Yanny, H. Newberg, X. Kang, M.C. Smith & D.P. Schneider
  SDSS maps the halo mass profile: $M_{vir} = 1.0 \times 10^{10} M_\odot$
L. Začs, J. Sperauskas, F.A. Musaev & O. Alksnis
  CH-like stars: The old disk population?

Author Index

Abrahamson, L. – 506
Aerts, C. – 483
Agertz, O. – 505, 507
Aguilar, L. – 507
Alksnis, O. – 508
Allen, D. – 505
Alves-Brito, A. – 153
Andersen, J. - 349, **497**
Andrievsky, S. – 349
Anguiano, B. – 505
Antoja, T. – 505
Arimoto, N. 505
Árnadottir, A. – 505
Asplund, M. – 153
Aumer, M. – 505
Azzollini, R. – 127

Bagetakos, I. – 301
Bailer-Jones, C. – **475**, 507
Bakos, J. – 127, 505
Banhatti, D. – 505
Barbuy, B. – **153**, 349
Barden, M. – 507
Barker, M. – 505
Battaglia, G. – **61**
Battinelli, P. – 505
Baumgardt, H. – 507
Beckman, J. – 127, 506
Beers, T. – 349, **461**, 507, 508
Bell, E.F. – 121, 507, 508
Benjamin, R. – **319**
Bensby, T. – 197, 507
Berentzen, I. – 165
Bergvall, N. – 507
Bessell, M. – 505
Bica, E. – 153
Bienaymé, O. – 507
Bigiel, F. – 301
Bijaoui, A. – 507
Bikmaev, I. – 505
Binney, J. – **145**, 505
Bland-Hawthorn, J. – 97, 133, **241**, 283, 393, 506
Blitz, L. – **307**
Boissier, S. – 506
Bonifacio, P. – 203, 349
Bono, G. – 507
Borthakur, S. – 507
Bournaud, F. – 429
Bragaglia, A. – **227**
Brinks, E. – **301**, 507

Bromm, V. – **337**
Brook, C. – 445
Buehler, S. – 505
Bullock, J.S. – **85**, 417
Burkert, A. – **437**

Caffau, E. – 349, 506
Callegari, S. – 506
Cano-Díaz, M. – 507
Carballo-Bello, J. – 505
Carollo, D. – 461
Carretta, E. – 227
Casagrande, L. – 505
Cavichia, O. – 505
Cayrel, R. – 349
Černiauskas, A. – 506
Chang, R. – 506
Chen, L. – 505, 506
Chiappini, C. – **191**, 325
Christlein, D. – **283**
Chun, M. – 508
Colavitti, E. – 505
Cole, A. – 505
Colpi, M. – 506
Costa, R. – 505, 506
Courty, S. – 445
Csabai, I. – 506
Curir, A. – 505

de Blok, W.J.G. 301, 507
de Jong, J.T.A. – **121**
de Jong, R. – 505
de Laverny, P. – 507
De Lucia, G. – **423**
De Silva, G.M. – **133**
de Mello, D. – 507
Debattista, V. – 507
Della Vecchia, C. – 508
Demers, S. – 505
Depagne, E. – 349
Dessauges-Zavadsky, M. – 41
Dobo, L. – 506
Dobrovolskas, V. – 506
Dolphin, A.E. – 121
Dottori, H. – 506
Drew, J. – 508
Driver, S. – **469**
Dubinski, J. – **165**

Edvardsson, B. – 506
Ekström, S. – 325

Elmegreen, B. – **289**
Engelbracht, C. – 506
Erwin, P. – 506

Fan, X. – 506
Faria, D. – 507
Feltzing, S. – **197**, 505, 507, 508
Ferguson, A – 505, 507
Fernández, D. – 505
Figueras, F. – 505, 507
Flynn, C. – 505, 506, 507
Forest, H. – 505
François, P. – 349
Fraternali, F. – **255**
Freeman, K.C. – 97, **111**, 133, 139, 393, 505
Friel, E. – 506
Fuchs, B. – 506, 507
Fynbo, J.P.U. – **41**

Galeev, A. – 505
Galli, D. – 203
Gao, X. – 505
García-Barreto, J. – 507
Gardner, E. – 506
Genzel, R. – **33**
Georgy, C. – 325
Gibson, B. – **445**
Gilmore, G. – **487**, 507
Giordano, L. – 506
Goldman, I – 506
Gómez, A. – 153
Gratton, R. – 203, 227
Grebel, E. – **49**, 483
Greimel, R. – 508
Grosbøl, P. – 506
Guittet, M. – 506
Guo, Q. – **411**
Gustafsson, B. – **3**
Gutiérrez, L. – 506

Hansen, C. – 506
Heiles, C. – 95
Heiter, U. – 506
Helmi, A. – 61, 139, 263, 423, 508
Hensler, G. – **269**
Herbert-Fort, S. – 506
Herczeg, G. – 506
Hernández-Toledo, H. – 507
Hill, V. – 153
Hirsch, R. – 325
Høg, E. – 506
Holmberg, J. – 507
Hou, J. – 506
House, E.L. – 445
Häusler, B. – 507

Ibata, R. – 505, 507
Innanen, K. – 506
Irwin, M. – 61, 505, 507
Ivanauskas, A. – 506
Ivanov, V. – 507

Jablonka, P. – 505
Jacobson, H. – 506
Jahreiss, H. – 506, 507
Jogee, S. – **67**
Johnson, R. – 507, 508
Johnston, K. – 507
Jordi, C. – 483
Jorgenson, R. – 95
Just, A. – 506

Kamiya, Y. – 355
Kang, X. – 506, 508
Karlsson, T. – **343**
Kaufmann, T. – 507
Kawata, D. – 445
Kazantzidis, S. – **417**
Kennicutt, R. – 301, 507
Kerekes, G. – 506
Kharchenko, N. – 221
Kim, J. – 507
Kim, W.-T. – **313**, 507
Klement, R. – 506
Klioner, S.A. – 483
Knude, J. – 506
Kobayashi, C. – **375**
Korn, A. – 506
Kroupa, P. – **209**, 507
Kučinskas, A. – 506

Lake, G. – 505
Lambert, D. – 507
Lee, H. – 507
Lee, Y. – 461, 507, 508
Leroy, A. – 301
Lewis, G. – 505, 507
Li, Y. – **263**
Lindegren, L. – 483
Luck, R. – 506
Ludwig, H.-G. – 349, 506
Lundström, I. – 505

Maciel, W. – 505, 506
Mackey, D. – 505
Maeder, A. – 325
Makaganiuk, V. – 506
Mapelli, M. – 506
Martínez-Delgado, D. – 505
Martig, M. – **429**
Mateu, C. – 506
Matteucci, F. – 505
Mattsson, L. – 506, 507

Mayer, L. – 505, 506
Mazzei, P. – 505
McIntosh, D. – 507
Meléndez, J. – 153
Melnikov, S. – 505
Menten, K. – 507
Meynet, G. – **325**
Mignard, F. – 483
Minniti, D. – 153
Molaro, P. – 349
Moore, B. – 505, 506, 507
Moreno, E. – 505, 507
Moustakas, J. – 506
Murante, G. – 505
Musaev, F. – 508
Møller, P. – 41

Naab, T. – 508
Navarro, J. – 508
Nedialkov, P. – 507
Newberg, H. – 508
Nissen, P. – **103**, 508
Nomoto, K. – **355**
Nordström, B. 349, 506
Norris, J. – 507

Oey, S. – 197, 507
Oh, S. – 507
Oh, S.H. – 507
Ortolani, S. – 153

Palouš, J. – **233**
Parrao, L. – 508
Pasquini, L. – **203**
Pedrosa, S. – 507
Pettini, M. – **21**
Pichardo, B. – 505
Pignatari, M. – 325
Pilachowski, C. – 506
Piontek, F. – 507
Piskunov, A. – 221
Plez, B. – 349
Pogge, R. – 506
Pohlen, M. – 127, 505
Ponman, T. – 507
Popescu, N. – 507
Portas, A. – 507
Portinari, L. – 507
Prantzos, N. – **381**, 506
Primas, F. 349, 508
Prochaska, J.X. – 41, 95
Prusti, T. – **483**
Puerari, I. – 507, 508
Purcell, C.W. – 85
Puzeras, E. – 508

Quinn, T. – 507

Radburn-Smith, D. – 505
Randich, S. – 203, 483
Rasmussen, J. – 507
Rawat, A. – 507
Re Fiorentin, P. – 507, 508
Recio-Blanco, A. – 507
Reddy, B. – 507
Renzini, A. – 153
Ribas, S. – 507
Richardson, J. – 507
Ripamonti, E. – 506
Rix, H.-W. – 121, 506, 508
Robertson, B.E. – **35**
Robishaw, T. – 95
Rockosi, C. – 508
Romeo, A. – 505, 507
Roškar, R. – 507
Ruhland, C. – 507
Ryan, S. – 505
Ryde, N. – **159**
Röser, S. – 221

Saha, K. – 508
Sakhibullin, N. – 505
Sale, S. – 508
Sales, L. – 508
Sánchez-Blásquez, P. – 445
Sarajedini, A. – 505
Scannapieco, C. – **369**, 507
Schaye, J. – 508
Schilbach, E. – 221
Schneider, D. – 508
Scholz, R.-D. – 221
Schuster, W. – 103, 508
Sellwood, J. – **73**
Shao, Z. – 508
Shen, S. – 508
Shlosman, I. – 165
Sick, J. – 505
Siebert, A. – 453
Siebert, A. – 505
Sil'chenko, O. – **173**
Silich, S. – 233
Silk, J. – **401**
Simmerer, J. – 508
Smecker-Hane, T. – 505
Smiljanić, R. – 203
Smith, M. – 508
Sommer-Larsen, J. – 41
Sperauskas, J. – 508
Spite, F. – 349
Spite, M. – **349**
Springel, V. – 369, 508
Stadel, J. – 507

Steinmetz, M. – **453**, 505, 507, 508
Stewart, K. – 85
Stinson, G. – 507

Takamiya, M. – 508
Tamm, A. – 506, 508
Tanabé, T. – 506
Tanvir, N. – 505, 507
Tautvaišiene, G. – 508
Taylor, E. – 507
Tempel, E. – 508
Tenjes, P. – 508
Tenorio-Tagle, G. – 233
Teyssier, R. – 429, 445, 505
Tissera, P.B. – 369
Tolstoy, E. – 61
Tominaga, N. – 355
Torra, J. – 505, 507
Tosi, M. – 227
Tossera, P. – 507
Trujillo, I. – **127**, 505
Tsangarides, S. – 505
Tsujimoto, T. – **393**

Umeda, H. – 355
Unruh, Y. – 508
Usero, A. – 301, 507

Valcheva, A. – 507
Valdez-Gutiérrez, M. – 508
Valenzuela, O. – 505, 507
Vallenari, A. – 507
van den Bosch, F. – 508
Vanzi, L. – 507
Verdes-Montenegro, L. – 507
Vidrih, S. – 506
Villalobos, Á. – 508

Vivas, K. – 506
Vlajić, M. – **97**

Wadadekar, Y. – 507
Wadsley, J. – 507
Walter, F. – 301, 507
Walton, N. – 483
Wanajo, S. – 355
Weiner, B. – 506
White, S. – **19**, 369, 411
Wilhelm, R. – 508
Wilkinson, M. – 507
Williams, M. – **139**
Willmer, C. – 508
Wolfe, A. – **95**
Wünsch, R. – 233
Wyder, T. – 508
Wylie de Boer, E. – 505
Wyse, R. – **179, 501**, 507

Xiao, Q. – 508
Xue, X. – 507, 508

Yanny, B. – 121, 508
Yin, J. – 506
You, M.S. – 507
Young, M. – 508

Zabludoff, A. – 506
Začs, L. – 508
Zaritsky, D. – 506
Zentner, A.R. – 417
Zhao, G. – 508
Zhao, J – 505
Zinn, R. – 506
Zinnecker, H. – **221**
Zoccali, M. – 153, 507
Zwitter, T. – 453, 507